ESO ASTROPHYSICS SYMPOSIA
European Southern Observatory

Series Editor: Bruno Leibundgut

T. Richtler · S. Larsen (Eds.)

Globular Clusters - Guides to Galaxies

Proceedings of the Joint ESO-FONDAP Workshop on Globular Clusters held in Concepción, Chile, 6–10 March 2006

 Springer

Volume Editors

Tom Richtler
Universidad de Concepción
Departamento de Astronomía
Concepción
Chile
tom@astro-udec.cl

Søren Larsen
Astronomical Institute
Utrecht University
Utrecht
The Netherlands
S.S.Larsen@uu.nl

Series Editor

Bruno Leibundgut
European Southern Observatory
Karl-Schwarzschild-Str. 2
85748 Garching, Germany

ISBN: 978-3-540-76960-6   e-ISBN: 978-3-540-76961-3

ESO Astrophysics Symposia ISSN: 1431-2433

Library of Congress Control Number: 2008938479

© Springer-Verlag Berlin Heidelberg 2009

This work is subject to copyright. All rights are reserved, whether the whole or part of the material is concerned, specifically the rights of translation, reprinting, reuse of illustrations, recitation, broadcasting, reproduction on microfilm or in any other way, and storage in data banks. Duplication of this publication or parts thereof is permitted only under the provisions of the German Copyright Law of September 9, 1965, in its current version, and permission for use must always be obtained from Springer. Violations are liable to prosecution under the German Copyright Law.

The use of general descriptive names, registered names, trademarks, etc. in this publication does not imply, even in the absence of a specific statement, that such names are exempt from the relevant protective laws and regulations and therefore free for general use.

Cover design: WMXDesign GmbH, Heidelberg

Printed on acid-free paper

9 8 7 6 5 4 3 2 1

springer.com

Preface

The idea to hold a workshop on globular clusters in Concepcion emerged during 2005 out of a variety of circumstances. Four years had passed since the IAU Symposium 207 on Extragalactic Globular Clusters in Pucón, a time span, which we thought to be long enough for justifying a new meeting with the intent to review the most recent developments in the field of extragalactic stars clusters. Originally intended to be a small-scale workshop, the response from the community was overwhelming so that only a full-scale international conference was able to cope with the numerous requests for talks and posters.

Finally, about 160 participants gathered in Concepción on March 6th, 2006. The venue was the university lecture hall located in the facultad de humanidades y artes of the Universidad de Concepción. Posters were exposed in the lobby of the faculty building. The weather was as good as one can reasonably expect from a late summer in Concepción. Although the programme was so tight that separate poster sessions other than those during coffee breaks could not be accomodated, posters received a lot of attention.

From the first to the last talk, the atmosphere was inspiring and the conference could keep its tension for five full days. This clearly shows that the attraction which globular clusters exercise on astrophysicists of quite different flavours, is as strong as ever.

Within the hierarchy of structure formation in the Universe, globular clusters link star formation with galaxy formation. Looking down from the globular cluster platform, one learns about stellar populations, looking up about galaxies. Sometimes it is not even necessary to look up, since the most massive clusters show properties which make their classification as star clusters or dwarf galaxies disputable. The concept of the conference was rather to look up and to put the emphasis on cluster systems and what they can tell us about their host galaxies. Strongly clustered star formation took or still takes place in almost all kinds of galactic environments, most spectacularly in ongoing mergers. Although the conditions for globular cluster formation must be realized locally in the interstellar medium, galaxy wide properties seem to be important to determine these conditions. The globular cluster systems of central galaxies may even be linked with the hosting galaxy cluster. Globular clusters can be used as dynamical probes for dark matter and thus enables a view onto the most fundamental conditions for galaxy formation, testing the CDM paradigm.

All these and more justify the conference title "Globular Clusters - Guides to Galaxies".

Enforced by limited time resources, we had to neglect the detailed view onto Milky Way globular clusters, which provide extremely valuable insight into cluster formation. We particularly regret that it was not possible to present a portrait of Omega Centauri.

We want to thank all sponsors which made the conference financially feasible: The FONDAP center of astrophysics, the European Southern Observatory, the Cerro Tololo Interamerican Observatory, the Universidad de Concepción, and the Facultad de Ciencias Físicas y Matematicas. The Facultad de Humanidades y Artes generously provided her lobby for poster presentations and for coffee and pastries.

The conference would not have been possible without the organisational skills and devotion of the Local Organising Committee: Doug Geisler, Matias Gómez, Aaron Romanowsky, and particularly Maria Eugenia Geisler.

Concepción, Chile *Tom Richtler*
and *Søren S. Larsen*
Utrecht, The Netherlands

Contents

Open Questions in the Globular Cluster – Galaxy Connection
Markus Kissler-Patig 1

Part I Detailed Studies of Individual Globular Clusters

Detailed Chemical Abundances of Extragalactic Globular Clusters
Rebecca Bernstein, Andrew McWilliam 11

Spectroscopic Abundances and Radial Velocities of the Galactic Globular Clusters 2MASS GC01 and 2MASS GC02: Preliminary Results
J. Borissova, V.D. Ivanov, A. Stephens, D. Minniti 17

Abundance Anomalies in Galactic Globular Clusters – Looking for the Stellar Culprits
C. Charbonnel, N. Prantzos 21

Globular Clusters in the Direction of the Inner Galaxy
J. Alonso-García, M. Mateo 25

Globular Cluster Research with Astronomical Archives
Carol A. Christian 27

Super-He-Rich Populations in Globular Clusters
Chul Chung, Young-Wook Lee, Suk-Jin Yoon, Seok-Joo Joo, and Sang-Il Han 29

Testing the BH 176 and Berkeley 29 Association with GASS/Monoceros
Peter M. Frinchaboy 31

New Yonsei-Yale (Y^2) Isochrones and Horizontal-Branch Evolutionary Tracks with Helium Enhancements
S.-I. Han, Y.-C. Kim, Y.-W. Lee, S.K. Yi, D.-G. Kim, and P. Demarque 33

Search for Candle Stars in Globular Clusters: Spectroscopic Analysis of Post-AGB Candidates
G. Jasniewicz, M. Parthasarathy 35

The Lack of Binaries Among Hot Horizontal Branch Stars: M80 and NGC5986
C. Moni Bidin, S. Moehler, G. Piotto, Y. Momany, A. Recio-Blanco, R.A. Méndez 37

Semi-Empirical Determination of the Mass Distribution of Horizontal Branch Stars in M3
A. Valcarce, M. Catelan 39

Part II The Most Massive Clusters

Globular Clusters, Galactic Nuclei and Supermassive Black Holes
Patrick Côté 43

UCDs – A Mixed Bag of Objects
Michael Hilker 51

Ultra-Compact Dwarf Galaxies and Globular Clusters: A Review of Their Spatial and Dynamical Properties
M.J. Drinkwater, E. Evstigneeva, P. Firth, C. Power 59

The Maximum Mass of Star Clusters
M. Gieles, S.S Larsen, M.R. Haas, R.A. Scheepmaker, N. Bastian ... 63

The Stellar Population of Ultra-Compact Dwarf Galaxies
Michael D. Gregg, Michael J. Drinkwater, Katya Evstigneeva, Arna Karick 69

News on Ultra-Compact Dwarfs and Blue Globular Clusters
Steffen Mieske 75

UCDs and GCs: Structural Differences from HST Imaging
E.A. Evstigneeva, M.J. Drinkwater, M. Hilker,, C.Y. Peng 79

Ultra-Compact Stellar Systems in the Fornax Galaxy Cluster
P. Firth, M.J. Drinkwater, E.A. Evstigneeva, A. Karick, M.D. Gregg, M. Hilker, K. Bekki, J.B. Jones, S. Phillips 81

Multi-Colour Imaging of Ultra-Compact Objects in the Fornax Cluster
A.M. Karick, M.D. Gregg, M.J. Drinkwater, M. Hilker, P. Firth 83

Part III Young Star Clusters

Hierarchical Formation of Galactic Clusters
Bruce G. Elmegreen 87

Young Massive Clusters – Formation Efficiencies and (Initial) Mass Functions
Søren S. Larsen... 95

The Radii of Thousands of Star Clusters in M51 with *HST/ACS*
R.A. Scheepmaker, M. Gieles, M.R. Haas, N. Bastian, S.S. Larsen.... 103

Extragalactic Star Clusters in Merging Galaxies
Gelys Trancho 107

The Environment of Young Massive Clusters
Leonardo Vanzi 111

Star/Cluster Formation in Complexes: Insights from IFUs and HST
Nate Bastian 115

Spectral Evolution of Blue Concentrated Star Clusters in the Large Magellanic Cloud
J.J. Clariá, J.F.C. Santos Jr., A.V. Ahumada, E. Bica, A.E. Piatti,
M.C. Parisi 117

Young Star Clusters in the SMC
Katharina Glatt, Eva K. Grebel, Andreas Koch 119

Molecular Clouds and Star Formation in the Magellanic System by NANTEN
A. Kawamura, T. Minamidani, Y. Mizuno, N. Mizuno, T. Onishi, A.
Mizuno, Y. Fukui, NANTEN team 121

Two Star Cluster Populations in NGC 45
M.D. Mora, S.S. Larsen, M. Kissler-Patig 123

Characterization of Open Cluster Remnants
D.B. Pavani, E. Bica.. 125

HST Photometry of the Binary Globular Cluster Sersic 13N-S in NGC5128[1]
D. Villegas, D. Minniti, J.G. Funes 127

Part IV Globular Cluster Systems in Dwarf and Irregular Galaxies

LMC Cluster Abundances and Kinematics
Doug Geisler, Aaron Grocholski, Ata Sarajedini, Andrew Cole, Verne Smith 133

Globular Clusters in Dwarf Galaxies
Bryan W. Miller 141

Globular Clusters in Dwarf and Giant Galaxies
Sidney van den Bergh 149

The Age-Metallicity Relation of the SMC
Andrea Kayser, Eva K. Grebel, Daniel R. Harbeck, Andrew A. Cole, Andreas Koch, John S. Gallagher, Gary S. Da Costa 157

Integrated Spectroscopic Analysis of Galactic and Small Magellanic Cloud Clusters
A.V. Ahumada, J.J. Clariá, E. Bica 161

Variable Stars in the Globular Clusters and in the Field of the Fornax dSph Galaxy
C. Greco[1], G. Clementini[1], E.V. Held[2], E. Poretti[3], M. Catelan[4], L. Dell'Arciprete[3], M. Gullieuszik[2], M. Maio[1], L. Rizzi[5], H.A. Smith[6], B.J. Pritzl[7], A. Rest[8], N. De Lee[6] 163

Physical Parameters of Intermediate-Age LMC Clusters from Modelling of HST CMDs
L.O. Kerber, B. Barbuy, E. Brocato 165

RGB Properties of the LMC/SMC Clusters in the Infrared
Radostin Kurtev, Valentin Ivanov, Jura Borissova, Márcio Catelan, and Douglas Geisler 167

WLM-1: A Non-Rotating, Gravitationally Unperturbed, Highly Elliptical Extragalactic Globular Cluster
Andrew W. Stephens, Márcio Catelan, Roxana P. Contreras 169

Part V Globular Cluster Systems in Spiral Galaxies

Star Clusters in M33 – Clues to Galaxy Formation and Evolution
Rupali Chandar 173

M31 and its Globular Clusters
Kathy Perrett 181

IR Integrated Light Colors For Galactic GCs and An Update on Young M31 Globular Clusters
Judith Cohen 189

Nuclear Star Clusters in Edge-on Galaxies
Anil C. Seth, Julianne J. Dalcanton, Paul W. Hodge, Victor P. Debattista 193

HST ACS Wide-Field Photometry of the Sombrero Galaxy Globular Cluster System
Lee Spitler 197

Intermediate-Age Globular Clusters in M31
Jay Strader 203

Metal-Poor Globular Clusters of the Galactic Bulge
B. Barbuy, M. Zoccali, S. Ortolani, Y. Momany, D. Minniti, V. Hill, A. Renzini, E. Bica, A. Alves-Brito, A. Goméz, L. Pasquini, R.M. Rich 207

Globular Cluster System and Milky Way Properties Revisited
C. Bonatto, E. Bica, B. Barbuy, S. Ortolani 209

RR Lyrae-Based Calibration of the Globular Cluster Luminosity Function
M. Di Criscienzo, F. Caputo, M. Marconi, and I. Musella 213

Globular Cluster Systems in Spiral Galaxies Using ACS Imaging
Kieran Forde 215

Laser Guide Star Imaging of M31 Globulars
Michael D. Gregg, Arna Karick, Bruce Macintosh 217

GALEX UV Observations of M31 Globular Clusters
Soo-Chang Rey, Sangmo T. Sohn, R. Michael Rich, Suk-Jin Yoon, Chul Chung, Sukyoung K. Yi, and Young-Wook Lee 221

Integrated Spectroscopy of Galactic Globular Clusters
Ray M. Sharples, Jaeil Cho 223

Part VI Globular Cluster Systems in Early-Type Galaxies

Globular Cluster Systems: Do They Really Trace Star Formation? (Or Rather: What Mode of Star Formation Do They Trace?)
Michael A. Beasley 227

Globular Clusters in Early Type Galaxies
Jean P. Brodie 235

Globular Clusters and Galaxy Formation
Duncan A. Forbes 245

Globular Cluster Systems in Giant Ellipticals: New and Old Patterns
William E. Harris 253

The ACS Virgo Cluster Survey
Andrés Jordán 263

Globular Clusters at the Centre of the Fornax Cluster: Tracing Interactions Between Galaxies
Lilia P. Bassino, Tom Richtler, Favio R. Faifer, Juan C. Forte, Boris Dirsch, Doug Geisler, Ylva Schuberth 271

Globular Cluster Bimodality Revisited (and the Globulars-Galaxy Halo Connection)
Juan C. Forte, Favio Faifer, Doug Geisler 275

Globular Cluster Systems, Diffuse Star Clusters, and Host Galaxies in the ACS Virgo Cluster Survey
Eric W. Peng 279

Hot Populations in M87 Globular Clusters
S.T. Sohn, R.W. O'Connell, A. Kundu, W.B. Landsman, D. Burstein, R.C. Bohlin, J.A. Frogel, J.A. Rose 283

A Subaru/Suprime-Cam Wide-Field Survey of Globular Cluster Populations around M87
Naoyuki Tamura, Ray M. Sharples, Nobuo Arimoto, Masato Onodera, Kouji Ohta, Yoshihiko Yamada 287

Stellar Populations of Globular Clusters in NGC 1407
A.J. Cenarro, J.P. Brodie, M.A. Beasley, J. Strader 293

The Globular Cluster System of NGC 5846 Revisited: Colours, Sizes and X-Ray Counterparts
Ana L. Chies-Santos, Basilio X. Santiago, Miriani G. Pastoriza, Duncan A. Forbes 295

Globular Cluster Systems in Shell Ellipticals
Jaeil Cho, Ray M. Sharples 297

GMOS Photometry of Five Globular Cluster Systems: NGC 4649, NGC 3923, NGC 524, NGC 3115 and NGC 3379
F.R. Faifer, J.C. Forte, M. Beasley, T. Bridges, D. Forbes, K. Gebhardt, D. Hanes, M. Norris, M. Pierce, R. Proctor, R. Sharples, and S. Zepf 299

Structural Parameters from Ground-based Observations of Globular Clusters in NGC 5128
M. Gómez, D. Geisler, W.E. Harris, T. Richtler, G.L.H. Harris, K.A. Woodley 301

Globular Cluster Populations in Early-Type Galaxies
M. Hempel, M. Kissler-Patig, T.H. Puzia, M. Hilker, S. Zepf, A. Kundu 303

The Low-Mass X-Ray Binary Globular Cluster Connection in the ACS Virgo Cluster Survey
A. Jordán, G.R. Sivakoff, C.L. Sarazin, J.P. Blakeslee, E.L. Blanton, P. Côté, L. Ferrarese, J.A. Irwin, A.M. Juett, S. Mei, E.W. Peng, M.J. West 305

The Globular Cluster System of NGC 5128: Combining Broad-Band Color and Lick Index Analysis
Thomas Lilly, Uta Fritze-v. Alvensleben, Richard de Grijs 307

The Galaxy – Globular Cluster Connection in NGC 3115
Mark A. Norris, Ray M. Sharples, Harald Kuntschner 309

Velocity Dispersions of Bright Globular Clusters in NGC 5128
Marina Rejkuba, Dante Minniti, Georges Meylan 311

Part VII Evolution of Cluster Systems and their Host Galaxies

Imprint of Galaxy Formation and Evolution on Globular Cluster Properties
Kenji Bekki 315

Formation of Globular Clusters in Hierarchical Cosmology: ART and Science
Oleg Y. Gnedin, José L. Prieto 323

Globular Cluster Formation in Mergers
François Schweizer 331

The Formation Histories of Metal-Rich and Metal-Poor Globular Clusters
Stephen E. Zepf 339

Globular Cluster System Evolution in Early Type Galaxies
R. Capuzzo-Dolcetta 347

Star Cluster Evolution: From Young Massive Star Clusters to Old Globulars
Richard de Grijs 353

A Wide-Field Survey of the Globular Cluster Systems of Giant Galaxies
Katherine L. Rhode 357

IGCs in the Virgo Cluster
Marianne Takamiya, Michael West, Patrick Côté, Andrés Jordán, Eric Peng, Laura Ferrarese, the ACS VCS Team 361

A New Explanation of Globular Cluster Color Distributions
Suk-Jin Yoon, Sukyoung Ken Yi, Young-Wook Lee 367

Formation of Intracluster and Intercluster Globular Clusters
Kenji Bekki, Hideki Yahagi 373

The Effect of Giant Molecular Clouds on Star Clusters
M. Gieles, S.F. Portegies Zwart, E. Athanassoula 375

Metal-rich Globular Clusters: An Unaccounted Factor Responsible for Their Formation?
V.V. Kravtsov 377

On the Globular Cluster Color Distributions
Suk-Jin Yoon, Chul Chung 381

Part VIII Dynamical Evolution of Star Clusters

Dissolution of Globular Clusters
Holger Baumgardt 387

Dynamical Masses of Young Star Clusters: Constraints on the Stellar IMF and Star-Formation Efficiency
Nate Bastian and Simon P. Goodwin 395

Dynamical Evolution of Rotating Globular Clusters with Embedded Black Holes
José Fiestas, Rainer Spurzem 399

The Dynamical Evolution of Young Clusters and Galactic Implications
Pavel Kroupa 403

Simulations of Globular Clusters Merging in Galactic Nuclear Regions
P. Miocchi, R. Capuzzo Dolcetta, P. Di Matteo 407

The Origin of the Gaussian Initial Mass Function of Globular Cluster Systems
Geneviève Parmentier, Gerard Gilmore 411

Evolution of Globular Cluster Systems
E. Vesperini 415

Tidal Disruption and the Tale of Three Clusters
Guido De Marchi, Luigi Pulone, Francesco Paresce 419

Tidal Tails Around Globular Clusters: Are they Good Tracers of Cluster Orbits?
P. Di Matteo, R. Capuzzo Dolcetta, P. Miocchi, M. Montuori 421

Modelling the Tidal Tails of NGC 5466
M. Fellhauer, V. Belokurov, M.I. Wilkinson, N.W. Evans, and G. Gilmore 423

The Search for Tidal Tails of Globular Clusters: NGC4147
Katrin Jordi, Eva K. Grebel for the SDSS Collaboration 425

Internal Rotation of Young Globular Clusters
E. Vesperini, S.E. Zepf 427

Mass Segregation in Young Star Clusters
E. Vesperini, S.L.W. McMillan, S.F. Portegies Zwart 429

Part IX Dynamics of Globular Cluster Systems

Kinematics of Globular Cluster Systems
Aaron J. Romanowsky 433

Dark Matter in the Elliptical Galaxies NGC 1399 and NGC 4636
Ylva Schuberth, Tom Richtler, Michael Hilker 445

Ages, Abundances, and Kinematics of Globular Clusters in NGC 3379 and NGC 4649 with Gemini/GMOS
T. Bridges, M. Beasley, F. Faifer, D. Forbes, J. Forte, K. Gebhardt, D. Hanes, M. Norris, M. Pierce, R. Proctor, R. Sharples, and S. Zepf . 449

The Dark Halo of NGC 1399 and MOND
Tom Richtler, Ylva Schuberth, A. Romanowsky 453

Dynamics of the Globular Cluster System of NGC 5128
Kristin A. Woodley 455

Participants

Aguayo, Gustavo	Universidad de Concepción, Chile
Ahumada, Andrea	Observatorio Astronómico, U. Nacional de Córdoba, Argentina
Alonso-García, Javier	University of Michigan, USA
Arevalo, Fabiola	Universidad de Concepción, Chile
Barbuy, Beatriz	University of São Paulo, Brazil
Barraza, Maria Eugenia	Universidad de Concepción, Chile
Bassino, Lilia	Universidad Nacional de La Plata, Argentina
Bastian, Nathan	Astronomy/Physics Department, University of College, London, UK
Baumgardt, Holger	Sternwarte der Universitaet Bonn, Germany
Beasley, Michael	UCO/Lick Observatory, Santa Cruz, USA
Bekki, Kenji	University of New South Wales, Sydney, Australia
Bernstein, Rebecca	University of Michigan, USA
Biswas, Indraneil	University of Virginia, Charlottesville, USA
Bonatto, Charles	IF/UFRGS, Porto Alegre, Brazil
Bono, Giuseppe	Rome Observatory, Italy
Borissova, Jura	ESO Santiago, Chile
Bridges, Terry	Astronomy Group, Queen's University, Kingston, Ontario, Canada
Brodie, Jean	University of California, Santa Cruz, USA
Burkert, Andreas	University of Munich, Germany
Cameron, Scott	University of Michigan, USA
Catelan, Marcio	P. Universidad Católica, Santiago, Chile
Capuzzo-Dolcetta, Roberto	Department of Physics, University of Roma "La Sapienza", Italy
Cenarro, Javier	Universidad Complutense de Madrid, Spain
Chandar, Rupali	Johns Hopkins University, Baltimore, USA
Charbonnel, Corinne	Geneva Observatory and CNRS, Sauverny, Switzerland
Chies-Santos, Ana Leonor	IF-UFRGS, Porto Alegre, Brazil
Cho, Jaeil	Durham University, UK
Christian, Carol	STScI, Baltimore, USA
Chung, Chul	Department of Astronomy, Yonsei University

XVIII Participants

Claria, Juan Jose	Observatorio Astronomico de Córdoba, Argentina
Cohen, Judith	California Institute of Technology, Pasadena, USA
Colucci, Janet	University of Michigan, USA
Contreras, Rodrigo	P. Universidad Católica, Santiago, Chile
Côté, Patrick	Herzberg Institute of Astrophysics, Victoria, Canada
de Grijs, Richard	Department of Physics & Astronomy, University of Sheffield, UK
de Marchi, Guido	European Space Agency, The Netherlands
di Matteo, Paola	Observatoire de Paris, LERMA, France
Di Criscienzo, Marcella	INAF-Osservatorio di Capodimonte, Napoli, Italy
Drinkwater, Michael	Department of Physics, University of Queensland, Australia
Elmegreen, Bruce	IBM Watson Research Center, USA
Escobar, Maria Eliana	P. Universidad Católica, Santiago, Chile
Evstigneeva, Ekaterina	Physics Department, University of Queensland, Brisbane, Australia
Faifer, Favio	Universidad Nacional de La Plata - CONICET, Argentina
Fellhauer, Michael	Institute of Astronomy, University of Cambridge, UK
Fiestas, Jose	Astronomisches Rechen-Institut Heidelberg, Germany
Firth, Peter	University of Queensland, Brisbane, Australia
Forbes, Duncan	Swinburne University, Australia
Forde, Kieran	National University of Ireland, Galway, Ireland
Forte, Juan Carlos	Universidad Nacional de La Plata, Argentina
Frinchaboy, Peter	University of Virginia, USA
Fuentes, Rodrigo	Universidad de Concepción, Chile
Garcia, Alejandro	Universidad de Concepción, Chile
Geisler, Doug	Universidad de Concepción, Chile
Gieles, Mark	Utrecht University, The Netherlands
Gieren, Wolfgang	Universidad de Concepción, Chile
Glatt, Katharina	Astronomical Institute, University of Basel, Switzerland
Gnedin, Oleg	Ohio State University, Columbus, USA
Gomez, Matias	Universidad de Concepción, Chile
Gonzalez, Rosa A.	Centro de Radioastronomía y Astrofísica, UNAM, Mexico
Goudfrooij, Paul	STScI, Baltimore, USA
Grebel, Eva	University of Basel, Switzerland
Gregg, Michael	University California, Davis, USA
Han, Sang-Il	Yonsei University, Seoul, Korea
Hanes, David	Astronomy Group, Queen's University, Kingston, Ontario, Canada

Harris, Gretchen	University of Waterloo, Canada
Harris, William	McMaster University, Hamilton, Canada
Hempel, Maren	Department Physics and Astronomy, Michigan State University, USA
Hesser, James E.	NRC/HIA, Victoria, Canada
Hilker, Michael	Sternwarte der Universitaet Bonn, Germany
Huxor, Avon	University of Hertfordshire, Hatfield, UK
Infante, Leopoldo	P. Universidad Católica, Santiago, Chile
Jasniewicz, Gerard	CNRS/University of Montpellier II, France
Johnson, Kelsey	University of Virginia, Charlottesville, USA
Jordan, Andres	ESO Garching, Germany
Jordi, Katrin	Astronomical Institute, University of Basel, Switzerland
Karick, Arna	University of California, Davis/ LLNL, Livermore, USA
Kawamura, Akiko	Nagoya University, Japan
Kayser, Andrea	University of Basel, Switzerland
Kerber, Leandro	IAG/USP, São Paulo, Brazil
Kissler-Patig, Markus	ESO Garching, Germany
Kravtsov, Valery	Universidad Catolica del Norte, Antofagasta, Chile
Kroupa, Pavel	Sternwarte der Universitaet Bonn, Germany
Kurtev, Radostin	University of Valparaiso, Chile
Larsen, Soeren	ESO Garching, Germany
Lee, Young-Wook	Yonsei University, Seoul, Korea
Leiton, Roger	Universidad de Concepción, Chile
Lilly, Thomas	Institut fr Astrophysik, University of Goettingen, Germany
Mast, Damian	Observatorio Astronomico UNC, Cordoba, Argentina
McLaughlin, Dean	Leicester University, UK
Mieske, Steffen	ESO Garching, Germany
Miller, Bryan	Gemini Observatory, La Serena, Chile
Miocchi, Paolo	Dipartimento di Fisica, Universita' di Roma "La Sapienza", Rome, Italy
Moni Bidin, Christian	Universidad de Chile, Santiago, Chile
Mora, Marcelo	ESO Garching, Germany
Muñoz, Priscilla	Universidad de Concepción, Chile
Norris, Mark	University of Durham, Department of Physics, UK
Nidever, David	University of Virginia, USA
Parisi, Celeste	Observatorio Astronomico de Cordoba, Argentina
Parmentier, Genevieve	Institute of Astronomy, Cambridge University, UK
Pavani, Daniela B.	IAG/USP, São Paulo, Brazil
Peng, Eric	NRC-HIA, Victoria, Canada

XX Participants

Perrett, Kathy	Department of Astronomy & Astrophysics, University of Toronto, Canada
Puzia, Thomas H.	STScI, Baltimore, USA
Reines, Amy	University of Virginia, Charlottesville, USA
Rejkuba, Marina	ESO Garching, Germany
Rey, Soo-Chang	Chungnam National University, Daejeon, Korea
Rhode, Katherine	Wesleyan University, Astronomy Department, Connecticut, USA
Richtler, Tom	Universidad de Concepción, Chile
Romanowsky, Aaron	Universidad de Concepción, Chile
Rubio, Monica	Universidad de Chile, Santiago, Chile
Sabogal, Beatriz	Universidad de Concepción, Chile
Salinas, Ricardo	Universidad de Concepción, Chile
Scheepmaker, Remco	Utrecht University, The Netherlands
Schuberth, Ylva	Sternwarte der Universitaet Bonn, Germany
Schweizer, Francois	Carnegie Observatories, Pasadena, USA
Seth, Anil	University of Washington, Seattle, USA
Sharples, Ray	Department of Physics, University of Durham, UK
Sohn, Tony	Astronomy and Space Science Inst., Daejon, South Korea
Solis, Miguel	Universidad de Concepción, Chile
Spitler, Lee	Swinburne University of Technology, Australia
Spitler, Lee	Swinburne University of Technology, Australia
Stephens, Andrew	Gemini Observatory, Hilo, USA
Strader, Jay	UCO/Lick Observatory, California, USA
Takamiya, Marianne	University of Hawaii at Hilo, USA
Tamura, Naoyuki	Department Physics, U. Durham, UK
Trancho, Gelys	Gemini Observatory, Hilo, USA
Valcarce, Aldo	P. Universidad Católica, Santiago, Chile, Santiago, Chile
Valenzuela, Rodrigo	Universidad de Concepción, Chile
Vanzi, Leonardo	ESO Santiago, Chile
Vesperini, Enrico	Drexel University, Philadelphia, USA
Villegas, Daniela	P. Universidad Católica, Santiago, Chile, & ESO Garching, Germany
Wallerstein, George	Univ. Washington, Seattle, USA
Weidner, Carsten	Sternwarte der Universitaet Bonn, Germany
West, Michael	Gemini Observatory, La Serena, Chile
Woodley, Kristin	McMaster University, Canada
Yoon, Suk-Jin	Yonsei University, Seoul, South Korea
Zepf, Stephen	Department of Physics and Astronomy, Michigan State Universty, USA
Zapata, Abner	Universidad de Concepción, Chile
Zoccali, Manuela	P. Universidad Católica, Santiago, Chile

Open Questions in the Globular Cluster – Galaxy Connection

Markus Kissler-Patig

European Southern Observatory, Karl-Schwarzschild-Str. 2, 85748 Garching, Germany mkissler@eso.org

Abstract. I present a short (and subjective) view of the open problems in the globular cluster – galaxy connection. I start by looking back onto the questions that mostly Canadians asked to each other in the 80's and how they (the questions, not the Canadians) evolved to the major questions that we ask ourselves today. In turn, I then look at todays pending problems on individual star clusters (e.g. measuring their integrated properties), on globular cluster systems, and finally on the link to galaxy formation and evolution. I end up being not very conclusive.

1 Globular Cluster Systems in the 80's

Systems of globular clusters were only occasionally studied before the 70's (e.g. [1, 2]). It took a few fearless Canadians to get the field to boom in the 80's, as illustrated by two landmark reviews ([3, 4]). The first conference dedicated to *systems* of globular clusters (the IAU Symposium 126, in 1986 [5], see Fig. 1) was the logical consequence and set the path to today's research field.

But what were the questions that were asked at that time? They were, of course, driven by the observational tools: ground-based imaging (see [4]) with only very few, brave excursions to spectroscopy in extragalactic systems (e.g. [6]). Thus, the morphologies (e.g. surface density profiles, ellipticities) of the system were of prime importance, as well as the puzzling universal globular cluster luminosity function, and the number of globular clusters ('specific frequency' when normalized to galaxy luminosity, [7]). The connection of these properties with the formation and evolution history of the host galaxy was recognized early, and the three types of globular cluster formation scenarios discussed today were all already thought of at that time (e.g. [4]): pre-galactic formation, formation in proto-galaxies, and formation in mergers (the latter with reference to [8]).

The IAU Symposium 126 gave a flavor of the hot topics of the 80's. The Milky Way System played a central role as it was the best studied one (in terms of kinematics, chemistry, ages). Local Group systems served as counter-examples to the Milky Way (intermediate ages, peculiarities in the integrated properties of massive star clusters). Globular cluster systems were already studied in galaxies out to the distance of the Coma galaxy cluster,

Fig. 1. Participants of the IAU Symposium 126, in 1986 – the forerunners of globular cluster *systems*.

but the amount of information gathered was sparse given the observational technology. Although links to stellar populations of the host galaxies were presented, no real attempt to strongly constrain the galaxy formation and evolution history was made.

2 Globular Cluster Systems in the 90's

In the 90's, the field slowly evolved to its current standing: namely being a powerful avenue to constraining the formation and evolution of galaxies, in particular of early-type galaxies. This is nicely illustrated by the title of the landmark meeting, the ASP Conference: "The Globular Cluster – Galaxy Connection" [9] (see Fig. 2). Whereas the questions of the 80's still played an important role, the proceedings of this conference now include two sections representative for the evolution in the field: "Extragalactic Globular Clusters Systems" and "The Formation of Globular Cluster Systems within the Context of their parent Galaxies".

In the early 90's, the discovery of 'bimodal' colors distributions in globular cluster systems ([10, 11]) kicked of an industry of papers still ongoing today. The clear distinction (morphological, kinematic, chemical abundance)

Fig. 2. Participants of the Santa Cruz meeting 'The Globular Cluster – Galaxy Connection', in 1992 – the pioneers of the sentence "Globular clusters are powerful probes of the formation and evolution of galaxies..." (and the gazillion of variations thereof).

between a blue and a red sub-population of globular clusters in (almost) all galaxies gave birth to many theories and speculations.

Further, the field fully profited in the 90's from the mini-revolution induced by HST. The space-based imaging allowed to reach in a moderate amount of time the turn-over of the globular cluster luminosity function and to spatially resolve the star clusters in nearby galaxies. Wide-field (ground-based) studies tend to get neglected since then.

3 Globular Cluster Systems in the 00's: Todays Open Questions

The open questions of today have been posed during two conference at the beginning of the new millennium ([12,13] and see Fig. 3). Investigators focus on the characteristics of red and blue sub-populations in their studies, trying to tie these different groups to various phases of galaxy formation and evolution. The ideas are heavily influenced by the fashionable hierarchical picture of galaxy formation that emerged from the ΛCDM paradigm. The

Fig. 3. Participants of the Pucon Meeting (IAU Symposium 207), in 2001. Believers of the Universal Bimodality.

formation of globular cluster sub-populations is tied to the predicted high galaxy merger rate during the lifetime of the universe.

As the star formation in galaxies is pushed back again in time by various observations (e.g. [14]), alternatives for the multiple globular cluster populations are sought for. In particular the blue, Milky Way halo-type globular clusters, known as the oldest objects in the universe, require a very early formation phase hard to reconcile with the current models of formation of early-type galaxies/bulges.

So, the key open question might be: how good are we at connecting the evolutionary history of globular clusters with the one of galaxies?

4 Understanding Our Tools: Individual Globular Clusters

As we focus on system aspects, we tend to forget that in order to use globular clusters as predictive tools, we need to understand them well. Especially the birth and death of star clusters must be understood if we want to infer anything about the *past* from the star clusters we see *today* (i.e. at a special time) in nearby galaxies.

While the dissolution of star clusters starts to be well understood (see several contributions in these proceedings), the formation process is still poorly studied. Ultimately, this prevents us from making any more detailed predictions on the formation/assembly of globular cluster systems. My guess is that the globular cluster system community would profit a lot from a closer contact with the star formation community. Especially at the epoch of ALMA, I predict globular cluster formation to become a booming subject.

Individual star clusters are snap-shots of stellar evolution. It is no surprise then, that detailed stellar isochrones are needed to describe the observed integrated properties of star clusters. This field is continuously progressing, but controversy still exists when it comes to describe the most difficult evolutionary phases such as the horizontal branch and AGB phases. The '2nd parameter' problem is not solved yet and major changes e.g. in the description of the horizontal branch could lead to a radically different interpretation of our observations (see e.g. Yoon et al.'s contribution in this volume and take it with a grain of salt).

5 On the Measurement of Integrated Properties

For another couple of decades, we will (unfortunately) have to rely on measuring *integrated* properties of star clusters outside the Local Group. We do this with great eager and are often (too?) swift with the interpretation of our measured properties. I will use this section to remind us that (i) both photometry as well as spectroscopy of the integrated light of extra-galactic globular clusters are difficult measurements, and (ii) the interpretation of these measurement relies on population synthesis models. Both points seem obvious of course, yet, too often I stumble over sections of (sometimes my own) articles in which it appears forgotten.

On the measurements: some recent very good examples of "things that can go wrong and nobody knows why" are listed in [15], that I encourage everybody to read. Data actually do contain systematic (probably instrumental) errors, some of which are very hard if not impossible to calibrate out. But it stresses, once again, that other errors must be minimized. Most important: good calibrations on the agreed standard systems (photometric, Lick indices, etc) are crucial. Especially the interpretation of spectroscopic data has become more subtle as we try to push the interpretation of the data and attempt to derive more parameters (now often α/Fe on top of age and metallicity).

The interpretation is done with respect to models that, even if they were perfect, currently are tied to the Lick system. If the data are not strictly on that system, a detailed comparison is vain. Again this sound obvious, but I could cite a few recent studies that did not bother to properly calibrate their data, yet feared not come forward with a wild interpretation. Along these lines, we must remind ourselves that models are not perfect: if the interpretation of a dataset relies on a particular population synthesis model but fails to be supported by another model, it loses quite of its appeal.

6 Globular Cluster Systems in the Nearby Universe

Most of us are currently trying to answer questions about the nature and origin of globular cluster systems in the nearby universe. I list below three of my favorite open questions that I would like to see answered in the next decade.

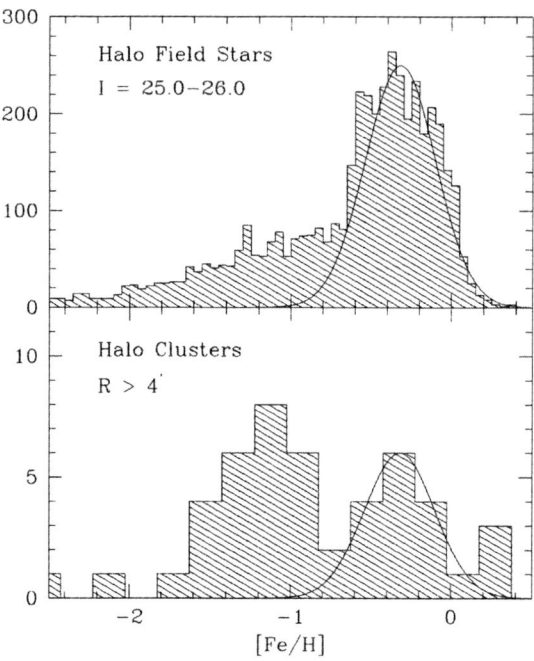

Fig. 4. Figure taken from [16], where it appeared for the first time. It was reproduced subsequently in updated versions and shows one of the main open questions, in my opinion: how are star and globular cluster formation coupled, yet exhibit such different metallicity distributions?.

The first one is the origin of the bi-modality (that I really do not believe to be "bi"). We have currently lots of ideas but still no clue. Current thoughts range from "Bi-modality is blue + lots of different reds" over "Bi-modality is a myth, e.g. an artifact of the colour-metallicity transformation" to "Bi-modality is natural and due to [insert here you favorite buzz word: re-ionization, downsizing, ... N.B. mergers are 'out']". One of my favorite graphs in that respect is shown in Fig. 4 (taken from [16]): while the globular clusters show the known bi-modality in metallicity, the associated stars do not?! Why are the chemical enrichment histories of these two component that we believe to be closely coupled so different?[1] I suspect that the answer will come from a collaboration with the galaxy formation modelers who, with the introduction of feedback mechanisms, start to derive interesting early phases of galaxy and star formation (e.g. [17] and reference therein).

Closely related to the bi-modality is the sequence of events in which globular clusters formed. What are the relative ages of globular clusters in a system? Given the difficulty to answer this in the Milky Way (see [18]) I doubt that we will soon answer this question in extragalactic systems, except maybe when intermediate age clusters are present. What might be interesting to trace, is whether a common chemical enrichment history can be found for the clusters of a system, or whether they are incompatible with having formed in a "closed box" and some accretion must be inferred?

Finally, a long standing assumption is that star clusters in early-type galaxies and spirals are similar . This allows us to use the population synthesis models calibrated locally (in spirals) to interpret star clusters in ellipticals and to extrapolate our local prejudices to nearby ellipticals. But is the assumption justified? Despite some work in that area (similar colours and spectroscopic indices, same fundamental plane, etc) I think that it is still an open question. Especially at the most massive end of the star cluster distribution, where star clusters slowly transit into the dwarf galaxy regime.

7 Galaxy Formation and Evolution

Finally, I will briefly share some thoughts about current clues on galaxy formation and evolution as constrained by globular cluster systems.

In my view, we currently have a rich literature of rather qualitative and incomplete scenarios describing the assembly of globular cluster systems rather then really constraining galaxy formation and/or evolution.

Having myself no brighter idea about this than is currently published, I peaked into the list of goals that galaxy formation theorist set to themselves. Among these goals, the ones that can be constrained in the near future by globular cluster system studies are the ones related to stellar populations in early-type galaxies (i.e. quite a small sub-set). These include: the distribution of stellar masses in galaxies, the distributions of chemical abundances (within and among galaxies and the IGM), the relation between galaxy properties and galactic environment, the evolution of all of the above as a function of redshift (i.e. look-back time). These are probably the areas in which we can help making some progress. The required observations will focus around the determination of ages, metallicities and other element abundances of individual globular clusters to a higher precision than currently achieved.

As a further addition to the above list: the build-up of the stellar halo appears to be (only?) traceable by globular clusters, and almost inaccessible to

---

[1] Whoever provides the answers gets one evening of free (Bavarian) beer.

studies of the integrated light. This might be an additional niche for globular cluster system studies.

8 Where to from Here?

In Summary, the list of open questions is (fortunately :-) still exhaustive and I certainly gave only a very subjective list. My best bet for the ones that will play a major role in the next decade is: How to get from a giant molecular cloud to a massive star cluster? Which parameters drive the CMDs? Are globular clusters in ellipticals really similar to the ones in spirals (integrated properties: ages, abundances)? How is globular cluster formation linked to star formation in galaxies?

There is enough work left for everybody, and the rest of these proceedings testify that we are well underway to answer some of the above questions.

References

1. H. Shapley: ApJ 48, 154 (1918)
2. W.A. Baum: PASP 67, 328 (1955)
3. W.E. Harris & R. Racine: ARA&A 17, 241 (1979)
4. W.E. Harris: ARA&A 29, 543 (1991)
5. *Globular cluster systems in galaxies*, ed by J.E. Grindlay & A.G.D. Philip (Dordrecht: Kluwer, 1988)
6. J. Huchra & J. Brodie: AJ 93, 779 (1987)
7. W.E. Harrisi & S. Van den Bergh: AJ 86, 1627 (1981)
8. F. Schweizer: Star formation in colliding and merging galaxies. In: *Nearly normal galaxies*, ed by S. Faber (New York: Springer, 1987) pp 19
9. *The globular clusters-galaxy connection*, ed by G.H. Smith & J.P. Brodie (San Francisco, CA: Astronomical Society of the Pacific (ASP), 1993) ASP conf. Series Vol. 48
10. K.M. Ashman & S.E. Zepf: ApJ 384, 50 (1992)
11. S.E. Zepf & K.M. Ashman: MNRAS 264, 611 (1993)
12. *Extragalactic star clusters* ed by D. Geisler, E.K. Grebel, D. Minniti (San Francisco, CA: Astronomical Society of the Pacific (ASP), 2002) IAU Symposium 207
13. *Extragalactic globular cluster systems* ed by M. Kissler-Patig (Berlin: Springer, 2003) ESO astrophysics symposium 14
14. *Multiwavelength mapping of galaxy formation and evolution* ed by R. Bender & A. Renzini (Berlin: Springer, 2005) ESO astrophysics symposium 33
15. S.S. Larsen, J.P. Brodie, J. Strader: A&A 443, 413 (2005)
16. G.L.H. Harris, W.E. Harris, G.B. Poole: AJ 117, 855 (1999)
17. D.J. Croton et al.: MNRAS 365, 11 (2006)
18. F. De Angeli et al.: AJ 130, 116 (2005)

Part I

Detailed Studies of Individual
Globular Clusters

Detailed Chemical Abundances of Extragalactic Globular Clusters

Rebecca Bernstein[1] and Andrew McWilliam[2]

[1] Astronomy Department, 500 Church St., University of Michigan, Ann Arbor, MI 48109, USA `rabernst@umich.edu`
[2] Carnegie Observatories, Santa Barbara St., Pasadena, CA 91101, USA `andy@ociw.edu`

Abstract. We are developing a technique for measuring the detailed chemical composition of unresolved, extragalactic globular clusters (GCs) from echelle spectra of their integrated light. To do this, we are using a "training set" of spatially resolved clusters. We scan these clusters to obtain integrated light spectra, and also take spectra of individual stars in these clusters to obtain "fiducial" abundances by the usual analysis methods. We briefly describe here the importance of obtaining detailed abundances, the technique we are developing to analyze integrated light spectra, and the accuracies that can be obtained with our technique.

1 The Advantage of Detailed Abundances

Low resolution spectroscopy and photometry of globular clusters (GCs) in other galaxies reveal that GCs form throughout the lifetimes of galaxies and in major star formation episodes. The abundance patterns and dynamics of the Milky Way GCs provided the first strong constraints on the formation of our own galaxy [6, 12]. The fact that extragalactic GC systems are known to be good tracers of the total star formation and host galaxy mass suggests that GC abundance patterns and kinematics provide strong constraints on galaxy formation in general.

In the Milky Way, especially useful formation diagnostics have come from the abundances ratios of key elements. Particularly useful are "α-elements" (e.g. Mg, O, Si, S, Ca, Ti), which come almost exclusively from Type II supernovae with progenitors that evolve on timescales of a few $\times 10^6$ yrs, and iron-peak elements (e.g. Cr, Mn, Fe, Co, Cu Zn), which have a major contribution from Type Ia supernovae with progenitors that evolve on timescales of $\sim 10^9$ yrs. The ratios of these elements, α/Fe, therefore trace the rate and duration of star formation. Unfortunately, high resolution abundances are very difficult to obtain for individual stars beyond the Milky Way and are just recently becoming available in the nearest dwarf satellite galaxies with 8-m class telescopes. A few luminous supergiants can be observed with Keck beyond the LMC, reaching to M31 (e.g. [16]); however, these massive, young stars only probe the current gas composition. Only red giant stars can provide the fossil evidence of a galaxy's enrichment history, and these are too faint to observe individually at the required spectral resolutions beyond distances of ~100 kpc.

It is possible, however, to obtain high resolution, high signal to noise spectra of GCs as far away as 4 Mpc with *current* telescopes given the low (roughly 5–25 km/s) velocity dispersion of GCs. It is important to note that the formation constraints available from detailed abundance ratios of GCs will never be obtainable from photometry or low-resolution line indices such as the Lick system for at least two reasons. First, abundances inferred from line indices depend strongly on the abundance ratios of the spectra used to calibrate the line system (e.g. [15,14]). Second, and more fundamentally, self-enrichment in GC stars can cause the Mg/Fe ratio to deviate from that with which the stars formed and from the abundance ratio of α-elements not affected by self-enrichment; unfortunately, estimates of the α/Fe abundance in the Lick system come mostly from strong Mg lines.

While the Milky Way clusters appear to be exclusively old and metal poor, there are examples of GC systems which appear to cover a range of ages and abundance patterns even within the Local Group. In addition to the young ($< 5\,\mathrm{Gyr}$) LMC clusters, some GCs in the disk of M31 also appear to be young, although their ages are difficult to determine (see [5] and references therein). Moreover, a range of GC ages may be common in galaxies with complex merger histories [13]. Note that for young GCs with unusual abundance ratios, the interpretation of Lick indices would depend on uncertain theoretical calibrations (e.g. [11]).

2 Abundance Analysis from Integrated Light

Chemical abundance analysis typically involves comparing the measured equivalent width (EW) of lines in the spectrum of a single star with predicted EWs that are based on spectrum synthesis calculations using model atmospheres. By contrast, abundance analysis of an unresolved GC must be based on measured EWs in the integrated light of the cluster (see Fig. 1). We have developed a technique that will let us interpret measured EWs in IL spectra by incorporating a CMD in the calculation of light-weighted, predicted EWs. To check the reliability of our results we are using a "training set" of Milky Way and LMC GCs, that span the full available range of age, metallicity, and α/Fe-ratios. This method is outlined below (see [1,2] for details).

Our analysis uses model CMDs, i.e. isochrones derived from stellar evolution tracks combined with a Kroupa IMF [9]. Observed CMDs can also be used to analyze the Milky Way clusters in the training set, but that does not present an interesting test-analysis for extragalactic, unresolved clusters. We treat these isochrones as possible CMDs without assuming that the ages or abundances are correct. Our analysis then begins with the Fe I and Fe II lines, of which there are many at a wide range of excitation potentials and wavelengths. For any isochrone, we produce stellar atmospheres for the stars

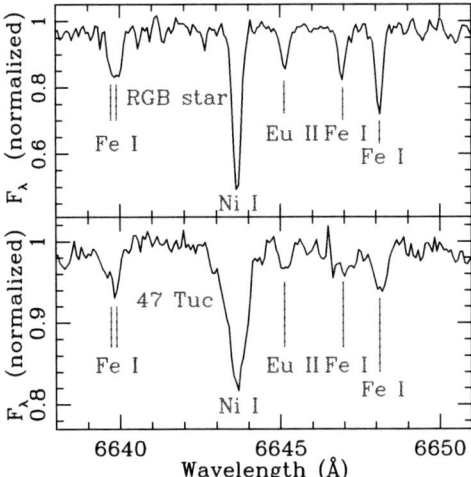

Fig. 1. A sample region of the integrated light spectrum of 47 Tuc compared with the spectrum of an RGB star in that cluster. Note that the lines marked in the RGB star can be readily detected in the integrated light of 47 Tuc, albeit broadened by the ~ 13 km sec^{-1} σ_v of the GC and weakened by the contribution of light from hotter stars. The isolated Eu II line at 6645 Å has an equivalent width of ~ 15 mÅ in the integrated light.

in the model CMD and synthesize each Fe line for those stars. We then vary [Fe/H] in the line synthesis until the light-weighted, combined EW for each Fe line matches the observed EW. This gives us the inferred value of [Fe/H] for each line. We average these together to obtain [Fe/H] for each isochrone.

As an example of this analysis, we show the inferred value of [Fe/H] for 47 Tuc from the observed Fe I and Fe II lines in its integrated light spectrum (see Fig. 2). By requiring that the [Fe/H] solution for all Fe I lines be independent of excitation potential, we can narrow the range of acceptable isochrones to those older than 5 Gyr. For those acceptable isochrones, the Fe I and Fe II lines give a unique solution at a small range of [Fe/H] around -0.65 dex. We use $\log gf$ values calibrated to Arcturus to reduce systematic errors. We then adjust the best-fit isochrone (10 Gyr, [A/H]$=-0.63$) for mass segregation in the core of the cluster and repeat the analysis. This changes the [Fe/H] abundance by only -0.05 dex. Using this adjusted isochrone, we then derive abundances by the same procedure for 18 other elements for which we can measure lines in the integrated light. We are able to obtain abundances for iron-peak elements (Sc, V, Cr, Mn, Fe, Ni), α-elements (Si, Ca, Ti), light elements (Na, Mg, Al), and neutron capture elements (Y, Zr, Ba, La, Nd, Eu) [2]. Our results are very similar to those obtained from standard analysis of single stars in 47 Tuc [4,3,8,10]. In addition, we detect significant (~ 0.2 dex) enrichment in Na and depletion of Mg relative Fe compared to the relative abundances of other α elements ([α/Fe]≈ 0.3 dex). This is consistent with

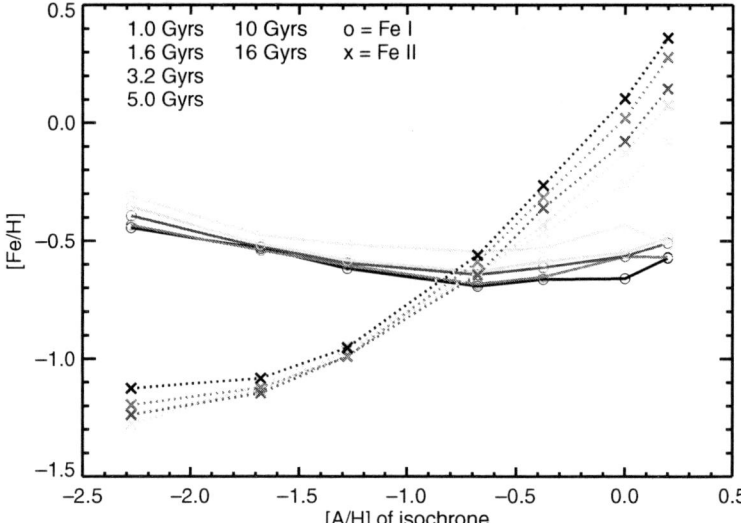

Fig. 2. [Fe/H] from the integrated light analysis of 47 Tuc. The x-axis shows the abundances of the Padova isochrones; the y-axis shows the inferred value of [Fe/H] from Fe I (*dots*) or Fe II (*boxes*) lines. The age (indicated by the *line color*) and [A/H] of the isochrone are not important for the analysis, but are used here to label the respective isochrones.

the known variation of Na and Mg from star to star in this and other Milky Way clusters due to self-enrichment within the cluster (see [7] and references therein).

Additional constraints on the isochrones will come from the Hα, Hβ, Hγ, and Hδ line profiles, which are sensitive to horizontal branch morphology; most of these lines cannot be accurately measured at low resolution. We can use these profiles to adjust the morphology of the horizontal branch in the isochrones in the cases where the presence of a blue horizontal branch is clearly indicated by the spectra.

This method will allow us to constrain the detailed chemical enrichment histories of stellar populations in normal galaxies within 4 Mpc for the first time. Moreover, each GC we observe and analyze at high resolution becomes an empirical calibrator for the line index systems, will always be applicable at larger distances than can be observed at echelle resolutions.

References

1. Bernstein, R. & McWilliam, A. 2006, in "Resolved Stellar Populations", ASP Conf. Ser., ed. D. Valls-Gabaud, M. Chevez (San Francisco: ASP), in press.
2. Bernstein, R. & McWilliam, A. 2006, ApJ (submitted).
3. Brown, J.A. & Wallerstein, G. 1989, AJ, 98, 1643.

4. Carretta, E., Gratton, R.G., Bragaglia, A., Bonifacio, P., & Pasquini, L. 2004, AA, 416, 925.
5. Cohen, J.G., Matthews, K., & Cameron, P.B. 2005, ApJ, 634, L45.
6. Eggen, O.J., Lynden-Bell, D., & Sandage, A. 1962, ApJ, 136, 748.
7. Gratton, R., Sneden, C., & Carretta, E. 2004, ARAA, 42, 385.
8. Kraft, R.P. & Ivans, I.I. 2003, PASP, 115, 143.
9. Kroupa, P. 2002, Science, 295, 82.
10. Lee, J.–W. & Carney, B.W. 2002, AJ, 124, 1511.
11. Puzia, T.H., Kissler-Patig, M., Thomas, D., Maraston, C., Saglia, R.P., Bender, R., Goudfrooij, P., & Hempel, M. 2005, AA, 439, 997.
12. Searle, L. & Zinn, R. 1978, APJ 225, 357.
13. Schweizer, F. 2006, "Globular Clusters – Guides to Galaxies", New York: Springer.
14. Tantalo, R. & Chiosi, C. 2004, MNRAS, 353, 917.
15. Thomas, D., Maraston, C., & Bender, R. 2003, MNRAS, 339, 897.
16. Venn, K.A., Tolstoy, E., Kaufer, A., Skillman, E.D., Clarkson, S.M., Smartt, S.S., Len non, D.J., & Kudritzki, R.P. 2003, ApJ, 126, 1326.

Spectroscopic Abundances and Radial Velocities of the Galactic Globular Clusters 2MASS GC01 and 2MASS GC02: Preliminary Results

J. Borissova[1], V.D. Ivanov[1], A. Stephens[2], and D. Minniti[3]

[1] European Southern Observatory, Ave. Alonso de Cordoba 3107, Casilla 19, Santiago 19001, Chile jborisso@eso.org,vivanov@eso.org
[2] Gemini Observatory, 670 N. Aóhoku Place, Hilo, HI 96720, USA stephens@gemini.edu
[3] Department of Astronomy, P. Universidad Católica, Av. Vicuña Mackenna 4860, Casilla 306, Santiago 22, Chile dante@astro.puc.cl

Abstract. We have presented the preliminary results of the first spectroscopic observations of the galactic globular clusters 2MASS GC01 and 2MASS GC02. The metallicity, radial velocities and distances to the clusters are determined.

1 Introduction

The known globular clusters (GC) in our Galaxy – 150 [2] were discovered mostly through optical searches, that are biased against highly obscured objects. The Galaxy is estimated to have 160±20 GCs [10]. We do not know how many clusters are still missing. The Two Micron All Sky (2MASS) and GLIMPSE Mid-IR Surveys offer an opportunity to carry a search for missing GCs, and Hurt et al. [3] and Kobulnicky et al. [7] reported three new GCs: 2MASS GC01, 2MASS GC02 and GLIMPLSE-C01. The first two clusters GC01 and GC02 are both located within 10 degrees to the Galactic center (Fig. 1).

Our goal is to address the following questions:

- What are the abundances of 2MASS GC01 (GC01) and 2MASS GC02(GC02)?
- What population do they belong to, based on their kinematics?

Here we present the preliminary metallicity and radial velocity measurements of GC01 and GC02.

2 Observations

We have obtained IR spectra at the ESO-VLT with ISAAC UT1 (ESO proposal 073.D-0158) with two setups: resolution 500, 0.3 arcsec wide slit centered at 2.2 μm, and resolution 3000, 1 arcsec wide slit, centered 1.5 microns.

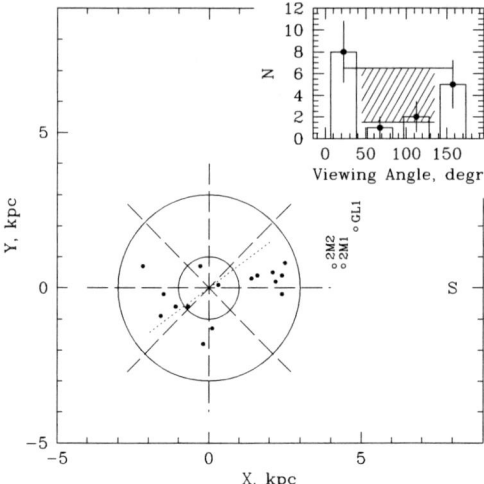

Fig. 1. The location of the globular clusters in the central 3 kpc region projected on the Galactic plane. The Sun is marked with S, the Galactic Center lies at the origin of the plot [6].

The exposure times were 10–15 min and 20–25 min, respectively. We have observed 6 slit positions per cluster, each containing at least two probable cluster members selected from (J-Ks, Ks) colour-magnitude diagrams [4] for both clusters. The data reduction were performed using standard *IRAF* routines and the ISAAC pipeline (Fig. 2).

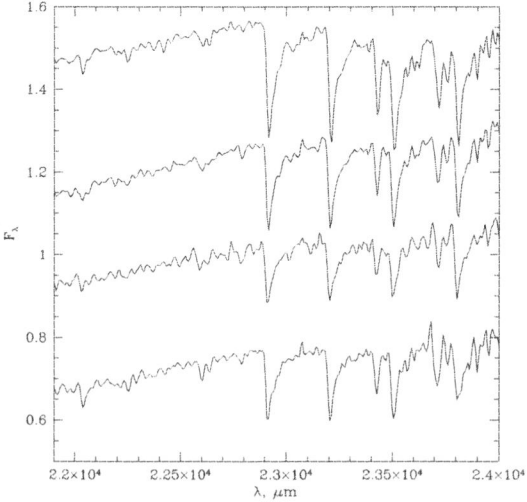

Fig. 2. A sample of reduced spectra of GC02.

3 Results

We determined the metallicities using the globular cluster metallicity indicator calibrated by Frogel et al. [1]. This technique uses the NaI (2.21 μm) and CaI (2.26 μm) and the band head of CO (2.29 μm). It has been calibrated with spectra of more than 100 giant stars in 15 Galactic globular clusters with well known [Fe/H] values in the range of -1.8 <[Fe/H]< -0.1, with an accuracy of 0.1 dex. We used the quadratic solution of the calibration and obtained the metallicities on the Zinn & West [11] metallicity scale. The average value of eight cluster members of GC01 is [Fe/H]=-1.18 ± 0.38, for GC02 (six stars) we calculated [Fe/H]=-1.06 ± 0.12.

The radial velocities were measured with *FXCOR* in *IRAF* using as a template the star BMB 181 (M7 III, [Fe/H]=-0.21, R_V=120.5 km/s) from the stellar library of Ivanov et al. [5]. The mean radial velocities of GC01 and GC02 are R_V=-374 ± 47 km/s (eight stars) and R_V=-310 ± 39 km/s (six stars), respectively.

The distance to the clusters was calculated by using intrinsic colors and magnitudes of Koornneef [8]. First, we chose our best signal to noise spectrum and carried out spectral classification by comparison with spectra of the stars with well determined spectral classes. We used for this comparison the spectra available in the ISAAC Library of Stellar Spectra for spectrophotometric calibration [9]. The best fit for GC01 was obtained for M2 III type star, which leads to the extinction A_v=15.65 mag and distance modulus $(m-M)_0$=12.18 mag ($D = 2.72$ kpc). For GC02 we have obtained M3/4 III, A_v=17.39 mag and distance modulus $(m-M)_0$=11.62 mag (2.12 kpc).

Our preliminary results suggest that most likely GC01 and GC02 belong to the Galaxy disk population, instead of the bulge. They are moderately metal rich and have relatively high radial velocities with respect to other globular clusters.

References

1. Frogel, J., Stephens, A., Ramirez, S., DePoy, D. 2001, AJ, 122, 1896
2. Harris, W.E. 1996, AJ, 112, 1487
3. Hurt, R., Jarrett, T., Kirkpatrick, D., Cutri, R. et al. 2000, AJ, 120, 1876
4. Ivanov, V.D., Borissova, J., Vanzi, L. 2000, A&A, 362, 1
5. Ivanov, V.D., Rieke, M., Engelbracht, C., Alonso-Herrero, A. et al. 2004, ApJS, 151, 387
6. Ivanov, V.D., Kurtev, R., Borissova, J. 2005, A&A, 442, 195
7. Kobulnicky et al. 2005, AJ, 129, 239
8. Koornneef, J. 1983, A&A, 128, 84
9. Pickles, A.J. 1998, PASP, 110, 863
10. van den Bergh, S. 1999, A&A Rev., 9, 273
11. Zinn, R., West, M. J. 1984, ApJS, 55, 45

Abundance Anomalies in Galactic Globular Clusters – Looking for the Stellar Culprits

C. Charbonnel[1,2] and N. Prantzos[3]

[1] Geneva Observatory, 51, chemin des Maillettes, CH-1290 Sauverny, Switzerland
Corinne.Charbonnel@obs.unige.ch
[2] LATT, CNRS UMR 5572, 14, av. E. Belin, 31400 Toulouse, France
[3] IAP, CNRS UMR 7095, 98b Bd. Arago, 75014 Paris, France prantzos@iap.fr

Abstract. Galactic globular cluster stars exhibit abundance patterns which are not shared by their field counterparts. It is clear from recent spectroscopic observations of GC turnoff stars that these abundance anomalies were already present in the gas from which the observed stars formed. This provides undisputed support to the so-called self-enrichment scenario according to which a large fraction of GC low-mass stars have formed from material processed through hydrogen-burning at high temperatures and then lost by more massive and faster evolving stars (and perhaps mixed with some original gas). Within this framework we present a new method to derive the Initial Mass Function of the polluter stars.

1 Abundance Anomalies in Galactic Globular Clusters

During the last three decades, an incredible amount of data has been collected on the chemical properties of galactic globular clusters (hereafter GCs) thanks to high spectral resolution abundance analysis (see [1] and [2] for recent reviews). The main results can be summarized as follows : (i) Individual GCs (with the notable exception of Omega Cen) appear to be fairly homogeneous as far as the iron peak elements (Ni, Cu) are concerned; (ii) They present very low scatter and the same trends as field stars for the neutron-capture elements (Ba, La, Eu) and the alpha-elements (Si, Ca); (iii) They exhibit however complex patterns and large star-to-star abundance variations for the lighter elements from C to Al which are not shared by their field counterparts. Among these anomalous patterns, the most striking ones are the so-called universal O—Na anticorrelation (first discovered by the Lick-Texas group) and the Mg—Al anticorrelation (see e.g. [3] and [4]). These patterns have been observed both in evolved stars and in fainter turnoff and subgiant cluster members.

It was soon recognized that the O—Na anticorrelation occurs thanks to the following coincidence: at a similar temperature ($\sim 2.5 \times 10^7$ K), proton-captures on ^{16}O and ^{22}Ne lead to the destruction of O and to the production of ^{23}Na ([5]). On the other hand the Mg—Al anticorrelation results from a sequence of proton-captures followed by β-decays that transforms ^{24}Mg into ^{25}Mg, ^{26}Mg and finally ^{27}Al ([6,7]); this chain is only effective at temperatures higher than $\sim 7 \times 10^7$ K due to the larger Coulomb barrier of Mg. However the internal temperature of the scarcely evolved GC stars which exhibit the abundance anomalies is too low for these abundance variations to be intrinsic. This sustains the so-called *self-enrichment* scenario according to which the abundance differences pre-existed in the material out of which the presently surviving stars formed. This requires the pollution of the intracluster gas by a first generation of more massive and faster evolving stars [8].[4] These features have been observed in all the GCs where they have been looked for and appear thus to be intrinsic properties related to the cluster formation process itself.

Although the nuclear mechanisms that build up the anticorrelations are clearly described, the identification of the astrophysical site were they took place still remains a challenge.

2 Polluting Agents? AGB Stars or Massive Rotating Stars?

It is claimed usually that massive AGB stars are responsible for the observed composition anomalies. However several custom-made detailed AGB models pointed out very severe drawbacks of the AGB pollution scenario. These difficulties stem from the subbtle competition between hot bottom burning and third dredge-up. This latter process does indeed contaminate the AGB envelope with the products of helium burning and creates abundance patterns in conflict with the ones observed [9, 10].

As an alternative [11] proposed the so-called *Winds of Massive Stars* scenario and suggested that hydrogen-burning in massive stars (i.e., with initial masses higher than ~ 10 $M_\odot$) is at the origin of the abundance patterns.[5] In this case, the material ejected in the interstellar matter by gently blowing winds of rapidly rotating massive stars does contribute to the formation of new low-mass stars. This idea is developed in great details and quantified by [14] who also discusses qualitatively other advantages of that idea.

3 Constraints on the IMF

In the framework of the self-enrichment scenario for GCs, we present a new method to constrain the initial mass function (IMF) of the polluters (we refer

---

[4]The fact that we see the same patterns in both scarcely and strongly evolved stars, which have respectively very thin and extremely deep convective envelopes, reveals primordial variations instead of pollution on already formed stars.

[5]The idea that massive stars may be at the origin of some anomalies in the composition of GCs has been discussed by [12] and [13] in order to explain the blue main sequence of the GC Omega Cen: the high helium content of the stars of that sequence could originate from the winds of massive stars, producing a large helium/metal ratio.

the reader to [11] for more details). We use the observed O/Na abundance distribution in NGC 2808 [15] to derive the amount of polluted material with respect to that of original composition. We find that $\sim 30\%$ of this GC stars have a pristine composition, while the remaining 70% has been contaminated to various degrees by H-burning products.

In view of the many uncertainties that enter this complex problem, we explore in some details two different types of self-enrichment scenarii differing in the composition of the polluter ejecta: Scenario I involves two clearly distinct stellar generation, the second of which is made exclusively from the nuclearly processed ejecta of the first one; in this case the ejecta of the polluter stars is processed at various degrees through H-burning. Scenario II involves only one stellar generation, the low-mass stars of which are contaminated on the making and to various degrees by extremely processed ejecta of their more massive and rapidly evolving sisters. Also, we explore both current possibilities for the polluters, namely AGB stars (4–9 $M_\odot$) and massive stars (10–100 $M_\odot$). In each case we take the mass of H-processed ejecta as large as possible, in order to constrain the polluter IMF on one side: For AGB stars, we assume that all the mass outside the white dwarf remnant is processed exclusively through H-burning. For massive stars, we assume that all the mass outside the He-core has the required composition.

We adopt a composite IMF, with an observationally derived part in the mass range 0.1–0.8 $M_\odot$ from [16] and a power-law for higher masses with a slope X that we aim at constraining. Scenario I and Scenario II require respectively slopes X< 0.8 and X< 1.25 if massive stars are the polluting agents, and X< 0.15 and X< 0.95 if AGB stars are the polluters. IMFs with the "classical" Salpeter slope X=1.35 fail to satisfy the observational requirements in any case.

The difficulty of the exercise stems on the fact that the parameter space is quite large. All our present assumptions are made in order to minimize the constraint on the IMF of the polluter stars since their ejecta are used in the most efficient way by forming exclusively stars still alive today. If stars with initial masses higher than 0.8 $M_\odot$ were assumed to be also formed from the polluter ejecta, the corresponding mass required would be still larger and the IMF of the polluting agents even flatter than the ones we derived.

4 Consequences for the Amount of Stellar Residues

Our study has also implications for the amount of dark objects (e.g., residues of stars with initial masses higher than 0.8 $M_\odot$) in GCs. The mass ratio of stellar residues to long-lived stars depends strongly on the assumption made about the mass range of the polluters, especially for flat IMFs as those required to explain the abundance distribution in GCs.

We find that the present number ratio of white dwarfs to long-lived stars, N_{WD}/N_{MS}, should be around 0.2 if the polluters were AGB stars, and much smaller if the polluters were massive stars. These values are lower than the N_{WD}/N_{MS} ratio infered by [17] in the case of the GC M4, and which is of the order of 1.

The low number ratio of white dwarfs over low-mass stars we obtain does not necessary point to a fatal flaw for the self-enrichment scenarii. It may well be that the ejecta mass and the resulting number of second generation stars is smaller than assumed (this is in fact certainly the case in reality), in which case a N_{WD}/N_{MS} ratio closer to the observationally infered one would be obtained.

References

1. R. Gratton, C. Sneden, E. Carretta: ARAA **42**, 385 (2004)
2. C. Sneden: IAU 228 on *From Li to U: Element tracers of early cosmic evolution*, Cambridge: Cambridge University Press, Eds. Hill, François, Primas, p. 337 (2005)
3. S.V. Ramirez, J.G. Cohen: AJ **123**, 3277 (2002)
4. I.I. Ivans, C. Sneden, R.P. Kraft, N.B. Suntzeff, V.V. Smith, G.E. Langer, J.P. Fullbright: AJ **118**, 1273 (1999)
5. P.A. Denissenkov, S.N. Denissenkova SvA Lett. **16**, 275 (1990)
6. G.E. Langer, R. Hoffman, C. Sneden: PASP **105**, 301 (1993)
7. G.E. Langer, R. Hoffman: PASP **107**, 1177 (1993)
8. P.L. Cottrell, G.S. Da Costa: ApJ **245**, L79 (1981)
9. Y. Fenner, S. Campbell, A.I. Karakas, J.C. Lattanzio, B.K. Gibson: MNRAS **353**, 789 (2004)
10. C. Charbonnel: IAU Symposium 228 on *From Li to U: Element tracers of early cosmic evolution*, Cambridge: Cambridge University Press, Eds. Hill, François, Primas, p. 347 (2005)
11. N. Prantzos, C. Charbonnel: A&A **458**, 135 (2006)
12. J.E. Norris: ApJ **612**, L25 (2004)
13. A. Maeder, G. Meynet: A&A **448**, L37 (2006)
14. T. Decressin, G. Meynet, C. Charbonnel, N. Prantzos, S. Eckström: A&A **464**, 1029 (2006)
15. E. Carretta, A. Bragaglia, R.G. Gratton, F. Leone, A. Recio-Blanco, S. Lucatello: A&A **450**, 523 (2006)
16. F. Paresce, G. De Marchi: ApJ **534**, 870 (2000)
17. H. Richer, J. Brewer, G. Fahlman, et al.: ApJ **574**, L151 (2002)

Globular Clusters in the Direction of the Inner Galaxy

J. Alonso-García and M. Mateo

Department of Astronomy, University of Michigan, 500 Church St, Ann Arbor, MI 48109, USA `jalonso,mmateo@umich.edu`

The age, chemical and kinematic distributions of stellar populations provide powerful constraints on models of the formation and evolution of the Milky Way. The globular clusters (GC) constitute a specially useful case because the stars within individual clusters are coeval and spatially distinct. But a serious limitation in the study of many globular clusters – especially those located near the Galactic Center – has been the existence of large and differential extinction by foreground dust. We have observed a sample of GC in the direction of the inner Galaxy (see Fig. 1) and now, applying the technique described in [1] to differentially deredden the cluster photometric data, we intend to derive precise relative ages, abundances, and distances from comparisons of the cluster color-magnitude diagrams (CMD) with modern stellar evolutionary models. The main goal of the work is to use such data to determine the relative mixtures, age sequence, and chemical properties of the cluster populations in the inner Milky Way. An important by-product of this process is a map of the differential extinction across a cluster to an angular resolution limited by the number of stars available to apply the method. This will allow us to achieve another of the goals of this work: to sample the ISM along numerous low latitude lines of sight, paying special attention to the low mass, low column density clouds of the cold ISM which are normally invisible at IR wavelengths.

We have already extracted the photometry and astrometry of the observed GC, and have been able to build their CMD. We have not applied our dereddening method yet, but from the CMD we can observe that the internal reddening is an important problem in almost all of these GC (see Fig. 2). Another important issue is the presence of a significant number of field stars in some of these CMDs. We, along with a group from the department of Statistics of the University of Michigan, are working on ways to simultaneously model the age, chemical and spatial distributions of stars along these sightlines for both the cluster and the field populations. We will also take into account any present reddening, which corresponds to local divergence between the observational and model CMDs as a function of position within a given field.

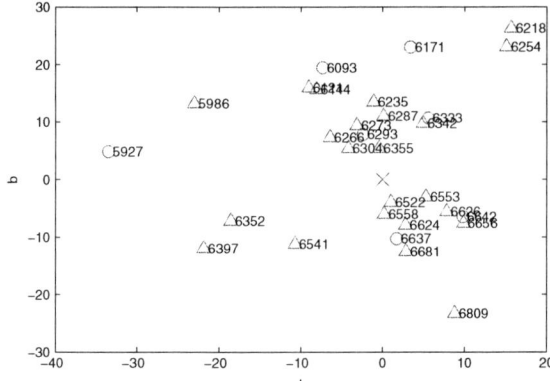

Fig. 1. Our sample of 30 globular clusters observed using the IMACS imager on the LCO Baade 6.5 m telescope, in B, V, and I. The GC observed in IMACS f/2 configuration ($27.5' \times 27.5'$) are shown as *circles* and the GC observed in IMACS f/4 configuration ($15' \times 15'$) are shown as *triangles*. The position of the Galactic center is shown with a *cross*. Time has been also awarded to this project to use the ACS camera on the HST during the present cycle, so we have the best expected photometry of the cores of the clusters.

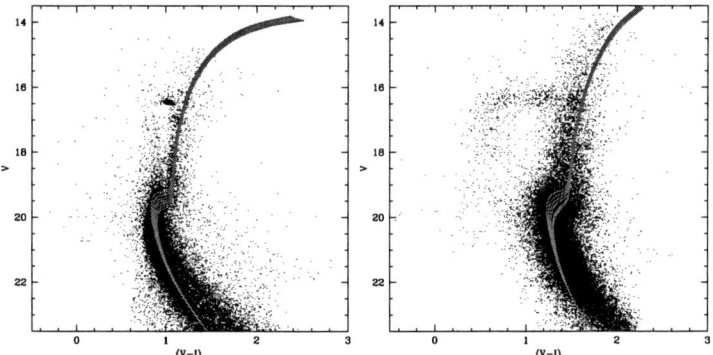

Fig. 2. CMDs of the inner $3'$ of NGC 6624 (*left*) and M62 (*right*), with the Yale isochrones with ages from 10 to 16 Gyr overplotted. Due to the internal reddening it is not yet possible to define a proper age for these clusters until we calculate the dereddening map for these fields.

Reference

1. von Braun, K. & Mateo, M.: AJ **121**, 1522 (2001)

Globular Cluster Research with Astronomical Archives

Carol A. Christian

Space Telescope Science Institute, 3700 San Martin Drive, Baltimore, MD 21218, USA carolc@stsci.edu

Abstract. As astronomical data archives advance, scientists have a plethora of diverse resources for research at their fingertips. Because these archives use standardized protocols, they can be useful for research through appropriate tools developed either by the host facility or by the science end users. These data holdings also have enormous potential for educational purposes from pre-college through non-majors courses and expert course curricula in universities. This paper focuses on software services and tools of use to researchers and will concentrate especially on opportunities presented by the National Virtual Observatory.

1 Literature Searches

The Astrophysics Data System (ADS) is commonly used to search astronomical literature. The National Virtual Observatory (NVO) provides an interface to catalogs through the *Registry* and the *Catalog Coverage Maps and Source Inventories*. These tools are useful for finding literature, diverse information, and data related to a target or type of target. By browsing the *Registry* output, links to catalogs can be found and the data retrieved.

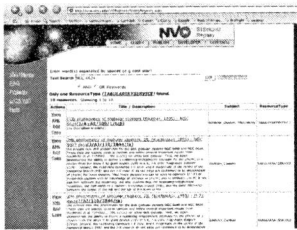

Fig. 1. NVO *Registry* search for cluster NGC 6624. Clicking on individual links allows the user to view individual *Registry* entries and also the URL for the catalog or resource listed.

The *Coverage Maps* can be useful for comparing a variety of sources of data and information, for example the sky coverage of observations taken by individual observatories or the distribution of a particular kind of object across the sky.

2 Data Location

The NVO *DataScope* tool is used to locate lists of observations and some preview imagery from a variety of registered sources. A search for data on NGC 6624, for example, yields listings of data from HST, 2MASS, ROSAT, XMM, SPITZER, Chandra and other observatories. By drilling down into the returned results in the *DataScope* interface, lists of observations can be located, and depending upon the services provided by the individual observatory or archive, links to data can be retrieved. For example the list of HST observations provides links to individual exposures with preview images (Fig. 2). Data from the *Sloan Digital Sky Survey* can be located through NVO or through the SDSS *SkyServer*, and various measurements and information about individual objects can be overplotted and displayed with the *Navigate* tool.

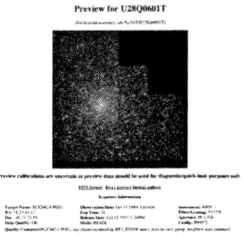

Fig. 2. Preview image of an HST observation retrieved through the NVO *DataScope* tool and the located list of HST observations for NGC 6624.

3 Data Viewing and Comparison

Data can be loaded directly from *DataScope* and/or stored to local disk. Tools like **Aladin** (local copy or served from the network) are useful tool for viewing data, combining data and overplotting other information such as catalog data. Users can also assemble several observations for comparison and creating color images from combinations of individual observations.

4 Summary

A variety of tools now exist for locating and acquiring data relevant to globular cluster research. For example, the NVO provides some useful tools primarily for locating the existence of catalog data and lists of observations pertaining to individual objects. Basic tools for retrieving data exist and other tools available over the network or as local copies (such as Aladin) prove useful for intercomparing retrieved observations and ancillary data. These tools continue in development and are intended to eventually provide mature research environments for scientists.

Super-He-Rich Populations in Globular Clusters

Chul Chung, Young-Wook Lee, Suk-Jin Yoon, Seok-Joo Joo, and Sang-Il Han

Center for Space Astrophysics, Yonsei University, Seoul 120-749, Korea
mitchguy@galaxy.yonsei.ac.kr

Abstract. Recently observations of the color-magnitude diagrams (CMDs) of the massive globular cluster ω Centauri have shown that it has a striking double main sequence (MS). Here we confirm, with the most up-to-date Y^2 isochrones, that this special feature can only be reproduced by assuming a large variation ($\Delta Y = 0.15$) of primordial helium abundance among several distinct populations in this cluster (Fig. 1). We further show that the same helium enhancement required for this special feature on the MS can by itself reproduce the extreme horizontal branch (HB) stars observed in ω Cen (Fig. 1). Similarly, the complex features on the HBs of other globular clusters, such as NGC 2808, NGC 6388 and NGC 6441, are explained by large internal variations of helium abundance (Fig. 2). The presence of super-helium-rich populations in some globular clusters suggests that a third parameter, other than metallicity and age, also influences the CMD morphology of these clusters.

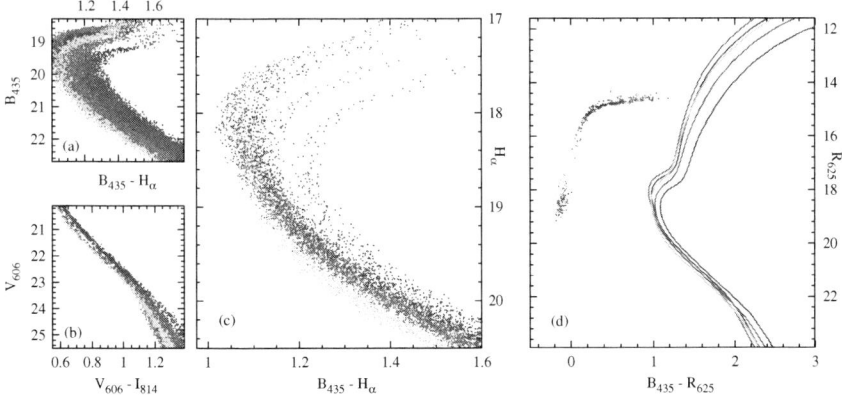

Fig. 1. Population models (synthetic HBs and new Y^2 isochrones) for ω Cen. Panel (**a**), (**b**) and (**c**) can be directly compared with the observed CMDs of Bedin et al. ([1]; see their Fig. 1). Panel (**d**) is for the case in which the helium abundances for the metal-rich populations are significantly enhanced. Only the case of $\Delta Y > 0.15$ can reproduce the observed features on the MS and HB simultaneously.

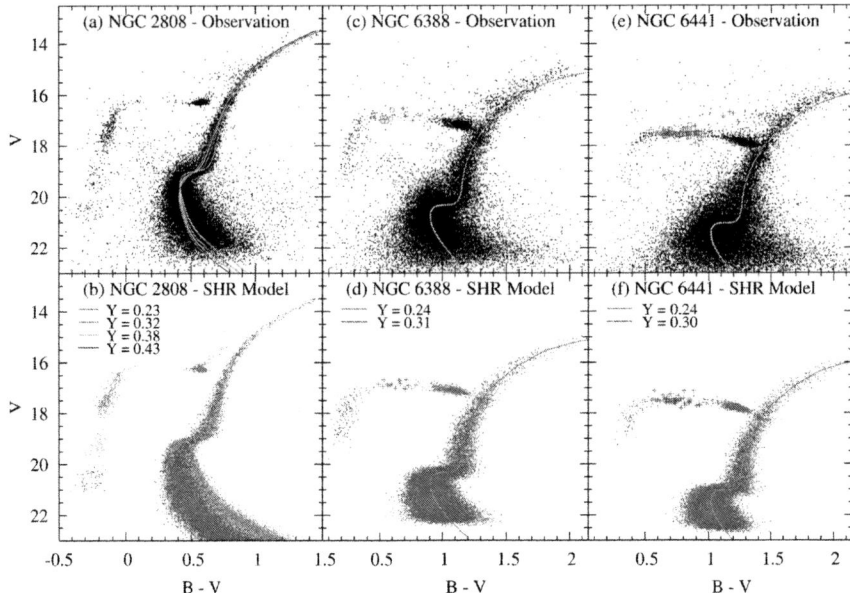

Fig. 2. Comparison of super-helium-rich models with the observed CMDs (**a**),(**c**) and (**e**) for NGC 2808, NGC 6388 & NGC 6441. While the CMDs are from Piotto et al. [2] and Rich et al. [6], RR Lyrae variables are from Pritzl et al. [3–5]. In (**b**), the synthetic CMD is constructed with the assumption that four distinct populations in this cluster have different helium abundance but the same metallicity (Z=0.0014) and age (10.1 Gyr). In (**d**) and (**f**), the Y= 0.24 and the Y= 0.30, 0.31 populations are respectively shown as red and blue dots. The spreads in helium abundance required to reproduce the CMD morphology are denoted. The observed bimodal HBs are explained with the sum of red HB with normal He abundance and blue HB with enhanced He abundance.

References

1. L. R. Bedin, G. Piotto, J. Anderson, S. Cassisi, I. R. King, Y. Momany, & G. Carraro: ApJ. **605**, L125 (2004)
2. G. Piotto, I. R. King, S. G. Djorgovski, C. Sosin, et al.: A&A. **621**, 777 (2002)
3. B. Pritzl, H. A. Smith, M. Catelan, A. V. Sweigart: ApJ **530**, 41 (2000)
4. B. Pritzl, H. A. Smith, M. Catelan, A. V. Sweigart: AJ **122**, 2600 (2001)
5. B. Pritzl, H. A. Smith, M. Catelan, A. V. Sweigart: AJ **124**, 949 (2002)
6. R. M. Rich et al.: ApJ **484**, 25 (1997)

Testing the BH 176 and Berkeley 29 Association with GASS/Monoceros

Peter M. Frinchaboy

University of Virginia, Department of Astronomy, P.O. Box 400325, Charlottesville, VA 22904-4325, USA `pmf8b@virginia.edu`

It has been previously noted that the outermost open clusters in the Milky Way seem to lie in a string-like configuration that is coincident with, and may be associated to, the Galactic anticenter stellar structure (GASS) or the "Monoceros Ring" [1,2]. Among the clusters that have been suggested to be associated with GASS are Berkeley 29 (Be29) and BH176, which have recently had their proper motion (μ) determined by [3] (hereafter D06). Matching the μ determinations from D06, to previously published radial velocities (RVs) for Berkeley 29 [4,5] and BH 176 [4] allows an attempt to derive their orbits.

D06 provided positions and 2MASS photometry [6] for stars in the fields of hundreds of open clusters derived from Tycho-2 and UCAC-2 μ, including BH176 and Be29. We matched the 2MASS photometry and positions from D06 to the photometry and RV samples for Be29 and BH176 from F06. We find that the stars used by D06 to determine a bulk μ for BH 176 are too bright to be cluster members, as shown in the $(V, V-I)$ color-magnitude diagram (CMD, Fig. 1a) from [4]. For Be29, we find a similar problem for most of the D06 μ stars (Fig. 1b); however, the tip of the Be29 red giant branch does overlap the faint red end of the UCAC-2 sample. Comparing the two samples finds only one star in common (i.e., having both a measured μ and an RV consistent with cluster membership), but this does allow exploration of a possible orbit for Be29.

The orbital motions of Be29 were calculated by using an orbit integrator and Galactic model from [7]. The orbit is followed backwards for a time interval equal to the age of each cluster (3.7 Gyr; [4]). We find that the resulting orbit, shown in Fig. 2, is roughly consistent with the orbit for GASS [8]. The preliminary orbit for Be29 shows that the cluster and GASS share a similar tilt with respect to the Galactic plane ($\sim 40 \pm 30$ deg for Be29 and ~ 17 deg for GASS; [4]), though of course further work is needed to improve the cluster motion. Moreover, whether GASS is an accretion product of just a flare of the disk is still debated [9]. A detailed 3-D kinematic follow-up on possible GASS clusters is needed to determine whether their origins due to are normal formation or accretion.

PMF acknowledges funding by the F.H. Levinson Fund of the Peninsula Community Foundation, a NASA GSRP, U. Virginia Faculty Senate Dissertation Fellowship, the Virginia Space Grant Consortium, and by a AAS-ITG.

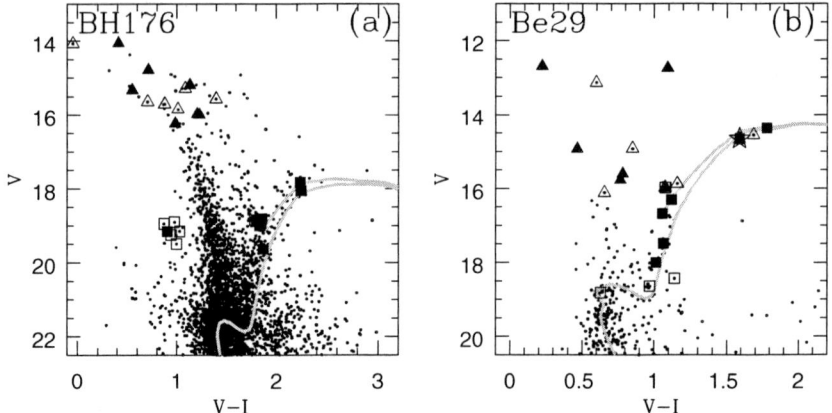

Fig. 1. (a) Color-magnitude diagram (CMD) for BH176 from F06. *Triangles* denote proper motion stars from D06, with *filled triangles* being stars that have probabilities > 60%. *Squares* are RV stars from F06, with filled squares denoting F06 RV members and *open squares* the non-members. The [10] isochrone match from F06 is overplotted in *grey*. (b) Same as (a) for Be29. The star denotes star used for orbit.

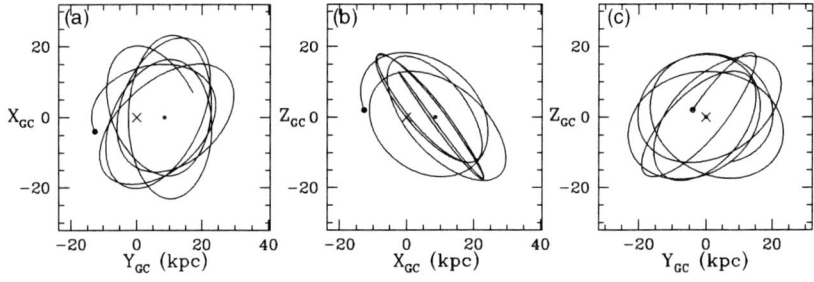

Fig. 2. Orbit for Be29 based on the one star in common between the RV membership of F06 and with proper motion from D06. The *large dot* denotes the cluster's current position, while the *cross* denotes the position of the Sun. This preliminary Be29 orbit shows the same orientation as GASS [8] in the Galactic X, Y plane (b).

References

1. N.F. Martin, R.A. Ibata, M.Bellazzini et al.: MNRAS **348**, 12 (2004)
2. P.M. Frinchaboy et al.: ApJ **602**, L21 (2004)
3. W.S. Dias, M. Assafin, V. Florio et al.: A&A **446**, 949 (2006)
4. P.M. Frinchaboy, R.R. Muñoz, R.L. Phelps et al.: AJ **131**, 922 (2006)
5. G. Carraro et al.: A&A, **128**, 1676 (2004)
6. M.F. Skrutskie, R.M. Cutrie, R. Stiening et al.: AJ **131**, 1163 (2006)
7. K.V. Johnston, D.N. Spergel, L. Hernquist: ApJ **451**, 598 (1995)
8. J. Peñarrubia, D. Martinez-Delgado, H.W. Rix et al.: ApJ **626**, 128 (2005)
9. Y. Momany, S. Zaggia, G. Gilmore et al.: A&A **451**, 515 (2006)
10. L. Girardi, A. Bressan, G. Bertelli, C. Chiosi: A&AS **141**, 371 (2000)

New Yonsei-Yale (Y^2) Isochrones and Horizontal-Branch Evolutionary Tracks with Helium Enhancements

S.-I. Han[1,2], Y.-C. Kim[2], Y.-W. Lee[1,2], S.K. Yi[1,2], D.-G. Kim[1,2], and P. Demarque[3]

[1] Center for Space Astrophysics, Yonsei University, Seoul 120-749, Korea
  sihan@csa.yonsei.ac.kr
[2] Department of Astronomy, Yonsei University, Seoul 120-749, Korea
[3] Department of Astronomy, Yale University, New Haven, CT 06520-8101, USA

Abstract. Recent studies [4,9,8,10,3] suggest that peculiar features observed in the horizontal-branch (HB) and main-squence (MS) of some Globular Clusters (GCs) are naturally reproduced by the presence of super-He-rich populations. However, there are no isochrones and self-consistent HB tracks available in the literature that are based on the up-to-date input physics and wide ranges of He and metal abundances. Here we present new sets of Y^2 isochrones and HB evolutionary tracks with the effects of He enhancements. The most up-to-date input physics are adopted, and the effects of α-enhancements ([α/Fe]=0.3) are also fully taken into account. These isochrones and tracks have been constructed for more realistic evolutionary population syntheses for old stellar populations, such as GCs and elliptical galaxies. The new isochrones and HB tracks well reproduce the observed color magnitude diagrams of Galactic GCs, including peculiar features observed in the HB and MS of ω Centauri and NGC 2808.

Table 1. Input physics

Input parameters	Description
Solar mixture	Grevesse and Noels [5]
OPAL Rosseland mean opacities	Rogers and Iglesias [11], Iglesias and Rogers [6]
Low temperature opacities	Alexander and Ferguson [1]
Equations of state	OPAL EOS; Rogers and Iglesias [11]
Energy generation rates	Bahcall and Pinsonneault [2], private communication
Neutrino losses	Itoh et al. [7]
Convective core overshoot	0.0$\sim$0.2 Hp when convective core develops
Helium diffusion	Thoul et al. [12]
Mixing length parameter	l/Hp=1.7431

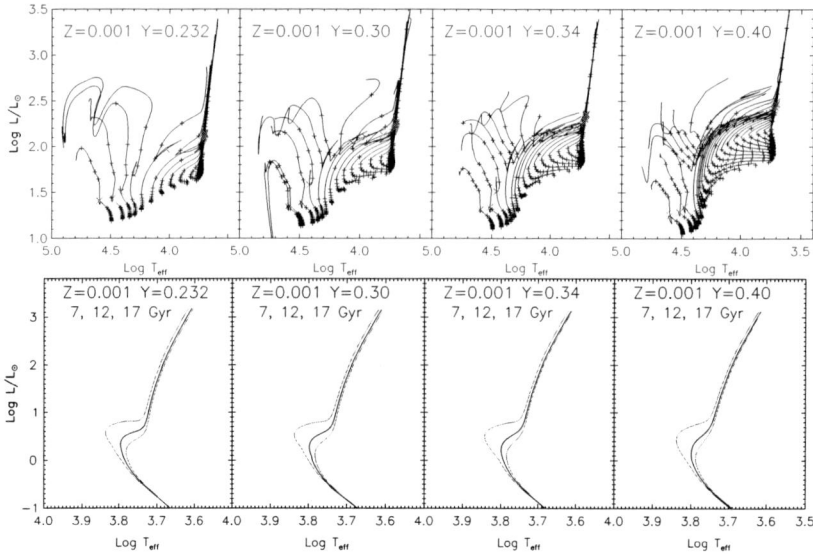

Fig. 1. New Y^2 isochrones and HB evolutionary tracks for standard He abundances ($\Delta Y/\Delta Z = 2.0$) and those for enhanced helium abundances.

References

1. D.R. Alexander and J.W. Ferguson: ApJ **437**, 879 (1994)
2. J.N. Bahcall and M.H. Pinsonneault: Rev. Mod. Phys. **60**, 297 (1992)
3. K. Bekki and J.E. Norris: ApJ **637**, L109 (2006)
4. F. D'Antona and V. Caloi: ApJ **611**, 871 (2004)
5. N. Grevesse and A. Noels: *Origin and Evolution of the Elements* (Cambridge University Press, Cambridge 1993) pp 15–25
6. C.A. Iglesias and F.J. Rogers: ApJ **464**, 943 (1996)
7. N. Itoh et al.: ApJ **339**, 354 (1989); erratum, ApJ 360, 741
8. Y.-W. Lee et al.: ApJ **621**, L57 (2005)
9. J.E. Norris: ApJ **612**, L25 (2004)
10. G. Piotto et al.: ApJ **621**, 777 (2005)
11. F.J. Rogers and C.A. Iglesias: The OPAL Opacity Code: New Results. In: *ASP Conf. Ser.*, vol 78, ed by S.J. Adelman, W.L. Wiese (Astronomical Society of the Pacific, San Francisco 1995) pp 31–50
12. A.A. Thoul, J.N. Bahcall, and A. Loeb: ApJ **421**, 828 (1994)

Search for Candle Stars in Globular Clusters: Spectroscopic Analysis of Post-AGB Candidates

G. Jasniewicz[1] and M. Parthasarathy[2]

[1] UMR 5024 CNRS/UM2, Université Montpellier II, France
    gjasniew@graal.univ-montp2.fr
[2] Indian Institute of Astrophysics, Koramangala, Bangalore 560034, India
    partha@iiap.res.in

1 Introduction

As satellites of galaxies, Globular Clusters (GCs) can provide extragalactic standard candles. An interesting approach was raised by Bond [1] who was the first to propose post-AGB stars as possible standard candles. Indeed, post-AGB are intrinsically bright within GCs, and theoretical tracks (Dorman et al. [2]) show a narrow luminosity function as they evolve through spectral types F and A. We give here the results of a first investigation among "UV-bright stars" carried out with the VLT-UVES at ESO by Jasniewicz et al. [3]. Targets are the so-called "UV-bright stars" in GCs. This term was introduced by Zinn et al. [5] for those stars lying above the horizontal branch and bluer than red giants. The survey for UV-bright and post-AGB stars in the GCs of our Galaxy is very far from being complete.

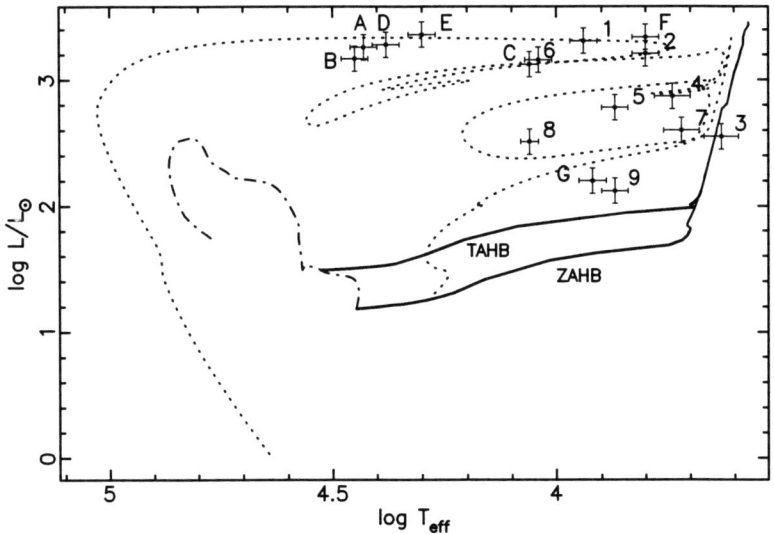

Fig. 1. The H-R diagram for UV-bright stars.

2 The H-R Diagram for UV-Bright Stars

All evolutionary sequences in Fig. 1 are from Dorman et al. [2] and drawn for [Fe/H]= -1.48, a core mass of $0.485 M_{sun}$, and a helium composition of 0.247. Solid thick lines: Zero Age HB (ZAHB) and Terminal Age HB (TAHB) sequences. The three post-HB evolutionary sequences correspond to three total masses (i.e core + envelope) at the ZAHB: $0.488 M_{sun}$ (dotted-dashed: AGB-manqué sequence), $0.520 M_{sun}$ (dotted line: post-early AGB sequence), $0.900 M_{sun}$ (solid thin line: AGB sequence). The highest luminosity at $L/L_{sun}=3.4$ is quasi-independent of metallicity.

Table 1.

NGC	5986	5986	6218	6656	6656	6712	7078	7078	7078
Star	ID6	ID7	ZNG7	V-4	ZNG5	ZNG1	K260	K996	K1082
Label	1	2	3	4	5	6	7	8	9
NGC	6254	5904	6712	5139	6205	5139	6205		
Star	ZNG1	ZNG1	ZNG1	ROA5701	BARN29	ROA24	ZNG4		
Label	A	B	C	D	E	F	G		

All stars plotted in Fig. 1 are identified in Table 1. Data for stars A to G are taken from the literature. Stars labelled 1 to 9 have been observed by Jasniewicz et al. [3], in using the blue and red arms of the UV-Visual Echelle Spectrograph (UVES) mounted on the VLT-UT2 (ESO-Chile). Spectral resolution was 40,000. Stars ID6 and ID7 in NGC 5986 are A–F supergiant stars with a post-AGB luminosity; ID7 shows a relative overabundance of some s-process elements whereas ID6 is O-rich and absorption lines of s-elements are very weak or not present at all. ID7 has probably evolved to the tip of the AGB and experienced the 3rd dredge-up. Both stars could be representative of the Have and the Have-Nots (s-process abundances) post-AGB stars discussed by Van Winckel [4]. Star NGC 6656 V-4 shows a significant overabundance of s-process elements but is not luminous enough to be in a post-AGB phase. This star could have suffered a 3rd dredge-up episode, or has been enriched in some way (binarity or primordial pollution).

References

1. H.E. Bond: IAU Symposium No 180, 460 (1997)
2. B. Dorman, R.T. Rood, & R.W. O'Connell: ApJ, 419, 596 (1993)
3. G. Jasniewicz, P. de Laverny, M. Parthasarathy, A. Lèbre, & F. Thévenin: A&A, 423, 353 (2004)
4. H. Van Winckel: ARAA, 41, 391 (2003)
5. R.J. Zinn, E.B. Newell, & J.B. Gibson: A&A, 18, 390 (1972)

The Lack of Binaries Among Hot Horizontal Branch Stars: M80 and NGC5986

C. Moni Bidin[1], S. Moehler[2], G. Piotto[3], Y. Momany[3], A. Recio-Blanco[4], and R.A. Méndez[1]

[1] Departamento de Astronomía, Universidad de Chile, Chile
[2] European Southern Observatory, Garching, Germany
[3] Dipartimento di Astronomia, Università di Padova, Italy
[4] Observatoire de la Côte d'Azur, Dpt. Cassiopée, France

1 Introduction

Extreme horizontal branch (EHB) stars play an important role in extragalactic astronomy, since they have been individuated as possibly being responsible for the UV upturn in elliptical galaxies and in the bulges of spiral galaxies, that has been proposed as an independent age indicator for this type of galaxies. In recent years the "binary scenario", in which EHB star formation is related to dynamical interactions inside binary systems, has been proposed as the main channel for their formation. In fact [1] indicated that 69±9% of field EHB stars should be close binary systems with short periods P≤10 days. Nevertheless, more recently [5] found a noticeably lower binary fraction (40–45%), and [2] found no evidence of binarity among 18 EHB stars in globular cluster NGC6752. They estimated that within a 95% confidence level the close binary fraction in EHB of this cluster should be lower than 20%.

Here we present preliminary results of the extension of the previous survey.

2 Results

Observations, data reduction, radial velocity (RV) measurements and error analysis were performed as in [2]. Systematic errors must still be rigorously measured, but the corrections applied here should be within 2–3 km/s from the true value. Our results are plotted in Fig. 1.

The most prominent result is that again we fail to detect the high RV variations observed among many field EHB stars. We conclude that there is no *clear* evidence of binarity in the samples, although we individuate a possible exception in M80. This star shows a modest (31 km/s) but statistically significant (nearly 5σ from zero) variation, and we consider it an interesting candidate, although the variation is not high enough to rule out the possibility that it is due to some distortion induced by noise. In NGC5986 we find one variation slightly higher than 3σ (26.1±8.3 km/s, i.e. 3.1σ), but it is not trustworthy due to the low S/N of the spectra. Such a variation is statistically

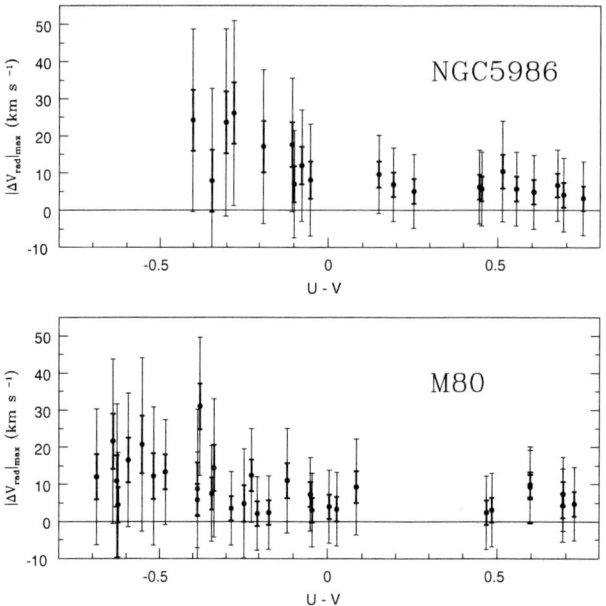

Fig. 1. Maximum RV variation observed for each target in 4 nights of observations, as a function of color U-V. Photometric data are from [3] and [4]. The thick errorbar indicates the 1σ interval, the thin one the 3σ. Each night we collected up to two spectra per star but observations on M80 were undersampled due to string wind from north.

reasonable among the great number of our measurements, it is hard to consider it significant. No further conclusion can be drawn at the moment.

In our sample we analyzed 11 EHB stars in M80. Our observations fix the best estimate for the close binary fraction $f=15\%$. Unfortunately it is not a strong constraint because of the poor temporal sampling, that lowered the sensitivity of the survey. Statistically considering our observations and results, the probability that $f \geq 48\%$ is lower than 10%, enough to rule out the high f observed by [1] but not the intermediate one found by [5].

In NGC5986 we observed 5 EHB stars. The sample is much too small to attempt any statistical consideration, but the result again points out a lack of close binaries, even in the presence of one candidate, because in this cluster our survey reaches a high detection probability (80% on average).

We can then conclude that in both clusters the general lack of close binaries among EHB stars is confirmed.

References

1. Maxted, P.F.L., Heber, U., Marsh, T.R., & North, R.C.: MNRAS **326**, 1391 (2001)
2. Moni Bidin, C., Moehler, S., Piotto, G., et al.: A&A **451**, 499 (2006)
3. Momany, Y., Cassisi, S., Piotto, G., et al.: A&A **407**, 303 (2003)
4. Momany, Y., Bedin, L.R., Cassisi, S., et al.: A&A **420**, 605 (2004)
5. Napiwotzki, R., Karl, C.A., Lisker, T., et al.: Ap&SS **291**, 321 (2004)

Semi-Empirical Determination of the Mass Distribution of Horizontal Branch Stars in M3

A. Valcarce and M. Catelan

Departamento de Astronomía y Astrofísica, Pontificia Universidad Católica de Chile, Av. Vicuña Mackenna 4860, 782-0436 Santiago, Chile `avalcarc@uc.cl`, `mcatelan@astro.puc.cl`

Abstract. We determine, by means of a semi-empirical study, the masses of horizontal branch stars in the globular cluster M3 (NGC 5272). We used the most recent and reliable observational datasets (broadband BVI photometry) available for the cluster, both for variable and nonvariable stars, to infer the most likely masses of individual horizontal branch stars by comparison against theoretical evolutionary tracks, suitably transformed to the observational planes. We found a mass distribution that is adequately described by a Gaussian, with $\langle M \rangle = 0.64\,M_\odot$ and $\sigma = 0.020\,M_\odot$.

1 Method and Result

In order to determine the mass distribution (MD) of non-variable horizontal branch (HB) stars in M3 we used the database of [7] in the filters (B, V) and (V, I) for the outer and innermost regions, respectively. For variable stars we used the mean magnitudes determined by [2] ($\langle B \rangle$ and $\langle V \rangle$) by integrating the light curves in intensity units and converting the result of this integration back to magnitude units. We used as color only $\langle B \rangle - \langle V \rangle$, since this is the same as the color of the equivalent static star to within 0.028 mag or better [1].

We used the theoretical evolutionary tracks for horizontal branch stars described in [5], adopting a metallicity of $[Fe/H] = -1.57$ [8] and an abundance of alpha elements of $[\alpha/Fe] = +0.3$ (e.g., [3]), as appropriate for M3. We transformed the evolutionary tracks to the observational plane using color transformations from [11], and interpolated additional evolutionary tracks for every $2 \times 10^{-4} M_\odot$ interval.

We adopted a distance modulus in the V-band equal to 15.0 mag to provide a better match between the predicted evolutionary time scales and the distribution of the HB stars in the CMD, and reddening values of $E(B-V) = 0.01$ mag [8] and $E(V-I) = 0.016$ mag [9].

In order to determine the MD of M3 horizontal branch stars as a whole, we had to adjust the relative proportions of variable and nonvariable stars according to the number ratios $B{:}V{:}R = 39{:}40{:}21$ [6], since the databases for variables and non-variables correspond to sectors of M3 with different sizes. Finally, we found that the mass distribution is essentially Gaussian (Fig. 1), with $\langle M \rangle = 0.64 M_\odot$ and $\sigma = 0.020 M_\odot$.

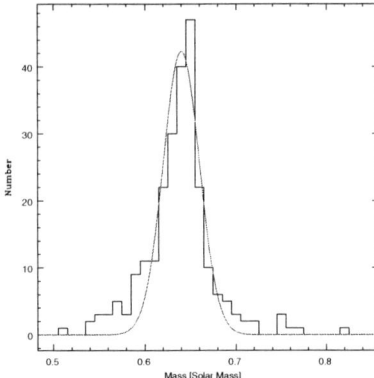

Fig. 1. MD for M3 HB stars, obtained from the combination of variable and non-variable stars. The best-fitting Gaussian has $\langle M \rangle = 0.64 M_\odot$ and $\sigma = 0.020 M_\odot$.

This result supports the semi-empirical MD previously found by [10] (who derived masses by minimizing the distance to the zero-age HB), and is at odds with the sharply bimodal distribution recently suggested by [4]. Therefore, the Castellani et al. suggestion that the peaked period distribution for the RR Lyrae stars in M3 can be naturally explained by a bimodal MD along the HB appears not to be supported by the available CMD data.

This work was supported by Proyecto Fondecyt Regular #1030954.

References

1. Bono, G., Caputo, F., & Stellingwerf, R. 1995, AJ, 99, 263
2. Cacciari, C., Corwin, T. M., & Carney, B. W. 2005, AJ, 129, 267
3. Carney, B. W. 1996, PASP, 108, 900
4. Castellani, M., Castellani, V., & Cassisi, S. 2005, A&A, 437, 1017
5. Catelan, M., Borissova, J., Sweigart, A. V., & Spassova, N. 1998, ApJ, 494, 265
6. Catelan, M. 2004, ApJ, 600, 409
7. Ferraro, F. R., Carretta, E., Corsi, C. E., Fusi Pecci, F., Cacciari, C., Buonanno, R., Paltrinieri, B., & Hamilton, D. 1997, A&A, 320, 757
8. Harris, W. E. 1996, AJ, 112, 1487 (Feb. 2003 version)
9. Rieke, G. H., & Lebofsky, M. J. 1985, ApJ, 288, 618
10. Rood, R. T., & Croker, D. A. 1989, in The Use of Pulsating Star in Fundamental Problems of Astronomy, IAU Colloq. 111, ed. E. G. Schmidt (Cambridge: Cambridge University Press), 103
11. VandenBerg, D. A., & Clem, J. L. 2003, AJ, 126, 778

Part II

The Most Massive Clusters

Globular Clusters, Galactic Nuclei and Supermassive Black Holes

Patrick Côté

Herzberg Institute of Astrophysics, National Research Council of Canada, 5071 W. Saanich Road, Victoria, BC V8S 1M2, Canada Patrick.Cote@nrc-cnrc.gc.ca

Abstract. I briefly review past efforts to characterize the properties of compact stellar nuclei in nearby galaxies, and describe a program – the ACS Virgo Cluster Survey (ACSVCS) – to observe 100 early-type members of the Virgo Cluster using the *Hubble Space Telescope*. The properties of the nuclei from this survey are compared with those of the globular clusters. A similar comparison between the nuclei and the supermassive black holes found in many massive galaxies points to an unforseen causal relationship between these two types of *central massive objects*.

1 Historical Background

It is well known that galaxies often contain compact nuclei near their centers. As the largest collection of early-type galaxies in the nearby universe, the Virgo Cluster has figured prominently in efforts to understand the properties and origin of these nuclei. In their landmark study of the Virgo Cluster, Binggeli, Sandage & Tammann [2] carried out a visual search for nuclei using blue-sensitive photographic plates, reporting the presence of a nucleus in about a quarter of Virgo's dwarf galaxies. Studies of the nuclei in Virgo galaxies include both ground-based (e.g., [5, 6, 1, 10]) and HST/WFPC2 imaging surveys [24] (Miller 2006, these proceedings). Recently, Geha et al. [16, 17] combined HST/WFPC2 imaging with Keck spectroscopy to examine the nuclear dynamics of a small number of nucleated dwarfs in Virgo.

Theories for the origin of these nuclei include mergers of star clusters through dynamical friction, mergers in the presence of an external tidal field, gas accretion with – and without – galaxy mergers, gas accretion in the presence of IGM confinement, two-body relaxation processes around a central black hole, and the fading of star clusters in evolving dIrr/BCD galaxies. Whatever their origin, there are strong reasons to believe that a modern, high-resolution survey of nuclei in early-type galaxies is in order. First and foremost, HST imaging of *late-type galaxies* has revealed that 50–70% of these systems have compact stellar clusters at, or near, their photocenters ([31, 7, 25, 3, 4, 34]; Seth 2006, these proceedings). While this is seemingly a larger fraction than found for early-type galaxies, few studies have had the depth and spatial resolution needed to characterize the nuclear properties of such galaxies.

2 The ACS Virgo Cluster Survey

The ACS Virgo Cluster Survey (ACSVCS; [8]) consists of HST imaging for 100 members of the Virgo Cluster, supplemented by imaging and spectroscopy from WFPC2, Chandra, Spitzer, Keck, KPNO, and CTIO. The program galaxies span a range of ≈ 460 in blue luminosity and have early-type morphologies: E, S0, dE, dE, N or dS0. All HST images were taken with the Advanced Camera for Surveys (ACS; [14]) using a filter combination roughly equivalent to the g and z bands in the SDSS photometric system. The images cover a $\approx 200'' \times 200''$ field with $\approx 0.1''$ resolution.

This article examines the connection between globular clusters, galactic nuclei and supermassive black holes (SBHs), summarizing results from the subset of ACSVCS papers which deal with the morphology, isophotal parameters and surface brightness profiles for early-type galaxies [12], their central nuclei [9] and scaling relations for nuclei and SBHs [13]. Other papers in this series have discussed the data reduction pipeline [20], the connection between low-mass X-ray binaries and globular clusters [21], the measurement and calibration of surface brightness fluctuation magnitudes [26, 27], the connection between globular clusters and ultra-compact dwarf galaxies [19], the color distributions of globular clusters [29], the half-light radii of globular clusters (Jordán et al. [22]) and diffuse star clusters in early-type galaxies [30].

3 Galactic Nuclei in the ACSVCS

3.1 Analysis of the Brightness Profiles and Identification of Nuclei

For each galaxy, azimuthally averaged surface brightness profiles were derived as described in Ferrarese et al. [12] and Côté et al. [9]. Nuclei were identified through a direct inspection of the ACS images and by fitting the surface brightness profiles, where nuclei appear as an "excess" over the inward extrapolation of the best-fitting galaxy model.

The $\sim$ one dozen ACSVCS galaxies brighter than $M_B \approx -20$ mag have surface brightness profiles that are well represented by "core-Sérsic" models [18, 33]. This model consist of a Sérsic [32] model outside a "break radius", r_b, and a shallower power-law profile inside r_b. None of these bright galaxies shows unambiguous evidence for a central stellar luminosity excess over the fitted profile. Galaxies fainter than $M_B \approx -20$ mag are almost always well fitted with a simple Sérsic model, with one notable exception: roughly two-thirds of these galaxies contain a compact central nucleus, which is visible as a sharp upturn in surface brightness within $\sim 1''$. For 51 galaxies in the ACSVCS, the nucleus is conspicuous enough that a measurement of its photometric and structural parameters is possible. When fitting the surface brightness profile of these nucleated galaxies, we add a central King model [23] to the underlying Sérsic component.

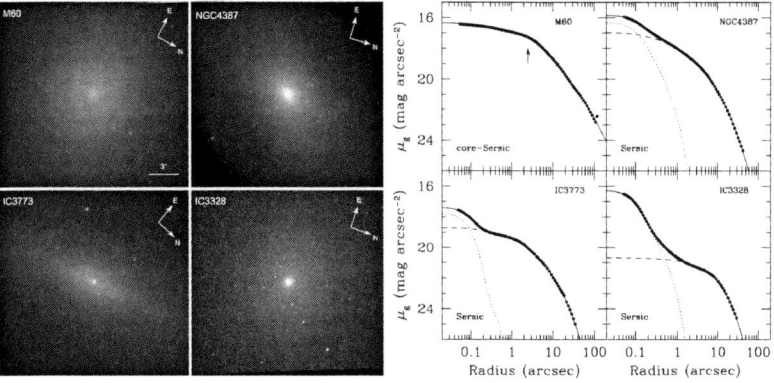

Fig. 1. *(Left Panels)* F475W (*g*) images of the central regions of four galaxies from the ACS Virgo Cluster Survey. *(Right Panels)* Azimuthally-averaged *g*-band surface brightness profiles for these same four galaxies. M60 is an example of a non-nucleated "core-Sérsic" galaxy; the best core-Sérsic model is shown as a *solid curve*. The *vertical arrow* shows the radius, r_b, at which the outer Sérsic profile "breaks" to an inner power-law. For the three other galaxies, we show the best-fit model which consists of a central King model for the nucleus (*dotted curve*) and a Sérsic model for the underlying galaxy (*dashed curve*). The *solid curve* shows the composite model.

Figure 1 shows images and surface brightness profiles for four representative galaxies from the ACSVCS. In the first panel, we show a "core-Sérsic" galaxy with $M_B \approx -21.4$ mag (M60), which also happens to have a dynamically measured SBH mass $\mathcal{M}_{\rm SBH} = 2 \times 10^8 \mathcal{M}_\odot$ [15]. There is no evidence for a stellar nucleus in this galaxy. Three fainter, nucleated Sérsic galaxies are shown in the remaining panels; only in IC3328 had a central nucleus been recognized from ground-based photographic imaging [2].

3.2 Incidence of Nucleation

At the outset of the ACSVCS, it was thought that ∼25% of the program galaxies contained nuclei, based on ground-based classifications from the VCC [2] (see Fig. 2). The excellent depth and angular resolution of the ACS images revealed that the true frequency of nucleation among early-type galaxies is in the range 60–80%, with the precise fraction varying as a function of galaxy magnitude (and depending on the specific criteria used to identify a nucleus). Our analysis shows that surface brightness selection effects are largely responsible for the large difference: i.e., in galaxies with central *g*-band surface brightnesses lower than ≈ 20.5 mag arcsec^{-2}, the agreement between the two surveys is nearly perfect; above 19.5 mag arcsec^{-2}, virtually all nuclei were missed by the ground-based survey. Needless to say, this comparison illustrates the importance of selection effects and suggests that the we too

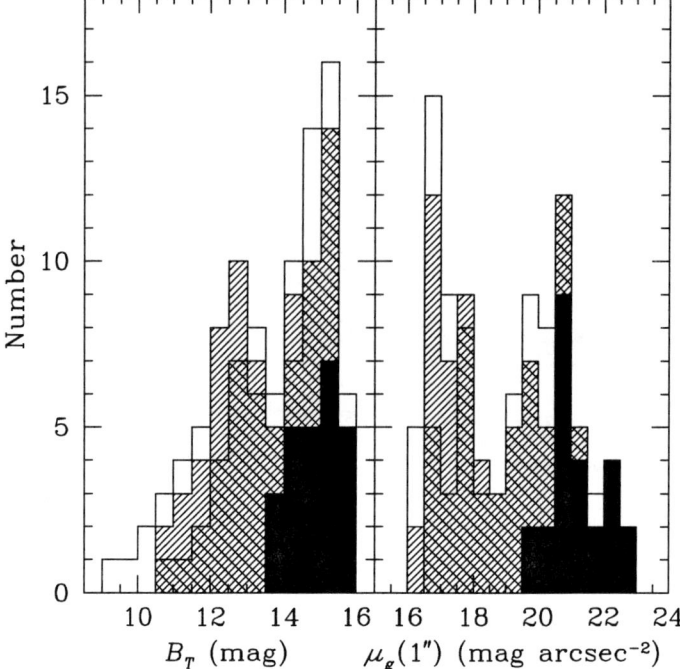

Fig. 2. *(Left Panel)* Luminosity distribution of galaxies in the ACS Virgo Cluster Survey for which a classification as nucleated or non-nucleated is possible. The double-hatched histogram shows the luminosity distribution for those galaxies which we classify unambiguously as nucleated; the hatched hatched histogram shows the results obtained using a less conservative definition for nuclei. The filled histogram shows the galaxies in our survey which were classified as nucleated in the Virgo Cluster Catalog [2]. *(Right Panel)* Distribution of g-band surface brightnesses for program galaxies measured at a geometric mean radius of $1''$. The samples and symbols are the same as in the previous panel.

may be missing nuclei in the highest-surface brightness galaxies. Thus, our estimate for the frequency of nucleation, $f_n \approx 60$–80%, is almost certainly a lower limit on the true frequency.

3.3 Offset Nuclei

Nuclei displaced from the photocenters of their host galaxies hold clues to the processes which trigger and/or regulate the formation of nuclei. Offsets may arise from ongoing merging of globular clusters through dynamical friction, the fading of stellar populations in dwarf irregular or blue compact dwarf galaxies as they evolve into dwarf ellipticals, recoil events following the ejection of a supermassive black hole from the nucleus, or counter-streaming instabilities that develop in flat and/or non-rotating systems. A search for offset nuclei reveals only five candidates with displacements of more than 0.5″, all in dwarf galaxies with transitional type (dIrr/dE) morphologies. In each case, their sizes, magnitudes and colors suggests that these "nuclei" are probably star clusters projected close to the galaxy photocenter.

3.4 The Luminosity Function

The globular clusters belonging to ACSVCS galaxies have a luminosity function that closely resembles a Gaussian distribution with turnovers at $\overline{g} \approx 24$ and $\overline{z} \approx 23$ mag. Working from the sample of 51 galaxies having nuclei with reliable photometric properties, we find that a broad ($1.5 \leq \sigma \leq 1.8$ mag) Gaussian distribution provides an adequate – though by no means unique – representation of the nuclei luminosity function. For the ACSVCS sample, the typical nucleus is $\approx 25\times$ brighter than a typical globular cluster. However, given that the nuclei are found to contain a roughly constant fraction of the host galaxy luminosity (see below), this ratio probably reflects the luminosity function of the sample galaxies.

3.5 A Size-Magnitude Relation

Images taken in the best ground-based conditions showed that the nuclei of Virgo Cluster galaxies were unresolved, with half-light radii $r_h \leq 0.4$–$0.5″$, or 30–40 pc [10]. Although a half-dozen or so nuclei appear unresolved in our images (i.e., $r_h \leq 2$ pc), the vast majority of the nuclei cannot be fitted with the ACS point-spread function. Working from the sample of ~ 45 resolved nuclei, we find a median half-light radius of $\langle r_h \rangle = 4.2$ pc, with the sizes of individual nuclei ranging from 62 pc down to the resolution limit of 2 pc. The nuclei sizes are found to scale with luminosity according to the relation $r_h \propto \mathcal{L}^{0.50 \pm 0.03}$. Because the majority of the nuclei are resolved, we can rule out low-level AGN as an explanation for the central luminosity excess in almost all cases.

3.6 A Color-Magnitude Relation

A comparison of the nuclei colors with the predictions of population synthesis models suggests that the nuclei have old to intermediate-age stellar populations (i.e., ages of 1 Gyr or more). Although their colors correlate with the luminosity of their host galaxy, they show a much tighter correlation with their own luminosities. This correlation is more evident among those galaxies fainter than $M_B \approx -17.6$ mag. The nuclei belonging to brighter galaxies frequently show very red colors, $(g-z) \sim 1.5$, and may constitute a separate type of object with a different formation route. The color-magnitude relation observed for the nuclei in the fainter galaxies suggest that their chemical enrichment was governed by local or internal factors. Monte Carlo simulations suggest that mergers of globular clusters through dynamical friction are unable to explain the observed color-magnitude relation.

3.7 The Connection Between Nuclei and Supermassive Black Holes

A comparison of the nuclei luminosities with those of their host galaxies reveals that the nucleus-to-galaxy luminosity ratio, η, is independent of galaxy luminosity. Our best estimate for the mean nucleus-to-galaxy luminosity ratio is $\langle \eta \rangle \approx 0.3\%$, albeit with considerable scatter about this value.

Merritt & Ferrarese [28] showed that the frequency function for galaxies with SBHs has a roughly normal distribution in $\log_{10}(\mathcal{M}_\bullet/\mathcal{M}_{gal})$. Remarkably, the mean of the frequency function for the nucleus-to-galaxy luminosity ratio (see Fig. 3) in our nucleated galaxies, $\langle \log_{10} \eta \rangle = -2.49 \pm 0.09$ dex

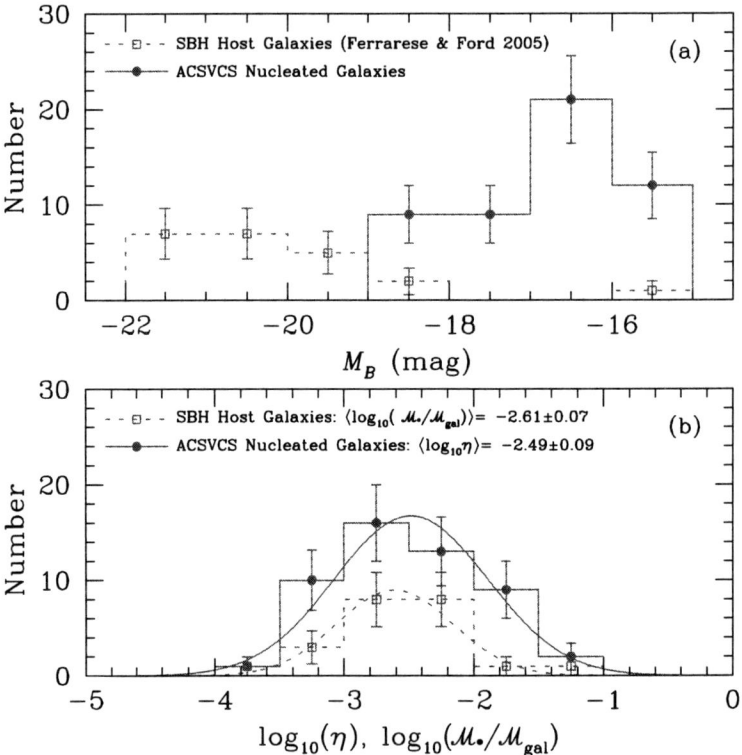

Fig. 3. *(Upper Panel)* Distribution of absolute blue magnitudes for 23 early-type galaxies with supermassive black holes (SBHs) taken from the compilation of Ferrarese & Ford [11] *(open squares)*. The distribution of 51 nucleated galaxies in the ACS Virgo Cluster Survey is shown by the *solid histogram*. *(Lower Panel)* Distribution of the SBH mass fraction for early-type galaxies compared to the luminosity fraction, η, for nucleated galaxies in the ACS Virgo Cluster Survey. In the former case, $\mathcal{M}_\bullet/\mathcal{M}_{gal}$ measures the dynamical mass of the SBH relative to host galaxy's bulge mass. For the nucleated galaxies, η is the ratio of the nucleus and galaxy luminosities, averaged in the g and z bandpasses. The *smooth curves* show the best-fit Gaussians.

($\sigma = 0.59 \pm 0.10$), is indistinguishable from that of the SBH-to-bulge mass ratio, $\langle \log_{10}(\mathcal{M}_\bullet/\mathcal{M}_{gal})\rangle = -2.61 \pm 0.07$ dex ($\sigma = 0.45 \pm 0.09$), calculated in 23 early-type galaxies with detected SBHs (Ferrarese & Ford 2005). In fact, the nuclei and SBHs share a number of other key similarities that are highly suggestive of a direct connection: e.g., they share a common location at the bottom of the gravitational potential wells defined by their parent galaxies and dark matter halos, and both are probably old components that formed during the earliest stages of galaxy evolution.

As noted in Côté et al. [9], the similarity of the frequency functions in Fig. 3 suggests that the compact stellar nuclei found in many of our program galaxies may be the low-mass counterparts of the SBHs detected in the bright galaxies (see also [35]). If so, then one should think in terms of Central Massive Objects (CMOs) – either SBHs or compact stellar nuclei – that accompany the formation of almost all early-type galaxies and contain a mean fraction $\approx 0.3\%$ of the total bulge mass.

Supporting evidence for this view has recently been presented by Ferrarese et al. (2006b), who used dynamical masses for the ACSVCS program galaxies to show that: (1) the masses of the nuclei are correlated with the masses of the host galaxies and; (2) the same correlation is obeyed by the SBHs detected in many giant galaxies. The challenge to theorists is to understand why different modes of CMO formation are favored in different regimes of galaxy mass.

I thank the conference organizers for their invitation and hospitality, and my collaborators on the ACS Virgo Cluster Survey for their hard work and enthusiasm during the past few years.

References

1. Binggeli, B., & Cameron, L.M. 1991, A&A, 252, 27
2. Binggeli, B., Sandage, A., & Tammann, G.A. 1985, AJ, 90, 1681 (BST85)
3. Böker, T., Laine, S., van der Marel, R.P., Sarzi, M., Rix, H.-W., Ho, L., & Shields, J.C. 2002, AJ, 123, 1389
4. Böker, T., Sarzi, M., McLaughlin, D.E., van der Marel, R.P., Rix, H.-W., Ho, L.C., & Shields, J.C. 2004, AJ, 127, 105
5. Caldwell, N. 1983, AJ, 88, 808
6. Caldwell, N., & Bothun, G. D. 1987, AJ, 94, 1126
7. Carollo, C.M., Stiavelli, M., & Mack, J. 1998, AJ, 116, 68
8. Côté, P., Blakeslee, J.P., Ferrarese, L., Jordán, A., Mei, S., Merritt, D., Milosavljević, M., Peng, E.W., Tonry, J.L., & West, M.J. 2004, ApJS, 153, 223
9. Côté, P., Piatek, S., Ferrarese, L., Jordán, A., Merritt, D., Peng, E.W., Haşegan, M., Blakeslee, J.P., Mei, S., West, M.J., Milosavljević, M., & Tonry, J.L. 2006, ApJS, in press (astro-ph/0603252)
10. Durrell, P.R. 1997, AJ, 113, 531
11. Ferrarese, L., & Ford, H. 2005, Space Science Reviews 116, 52333

12. Ferrarese, L., Côté, P., Jordán, A., Peng, E.W., Blakeslee, J.P., Piatek, S., Mei, S., Merritt, D., Milosavljević, M., Tonry, J.L., & West, M.J. 2006a, ApJ, in press, (astro-ph/0512474)
13. Ferrarese, L., Côté, P., Blakeslee, J.P., Dalla Bontá, E., Peng, E.W., Merritt, D., Jordán, A., Blakeslee, J.P., Haşegan, M., Mei, S., Piatek, S., Tonry, J.L., & West, M.J. 2006b, ApJL, in press (astro-ph/0603840)
14. Ford, H.C., et al. 1998, Proc. SPIE, 3356, 234
15. Gebhardt, K., et al. 2003, ApJ, 583, 92
16. Geha, M., Guhathakurta, P., & van der Marel, R.P. 2002, AJ, 124, 3073
17. Geha, M., Guhathakurta, P., & van der Marel, R.P. 2003, AJ, 126, 1794
18. Graham, A.W., Erwin, P., Trujillo, I., & Asensio Ramos, A. 2003, AJ, 125, 2951
19. Haşegan, M., Jordán, A., Côté, P., Djorgovski, S.G., McLaughlin, D.E., Blakeslee, J.P., Mei, S., West, M.J., Peng, E.W., Ferrarese, L., Milosavljević, M., Tonry, J.L., & Merritt, D. 2005, ApJ, 627, 203
20. Jordán, A., Blakeslee, J.P., Peng, E., Mei, S., Côté, P., Ferrarese, L., Merritt, D., Milosavljević, M., Tonry, J.L., & West, M.J. 2004a, ApJS, 154, 509
21. Jordán, A., Côté, P., Ferrarese, L., Blakeslee, J.P., Mei, S., Merritt, D., Milosavljević, M., Peng, E.W., Tonry, J.L., & West, M.J. 2004b, ApJ, 613, 279
22. Jordán, A., Côté, P., Blakeslee, J.P., Ferrarese, L., McLaughlin, D.E., Mei, S., Peng, E.W., Tonry, J.L., Merritt, D., Milosavljević, M., Sarazin, C.L., Sivakoff, G.R., West, M.J. 2005, ApJ, 634, 1002
23. King, I.R. 1966, AJ, 71, 64
24. Lotz, J.M., Miller, B.W., & Ferguson, H.C. 2004, ApJ, 613, 262
25. Matthews, L.D., et al. 1999, AJ, 118, 208
26. Mei, S., Blakeslee, J.P., Tonry, J.L., Jordán, A., Peng, E.W., Côté, P., Ferrarese, L., West, M.J., Merritt, D., Milosavljevic, M. 2005a, ApJS 165, 113
27. Mei, S., Blakeslee, J.P., Tonry, J.L., Jordán, A., Peng, E.W., Côté, P., Ferrarese, L., West, M.J., Merritt, D., Milosavljevic, M. 2005b, ApJS 165, 113
28. Merritt, D., & Ferrarese, L. 2001, MNRAS, 320, L30
29. Peng, E.W., Jordán, A., Côté, P., Blakeslee, J.P., Ferrarese, L., Mei, S., West, M.J., Merritt, D., Milosavljevic, M., & Tonry, J.L. 2006a, ApJ, 639, 95
30. Peng, E.W., Côté, P., Jordán, A., Blakeslee, J.P., Ferrarese, L., Mei, S., West, M.J., Merritt, D., Milosavljevic, M., & Tonry, J.L. 2006b, ApJ, 639, 838
31. Phillips, A.C., Illingworth, G.D., MacKenty, J.W., & Franx, M. 1996, AJ, 111, 1566
32. Sérsic, J.-L. 1968, Atlas de Galaxias Australes (Córdoba: Obs. Astron., Univ. Nac. Córdova)
33. Trujillo, I., Erwin, P., Asensio Ramos, A., & Graham, A.W. 2004, AJ, 127
34. Walcher, C.J., van der Marel, R.P., McLaughlin, D., Rix, H.-W., Böker, T., Häring, N., Ho, L.C., Sarzi, M., & Shields, J.C. 2005, ApJ, 618, 237
35. Wehner, E.H., & Harris, W.E. 2006, ApJ, in press (astro-ph/0603801)

UCDs – A Mixed Bag of Objects

Michael Hilker

Argelander-Institut für Astronomie, Universität Bonn, Auf dem Hügel 71, 53121 Bonn, Germany mhilker@astro.uni-bonn.de

1 Introduction

The realm of dwarf galaxies was extensively studied only in the last three decades. Dwarf spheroidals are considered to be the faintest galaxies, having baryonic masses comparable to those of bright globular clusters ($\sim 10^5 M_\odot$), but being 50–200 times more extended. Thus far one thought that dwarf galaxies were diffuse structures, with the exception of the compact elliptical M32 which is $\sim$8–10 times smaller than dwarf ellipticals of comparable luminosities but about 150 times more luminous than the brightest globular clusters of the Local Group. This gap in luminosity of compact stellar systems started to be filled in observationally during the last decade thanks to several large spectroscopic surveys in nearby galaxy clusters. The question arises what are the physical parameters and what is the origin of objects in the transition region between dwarf galaxies and star clusters? What is the smallest compact galaxy? What is the largest globular cluster? How can a massive *cluster* be descerned from a low-mass compact *galaxy*?

2 The Discovery of Compact Objects

2.1 Celebrating the 10-years Anniversary of "UCDs"

The discovery history of very massive compact star clusters started about 10 years ago. In a small spectroscopic survey of the globular cluster system of NGC 1399, the central galaxy of the Fornax cluster, Minniti et al. [1] confirmed a bright compact object as radial velocity member of the cluster: '... *Note that the objects at $V = 18.5$, $V - I = 1.48$ (our reddest "globular cluster"), which has $M_V = -12.5$, was identified as a compact dwarf galaxy on the images after light-profile analysis (M. Hilker, 1996, private communication [2]) ...*". In another spectroscopic survey, designed as a follow-up of a photometric investigation of the surface brightness-magnitude relation of dwarf ellipticals in the Fornax cluster, Hilker et al. [3] confirmed in 1999 two bright compact objects (including the one mentioned before) as Fornax members. They proposed that they '... *can be explained by a very bright GC as well as by a compact elliptical like M32. Another explanation might be that* *these objects represent the nuclei of dissolved dE,Ns ...'*. Further they suggested that *'... It would be interesting to investigate, whether there are more objects of this kind hidden among the high surface brightness objects in the central Fornax cluster ...'*.

Indeed, only one year later, in 2000, a systematic all-object spectroscopic survey within in a 2-degree field centred on the Fornax cluster revealed five compact Fornax members in the magnitude range $-13.5 < M_V < -12.0$ [4] which later, in 2001, were dubbed "Ultracompact Dwarf Galaxies" (UCDs) by Phillipps et al. [5]. Their physical properties were presented in a *Nature* article by Drinkwater et al. in 2003 [6].

2.2 A Word on Nomenclature

Before the term "ultracompact dwarf galaxy" (UCD) was invented, the new type of objects was circumscribed in different ways: e.g. compact stellar object, compact object, compact stellar system, (extremely) compact dwarf galaxy, super-massive star cluster, extremely large star cluster, etc.

After its introduction, the denomination "ultracompact dwarf galaxy" became widely used, but also provoked severe critisism, since it suggests that these objects have a galaxian origin. The term "ultra-diffuse star cluster" was opposed by Kissler-Patig [7] to demonstrate their link to massive star cluster formation. In an attempt to find a neutral description Haşegan et al. [8] named newly discovered compact objects in the Virgo cluster "dwarf-globular transition objects" (DGTOs).

One should note that star clusters of similar luminosities and sizes as UCDs/DGTOs are known in present-day galaxies, like the nuclear (star) clusters (NCs) of late-type spirals and dwarf ellipticals and the super-star clusters (SSCs) or young massive clusters (YMCs) of merger/starburst galaxies.

In this contribution, I will stay for simplicity with the term UCD to describe objects with properties as summarized in the next section. Note however that I don't suggest that they are of galaxian origin. I will argue instead that they may be an inhomogeneous group of objects perhaps with different origins.

2.3 UCD Properties

Once the existence of UCDs in the Fornax and Virgo clusters was proven by radial velocity measurements, e.g. [3, 4, 8, 9], their physical parameters became the subject of much study. In particular, their sizes, shapes, metallicities, ages, internal kinematics, masses and mass-to-light ratios are of interest. Recent and ongoing observing programmes, employing high-resolution imaging (HST) and spectroscopy (VLT and Keck) as well as high signal-to-noise spectroscopy and deep imaging to faint surface brightnesses, revealed most of those parameters for Fornax and Virgo UCDs [6, 8, 10–12]. The general properties are listed in Table 1.

Table 1. General properties of UCDs (LSB = low surface brightness)

Luminosity:	$-13.5 < M_V < -11.5$ mag
Colour:	Fornax: mainly red; Virgo: mainly blue
LSB features:	some have LSB envelopes with $80 < R_{\text{eff}} < 120$ pc
Shape:	best represented by King+Sérsic or Nuker profile
Effective radius:	$10 < R_{\text{eff}} < 30$ pc
Velocity dispersion:	$25 < \sigma_0 < 45$ km s^{-1} (central value)
Mass:	$M = 2\text{--}9 \times 10^7 M_\odot$
Mass-to-light ratio:	$M/L_V = 2\text{--}5$; Virgo DGTOs: 2–9 (global value)
Metallicity:	Fornax: -0.5 dex; Virgo: -1.2 dex, $[\alpha/\text{Fe}] = 0.3$
Age:	old: >8–10 Gyr

3 "The UCD Rush"

After the first discovery of UCDs in the Fornax cluster, many surveys followed to search for UCDs in different environments and towards fainter magnitudes.

In the Fornax cluster, Mieske et al. [13] identified compact objects in the brightness range $-12.0 < M_V < -10.0$ mag. They found that their distribution over luminosity is consistent with an extrapolation of the GC luminosity function. Jones et al. [9] discovered a sixth bright UCD in the central two degrees of the cluster, but could not find any UCD brighter than $M_V = -12.0$ in an all-object spectroscopic survey of two adjacent 2-degree fields. The spatial distribution of GCs and UCDs in Fornax shows that UCDs brighter than $M_V \simeq -11.5$ do not seem to belong to certain galaxies, but rather live in the intra-cluster space of the core region of the cluster.

In the Virgo cluster, Haşegan et al. [8] identified close to the central galaxy M87 six DGTOs and 13 DGTO candidates in the magnitude range $-11.8 < M_V < -10.8$. Three of the DGTOs have global M/L_V of the order 6–9 that cannot be explained by stellar population models. Jones et al. [9] discovered 9 UCDs with $-13.7 < M_V < -11.5$ in a 2-degree field around M87, widely distributed throughout intra-cluster space.

In other surveys, nearby groups and distant massive clusters were searched for UCDs. Around Centaurus A several bright GCs with $M_V > -11.2$ were identified, e.g. [14], but no massive UCDs. Those seem also to be absent in other nearby groups. In the very massive lensing cluster Abell 1689, two twins of M32 were discovered and several massive UCD candidates [15].

4 Filling the Parameter Space of Hot Stellar Systems

The physical parameters of UCDs have been compared to those of bright star clusters (young and old) and galactic nuclei by many authors. It is of special interest whether UCDs lie on known scaling relations in the parameter space of hot stellar systems. The most commonly used parameters for comparison are the absolute magnitude or mass (if available), the central or effective surface brightness, the effective (= half-light) radius, the central velocity dispersion, or combinations of such parameters (e.g. κ-space).

Here, the luminosity-size and luminosity-velocity dispersion plane, as well as a colour magnitude diagram of Fornax and Virgo UCDs are presented.

Figure 1 shows that Galactic GCs with $M_V < -7.5$ and GCs in the Fornax cluster (dotted line) scatter around a luminosity-independent size of about 2.7 parsec. Objects brighter than $M_V \simeq -10.5$, however, follow a luminosity-size relation, approximately along a line of constant surface density. M32-type galaxies lie on the extension of this relation. Also nuclei of early-type galaxies exhibit a luminosity-size relation, shifted towards smaller sizes at a given luminosity. Note that young massive star clusters in starburst/merger galaxies follow a mass-size relation that is consistent with that of UCDs [16].

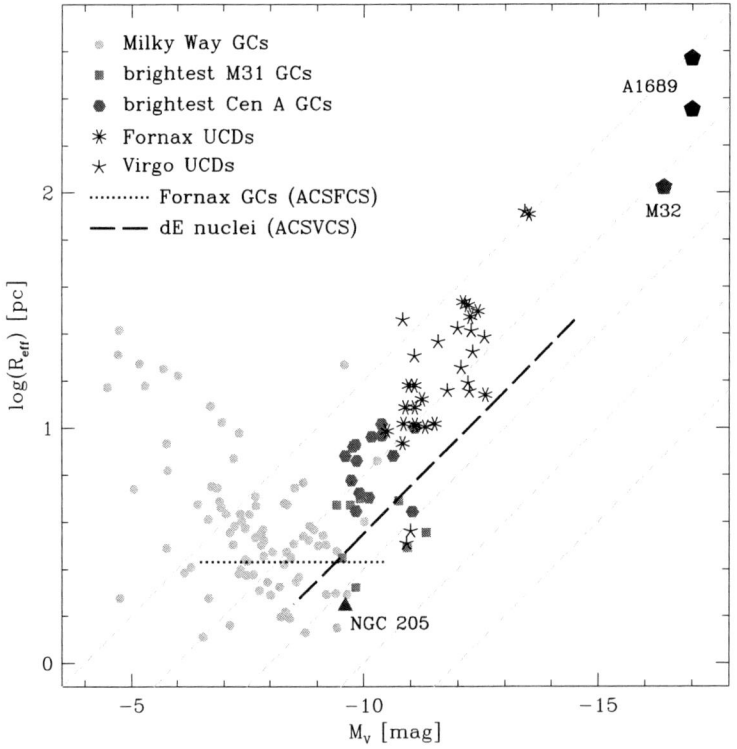

Fig. 1. Luminosity-size relation for GCs, UCDs, compact ellipticals and dwarf ellipical nuclei as indicated in the plot (data taken from the literature). The *diagonal dashed lines* are lines of constant surface intensity. The *horizontal dotted line* shows the average effective radius of "normal" GCs in the Fornax cluster, $\langle r_{\rm eff} \rangle = 2.7$ pc.

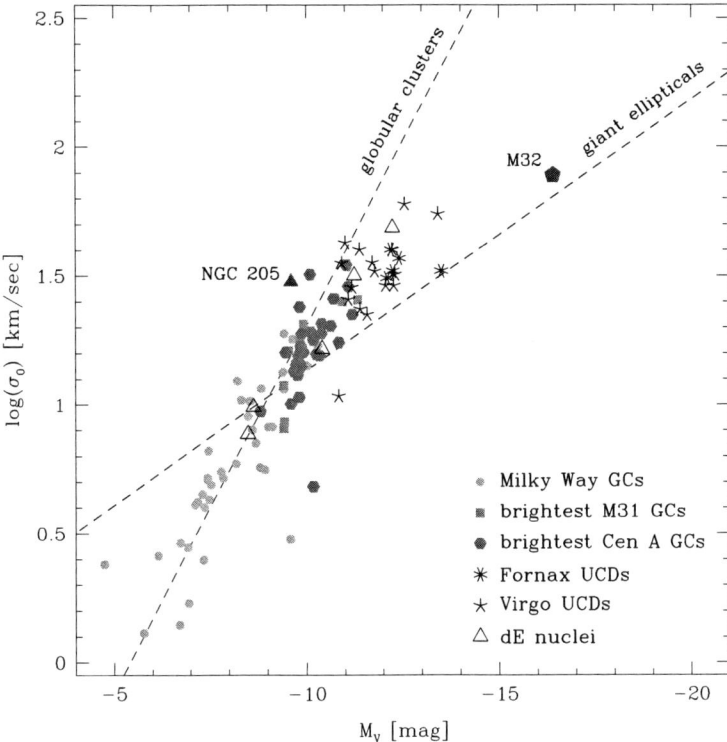

Fig. 2. Luminosity versus central velocity dispersion for GCs, UCDs, M32 and dwarf elliptical nuclei as indicated in the plot. The *dashed lines* show the known relations for Galactic globular clusters and giant ellipticals (Faber-Jackson relation).

Concerning their internal kinematical properties, compact objects brighter than $M_V \simeq -10.5$ also seem to deviate from the well defined relation of "normal" GCs (see Fig. 2). In the luminosity-velocity dispersion diagram, UCDs seem to bend over towards the Faber-Jackson relation of early-type galaxies.

The colour magnitude diagram in Fig. 3 exhibits some interesting differences between Fornax and Virgo UCDs. In the magnitude range $-13.0 < M_V < -11.5$, the Fornax UCDs are dominantly red ([Fe/H]$\simeq -0.5$), whereas the colour of most Virgo UCDs is blue ([Fe/H]$\simeq -1.2$) and consistent with that of dE nuclei. Note that the two brightest UCDs are at least twice as luminous as the second brightest UCD in their respective clusters. Both are metal-rich and possess a small envelope of low surface brightness.

5 Formation Scenarios for UCDs

Various formation scenarios have been suggested to explain the origin of UCDs. The three most promising and their implications are:

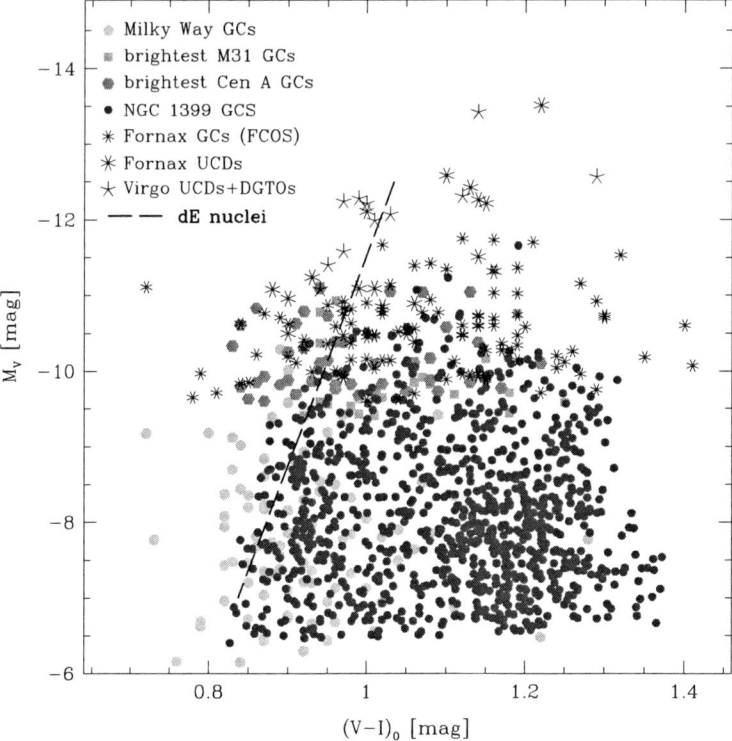

Fig. 3. Colour magnitude diagram for GCs and UCDs as indicated in the plot. The *dashed line* shows the colour-magnitude relation of dwarf elliptical nuclei.

(1) UCDs are the remnant nuclei of galaxies that have been significantly stripped in the cluster environment, also referred to as the "threshing scenario", e.g. [17, 18]. Numerical simulations have shown that nucleated dEs can be disrupted in a galaxy cluster potential under specific conditions and that the remnant nuclei resemble UCDs in their structural parameters [18]. Good candidates for isolated nuclei are the Galactic globular clusters ω Centauri [19] and M54 as the nucleus of the Sagittarius dSph.

In Fornax and Virgo, the present-day nuclei of dwarf ellipticals are less massive and more compact than the UCDs. This implies that the progenitor galaxies must have been very massive dE, Ns or maybe late-type spirals. The small number of UCDs in both clusters points to a rather selective "threshing" process. The high metallicity of the Fornax UCDs seems to disfavour this scenario for their origin.

(2) UCDs have formed from the agglomeration of many young, massive star clusters that were created during merger events, e.g. [20, 21, 16]. Inspired by the massive star forming knots in the Antennae galaxies, Kroupa [20]

first showed that the individual star clusters in complexes can merge to form a large massive, single star cluster. This work thus constitutes a prediction of UCD-type objects made right at the time when these were discovered. An evolved example of such a merged star cluster complex might be the 300 Myr old, super-star cluster W3 in NGC 7252 [22, 23].

If the old UCDs in Fornax and Virgo formed like this, the galaxy mergers must have happened early in the galaxy cluster formation history when the merging galaxies were still gas-rich. However, in the case of Fornax, these early mergers must have already possessed close to solar metallicity gas. The small number of UCDs would imply that only the most massive star cluster complexes survived as bound systems.

(3) UCDs are the brightest globular clusters and were formed in the same GC formation event as their less massive counterparts, e.g. [13, 14]. The smooth shape of the bright end of the GC luminosity function (no excess objects!) might support this scenario.

The most massive GCs then supposedly formed from the most massive molecular clouds (MCs) of their host galaxy, where more massive galaxies were able to form higher mass MCs (as e.g. M87) than lower mass galaxies (as e.g. M31). The luminosity-size relation of the most massive clusters suggests that there is a break of the formation/collapse physics at a critical MC mass. In Fornax, the formation of the most massive GCs was connected to that of the metal-rich bulge GCs, whereas in Virgo they must have formed with the metal-poor GCs, if this scenario would be correct.

In an attempt to unify the ideas of the different formation scenarios one might think of the following generalized scheme of massive star cluster formation: GCs with a mass of $\leq 5 \times 10^6 M_\odot$ are "single-collapse" GCs (SCGCs), i.e. their stars share a single age and metallicity. At a critical mass of $\simeq 5 \times 10^6 M_\odot$ the formation physics changes to "multiple-collapse" GCs (MCGCs) because the giant MC fragments into massive clusters as it contracts. The growth of SCGCs to MCGCs through merging can happen on different time scales. An immediate growth (~ 10–100 Myr) would correspond to the situation in super-star cluster complexes in mergers. The stars formed in the resulting MCGC then would have the same age and probably similar metallicities, although MCGCs would also be able to capture a substantial amount of older field stars from the host galaxy [20]. A successive growth over gigayears reflects the situation in nuclear star clusters. Nuclei of dwarf ellipticals could have formed via the merging of GCs which not necessarily had all the same age and/or metallcity. Another way of growing a nuclear star cluster is via episodic star formation of infalling gas in the centre of a gas-rich galaxy. The stars of those MCGCs are supposed to show a spread in ages and metallicities. Finally, through whichever channel the MCGCs formed, their evolution in the tidal field of a dense galaxy cluster over a Hubble time could have given rise to the population of old, isolated, massive, compact stellar systems we nowadays call UCDs.

6 Summary and Outlook

The name "ultracompact dwarf galaxies" (UCDs) collects/paraphrases old stellar systems in the transition region of globular clusters and compact dwarf galaxies ($-13.5 < M_V < -11.5$, 2–$9\times10^7 M_\odot$, $10 < r_h < 30$ pc, $25 < \sigma_0 < 45$ km s^{-1}). The known UCDs are found in the cores of galaxy clusters and are not concentrated towards individual galaxies unlike most of the GCs.

While we have good ideas on their possible origin, there are many questions left to answer concerning the nature of UCDs. Some important ones are: Do UCDs have multiple stellar populations? Why do some UCDs have high M/L ratios? Is this due to tides? Or do they contain dark matter? Is there tidal structure around UCDs? Do UCDs harbour black holes? What are the kinematics of UCDs within their host clusters?

Some of these questions will be answered in the next years with the help of ongoing and future observing programmes. The results will bring more light into the nature of these enigmatic objects.

References

1. D. Minniti, M. Kissler-Patig, P. Goudfrooij, G. Meylan: AJ **115**, 121 (1998)
2. M. Hilker: The Center of the Fornax Cluster: Dwarf Galaxies, cD Halo, and Globular Clusters. PhD Thesis, Univ. of Bonn (1998)
3. M. Hilker, L. Infante, G. Vieira et al: A&AS **134**, 75 (1999)
4. M.J. Drinkwater, J.B. Jones, M.D. Gregg, S. Phillipps: PASA **17**, 227 (2000)
5. S. Phillipps, M.J. Drinkwater, M.D. Gregg, J. Jones: ApJ **560**, 201 (2001)
6. M.J. Drinkwater, M.D. Gregg, M. Hilker et al: Nature **423**, 519 (2003)
7. M. Kissler-Patig: Extremely Massive Clusters and their link to Ultra Compact Dwarf Galaxies. In: *The Formation and Evolution of Massive Young Star Clusters*, ASP Conf. Ser. vol 322, ed by H.J.G.L.M. Lamers, L.J. Smith, A. Nota (San Francisco: Astronomical Society of the Pacific, 2004) pp. 535–539
8. M. Haşegan, A. Jordán, P. Côté et al: ApJ **627**, 203 (2005)
9. J.B. Jones, M.J. Drinkwater, R. Jurek et al: AJ **131**, 312 (2006)
10. T. Richtler, B. Dirsch, S. Larsen et al.: A&A **439**, 533 (2005)
11. R. De Propris, S. Phillipps, M.J. Drinkwater et al.: ApJ **623**, L105 (2005)
12. S. Mieske, M. Hilker, L. Infante, A. Jordán: AJ **131**, 2442 (2006)
13. S. Mieske, M. Hilker, L. Infante: A&A **418**, 445 (2004)
14. P. Martini, L.C. Ho: ApJ **610**, 233 (2004)
15. S. Mieske, L. Infante, M. Hilker: A&A **430**, L25
16. M. Kissler-Patig, A. Jordán, N. Bastian: A&A **448**, 1031 (2006)
17. L. Bassino, J.C. Muzzio, M. Rabolli: ApJ **431**, 634 (1994)
18. K. Bekki, W.J. Couch, M.J. Drinkwater, Y. Shioya: MNRAS **344**, 399 (2003)
19. M. Hilker, T. Richtler: A&A **362**, 895 (2000)
20. P. Kroupa: MNRAS **298**, 231 (1998)
21. M. Fellhauer, P. Kroupa: MNRAS **330**, 642 (2002)
22. C. Maraston, N. Bastian, R.P. Saglia et al.: A&A **416**, 467 (2004)
23. M. Fellhauer, P. Kroupa: MNRAS **359**, 223 (2005)

Ultra-Compact Dwarf Galaxies and Globular Clusters: A Review of Their Spatial and Dynamical Properties

M.J. Drinkwater[1], E. Evstigneeva[1], P. Firth[1], and C. Power[2]

[1] Department of Physics, University of Queensland, QLD 4072, Australia
   m.drinkwater@uq.edu.au, katya,firth@physics.uq.edu.au
[2] Centre for Astrophysics and Supercomputing, Swinburne University of Technology, PO Box 218, Hawthorn, VIC 3122, Australia
   cpower@astro.swin.edu.au

1 Introduction

The discovery of ultra-compact dwarf (UCD) galaxies has been very well reviewed by Michael Hilker in these proceedings. Here we wish to focus on the large-scale distribution of UCDs. Our study was originally motivated by a search for "M32-type" compact elliptical galaxies in the Fornax Cluster. We didn't find any of these, but we did demonstrate that morphological membership classification can overlook many compact dwarf cluster members [3]. This motivated the 2dF Fornax Cluster Spectroscopic Survey (FCSS, [4]), which measured redshifts of all objects, both "stars" and "galaxies" in the cluster centre, resulting in the detection of 5 UCDs ([5] and also [9]). High resolution imaging and spectroscopy have since established that the most luminous UCDs are a new type of object [6]. We have extended our UCD searches to fainter limits [7], finding a total of 62 compact objects filling a large region of intra-cluster space (see Fig. 1). We discuss the distribution of these objects below, noting that our working UCD definition (Fornax Cluster members barely resolved in photographic survey data with $m_{Bj} < 21.5$ mag) may also include genuine globular clusters (GCs).

2 Spatial Distribution

Given the similarity between UCDs and globular clusters, an obvious hypothesis is that the cluster UCDs are simply the high-luminosity end of the existing populations of GCs around the central cluster galaxies. Mieske et al. [10] compared the original UCDs found in the Fornax Cluster to the luminosity function of known GCs associated with the central galaxy NGC1399. They found there was no evidence for any gap between the distributions, except that the brightest UCD was too luminous to be part of the GC population. With a much larger sample of UCDs we can now compare both the luminosity *and* the spatial distributions of the objects as shown in Fig. 2. The figure

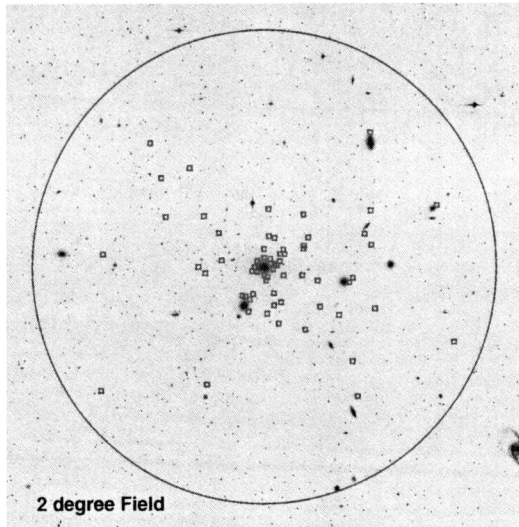

Fig. 1. The central field of the Fornax Cluster showing the locations of the 62 UCDs we have detected in our wide-field 2dF survey.

compares the bivariate radial and luminosity distribution of UCDs with a model for the globular cluster population (derived from [2]). The model is simply the product of the separate radial distribution and luminosity function, normalised to have the same total counts within a radius of 3.4 arc min. The integrated counts in the top panel of Fig. 2 show that the brighter ($b_J < 21$ mag) are about twice as numerous as GCs predicted by the model, so they clearly come from a different population.

Although many simulations have now shown that dwarf elliptical galaxies can be tidally stripped to form UCDs (e.g. [1]), we also need to show that the process actually takes place. We have started a detailed study of cosmological simulations of galaxy cluster formation with the aim of identifying objects that have sufficiently central orbits for stripping to occur (see Fig. 3).

3 Kinematics

We compare the kinematics of the dwarf galaxies, UCDs and globular clusters in Fig. 4. There are significant (99% confidence) differences between all three. The mean velocity of the UCDs is higher than that of the globular clusters—the UCDs follow the mean cluster potential whereas the GCs are tied to the lower-velocity of NGC1399. Furthermore, the UCDs have a much lower velocity dispersion than the dwarf elliptical galaxies, consistent with their being a more central population on lower-energy orbits.

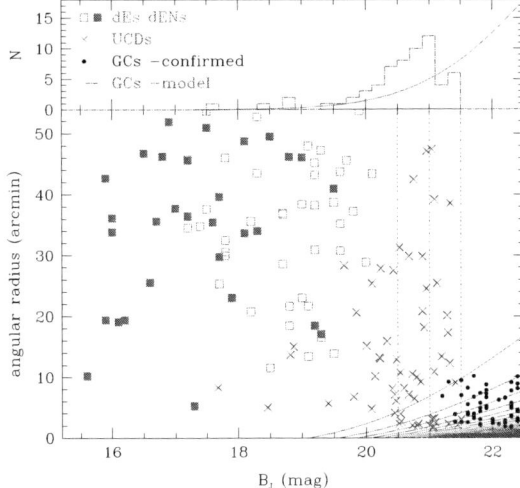

Fig. 2. Comparison of the distributions of Fornax Cluster dwarf galaxies [8], UCDs and globular clusters [2]. *Lower panel*: the distributions of the objects as a function of radius (from the cluster centre) and luminosity, shown by the different symbols. The *lines* at the bottom right are iso-density contours of the model distribution of globular clusters; the *vertical lines* indicate where the UCD survey is 96%, 82% and 36% complete. *Top panel*: total counts of UCDs and globular clusters as a function of magnitude. The GC counts are obtained by integrating the model in the lower panel. UCDs are clearly much more numerous than the GC population.

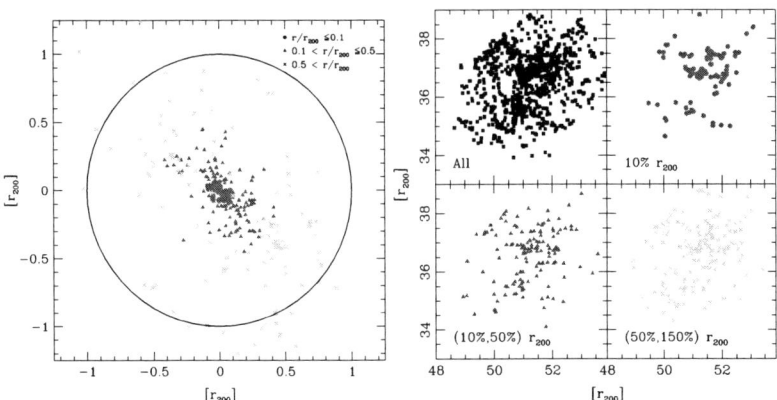

Fig. 3. Cosmological simulation of the formation of a galaxy cluster. The simulation has 10^6 particles; the virial mass is $5 \times 10^{14} M_\odot$ and the resolution is $5 \times 10^8 M_\odot$. *Left*: the final cluster ($z = 0$) with particles classified by their distance from the centre. *Right*: the proto-cluster (at $z = 8$) showing the original locations of the particles in each classification. The particles destined for the cluster centre ("10% r_{200}") are more clustered than average: these are the ones we will test for tidal stripping and possible UCD formation.

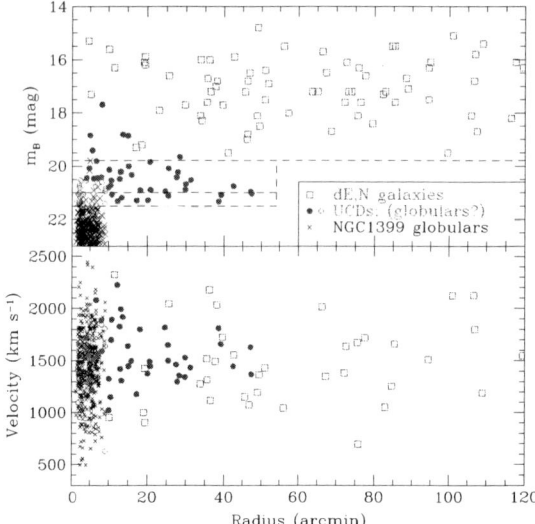

Fig. 4. Comparison of dwarf galaxy, UCD and globular cluster kinematics as a function of radius. *Top*: the distributions of luminosity. *Bottom*: the distributions in velocity. Note that the GC sample is only complete to a radius of 10 arc min.

4 Summary

We conclude the following. First, the 62 UCDs in our sample are clearly not distributed like the classical GC population around the central galaxy, being much more extended and more luminous. This does not preclude them being *intra-cluster* GCs [11]. Secondly, the UCDs are distinct from the current dE galaxy population, so if they did form by a tidal stripping process, it was from objects formed early at the cluster centres or by the selective disruption of a more extended parent population.

References

1. K. Bekki, W. Couch, M. Drinkwater: ApJ **522**, L105 (2001)
2. B. Dirsch, et al.: AJ **125**, 1908 (2003)
3. M. Drinkwater, M. Gregg: MNRAS **296**, L15 (1988)
4. M. Drinkwater, et al.: A&A **355**, 900 (2000a)
5. M. Drinkwater, J.B. Jones, M. Gregg, S. Phillipps: PASA **17**, 227 (2000b)
6. M. Drinkwater, M. Gregg, M. Hilker et al.: Nature **423**, 519 (2003)
7. M. Drinkwater, et al.: PASA **21**, 375 (2004)
8. H. Ferguson: AJ **98**, 367 (1989)
9. M. Hilker, et al.: A&AS **134**, 75 (1999)
10. S. Mieske, M. Hilker, L. Infante: A&A **383**, 823 (2002)
11. M. West, et al.: ApJ **453**, L77 (1995)

The Maximum Mass of Star Clusters

M. Gieles[1], S.S Larsen[1,2], M.R. Haas[1], R.A. Scheepmaker[1], and N. Bastian[3]

[1] Astronomical Institute, Utrecht University, Princetonplein 5, 3584 cc Utrecht, The Netherlands gieles@astro.uu.nl
[2] European Southern Observatory, Garching, Germany larsen@astro.uu.nl
[3] Department of Physics and Astronomy University College London, London, UK bastian@star.ucl.ac.uk

1 A Maximum from Size-of-Sample Effects or Physics?

If an universal untruncated cluster initial mass function (CIMF) of the form $N(M)\mathrm{d}M = C\,M^{-2}\mathrm{d}M$ is assumed, the mass of the most massive star cluster in a galaxy (M_{max}) is the result of the size-of-sample (SoS) effect. This implies a dependence of M_{max} on the total number of clusters (N). For a power-law index of -2, the constant $C = M_{\mathrm{max}}$ and N follows from integrating the CIMF from M_{min} to M_{max}, resulting in $N = M_{\mathrm{max}}/M_{\mathrm{min}}$. Since the cluster luminosity function (CLF) is also a power-law distribution, with a comparable index, a similar relation holds for the luminosity of the brightest cluster in a galaxy (L_{max}) and N, which has been observed [13, 11]. An attempt to compare M_{max} in a sample of galaxies with the star formation rate (SFR) has shown a similar relation [12]. However, finding the most massive cluster in a galaxy is not trivial, since star clusters fade rapidly due to stellar evolution. For example, a 1 Gyr old cluster of $10^6\,M_\odot$ has about the same luminosity as a 4 Myr old cluster of $10^4\,M_\odot$. The SoS effect also implies that M_{max} within a cluster population increases with equal logarithmic intervals of age. This is because the number of clusters formed in logarithmic age intervals increases (assuming a constant cluster formation rate). This effect has been observed in the SMC and LMC [10]. The observations of this increase argues for a M_{max} (in the LMC and SMC) that is determined by sampling statistics, *or* a physical upper limit that is higher than the M_{max} following from statistics.

Based on the maximum pressure (P_{int}) inside molecular clouds, it has been suggested that a physical maximum mass ($M_{\mathrm{max}}^{\mathrm{phys}}$) should exist, which scales as $M_{\mathrm{max}}^{\mathrm{phys}} \propto P_{\mathrm{int}}^{1/2}$ [6]. The ISM pressure in a galaxy scales approximately as the square of the column density of molecular gas (Σ^2), and when assuming that P_{int} is determined by the ISM pressure (i.e. pressure equilibrium), then $M_{\mathrm{max}}^{\mathrm{phys}} \propto \Sigma$. Since the star formation rate (SFR) scales in another way with Σ, namely SFR $\propto \Sigma^{1.4}$, and since $M_{\mathrm{max}}^{\mathrm{phys}}$ is independent of the size of the galaxy (A), for a certain minimum A and SFR a Σ_{crit} should exist where $M_{\mathrm{max}} = M_{\mathrm{max}}^{\mathrm{phys}}$. For galaxies where $\Sigma > \Sigma_{\mathrm{crit}}$, $M_{\mathrm{max}}^{\mathrm{phys}}$ is lower than the M_{max} determined by sampling statistics. To observe signatures of the presence of $M_{\mathrm{max}}^{\mathrm{phys}}$, one should look in big galaxies where Σ (or the SFR) is high.

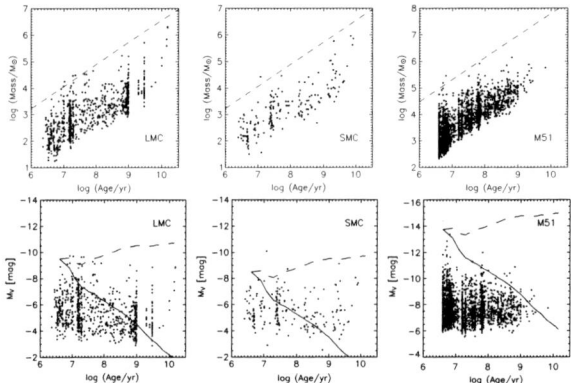

Fig. 1. *Top:* Age-mass diagrams of clusters in the LMC, SMC and M51. *Bottom:* Age-luminosity diagrams for the same clusters. The SoS relations are shown as *dashed lines*. Fading lines from SSP models are shown as *full lines* (from [7]).

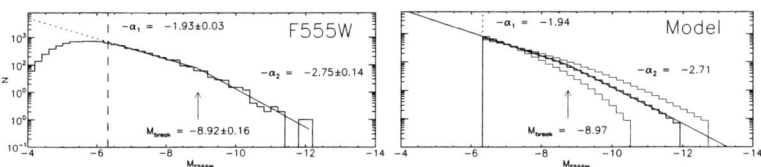

Fig. 2. *Left:* Observed CLF of ~ 6000 star clusters in M51. *Right:* Modelled CLF of a population with $M_{\max}^{\rm phys} = 5 \times 10^5\,M_\odot$ (from [8]).

2 The Size-of-Sample Effect in M51

A good candidate galaxy, which is big and has a high SFR, is M51. We used the 1052 star clusters identified by [2] to study the SoS relation of $M_{\max}$ with log(age). In Fig. 1 we show a comparison between the clusters in the LMC (left), the SMC (middle) and M51 (right). In the top panel we compare the age-mass diagrams, where we have overplotted the predicted SoS increase of $M_{\max}$ with log(age) as dashed lines, based on a power-law CIMF with index -2. As was shown by [10], $M_{\max}$ in the LMC and SMC follows this prediction quite well. In M51, however, there is a lack of old ($>\sim 10^8$ yr), massive ($>\sim 10^6\,M_\odot$) clusters. In the bottom panels we show the luminosity (magnitude) *vs.* log(age). The SoS relation for $M_{\max}$ is converted to $L_{\max}$ using the GALEV SSP models [1] and is almost a horizontal line. Fading lines, scaled to the brightest clusters at young ages, are shown as full lines. The brightest cluster *vs.* log(age) in M51 follow this fading line of a $5 \times 10^5\,M_\odot$ cluster quite well, similar to what was found for the "Antennae" galaxies (for a $10^6\,M_\odot$ cluster) [15]. *This suggests that the cluster mass function in M51 and the "Antennae" galaxies is truncated around ~ 0.5–$1.0 \times 10^6\,M_\odot$.*

3 The Integrated Star Cluster Luminosity Function

Since the age determination from broad-band colours has limitations, we want to have an independent check of the *truncated mass function scenario*, without relying on age determination. Therefore, we model for two scenarios the integrated cluster luminosity function (CLF) of a population which has formed with a constant cluster formation rate (CFR): (1) $M_{\rm max}$ is determined by SoS effects and increases with log(age) and (2) $M_{\rm max} = M_{\rm max}^{\rm phys}$ is constant with log(age). The CLF in case (1) is a power-law distribution, with an index similar to the underlying mass function. This has been observed for various spiral galaxies and the LMC and SMC [7, 11]. The resulting CLF of scenario (2) is better described by a double power-law distribution, for which the location of the break is determined by $M_{\rm max}^{\rm phys}$. On the bright side of the CLF the index is smaller than -2 (i.e. steeper), and on the faint side it is $\simeq -2$. The steeper bright side is because a truncation in the mass function will be spread out over a range of luminosities due to the age spread in the population and fading of clusters in time (e.g. young clusters with $M_{\rm max}^{\rm phys}$ are brighter than old clusters with $M_{\rm max}^{\rm phys}$). Tentative evidence for a double power-law CLF was observed for NGC 6946 and M51 [7].

Recently, the Hubble Heritage project released new *HST/ACS* data of M51, covering the entire disc with 6 pointings. We used this dataset and selected clusters based on the size. All sources found with SExtractor (~70 000), were compared to (extended) cluster profiles convolved with the camera PSF. Around 6 000 sources, above a conservative completeness limit, were found to be more extended than the instrumental PSF. The resulting CLF of this sample shows a pronounced double power-law behaviour and is very similar to what was found from the models (see Fig. 2).

Several predictions from the CLF model are found back in the observations: (1) The power-law index on the bright side $(-\alpha_2)$ increases when going to bluer filters. This is because clusters fade more rapidly in the bluer filters, which spreads out the luminosity of $M_{\rm max}^{\rm phys}$ over a larger range of magnitudes; (2) The break in the CLF shifts to brighter luminosities when going to redder filters. This is because the majority of the clusters with the break luminosity is red (see [7] for details). The best agreement between data and model, taking into account cluster disruption and extinction, is for $M_{\rm max}^{\rm phys} = 5 \times 10^5 \, M_\odot$. A similar double power-law CLF was observed for the "Antennae" clusters [14], although with a break 1.4 mag brighter, implying that $M_{\rm max}^{\rm phys}({\rm Antennae}) = 4 \times M_{\rm max}^{\rm phys}({\rm M51}) \simeq 2 \times 10^6 \, M_\odot$. We note that a direct comparison between the CLF of "Antennae" clusters and the one following from our model is dangerous because of the non-constant CFR in the "Antennae" galaxies. Nevertheless, *the observed break in the CLF is an independent confirmation of the truncated mass function scenario, confirming the results from the SoS comparison of Sect. 2.*

4 The Environmental Dependency of $M_{\rm max}^{\rm phys}$

The difference between $M_{\rm max}$ in the "Antennae" galaxies and in M51 and the recently discovered super-massive star clusters [3] (also Bastian in these proceedings), suggest an environmental dependent $M_{\rm max}^{\rm phys}$. We looked for variations of the bend location *within* M51 at different galactocentric radii (R). If $M_{\rm max} \propto \Sigma$, and $\Sigma \propto \exp(-R/R_{\rm h})$, then $M_{\rm max} \propto \exp(-R/R_{\rm h})$, with $R_{\rm h}$ the disc scale length of molecular gas. We found a correlation, since in three radial bins with $\bar{R}/{\rm kpc} = [1.5, 4.5, 7]$ we find the bend at $M_V = [-8.6, -8.5, -7.7]$ [9]. Although the errors in the fit are large (± 0.2 mag), the decreasing $M_{\rm max}^{\rm phys}$ with R is a third argument supporting the truncated mass function scenario in M51.

5 Final Thoughts

Our observations of a truncation of the *integrated* mass function does not necessarily imply that a truncation is visible in the CIMF, since there N is much lower. Therefore, the observations of an untruncated CIMF in M51 [4] and the "Antennae" galaxies [15] are not in disagreement with what we discuss here. In addition, the scaling of $L_{\rm max}$ with N is expected to be determined by the SoS effect, since the brightest cluster is generally young (< 10 Myr). The number of clusters in a young sample is too small to sample the mass function up to $M_{\rm max}^{\rm phys}$.

Acknowledgements. I thank Bruce Elmegreen for interesting discussions during the meeting in Concepción and the organisers for a great conference and a nice asado!

References

1. P. Anders, & U. Fritze-v. Alvensleben, A&A **401**, 1063 (2003)
2. N. Bastian, M. Gieles, H.J.G.L.M. Lamers, et al., A&A **431**, 905 (2005)
3. N. Bastian, R.P. Saglia, P. Goudfrooij, et al., A&A **448**, 881 (2006)
4. A. Bik, H.J.G.L.M. Lamers, N. Bastian, et al., A&A **397**, 473 (2003)
5. B.G. Elmegreen, in ASP Conf. Ser. 322: The Formation and Evolution of Massive Young Star Clusters **322**, 277 (2004)
6. B.G. Elmegreen, D.M. Elmegreen, AJ **121**, 1507 (2001)
7. M. Gieles, S.S. Larsen, N. Bastian, I.T. Stein, A&A **450**, 129 (2006)
8. M. Gieles, S.S. Larsen, R.A. Scheepmaker, et al., A&A **446**, L9 (2006)
9. M.R. Haas, M. Gieles, R.A. Scheepmaker et al., A&A **487**, 937 (2008)
10. D.A. Hunter, B.G. Elmegreen, T.J. Dupuy, M. Mortonson, AJ **126**, 1836 (2003)
11. S.S. Larsen, AJ **124**, 1393, (2002)

12. C. Weidner, P. Kroupa, S.S. Larsen, MNRAS **350**, 1503 (2004)
13. B.C. Whitmore, in A Decade of HST Science, 153–178 (2003)
14. B.C. Whitmore, Q. Zhang, C. Leitherer, et al., AJ **118**, 1551 (1999)
15. Q. Zhang, S.M. Fall, ApJL **527**, L81 (1999)

The Stellar Population of Ultra-Compact Dwarf Galaxies

Michael D. Gregg,[1,2] Michael J. Drinkwater,[3] Katya Evstigneeva,[3] and Arna Karick[1,2]

[1] Physics Department University of California, Davis, California 95616, USA
[2] IGPP/LLNL, L-413, Livermore, CA 94550, USA
   gregg,akarick@igpp.ucllnl.org
[3] Department of Physics University of Queensland, QLD 4072, Australia
   m.drinkwater,katya@physics.uq.edu.au

1 A Compact Dilemma

The burgeoning field of research on ultra-compact dwarf (UCD) galaxies faces a fundamental dilemma: what exactly is a UCD, anyway? We all feel we know one when we see one, but where some of us see many, others see just two. The operational definition varies from astronomer to astronomer, and, lacking detailed data, UCDs can be difficult or impossible to distinguish from globular clusters; some would claim that there is no difference. Originally, UCDs were objects classified as stellar in the APM catalog but having radial velocities placing them in the Fornax or Virgo clusters. This simple definition breaks down at other distances, nearer and farther, and as better imaging data become available. We need an objective definition.

There are now close to 100 compact objects spread liberally around the intergalactic spaces of the Fornax cluster, most with $19 < B < 22.5$ or $-9.5 < M_B < -12$ (Firth et al., these proceedings). The question arises, are they UCDs or true intergalactic globular clusters which are expected to accumulate during the formation of galaxy clusters [6]? This depends on the definitions of UCDs and globulars, of course. Milky Way globular clusters form a distinctly simple set: with a handful of exceptions, each is a single-aged, single-abundance, old population, and no dark matter is required to account for the internal dynamics. If UCDs are galaxies, then they must differ from the simple nature of globular clusters.

Determining the presence of dark matter presents a serious observational challenge in relatively faint, low internal velocity dispersion objects because the requisite spectra must be both high (echelle) resolution and high signal-to-noise. This has been done for a handful of UCDs [2,4]; the resulting M/L$\approx$ 4, about twice that of a globular, requiring some dark matter. So UCDs must be classed as bona fide galaxies.

It is prohibitively expensive to obtain echelle data for large samples of compact objects in Fornax or Virgo, and beyond. An alternative is to use population diagnostics from lower resolution spectra or broad band imaging. Although the usual age/metallicity degeneracy caveats apply, it is possible to determine if an object has a spread of age or abundance, especially if applied differentially, comparing globulars and larger ellipticals to UCD samples, thereby distinguishing UCDs from simple globular clusters.

2 UCD Stellar Populations from Optical Spectra

The Low Resolution Imaging Spectrograph (LRIS) at Keck was used to obtain 6Å resolution spectra of the original six Fornax UCDs. These reveal that the UCDs are at least moderately old and have a range of absorption line strengths, in some cases surprisingly strong. UCD01 in particular has very strong CN4200 and CN3850 absorption in addition to strong Mg and Fe, approaching the line strengths of the largest ellipticals, such as NGC1399 (Fig. 1). A quantitative comparison shows that the UCDs have moderately enhanced Balmer lines for their Mg2 strength, compared to old stellar population objects, such as globular clusters and the oldest ellipticals (Fig. 2). The most straightforward interpretation of this is that UCDs contain an intermediate age population, 3–5 billion years old. As such these few Fornax UCDs are then *not* globular clusters, but galaxies.

After the discovery of UCDs in the Virgo cluster [5], we observed them with the Echelle Spectrograph Imager at Keck Observatory. The Virgo UCDs also exhibit relatively strong absorption features indicative of high abundance. Unlike the Fornax UCDs, however, the Virgo objects show no evidence for enhanced Balmer absorption and, if anything, are slightly older than the Virgo dE, N galaxy sample of Geha et al. [3]; (Fig. 2). At the same time, the

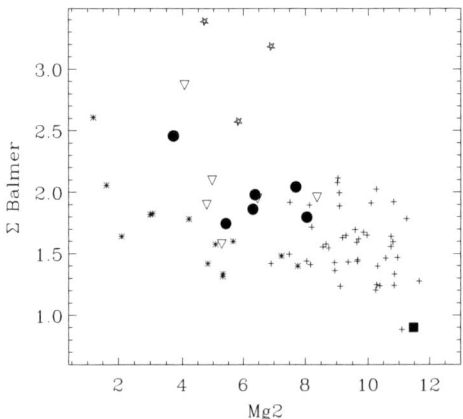

Fig. 1. Comparison of early-type objects in Fornax from medium resolution spectra. The plus signs are ellipticals, asterisks are Milky Way globulars, stars are starburst objects, *open triangles* are dE,N nuclei, and *filled circles* are UCDs. The UCDs and dEs have enhanced Balmer line strength relative to the purely old globular sequence, evidence that they house intermediate age populations.

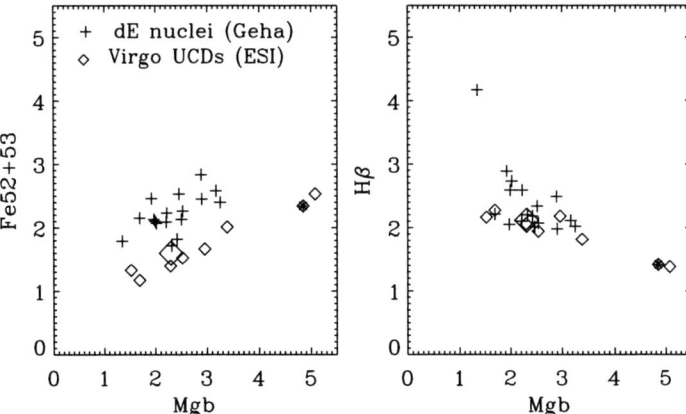

Fig. 2. Comparison of Virgo UCDs and dE, N line strengths. Unlike the Fornax UCDs, the Virgo objects do not have enhanced Balmer lines, so are perhaps entirely old. But the Virgo UCDs have weaker Fe at a given Mgb strength compared to the dEs, indicating a different enrichment history.

UCDs have lower Fe line strengths at a given Mgb index. At least in Virgo, the UCDs and dE, N galaxies appear to have different enrichment histories, and therefore probably also different formation mechanisms, making galaxy threshing difficult to accommodate [1].

3 UCD Stellar Populations from GALEX Imaging

Turning to other resources, we matched the positions of the known UCDs in Fornax to the GALEX satellite catalog made from the 1500s archival image. A dozen matches within 2″ were found in the NUV, about half that many in the FUV. From our own deep imaging with the CTIO 4m 8K Mosaic, we have r'-band photometry of the UCDs, so we can examine optical/UV color–color behavior of the UCDs (Fig. 3). In the $g - r$ vs. $r - NUV$ plane, the dE, N galaxies lie entirely within the normal galaxy locus, while for the most part, the UCDs bridge the region between the galaxies and the stellar locus. The Fornax UCDs and dE, N galaxies almost do not overlap, a little surprising given their coincidence in Fig. 2. This could be consistent with the LRIS result if the apparent intermediate age population in UCDs is older, ∼5 Gyr, than that in the dE, N nuclei, perhaps ∼2–3 Gyr.

4 Summary

- From our low resolution Keck/LRIS spectra, the 6 original Fornax UCDs appear to have intermediate age populations, similar to dE nuclei.

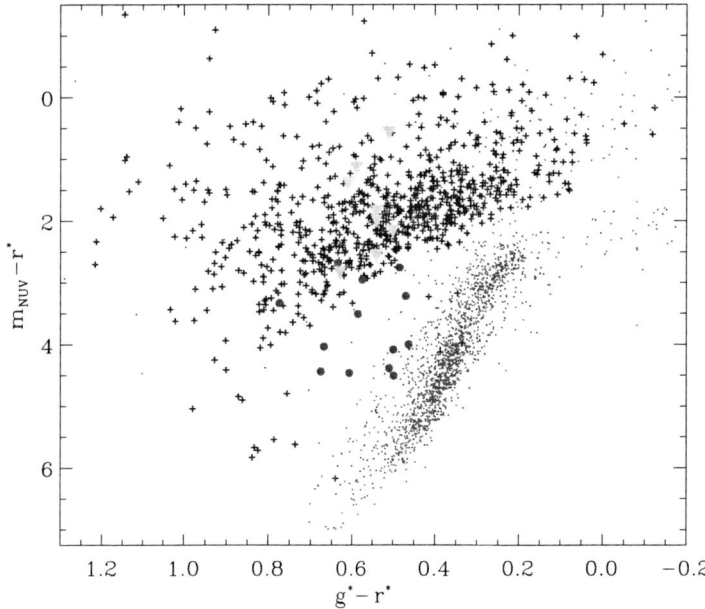

Fig. 3. Optical–NUV color–color diagram using the GALEX catalog of Fornax cluster objects. The UCDs (*large dots*) separate from the dE, N galaxies (*triangles*), which lie among the ordinary galaxies (+). The GALEX data can also efficiently separate UCDs from stars, even though the UV images have 4″ spatial resolution.

- Higher resolution Keck/ESI spectroscopy of indicates that the Virgo UCDs are uniformly old and dissimilar to Virgo dE nuclei.
- The UV properties of UCDs and dEs are distinct, with the UCDs being redder, presumably older.
- The UCD population can vary from cluster to cluster, implying different evolutionary histories for both the UCDs and the clusters.

The results described here need to be extended to larger samples of UCDs, now available in Fornax and, soon, Virgo. High quality medium resolution spectra, deeper GALEX imaging, and infrared photometry or spectroscopy could all be applied to comparing the stellar populations of UCDs, yielding clues to their origin and to the evolution of their parent galaxy clusters.

This work is supported by NSF grant 0407445 and carried out partly at IGPP under the auspices of the US DOE by LLNL, contract no. W-7405-Eng-48. This project was also supported by grants from the ARC and ANSTO Access to Large Research Facilities Scheme.

References

1. K. Bekki, et al.: Ap. J. L. **552**, 105 (2001)
2. M. Drinkwater, et al.: Nature, **423**, 519 (2003)
3. M. Geha, et al.: A. J. **124**, 3073 (2002)
4. M. Hasegan, et al.: Ap. J. **627**, 203 (2005)
5. B. Jones, et al.: A. J. **131**, 312 (2006)
6. M. West, et al.: Ap. J. **453**, 77 (1995)

News on Ultra-Compact Dwarfs and Blue Globular Clusters

Steffen Mieske

ESO Garching, Karl-Schwarzschild-Str.2, 85748 Garching b. München, Germany
smieske@eso.org

This contribution consists of two parts. In the first part, we discuss the separation between ultra-compact dwarf galaxies (UCDs) and globular clusters (GCs) in the Fornax cluster. Based on spectroscopic and structural information, we find that the limit between both classes occurs at about $M_V = -11$ mag. In the second part of the contribution we analyse color-magnitude relations in Globular Cluster Systems observed in the ACS Virgo Cluster Survey. We find that the blue GC population exhibits a clear trend of redder color with increasing luminosity. This trend persists over the entire investigated luminosity range of Virgo cluster galaxies, and is more pronounced for GCs at smaller galactocentric distances.

1 The Limit Between UCDs and GCs in Fornax

Ultra-compact dwarf galaxies (UCDs) have been proposed as a new class of galaxies, populating the central regions of the Fornax, Virgo and Abell 1689 clusters [1–5]. UCDs are placed between the sequence of globular clusters and dwarf elliptical galaxies in the fundamental plane [6], having absolute magnitudes $M_V > -13.5$ mag. Two UCD formation channels are discussed: (1) "Super star clusters" formed in galaxy mergers [7]. (2) Remnant nuclei of dE,Ns stripped in the potential well of their host cluster [8].

In order to properly discuss those scenarios, an important pre-requisite is to cleanly separate UCDs from the morphologically similar globular clusters (GCs). In Mieske et al. ([9], see Fig. 1) we investigate this issue, and find two distinctive breaks in the properties of Fornax compact objects at $M_V \simeq -11$ mag ($3 * 10^6 M_*$):

1. In the metallicity-luminosity plane, objects with $M_V < -11$ mag have a narrow metallicity distribution centred around [Fe/H]$= -0.6$ dex. Objects with $M_V > -11$ mag have a broader metallicity distribution centred around -1.2 dex.
2. In the size-luminosity plane, objects with $M_V < -11$ mag have sizes that positively correlate with luminosity. Objects with $M_V > -11$ mag have luminosity independent sizes.

This makes us conclude that in Fornax, the separation between GCs and UCDs occurs at $3*10^6 M_*$ ($M_V = -11$ mag):

GCs have masses $m < 3*10^6 M_*$, luminosity independent sizes and a broad metallicity distribution. **UCDs** have masses $m > 3*10^6 M_*$, sizes correlating with luminosity and a narrow metallicity distribution.

Since merged star clusters are more extended than single clusters of same mass (e.g. [10,11]), we conclude in more general terms: the transition between single and merged compact stellar systems occurs at $\approx 3*10^6 M_*$.

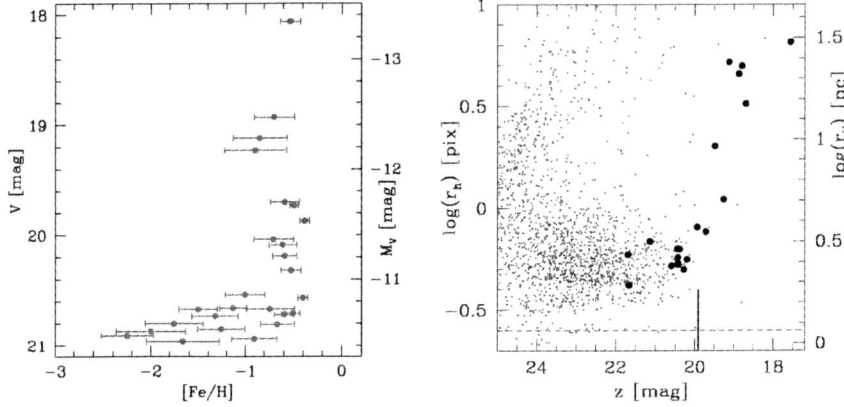

Fig. 1. *Left*: [Fe/H]-magnitude diagram for a sample of Fornax compact objects [9]. Note the change in metallicity distribution at $M_V \simeq -11$ mag. The [Fe/H] difference between $M_V < -11$ mag and $M_V > -11$ mag is significant at the 3.7σ level. *Right*: *Filled circles* indicate instrumental magnitude vs. half-light radius r_h for Fornax compact objects with HST-ACS imaging ([9] and references therein). *Dots* show *all* the objects in the two ACS pointings containing compact objects. Note the uprise in the size distribution at around the vertical tick, which corresponds to the magnitude of the metallicity break in the left panel. The *dashed horizontal line* indicates the resolution limit.

2 The Color-Magnitude Relation for Blue GCs

While the relation between mean color of a GC system and host galaxy luminosity is now well established [12], an internal color-magnitude (CM) relation in the sense of redder average color of single GCs with increasing average single GC luminosity had not been reported until quite recently. Based on their HST/ACS imaging program, Harris et al. [13] investigate the GC color distribution for eight nearby brightest cluster galaxies (BCGs). For several of the investigated galaxies they find a clear trend of redder color with increasing luminosity for the blue GC population. Strader et al. [14] used the public archive data of the ACS Virgo Cluster Survey (ACSVCS,

Côté et al. [15]) to investigate a possible CM relation for the three brightest Virgo galaxies M49, M87 and M60. They find significant CM trends for the blue GC population of M87 and M60, but not for M49.

Those two studies do not investigate how the blue peak slope varies as a function of host galaxy luminosity or galactocentric distance.

Therefore, we have investigated the CM trend of GCs in the full ACS Virgo Cluster Survey data set with special emphasis on the environmental dependence of the slope [17]. We analyse the globular cluster populations in those ACS Virgo Cluster Survey galaxies that have a significant number of GC candidates [12] and an SBF distance available [16]. Those are 78 galaxies in the range $-21.7 < M_B < -15.2$ mag [15,17].

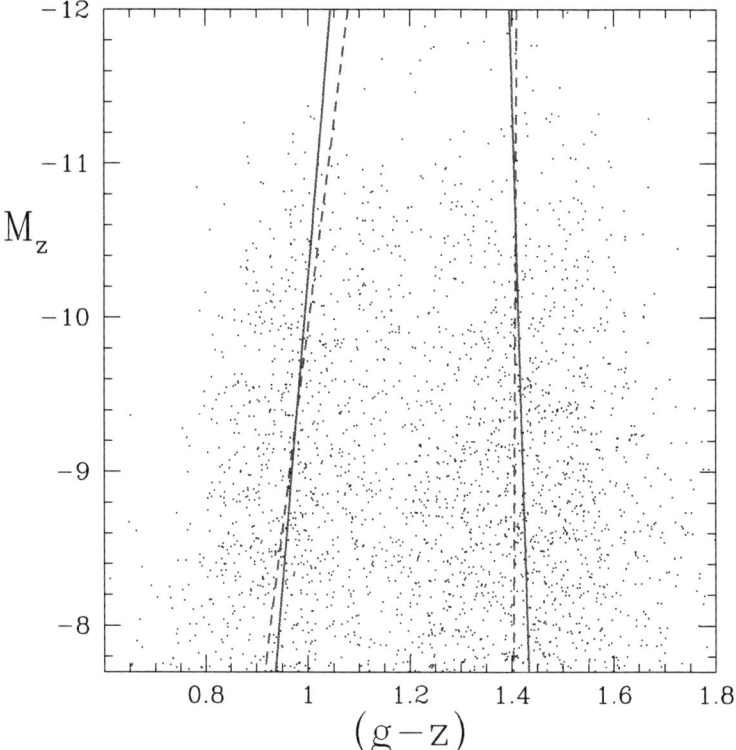

Fig. 2. Color-magnitude diagram of the joint GC population of the three brightest Virgo cluster galaxies M49, M87 and M60 [12,17]. The *solid lines* indicate fits obtained from biweight estimators of the mean color as a function of magnitude in either peak. The *dashed lines* give the linear fit obtained from KMM fitted peak positions as a function of magnitude. In both cases, a significant slope for the blue GC population is found, while the red population does not exhibit a notable trend.

These are our findings:

1. From fits to the blue peak position based both on KMM [18] and biweight estimators [19], we find a 10σ correlation between luminosity and color for the blue GCs of the three brightest Virgo Cluster galaxies (M49, M87, M60): brighter GCs are redder than their fainter counterparts (see Fig. 2). For M87 and M60 alone we also find comparable slopes, while M49 does not exhibit a measurable trend. We find no significant trend for the red sub-population.
2. The slope for the blue population gradually decreases with decreasing average host galaxy luminosity, but remains detectable over the entire investigated luminosity range.
3. The color-magnitude trend of the blue GC population is significantly stronger for GCs with smaller projected galactocentric distances.

In Mieske et al. [17] we discuss several mechanisms that might create the observed dependencies:

I. Presence of super-star-clusters or stripped nuclei. II. GC self-enrichment. III. Accretion of GC systems from lower mass galaxies. IV. Capture of field stars. V. Cluster star evaporation.

From the discussion we conclude that none of those scenarios can explain the observations in a straightforward manner. A combination of several mechanisms may be at work. More theoretical investigations will be needed to better constrain their respective contributions.

References

1. Hilker, M. et al. 1999, A&AS, 134, 75
2. Drinkwater, M. J. et al. 2000, PASA, 17, 227
3. Hasegan, M. et al. 2005, ApJ, 627, 203
4. Jones, L. R. et al. 2006, AJ, 131, 312
5. Mieske, S. et al. 2005, A&A, 430L, 25
6. Drinkwater, M. J. et al. 2003, Nature, 423, 519
7. Fellhauer, M. & Kroupa, P. 2002, MNRAS, 330, 642
8. Bekki, K. et al. 2003, MNRAS, 344, 399
9. Mieske, S., et al. 2006, AJ, 131, 2442
10. Bekki, K. et al. 2004, ApJ, 610L, 13
11. Kissler-Patig, M. et al. 2006, A&A, 448, 1031
12. Peng, E. et al. 2006, ApJ, 639, 95 (ACSVCS paper IX)
13. Harris, W. E. et al. 2006, ApJ, 636, 90
14. Strader, J. et al. 2005, AJ submitted, astro-ph/0508001
15. Côté, P. et al. 2004, ApJS, 153, 223 (ACSVCS paper I)
16. Mei, S. et al. 2005, ApJ, 625, 121 (ACSVCS paper V)
17. Mieske, S. et al. 2006, ApJ submitted (ACSVCS paper XIV)
18. Ashman, K. M. et al. 1994, AJ, 108, 2348
19. Beers, T. C. et al. 1990, AJ, 100, 32

UCDs and GCs: Structural Differences from HST Imaging

E.A. Evstigneeva[1], M.J. Drinkwater[1], M. Hilker[2], and C.Y. Peng[3]

[1] Physics Department, University of Queensland, QLD 4072, Australia
   katya,mjd@physics.uq.edu.au
[2] Sternwarte der Universitat Bonn, Auf Dem Hugel 71, 53121 Bonn, Germany
   mhilker@astro.uni-bonn.de
[3] Space Telescope Science Institute, 3700 San Martin Drive, Baltimore, MD 21218, USA cyp@stsci.edu

1 Introduction

Ultra-compact dwarf (UCD) galaxies are a new type of galaxy recently discovered in the central regions of the Fornax [5, 1, 2] and Virgo [6] galaxy clusters. Possible origins of UCDs include the following hypotheses: (1) they are the remnant nuclei of stripped dwarf galaxies which have lost their outer parts in the course of tidal interaction with the galaxy cluster potential; (2) they are stellar superclusters formed in galaxy merger events; (3) they are simply very luminous globular clusters (GCs).

Here we use HST imaging of 5 Fornax and 6 Virgo UCDs in the absolute magnitude range $-13.6 < M_V < -12.0$ to test the 3rd hypothesis.

2 Data and Models

The data for the Virgo UCDs were taken with the Advanced Camera for Surveys (ACS), High Resolution Channel (HRC). Exposure times were 870 sec in the F606W filter and 1050 sec in the F814W filter. The Fornax UCD data consist of 40 min exposures taken with the Space Telescope Imaging Spectrograph (STIS) in unfiltered mode. The images of Virgo and Fornax UCDs were modeled using the two-dimensional fitting algorithm GALFIT [9] and assuming King [7, 4] and Nuker [8] models for the luminosity profile.

3 Results and Conclusions

UCDs are poorly fitted with the King model, but are very well fitted with the Nuker law – a double power law (except two UCDs, which require two components: a core and a halo). Figure 1 compares King and Nuker models for one of the UCDs: note the differences in the central and outer parts.

How do UCDs differ from GCs?

(1) Three UCDs have small law surface brightness envelopes. "Normal" GCs do not.
(2) Outer structure: King models were specifically introduced to fit globular clusters [7]. King models predict a truncation radius, beyond which stars are stripped from the cluster by the galactic tidal field.
UCDs do not show this tidal cutoff.
(3) Core structure: A main feature of King models is their central cores – the regions of constant surface brightness. We have found from Nuker models that only one UCD has a flat core (the inner power law slope $\gamma \sim 0$), all other UCDs have shallow or steep cusps in the center ($0.2 < \gamma < 1.5$).

The CONCLUSION is: the UCDs are NOT like typical GCs.

Since we make the comparisons based on King models only, there is a CAVEAT. Although King models provide very good representations of GC profiles in many cases, there exist GCs where the models deviate from the profiles [3]: core collapse may result in a central surface brightness exceeding the one predicted by cored King models, some clusters possess stars beyond the truncation radius predicted by theory, etc.

Fig. 1. From *left* to *right*: (1) SB profiles for the King and Nuker models (NOT convolved with the PSF); (2) Residuals after subtracting King model (PSF convolved) from the image; (3) Residuals after subtracting Nuker model (PSF convolved).

References

1. M. Drinkwater, J.B. Jones, M. Gregg, S. Phillipps: PASA **17**, 227 (2000)
2. M. Drinkwater, M. Gregg, M. Hilker et al.: Nature **423**, 519 (2003)
3. R. Elson, P. Hut, S. Inagaki: ARA&A **25**, 565 (1987)
4. R. Elson, Stellar dynamics of globular clusters. In: Globular Clusters, ed. by C.M. Roger, F. Sanchez, P. Fournon (Cambridge Contemp., Cambridge University Press, 1999)
5. M. Hilker, L. Infante, T. Richtler: A&AS **138**, 55 (1999)
6. J.B. Jones, M. Drinkwater, R. Jurek et al.: AJ **131**, 312 (2006)
7. I. King: AJ **71**, 64 (1966)
8. T. Lauer, E. Ajhar, Y.-I. Byun et al.: AJ **110**, 2622 (1995)
9. C. Peng, L. Ho, C. Impey, H.-W. Rix: AJ **124**, 266 (2002)

Ultra-Compact Stellar Systems in the Fornax Galaxy Cluster

P. Firth[1], M.J. Drinkwater[1], E.A. Evstigneeva[1], A. Karick[2], M.D. Gregg[2], M. Hilker[3], K. Bekki[4], J.B. Jones[5], and S. Phillips[6]

[1] Department of Physics, University of Queensland, Qld 4072, Australia
 firth@physics.uq.edu.au
[2] University of California, Davis, California 95616, USA
[3] Sternwarte der Universitat Bonn, Auf Dem Husel 71, 53121 Bonn, Germany
[4] School of Physics, University of New South Wales, Sydney 2052, Australia
[5] Queen Mary, University of London, Mile End Road, London E14NS, UK
[6] University of Bristol, Tyndall Avenue, Bristol BS8 ITL, UK

1 Observing the UCD/GC Interface

Ultra-compact dwarfs (UCDs) are massive but compact gravitationally-bound stellar systems discovered in the nearby Fornax [1–5] and Virgo [6] galaxy clusters. Several UCD formation theories have emerged – that they are ultra-massive examples of globular clusters (GCs) [4]; or stellar super-clusters created in gas-rich galaxy mergers [7]; or the remnant cores of tidally-stripped nucleated dwarf galaxies [8].

We completed spectroscopic observations in November 2004 of colour-selected point source targets ($18.00 <$r$' <22.75$) in four $25'$ diameter VLT fields surrounding NGC1399, the massive cD galaxy at the core of the Fornax galaxy cluster (Fig. 1:LEFT). Targets were selected from g$'$r$'$i$'$ imaging with the CTIO Blanco 4 m telescope [9]. We have discovered 30 new compact stellar systems at the cluster redshift, adding to 62 previously catalogued UCDs [1–3]. Our observations extend to the absolute magnitude range of globular clusters, enabling us to explore the UCD/GC interface ($-12 <M_{r'} <-9$) in a forthcoming paper.

2 UCD Radial Distribution and Kinematics

In Fig. 1:RIGHT the radial distribution of 51 previously known Fornax UCDs for which we have u$'$g$'$r$'$i$'$z$'$ photometry, and 30 new compact stellar systems from our VLT observations, is plotted against $M_{r'}$ magnitude. The new data show that compact stellar systems over a range of magnitudes are found extensively in intra-cluster space.

The UCD system has a mean velocity of 1478 km s^{-1} and a dispersion (σ_0) of 244 km s^{-1}. Our redshift data show weak evidence for a net rotation of the 92-member UCD system about NGC1399. The velocity gradient is 57 ± 42 km s^{-1} deg^{-1} in R.A. from NGC1399 (V/$\sigma_0 = 0.23 \pm 0.17$). This

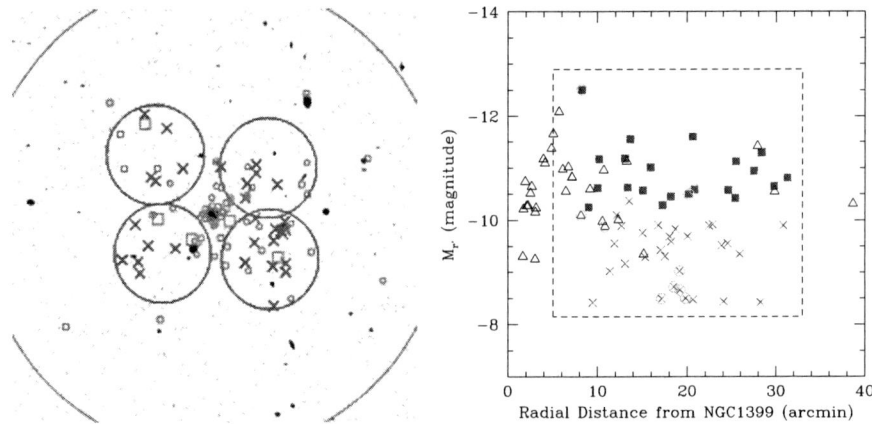

Fig. 1. LEFT: Approximately 1.5° square image of UCDs surrounding NGC1399. The 2dF field and four smaller VLT fields shown here contain previously catalogued UCDs (*open small circles/squares*) and newly discovered stellar systems (*crosses*). Our results suggest a UCD bar/filament structure stretches across NGC1399. RIGHT: $M_{r'}$ magnitude against radial distance from NGC1399 (m−M = 30.9) of 51 previously known UCDs, being those observed only with 2dF (*triangles*) or re-observed with VLT (*squares*), together with 30 newly-discovered UCDs (*crosses*). UCDs close to prominent cluster member galaxies other than NGC1399 are circled. The *dashed rectangle* shows the VLT field/magnitude outer limits.

rotation contrasts with the finding [10] that the inner GC population (2′–9′ radius) shows little or no rotation about NGC1399.

References

1. M. Hilker, L. Infante, G. Vieira, M. Kissler-Patig, T. Richtler: A&AS **134**, 75 (1999)
2. M.J. Drinkwater, J.B. Jones, M.D. Gregg, S. Phillipps: PASA **17**, 227 (2000)
3. M.J. Drinkwater, W.J. Couch, H.C. Ferguson, M. Hilker, J.B. Jones, A. Karick, S. Phillipps: PASA **21**, 375 (2004)
4. S. Mieske, M. Hilker, L. Infante: A&A **383**, 823 (2002)
5. S. Phillipps, M.J. Drinkwater, M.D. Gregg, J.B. Jones: ApJ **560**, 1, 201 (2001)
6. J.B. Jones, M.J. Drinkwater, R. Jurek, S. Phillipps, M.D. Gregg, K. Bekki, W.J. Couch, A. Karick, Q.A. Parker, R.M. Smith: AJ **131**, 312 (2006)
7. M. Fellhauer, P. Kroupa: MNRAS **330**, 642 (2002)
8. K. Bekki, W.J. Couch, M.J. Drinkwater: ApJ **552**, 105 (2001)
9. A. Karick et al.: PhD Thesis (2005)
10. T. Richtler et al.: AJ **127**, 2094 (2004)

Multi-Colour Imaging of Ultra-Compact Objects in the Fornax Cluster

A.M. Karick[1], M.D. Gregg[1], M.J. Drinkwater[2], M. Hilker[3], and P. Firth[2]

[1] University of California, Davis, USA akarick@igpp.ucllnl.org
[2] University of Queensland, Australia
[3] Sternwarte der Universitat Bonn, Germany

Ultra-compact dwarf galaxies (UCDs) are a new type of galaxy we have discovered in the central region of the Fornax and Virgo clusters. Unresolved in ground-based imaging, UCDs have spectra typical of old stellar systems. Ninety-two have been found in Fornax [1,3], making them *the most numerous galaxy type in the cluster*. Although they form a cluster wide population, an over-density surrounds the central cluster galaxy, NGC1399, fueling controversy over their nature and origin. Several formation scenarios have been proposed. UCDs may be the remnant nuclei of tidally stripped dE,N galaxies [2] or they may be the bright tail of the globular cluster (GC) population associated with NGC1399 [3,"UCOs"]. Alternatively they may be the first spectroscopically confirmed intracluster globular clusters (IGCs) in Fornax, resulting from hierarchical star cluster formation [4] and merging in intracluster space. The 5 brightest Fornax UCDs have M/L ratios indicating at least some dark matter, unlike typical GCs.

1 Multicolour Imaging of the Fornax Cluster

We obtained deep multicolour (u'g'r'i'z') imaging of the central region of the Fornax Cluster using the CTIO 4 m Mosaic Telescope. Figure 1 (LEFT) shows the radial distribution of the present population of UCDs, GCs and dE,Ns. Sixty-two UCDs were discovered in our 2 dF spectroscopic surveys of Fornax. An additional 30 were discovered in our recent VLT survey, from which candidates were pre-selected using this multicolour data [5]. Figure 1 (RIGHT) shows the g'−i' colour-magnitude relation (CMR) for the UCDs, UCOs [3], NGC1399 GCs [6–8], and cluster dwarf galaxies [9]. All objects are spectroscopically confirmed cluster members.

Bright UCDs and dEs follow similar CMRs: both populations become redder with increasing luminosity. The slope for the CMR for bright UCDs and fainter UCD candidates is −0.05 mag, similar to the slope of the CMR of candidate UCDs (−0.07 mag) in Abell 1689 [10]. This is qualitatively consistent with UCDs being the stripped nuclei of dE,Ns. UCDs have similar colours to NGC1399 GCs, however they exhibit a larger spread in colour at 18< i'<20, possibly reflecting a mixed metallicity or age spread. These results will be discussed in more detail in a forthcoming paper.

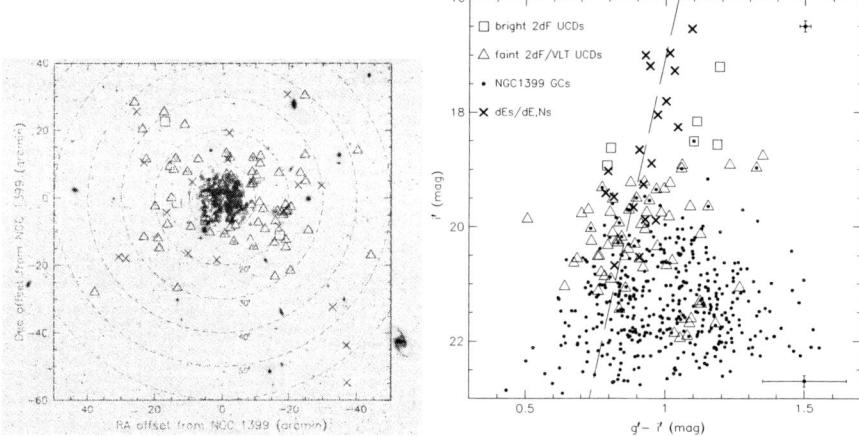

Fig. 1. LEFT: Distribution of compact objects surrounding NGC1399: bright 2dF UCDs (*boxes*), fainter 2dF/VLT UCDs (*triangles*), dE,Ns (*crosses*), NGC1399 GCs and UCOs (*small circles*). The NGC1399 GCs extend out to 10′ (survey limited) and overlap a small fraction (7%) of the innermost UCDs. A significant number of the 2dF/VLT UCDs (25%) overlap the UCOs which extend out to 20′. RIGHT: $g'-i'$ CMR for compact objects in Fornax: 5 brightest 2dF UCDs (*boxes*), fainter 2dF/VLT UCDs (*triangles*), cluster dEs and dE,Ns (*crosses*) and NGC 1399 globular clusters (*points*). The *dashed line* shows the fit to the dwarf galaxy CMR. Typical error bars for $i'\sim16.5$ mag and $i'\sim22.5$ mag objects are also shown.

At faint magnitudes, GC and UCDs cannot be distinguished by colour alone. High resolution spectroscopy to measure their internal velocity dispersions and metallicities, is needed to distinguish between GCs and UCDs. If dark matter is required to explain their dynamics then these faint compact objects are real galaxies or UCDs.

This project was supported by grants from the ARC and ANSTO Access to Large Research Facilities scheme. This material is based on part upon work supported by NSF grant no. 0407445 and carried out at IGPP under the auspices of the US DOE by LLNL under contract no. W-7405-Eng-48.

References

1. M. J. Drinkwater et al.: PASA **21**, 375 (2004)
2. K. Bekki, W.J. Couch, M.J. Drinkwater: ApJ **552**, 105 (2001)
3. S. Mieske, M. Hilker, L. Infante: A&A **418**, 445 (2004)
4. M. Fellhauer, P. Kroupa: MNRAS **330**, 642 (2002)
5. A. M. Karick: PhD Thesis, University of Melbourne, Melbourne (2005)
6. B. Dirsch, et al.: AJ **127**, 2114 (2004)
7. M. Kissler-Patig: AJ **115**, 105 (1998)
8. D. Minniti, M. Kissler-Patig, P. Goudfrooij, G. Meylan: AJ **115**, 121 (1998)
9. H. C. Ferguson: AJ **98**, 367 (1989)
10. S. Mieske, L. Infante: AJ **128**, 1529 (2004)

Part III

Young Star Clusters

Hierarchical Formation of Galactic Clusters

Bruce G. Elmegreen

IBM Research Division, T.J. Watson Research Center, 1101 Kitchawan Road, Yorktown Hts., NY 10598, USA bge@watson.ibm.com

Young stellar groupings and clusters have hierarchical patterns ranging from flocculent spiral arms and star complexes on the largest scale to OB associations, OB subgroups, small loose groups, clusters and cluster subclumps on the smallest scales. There is no obvious transition in morphology at the cluster boundary, suggesting that clusters are only the inner parts of the hierarchy where stars have had enough time to mix. The power-law cluster mass function follows from this hierarchical structure: $n(M_{cl}) \propto M_{cl}^{-\beta}$ for $\beta \sim 2$. This value of β is independently required by the observation that the summed IMFs from many clusters in a galaxy equals approximately the IMF of each cluster.

1 Introduction

One way to study the origin of clusters is to observe the general morphology of star formation, how it fits with the surrounding gas distribution, and how it responds to various energy sources like other star formation or galactic-scale processes like spiral waves. A startling revelation in ISM structure came 20 years ago following Infrared Astronomical Satellite surveys of dust emission and CO and HI surveys of the Galactic plane. These surveys demonstrated that interstellar gas is not a random arrangement of round clouds with a smooth intercloud medium. It is a scale-free continuum of structures with a power-law power spectrum, correlated velocities and densities, and no obvious limits on either large or small scales except those defined by the galaxy itself. This view was present as far back as the 1950s, but dropped from ISM models in the intervening two decades (see review in [29]). Particularly important papers were those by Beech [5], in which fractal structure was found for the first time in the Lynds dark clouds, Low et al. [48], in which the pervasive and highly structured IR cirrus clouds were discovered, Croisier & Dickey [8], in which the power-law power spectrum of HI was discovered, and Stutzki et al. [58], in which the power-law power spectrum of CO emission was discovered. All of these followed the influential paper by Larson [47] in which giant molecular cloud (GMC) properties like size, density, and velocity dispersion were shown to be correlated in a manner reminiscent of turbulence. Whole galaxies were eventually found to have correlated properties too [57, 61, 26].

These papers and others led to a change in thinking about the structure, energetics, and evolution of interstellar gas. It did not take long for the theory of star formation to change with it.

Star formation is now seen to follow scale-free patterns like the gas, suggesting that stars form in turbulent gas wherever the density is large, making clusters and loose groups with a power-law distribution of masses. The result is a hierarchy of clouds and young stellar groupings with HI "superclouds" [15] and star complexes [12,13] on the largest scale, GMCs and OB associations generally clustered together inside of them [23,24,37], and molecular cloud cores with galactic clusters inside. An early review is in Scalo [52]. GMCs are not isolated regions of star formation, nor are they the largest scale of cloud structure. GMCs are not ballistic objects, nor long-lived objects, although their pieces may shuffle around and form new clouds after star formation breaks them apart [15]. Self-gravitating clouds more massive than GMCs are observed but they are not molecular. Virialized density decreases with increasing mass at constant pressure, so the largest self-gravitating clouds do not self-shield against dissociative radiation in our Galaxy and are mostly atomic [24]. The largest clouds are more highly molecular in a high-pressure galaxy like M51 [51]. Generally the molecular fraction in a galaxy follows the pressure [16,64]. Thus the largest GMC mass in our Galaxy [63] is not the largest self-gravitating cloud mass. Neither is the largest GMC mass related to the largest cluster mass, which should have a different dependence on ISM properties like pressure P and core density n: $M_{cl,max} \sim 6 \times 10^3 \left(P/3 \times 10^8 \, k_\mathrm{B}\right)^{1.5} \left(n/10^5 \, \mathrm{cm}^{-2}\right)^{-2}$ [25].

GMCs are not special cloud structures, they are only the dense self-shielded parts of the ISM hierarchy [1,24], and among them, only the most massive tend to be self-gravitating in large-scale surveys [42]. Molecular cores inside GMCs can be self-gravitating too. Density peaks in the diffuse ISM should be viewed as transient, lasting only a few internal turbulent crossing times before shear and random motions from inside and outside change their identities. Even GMCs and their star-forming cores are probably somewhat transient, although perhaps not for the same reasons as diffuse clouds. GMC cores appear to begin star formation very quickly after they form, and the pressure from this star formation disrupts them [18,41]. Short cloud lifetimes appear to be the norm over a wide range of scales, from individual clusters to whole star complexes, with the actual time scale for formation increasing with size, in proportion to the turbulent crossing time [14].

Hierarchical star formation has been recognized for a long time. In a series of papers in the 1980's, Feitzinger and collaborators quantified the hierarchical structure and fractal dimension in several nearby galaxies, including the LMC [32–34]. Other recent studies of the top two levels in the hierarchy, from star complexes to OB associations, were made for the LMC [49,40,36], the SMC [7], M31 [4], NGC 300 [50], M51 [3], and M33 [43]. Large surveys of galaxies were in Bresolin [6] and Gusev [39]. These studies found collections of OB stars in OB associations, typically 80 pc in diameter, which are themselves collected into star complexes several hundred pc in diameter. A review of star complexes is in Efremov [13].

The top of the hierarchy consists of 10^7 $M_\odot$ clouds and star complexes that are most likely formed by gravitational instabilities in either the ambient medium, forming flocculent spirals, or in the dense shocks (dust lanes) of spiral arms when a stellar spiral wave is strong. Inside density wave spirals, giant clouds and star complexes are nearly equally spaced with a separation of about 3 times the arm width [23,31], similar to the relative separation of clumps in other filamentary clouds [55]. This is the characteristic length of the threshold gravitational instability for the tube-like concentration of gas that is a dust lane [17]. The instability also produces feathery clouds that trail off into the interarm region [2,44]. There might be a characteristic mass for these largest clouds, comparable to the theoretical Jeans mass of $\sim c^4/\left(\pi^2 G^2 \Sigma\right)$ for velocity dispersion c and gas mass column density Σ. Such a peaked mass function is observed in the highly compressed parts of the interacting galaxy IC 2163 [31], but it is not known whether such a peaked function is a general feature of giant spiral arm clouds.

OB associations, loose stellar groupings and star clusters are evidently fragments of these giant clouds. The associations and groups usually cluster inside the giant clouds [37]. On smaller scales, there are universal power-law mass functions, which are presumably the result of scale-free processes such as turbulence and gravitational fragmentation.

The hierarchical structure does not stop at OB subgroups. It continues down to the sub-parsec scale of individual clusters and inside the clusters, which are often hierarchical themselves (e.g., ρ Oph [56], NGC 2264 [9], Serpens [59]). Kiss [45] found 3872 T Tauri stars from 2MASS using colors as a guide. There were 64 possible T associations among these. These groupings have a star-number distribution that continues to increase like a power law from 138 stars down to 4 stars. This distribution implies that hierarchical stellar groupings may continue down to a few stars.

Between the top and bottom of the hierarchy there is a broad range of correlated scales. As just mentioned, there is hierarchical structure in stellar groupings, autocorrelation of clusters [66], power-law power spectra of optical light in galaxies [27,28,62]; power-law size distributions of star fields [25,30], and fractional powers in the run of star-counts versus distance [7]. In general, young star fields are hierarchical, even if the groupings are unbound (e.g. [36] for the LMC). Local groupings of low mass x-ray stars (T Tauri stars) in Gould's Belt also have a hierarchical structure [38].

2 Cluster Mass Functions

The power-law mass functions for atomic and molecular clouds and for star clusters and OB associations are most likely the result of the hierarchical nature of the ISM. Clouds like these are not isolated objects but are interconnected by diffuse and molecular gas in a widespread network. The network has a power-law power spectrum, as discussed above, and this means there is no characteristic cloud size or mass. There may be only upper and lower limits. Because clusters form in scale-free gas clouds with about constant efficiency (to within a factor of ~ 3), their initial stellar masses are also scale free. The sizes of clusters are apparently not scale-free as there seems to be a characteristic cluster radius [54].

The mass distribution of clusters in spiral galaxy disks is a power law with a negative slope of around $\beta = 2$ on a log–log plot with linear intervals of cluster mass (or a negative slope of 1 on a log–log plot with log intervals of mass). In the Antennae galaxy, $\beta = 1.95 \pm 0.03$ for young clusters and $\beta = 2.00 \pm 0.08$ for old clusters [65]. For the LMC, $\beta = 1.85 \pm 0.05$ [11]. For M51, $\beta = 2$ [35]. For NGC 3310, $\beta = 2.04 \pm 0.23$ and for NGC 6745, $\beta = 1.96 \pm 0.15$ [10]. A mass function with a slope of $\beta = 2$ follows from an ISM that is fractal with a three-dimensional power spectrum slope of -3.66 [20, 30], the same as for velocity in 3D Kolmogorov turbulence.

Summed IMFs from clusters can produce a global IMF that is nearly the same as the individual IMFs if the cluster mass function slope is the observed value of $\beta = 2$. Galaxy-wide IMFs are in fact very close to the Salpeter IMF. These IMFs are determined in a variety of ways, including metallicity, colors, Hα equivalent widths, and color-magnitude diagram star counts. Cluster IMFs have an average slope comparable to the Salpeter value also [53]. This agreement between galaxy and cluster IMFs implies that the cluster mass function is close to $\beta = 2$. A slightly steeper cluster mass function makes the summed IMF significantly steeper [46,60], and in clear disagreement with the IMF observations, reviewed in [22]. For example, $\beta = 2.3$ implies the galaxy IMF should have a slope of -2.9 [60], which is much steeper than the commonly observed Salpeter slope of approximately -2.4. Evidently, galaxy-wide IMFs are a sensitive measure of the slope of the mass function of clusters and of general regions of star formation.

The approximate agreement between summed IMFs and cluster IMFs implies, in practical terms, that stars of any mass can form in clusters of any mass. A cluster seems to choose its stars randomly from a universal IMF. Such random sampling would be illogical if a cluster were to choose a star more massive than itself, but this event has negligible occurrence in practice [22]. A related property is that a power-law IMF implies that the maximum likely star mass out of N clusters of mass M equals the maximum likely star mass in one cluster of mass NM [22].

Hierarchical structure with $\beta = 2$ is also required by the observation that cluster IMFs are independent of cluster mass. More massive clusters have more sub-units for the local sum of sub-unit IMFs, and this would produce steeper IMFs for more massive clusters than low mass clusters if $\beta > 2$.

3 Summary

The hierarchy of star formation extends from "star complexes" on ~ 500 pc scales to the interiors of embedded clusters on sub-pc scales. This smooth continuation is evident from power spectra, mass functions, autocorrelation functions, and other studies. There is apparently no threshold or change at a "cluster" boundary other than the change from unmixed hierarchical structure outside the boundary to mixed stellar orbits inside the boundary. Clusters are best defined dynamically where the cluster age equals the dynamical time at a density much higher than the tidal limit. Cluster self-boundedness is a separate issue. In fact, the efficiency of star formation is automatically large at high density in a power law ISM [19,21], so self-boundedness is somewhat inevitable for very young clusters once the local star formation process ends.

Hierarchical structure produces a cluster mass spectrum with equal mass in equal logarithmic intervals. This means $\beta \sim 2$, particularly with a Kolmogorov-like power spectrum for structure. In fact, $\beta \sim 2$ is observed directly for many cluster systems. $\beta \sim 2$ is also required by the observation that galaxy-wide IMFs are equal to individual cluster IMFs. One important implication of this is that stars of any mass can be associated with clusters with any number of stars.

The top of the hierarchy (flocculent spiral arms and star complexes) should have a characteristic cloud or cluster mass comparable to the ambient Jeans mass, although there are few observations to confirm this point. A peaked mass function for star complexes has been observed for only one galaxy so far; the bound clusters inside these star complexes have a power-law mass function [30]. If halo globular clusters had a peaked mass function at birth, then it seems possible they were at the top of a hierarchy of cloud structures also, but in ultra-high pressure regions to make them compact.

References

1. R.J. Allen, P.D. Atherton, R.P.J. Tilanus: Nature **319**, 296 (1986)
2. S.A. Balbus: ApJ **324**, 60 (1988)
3. N. Bastian, M. Gieles, Y.N. Efremov et al.: A&A **443**, 79 (2005)
4. P. Battinelli, Y.N. Efremov, E.A. Magnier: A&A **314**, 51 (1996)
5. M. Beech: Ap. Sp. Sci. **133**, 193 (1987)
6. F. Bresolin, R.C. Kennicutt, L. Ferrarese, et al.: AJ **116**, 119 (1998)
7. Y. Chen, M.M. Crone: AAS **207**, 113.05 (2005)
8. J. Crovisier, J.M. Dickey: A&A **122**, 282 (1983)
9. S.E. Dahm, T. Simon: AJ **129**, 829 (2005)
10. R. de Grijs, P. Anders, N. Bastian et al.: MNRAS **343**, 1285 (2003)
11. R. de Grijs, P. Anders: MNRAS **366**, 295 (2006)
12. Y.N. Efremov: Soviet Astron. Lett. **4**, 66 (1978)
13. Y.N. Efremov: AJ **110**, 2757 (1995)

14. Y.N. Efremov, B.G. Elmegreen: MNRAS **299**, 588 (1998)
15. B.G. Elmegreen: ApJ **231**, 372 (1979)
16. B.G. Elmegreen: ApJ **411**, 170 (1993)
17. B.G. Elmegreen: ApJ **433**, 39 (1994)
18. B.G. Elmegreen: ApJ **530**, 277 (2000)
19. B.G. Elmegreen: ApJ **577**, 206 (2002)
20. B.G. Elmegreen: Triggering the formation of massive clusters. In: *The Formation and Evolution of Massive Young Star Clusters*, ASP Conf. Ser. vol 322, ed by H.J.G.L.M. Lamers, L.J. Smith, A. Nota (Astronomical Society of the Pacific, San Francisco 2004) pp 277–288
21. B.G. Elmegreen: Star-forming complexes in galaxies. In: *The Many Scales in the Universe*, ed by Jose Carlos del Toro Iniesta et al. (Springer, Dordrecht 2006) pp 99
22. B.G. Elmegreen: ApJ (2006) submitted
23. B.G. Elmegreen, D.M. Elmegreen: MNRAS **203**, 31 (1983)
24. B.G. Elmegreen, D.M. Elmegreen: ApJ **320**, 182 (1987)
25. B.G. Elmegreen, D.M. Elmegreen: AJ **121**, 1507 (2001)
26. B.G. Elmegreen, S. Kim, L. Staveley-Smith: ApJ **548**, 749 (2001)
27. B.G. Elmegreen, D.M. Elmegreen, S. Leitner: ApJ **590**, 271 (2003)
28. B.G. Elmegreen, S. Leitner, D.M. Elmegreen et al.: ApJ **593**, 333 (2003)
29. B.G. Elmegreen, J. Scalo: ARAA **42**, 211 (2004)
30. B.G. Elmegreen, D.M. Elmegreen, R. Chandar et al.: ApJ **644**, (2006) in press
31. D.M. Elmegreen, B.G. Elmegreen, M. Kaufman et al.: ApJ **642**, 158 (2006)
32. J.V. Feitzinger, E. Braunsfurth: A&A **139**, 104 (1984)
33. J.V. Feitzinger, T. Galinski: A&A **179**, 249 (1987)
34. J.V. Feitzinger, T. Galinski: Bull. Inf. Centre Donnees Stellaires **34**, 231 (1988)
35. M. Gieles, S.S. Larsen, R.A. Scheepmaker et al.: A&A **446**, L9 (2006)
36. D. Gouliermis, M. Kontizas, R. Korakitis et al.: AJ **119**, 1737 (2000)
37. D.A. Grabelsky, R.S. Cohen, L. Bronfman et al.: ApJ **315**, 122 (1987)
38. P. Guillout, M.F. Sterzik, J.H.M.M. Schmitt et al.: A&A **337**, 113 (1998)
39. A.S. Gusev: A&AT **21**, 75 (2002)
40. J. Harris, D. Zaritsky: AJ **117**, 2831 (1999)
41. L. Hartmann, J. Ballesteros-Paredes, E.A. Bergin: ApJ **562**, 852 (2001)
42. M.H. Heyer, J.M. Carpenter, R.L. Snell: ApJ **551**, 852 (2001)
43. G.R. Ivanov: Pub. Astr. Soc. Rudjer Boskovic **5**, 75 (2005)
44. W.-T. Kim, E.C. Ostriker: ApJ **570**, 132 (2002)
45. Z.T. Kiss, L.V. Tóth, D. Ward-Thompson et al.: A&A (2006) submitted
46. P. Kroupa, C. Weidner: ApJ **598**, 1076 (2003)
47. R.B. Larson: MNRAS **194**, 809 (1981)
48. F.J. Low, E. Young, D.A. Beintema et al.: ApJL **278**, 19 (1984)
49. F. Maragoudaki, M. Kontizas, E. Kontizas et al.: A&A **338**, L29 (1998)
50. G. Pietrzyn'ski, W. Gieren, P. Fouqué et al.: A&A **371**, 497 (2001)
51. R.J. Rand, S.R. Kulkarni: ApJL **349**, 43 (1990)
52. J. Scalo: Fragmentation and hierarchical structure in the interstellar medium, In *Protostars and Planets II*, ed by D.C. Black, M. S. Matthews (University of Arizona Press, Tucson 1985) pp 201–296
53. J. Scalo: The IMF revisited: a case for variations, In: *The Stellar Initial Mass Function* ed by G. Gilmore, I. Parry, S. Ryan (Cambridge University Press, Cambridge 1998) pp 201–236

54. R.A. Scheepmaker, M. Gieles, M. R. Haas et al.: astroph/0605022 (2006)
55. S. Schneider, B.G. Elmegreen: ApJS **41**, 87 (1979)
56. M.D. Smith, R. Gredel, T. Khanzadyan et al.: MmSAI **76**, 247 (2005)
57. S. Stanimirovic, L. Staveley-Smith, J.M. Dickey et al.: MNRAS **302**, 417 (1999)
58. J. Stutzki, F. Bensch, A. Heithausen, et al.: A&A **336**, 697 (1998)
59. L. Testi, A.I. Sargent, L. Olmi et al.: ApJL **540**, 53 (2000)
60. C. Weidner, P. Kroupa: ApJ **625**, 754 (2005)
61. D.J. Westpfahl, P.H. Coleman, J. Alexander et al.: AJ **117**, 868 (1999)
62. K.W. Willett, B.G. Elmegreen, D.A. Hunter: AJ **129**, 2186 (2005)
63. J.P. Williams, C.F. McKee: ApJ **476**, 166 (1997)
64. T. Wong, L. Blitz: ApJ **569**, 157 (2002)
65. Q. Zhang, S.M. Fall: ApJL **527**, 81 (1999)
66. Q. Zhang, S.M. Fall, B.C. Whitmore: ApJ **561**, 727 (2001)

Young Massive Clusters – Formation Efficiencies and (Initial) Mass Functions

Søren S. Larsen

Astronomical Institute, Utrecht University, Princetonplein 5, NL-3584 CC, The Netherlands `larsen@astro.uu.nl`

Abstract. Globular clusters are often assumed to be good tracers of major star formation episodes in their host galaxies. While observations over the past 2 decades have confirmed the presence of young objects with globular cluster-like properties in many galaxies, it is still not well understood exactly how the formation efficiency of bound star clusters relative to field stars and their mass spectrum depend on external factors. The cluster initial mass function typically appears to be consistent with a power-law with a slope ~ -2, but most attempts to constrain any upper limit on the CIMF have been limited by size-of-sample effects. However, evidence is starting to accumulate for possible truncation of the cluster mass function. It is tentatively suggested that the upper mass limit may currently be at $\sim 10^5$ $M_\odot$ in the Milky Way disk, while there are indications that it is $\sim 5 \times 10^5$ $M_\odot$ in M51 and about 2×10^6 $M_\odot$ in the Antennae. Some extreme starbursts (e.g. Arp 220, NGC 7252) are (or were) able to form clusters as massive as 10^7 $M_\odot$. The overall formation efficiency of star clusters (relative to field stars) in the Galactic bulge may not have been much different from that in the disk today, but was probably significantly higher for metal-poor GCs in halos.

1 YMCs – Guides to Young Stellar Populations?

Three or four decades ago, globular cluster (GC) formation was thought by many to be a phenomenon occurring only in the early Universe (e.g. [20]). In the meantime, young, compact star clusters with masses in the range 10^4–10^6 $M_\odot$ have been found in many different galaxies (e.g. [18, 21, 24]), and there is a growing concensus that these "Young Massive Clusters" (YMCs) may well be young analogues of the old GCs associated almost universally with the spheroidal components of galaxies. By implication, GCs have become potentially interesting not just as fossil left-overs from the early Universe, but more generally as test particles for studies of extragalactic stellar populations.

In keeping with this spirit, contemporary observing proposals or papers on GCs often include an introductory remark along the lines of: *"GCs are thought to be good tracers of the major star forming episodes in their host galaxies"*. YMCs/GCs may indeed trace star formation quite generally, but it is also clear that some caution must be exercised. For example, the number of GCs per field star varies from galaxy to galaxy (the classical "specific frequency problem"), as well as between stellar populations within galaxies [14,11]. This must be due to differences in the formation efficiency of GCs relative to field stars, in the GC survival rates, or (more likely) some combination of the two. Of the roughly 150 catalogued GCs in our own Galaxy [13], about 2/3 are associated chemically, spatially and kinematically with the halo, while this is true for only ∼1% of the stars. Conversely, some 90% of the stellar mass resides in the thin disk of our Galaxy, while no GCs are currently known to be associated with this component. Even though some quite massive clusters might be located in remote parts of the Galactic disk [7, 10], a simple scaling by stellar mass of the GCs in the bulge or halo would predict hundreds or thousands of GCs in the disk, seemingly at odds with the observations.

All this is of course well known, and is the reason why GCs are often assumed to trace only "*major* star forming episodes". The problem then remains to define when a star forming episode qualifies as "major". Perhaps GCs mainly trace spheroid formation [4], but some stars currently residing in spheroids may originally have formed in disks. As an example, it is illustrative to consider the outcome of merging two Milky Way galaxies: The merger product would contain about 200 metal-poor and 100 metal-rich pre-existing GCs from the progenitor galaxies, assuming no GCs are destroyed. The current gas mass in the Milky Way is about 0.5×10^{10} $M_\odot$ [6], which is about half the mass of the bulge. Assuming that the merger would form GCs with the same efficiency as the bulge, using all the available gas, about 50 new metal-rich GCs would form (and survive). The resulting GC population would then consist of three major sub-components: metal-poor, old GCs originating in the progenitor galaxy halos, moderately metal-rich GCs from the pre-existing bulges, and very metal-rich GCs formed in the merger. The estimate of the number of new GCs is obviously very crude. However, the main point here is that while the majority of the *stars* in the resulting spheroid (about 90%) would have formed in the disks already before the merger took place, their formation history might not be reflected in the GC system.

Mergers at higher z were likely more gas-rich, and the discrepancy between the star- and GC formation histories may have been less extreme than in the example outlined above. Nevertheless, since GCs play such an important role in attempts to constrain the star formation histories of early-type galaxies, there is a clear need to also quantify the limitations better.

2 What Can We Learn from Studies of YMCs?

In order to understand the differences between properties of star cluster populations in different galaxies better, it is useful to divide the problem into three sub-problems which can, to a large extent, be addressed separately: (1) Understanding the *cluster (initial) mass function* (CIMF) – Is it universal, or do some parameters (e.g. the slope or upper mass limit) vary as a function of external parameters (star formation rate, gas pressure/density)? (2) The cluster *formation efficiency* relative to "field" stars: Again, how does this depend on external factors? (3) *Disruption effects:* these are driven both by internal mechanisms (two-body relaxation, binaries, black holes) and external factors (shocks, interactions with giant molecular clouds, tidal fields). However, it is a hard problem and progress has been slow in making reliable, quantitative predictions for the time scales involved, at least until very recently.

In the following I will concentrate mainly on the first item in the list, with only a few remarks about formation efficiencies. Disruption effects are covered elsewhere in this volume (e.g. Baumgardt, De Marchi, Vesperini).

2.1 The Cluster Initial Mass Function

The number of galaxies with reliable constraints on the CIMF remains small. Probably the best-known example is the Antennae, where the CIMF appears well approximated by a power-law $dN/dM \propto M^\alpha$ with $\alpha \approx -2$ over the range $10^4 < M/M_\odot < 10^6$ [26]. Similar mass functions have been found in M51 [2], NGC 3310 and NGC 6745 [8], the Milky Way [9] and in the LMC [15], although not all studies cover the same mass range. It seems reasonable to conclude that star clusters typically form with a mass spectrum that can be well approximated by a uniform power-law over some mass range. It should be mentioned that some dwarf galaxies contain a few clusters which are much brighter than one would expect from the total number of star clusters in those galaxies [3]. Here, however, I mainly focus on the opposite problem, i.e. whether there might be a *truncation* of the CIMF at some upper mass $M_{\rm trunc}$, thus inhibiting efficient formation of clusters above a certain mass limit ($M_{\rm trunc}$) under certain circumstances (e.g. in the Milky Way disk today).

2.2 Some Considerations on the Milky Way

Starting again with the Milky Way, it is interesting to consider the consequences of postulating that young clusters are drawn purely at random from a power-law distribution with $\alpha = -2$. The current star formation rate in bound star clusters in the solar neighbourhood is estimated to be around 5.2×10^{-10} M$_\odot$ yr^{-1} pc^{-2} [16]. Assuming for simplicity a constant cluster formation rate in the Galactic disk over the past 10 Gyr and that the cluster formation rate in the Solar neighbourhood is representative for the disk as a whole, this corresponds to $\sim 10^9$ M$_\odot$ formed in bound clusters within the Solar circle (8.5 kpc). This is most likely a conservative estimate and the true number may well be significantly higher. Sampling these clusters from a power-law CIMF for $10^2 < M/M_\odot < 10^7$, one predicts a total of close to 9×10^5 clusters formed over the lifetime of the disk, of which 9000, 900 and 80 have $M > 10^4 M_\odot$, $M > 10^5 M_\odot$ and $M > 10^6 M_\odot$. These numbers do not depend strongly on the adopted upper and lower integration limits, although they do depend on the CIMF slope – a steeper slope implies a more bottom-heavy CIMF, with fewer high-mass clusters.

There are two consistency checks worth making: first, the stellar mass of the Milky Way bulge is about 10% of that of the disk. If clusters formed with the same efficiency in the bulge as they do in the disk now, one might expect about 90 clusters with $M > 10^5 M_\odot$ in the bulge, of which 8 have $M > 10^6 M_\odot$. The actual observed numbers of GCs in the bulge are smaller by a factor of 4, which may be partly due to disruption effects. However, there is no indication that clusters formed with a *higher* efficiency in the bulge, and the number of GCs observed in the bulge may be roughly consistent with a formation efficiency (relative to field stars) similar to that observed today in the Galactic disk. The numbers for the halo are more difficult to reconcile with this picture, since the halo has about twice as many GCs as the bulge but an order of magnitude fewer stars. This suggests a higher formation efficiency of metal-poor halo GCs, consistent with observations of early-type galaxies [14, 11]. The distinction between *halo* (metal-poor) GCs and all other star clusters may be more fundamental than the one between old GCs in general and present-day star formation. Note that these arguments are slightly different from those of McLaughlin [19] who argued for a universal cluster formation efficiency relative to the total available *gas* mass.

Second, we can compare with the number of massive star clusters actually observed in the disk. By construction, the current formation rate agrees with the number of low-mass clusters ($M < 10^3$ $M_\odot$) observed locally, but it is interesting to see what happens when extrapolating to higher masses. Recently, at least two young clusters with masses in the range 10^4–10^5 $M_\odot$ and ages $< 10^7$ years have been identified [7, 10]. If they are taken as representative of the formation rate of such objects within a distance of 5 kpc, this corresponds to 2 kpc^{-2} Gyr^{-1}, or 4500 such clusters formed within the Solar circle over 10 Gyr. Again, this is roughly consistent with the order-of-magnitude estimates above (in fact, 8000 clusters with $10^4 < M/M_\odot < 10^5$ are predicted), suggesting that clusters like Westerlund 1 with masses up to about 10^5 $M_\odot$ occur naturally (albeit rarely) as part of the normal hierarchy of star formation in the Milky Way disk today.

A typical disruption time for a cluster with mass M in the solar neighbourhood is $t_{\rm dis} = 1.3\,{\rm Gyr}\,(M/10^4 M_\odot)^{0.62}$ [16]. Assuming that this scaling remains valid for $M > 10^4$ $M_\odot$, a cluster with an initial mass of 10^5 $M_\odot$ is expected to disrupt in about 5 Gyrs while a 10^6 $M_\odot$ cluster has a lifetime well in excess of a Hubble time. The disk should then contain about 500 clusters with masses greater than 10^5 $M_\odot$ and still virtually all of the 80 clusters with $M > 10^6$ $M_\odot$ formed over its lifetime. Of these objects, 7 should be found within a distance of 1 kpc. This estimate makes the crude assumption that clusters disrupt instantaneously, while in practice mass is lost at a nearly constant rate over the lifetime of the cluster. Although current catalogs of Milky Way open clusters are highly incomplete beyond 1 kpc, it seems unlikely that a large population of clusters with masses in the range 10^5–10^6 $M_\odot$ could have been missed. It appears plausible that the CIMF in the Milky Way is truncated somewhere in the vicinity of 10^5 M$_\odot$, or at least becomes much steeper than a power-law with a slope of -2. However, it would be highly desirable to quantify better the completeness of current cluster surveys in the disk out to large distances (several kpc).

2.3 Constraints on the CIMF in Other Galaxies

Studies of the CIMF in external galaxies are complicated by the rapid change in mass-to-light ratio that characterizes simple stellar populations. Observed luminosities cannot be converted to masses unless the age of each individual cluster is known with some accuracy, which generally requires U-band imaging. Several studies have shown that the *luminosity* function of young star clusters generally appears to be sampled all the way up to its statistical upper limit [24, 3, 17]. Even the scatter around the predicted relation is consistent with random sampling [17]. No direct evidence for truncation of the LF has been found so far, suggesting that most large galaxies are physically able to form star clusters with masses *at least* up to about 10^5 M$_\odot$.

In general, there is no straight-forward way to infer the mass function of a cluster sample from the observed luminosity function [26]. Only in the special case where the MF is a simple, untruncated power-law, or the age distribution is a delta function, will the LF and MF have the same shape. This point is illustrated in Fig. 1, which compares the mass- and luminosity functions for simulated cluster samples with power-law mass functions truncated at 10^8 M$_\odot$ and 10^5 M$_\odot$. The clusters were assigned random ages uniformly distributed between 10^6 years and 10^9 years and masses were converted to M_V magnitudes using SSP models [5]. Mass loss and cluster disruption were ignored. In the left-hand panel, the MF truncation occurs as a simple cut-off, while in the LF (right) the mass cut results in a steepening of the slope at the

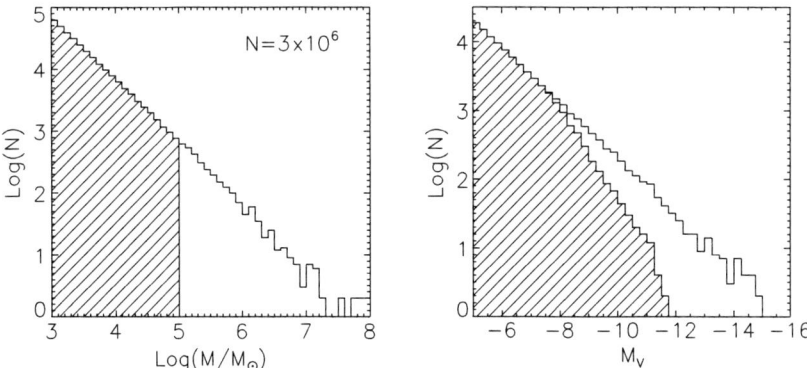

Fig. 1. Mass and M_V distribution for a simulated cluster population with 3×10^6 clusters and $6 < \log(\mathrm{age/yr}) < 9$. Hashed and outlined histograms are for mass distributions truncated at 10^5 M$_\odot$ and 10^8 M$_\odot$.

bright end rather than any distinct cut-off in luminosity. For a real cluster population, the maximum mass may be inferred from the "bend" that occurs at $M_V \approx -8$ in Fig. 1, assuming that the MF is a power-law up to M_{trunc} [12]. Such a bend has been observed in the Antennae and M51, where it may be explained by a MF truncated near 2×10^6 $M_\odot$ and $\sim 5 \times 10^5$ $M_\odot$ [26,12]. In still more active galaxies such as Arp 220, NGC 7252 and NGC 1316, there are clusters with masses as high as 10^7 $M_\odot$ [1,25].

Let us now consider the behaviour of the following quantities as a function of the total number of clusters (N) in a population with a truncation at an upper mass limit M_{trunc}: (1) the maximum mass M_{max} occurring in the population, (2) the mass of the brightest cluster $M_{\text{brightest}}$, and (3) the magnitude of the brightest cluster, $M_V^{\text{brightest}}$. If N clusters are sampled at random from a power-law $dN/dM \propto M^{-2}$ with lower mass limit M_{min}, then statistically the most massive cluster will have $M_{\text{max}} = N\, M_{\text{min}}$ [24,3]. From Monte-Carlo simulations of various cluster formation histories, Weidner et al. [23] found that when clusters are sampled at random from a power-law MF, the most massive cluster is also the brightest in about 95% of the cases. If the mass function has a real physical upper limit, this is not necessarily the case.

The left panel in Fig. 2 shows the results of Monte-Carlo simulations for M_{max} and $M_{\text{brightest}}$ in cluster samples with various M_{trunc} limits. Clusters were drawn at random from the same population used in Fig. 1, truncated at M_{trunc} values between 10^4 $M_\odot$ and 10^7 $M_\odot$. The median values of M_{max} and $M_{\text{brightest}}$ in 1000 experiments are shown with solid and dashed lines as a function of the number of clusters sampled. The relation predicted by

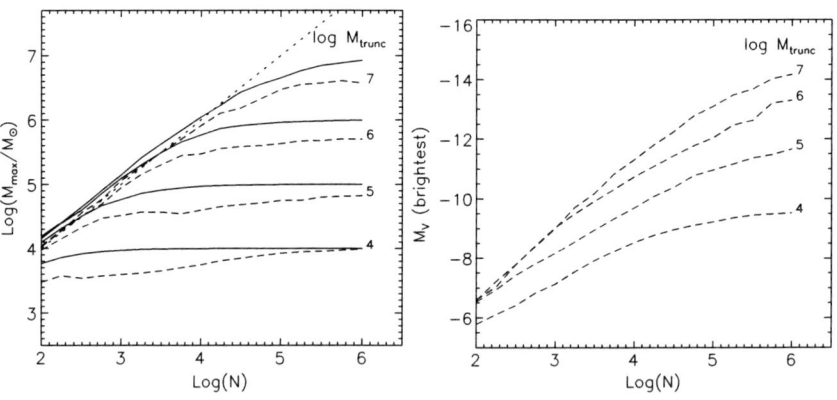

Fig. 2. *Left*: Mass of the most massive (*solid lines*) and most luminous (*dashed lines*) cluster as a function of total number of clusters in simulated cluster samples with various fixed upper mass limits. The *dotted line* is the maximum cluster mass expected from random sampling from an untruncated power-law. *Right*: The luminosity of the brightest cluster as a function of N and M_{trunc}.

pure sampling statistics is indicated as a dotted line. For small N and large $M_{\rm trunc}$, the truncation of the MF is not "felt" and the $M_{\rm max}$ and $M_{\rm brightest}$ vs. N relations approach the curve for random sampling, i.e. the Weidner et al. result is reproduced. Conversely, for small $M_{\rm trunc}$ and large N, statistical effects become unimportant and $M_{\rm brightest} \sim M_{\rm max} \sim M_{\rm trunc}$. In between these extremes there is a regime where the mass function is likely to be sampled up to higher masses than the luminosity function, or $M_{\rm max} > M_{\rm brightest}$. This turns out to be the situation encountered in many real galaxies.

For both $M_{\rm max}$ and $M_{\rm brightest}$, Fig. 2 shows a fairly rapid transition between the regimes where size-of-sample effects and truncation dominate. Thus, it might seem that the observed strong relation between total number of clusters in galaxies and the luminosity of the brightest cluster [24,3,17] is incompatible with truncation of the CIMF playing any important role. However, as shown in the right-hand panel of Fig. 2, $M_V^{\rm brightest}$ has a steep dependency on N over a much wider range in N than $M_{\rm max}$ and $M_{\rm brightest}$. This is because the mean *age* of the brightest cluster shifts towards younger ages for higher N, as it becomes increasingly likely to encounter a cluster with mass near $M_{\rm trunc}$ in the brief phase where the M/L ratio is very low.

From the preceding discussion it should be clear that the LF can be dominated by sampling effects, even if the MF does have a physical truncation. For example, for $M_{\rm trunc} = 10^5$ M$_\odot$, the $M_{\rm max}$ curve starts to flatten at $N \sim 10^3$, implying that the MF would be sampled up to its physical upper limit already in a relatively cluster-poor galaxy. The $M_V^{\rm brightest}$ curve, on the other hand, would continue to rise beyond $N = 10^5$, corresponding roughly to a Milky Way-sized galaxy (cf. Sect. 2.2).

The key to detecting a physical upper limit of the CIMF will be to cover a dynamic range large enough to study the LF in detail and detect signatures such as the "bend" in Fig. 1. Ultimately, however, inferences about the MF based on the LF remain indirect and dependent on assumptions about the shape of the MF, star formation history etc., and ideally it would be desirable to have direct information about the mass functions in several more galaxies. Several such studies are now underway, and may provide important insight into the MF in the not too distant future.

3 Concluding Remarks

While research in extragalactic star clusters remains a very active field, we are still facing a number of important questions. The ubiquity of YMCs in external galaxies is now well established, but the apparent absence of a large population of clusters with masses in the range 10^5–10^6 M$_\odot$ in the Galactic disk remains somewhat of a puzzle. This may suggest an upper limit to the CIMF near 10^5 M$_\odot$ in the Milky Way, as noted already by van den Bergh & Lafontaine [22], although studies of disk clusters are still hampered by our own location close to the Galactic plane. However, the recent realization that the Milky Way is still forming clusters with masses near 10^5 M$_\odot$ hints that the CIMF in the Milky Way may not be very different from that in the LMC, the main difference being that the census of YMCs is more complete in the LMC. There is some evidence for truncation at a somewhat higher mass in M51 ($\sim 5 \times 10^5$ M$_\odot$) and the Antennae ($\sim 2 \times 10^6$ M$_\odot$), while Arp 220, NGC 1316 and NGC 7252 host clusters as massive as $\sim 10^7$ M$_\odot$. These galaxies also define a sequence of increasing star formation rate, suggesting that galaxies with higher SFRs are physically able to form more massive clusters. This may provide a hint as to why GC formation was common at high z, when SFRs and gas densities were generally higher.

References

1. N. Bastian, R.P. Saglia, P. Goudfrooij, et al.: A&A, 448, 881 (2006)
2. A. Bik, H.J.G.L.M. Lamers, N. Bastian, et al.: A&A, 397, 473 (2003)
3. O.H. Billet, D.A. Hunter, & B.G. Elmegreen: AJ, 123, 1454 (2002)
4. J.P. Brodie & J. Strader: Ann. Rev. Astron. Astrophys., 44, 143 (2006)
5. G.A. Bruzual & S. Charlot: MNRAS, 344, 1000 (2003)
6. B.W. Carroll & D.A. Ostlie: An Introduction to Modern Astrophysics, San Francisco, CA: Addison-Wesley Publishing Company (1996)
7. J.S. Clark, I. Negueruela, P.A. Crowther, & S.P. Goodwin: A&A, 434, 949 (2005)
8. R. de Grijs, P. Anders, N. Bastian, et al.: MNRAS, 343, 1285 (2003)
9. B.G. Elmegreen & Yu.N. Efremov: ApJ, 480, 235 (1997)
10. D.F. Figer, J. MacKenty, M. Robberto, et al.: ApJ, 643, 1166 (2006)
11. J.C. Forte, F. Faifer, & D. Geisler: MNRAS, 357, 56 (2005)
12. M. Gieles, S.S. Larsen, N. Bastian, & I.T. Stein: A&A, 450, 129 (2006)
13. W.E. Harris: AJ, 112, 1487 (1996)
14. W.E. Harris & G.L.H. Harris: AJ, 123, 3108 (2002)
15. D.A. Hunter, B.G. Elmegreen, T.J. Dupuy, & M. Mortonson: AJ, 126, 1836 (2003)
16. H.J.G.L.M. Lamers, M. Gieles, N. Bastian, et al.: A&A, 441, 117 (2005)
17. S.S. Larsen: AJ, 124, 1393 (2002)
18. S.S. Larsen: In: Planets to Cosmology: Essential Science in Hubble's Final Years, ed. M. Livio, Santa Cruz, CA: STScI (2005)
19. D.E. McLaughlin: AJ, 117, 2398 (1999)
20. P.J.E. Peebles & R.H. Dicke: ApJ, 154, 891 (1968)
21. F. Schweizer: In: Extragalactic Star Clusters, eds. D. Geisler, E.K. Grebel, D. Minniti, San Francisco, CA: ASP (2002)
22. S. van den Bergh & A. Lafontaine: AJ, 89, 1822 (1984)
23. C. Weidner, P. Kroupa, & S.S. Larsen: MNRAS, 350, 1503 (2004)
24. B.C. Whitmore: In: A Decade of HST Science, eds. M. Livio, K. Noll, M. Stiavelli, Cambridge: Cambridge University Press, p. 153 (2003)
25. C.D. Wilson, W.E. Harris, R. Longden, N.Z. Scoville: ApJ, 641, 763 (2006)
26. Q. Zhang & S.M. Fall: ApJ, 527, L81 (1999)

The Radii of Thousands of Star Clusters in M51 with *HST/ACS*

R.A. Scheepmaker[1], M. Gieles[1], M.R. Haas[1], N. Bastian[2], and S.S. Larsen[1]

[1] Astronomical Institute, Utrecht University, Princetonplain 5, 3584 CC Utrecht, The Netherlands `scheepmaker@astro.uu.nl`
[2] Department of Physics and Astronomy University College London, London, UK

1 M51 – A Star Cluster Laboratory

The young (<1 Gyr) star cluster population of M51 (NGC 5194, Hubble-type Sc) is, due to its relative proximity and high number of star clusters, an excellent candidate for studies to the formation and evolution of young star clusters (YCs) in spiral disks. M51 has many star clusters because of its high star formation rate, which may have been caused by its interaction with NGC 5195, and because the disk is almost face-on, we can observe the clusters throughout the disk without being hampered by extinction too much. The new *HST/ACS* mosaic image of M51, taken as part of the Hubble Heritage program [13], covers the entire disk in 6 pointings in $F435W$ ($\sim B$), $F555W$ ($\sim V$), $F814W$ ($\sim I$) and $F658N$ ($H\alpha$), with a resolution of ~ 2 pc per pixel. Previous studies on the star cluster population of M51 used lower resolution *WFPC2* data that did not cover the entire disk (see e.g. [10, 1, 11]). In the study of [1], clusters were selected based on their fit to SSP models, which led to a possible contamination of individual stars. In a different study clusters were selected based on their sizes, which made it difficult to select a large sample due to the limited resolution [11]. However, by exploiting the superb resolution of the *ACS* camera and the large field of view of $\sim 17.5 \times 24.8$ kpc, we now *can* select a large sample of clusters based on their sizes. This sample can then be used to study the formation history and evolution of the star cluster population and to derive infant mortality rates and study how these depend on the environment. Since this is work in progress, we here select a smaller preliminary sample of clusters and we present some first results that can already be derived by looking at the sizes and the positions of the star clusters in the disk of M51.

2 Selecting a Reliable Cluster Sample Based on Radii

We measure the effective radii of 75 436 sources in our data with the *Ishape* routine [9], which convolves the PSF of the telescope with Moffat 15 profiles [12] of different sizes and then selects the fit with the lowest χ^2. In this way *Ishape* can determine an accurate radius of a marginally extended source down to ~ 0.5 pc at the distance of M51. We assume that a source is a cluster when its radius is > 0.5 pc and when a fit using a Moffat profile gives a lower χ^2 than a fit using just the PSF (i.e. assuming the source is a point source). This gives us a sample of 14 950 clusters. If we apply a 90% completeness limit of B & $V < 23.3$ mag, which we determined for a 3 pc source in a high background region, we find a sample of 4357 resolved clusters above the completeness limit. This sample was used to study the cluster luminosity function and the related maximum mass for star clusters in M51 ([6], also Gieles et al. in these proceedings).

This sample can also be used to study the distribution of the clusters in the disk of M51 (Sect. 3). However, it turns out that many of the clusters in this sample are in crowded regions or in regions with a highly varying background, making the value of the measured radius unreliable due to blending effects and fits to parts of the background light. We therefore also select a smaller sample of 769 clusters which are isolated and in low background regions and therefore have more accurate size estimates. We use this accurate cluster sample to study the radii themselves (Sects. 4 and 5).

3 The Distribution of Star Clusters in the Disk

We use the sample of 4357 resolved clusters to plot the surface density distribution of the clusters in the disk of M51. Figure 1a shows a bump in the surface density distribution at a galactocentric distance of $\sim$6 kpc. This distance is remarkably similar to the corotation radius of $\sim 5.5 \pm 1.0$ kpc, which is the distance where the rotational velocity of the stellar and gaseous component is the same as the one of the spiral arms [17, 15]. The bump at this location indicates that the corotation radius is a preferred site for cluster formation. We note that one of the biggest complexes of clusters, named G2, is also found near the corotation radius [2].

Assuming the surface density distribution of clusters is an exponential distribution of the form $\Sigma \propto e^{-D/R_h}$, we can derive the scalelength R_h. Stars in the disk of the Milky Way (MW) follow an exponential distribution with a scalelength of $\sim$3.5 kpc [3]. For M51 this scalelength is expected to be larger, since the disk of M51 is bigger than the disk of the MW. Due to the dissipative nature of molecular gas, the distribution of GMCs is in general more centrally concentrated than the distribution of the stars. The surface density of GMCs in M51 has a scalelength of $R_h = 2.4$ kpc [5]. Since we are looking at *young* clusters, we would expect them to be correlated with the GMCs more than the total field population. In Fig. 1a we overplotted the surface density distributions of GMCs and stars. Obviously, because of the presence of the bump, the surface density distribution of the clusters can not be approximated by a straight line. However, the clusters seem to be more correlated with the GMCs, especially for distances <3 kpc. In a future paper we will address this issue in much more detail, also by measuring the surface distribution of the stars in M51 ourselves [14].

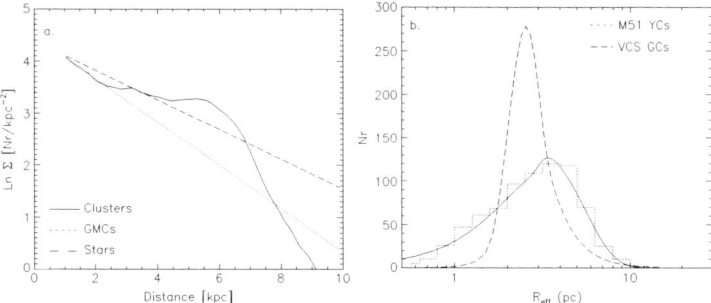

Fig. 1. (a) The surface density versus galactocentric distance for the complete sample of 4357 resolved clusters in M51 (*solid line*). The *dashed line* shows the density distribution of stars in the Milky Way ($R_h = 3.5$ kpc) and the *dotted line* shows the density distribution of GMCs in M51 ($R_h = 2.4$ kpc). (b) The effective radius distribution of 769 star clusters in M51 (*histogram*). The *solid line* is the best fit of Eq. 22 of [8]. The *dashed line* is the best fit of the same function to the GCs of the *ACS* Virgo Cluster Survey.

4 A Preferred Radius for Both Young and Globular Clusters

In Fig. 1b we show the effective radius distribution of our smaller sample of 769 clusters, for which we have determined more accurate radii (Sect. 2). Also plotted is the radius distribution of the much older GCs of the *ACS* Virgo Cluster Survey (*VCS*) of [8]. Both distributions, concerning completely different populations, have their peak at ∼3 pc, while the general shapes of the distributions are different. For early-type galaxies, the mean effective radius scales inversely with galaxy color [8]. It is not clear if the same relation also holds for late-type spiral galaxies, but when we correct the radii of the *ACS VCS* GCs to the color of M51, the peak shifts to the right and the similarity in the location of both peaks is even stronger. The effective radius distribution of the galactic GCs also peaks at 3 pc [7]. This suggests that for both young and old populations of star clusters there exists a preferred effective radius of ∼3 pc. The preferred radius also implies that the procedure for deriving distances, based on the median radius and described in [8], could also work for spiral galaxies with young cluster populations. If we apply the procedure to our cluster sample we find a distance of 8.6 ± 0.9 Mpc, very close to the assumed distance to M51 of 8.4 ± 0.6 Mpc based on the planetary nebulae luminosity function [4].

5 Radius Versus Distance

The GCs in the Milky Way show a relation between their effective radius and galactocentric distance of the form $R_{\text{eff}} \propto D^{\sim 1/2}$ [16], which can be explained by both a remnant of the formation process (denser clouds near the Galactic centre) and by tidal relaxation of the clusters in the Galactic potential, since the tidal radius scales as $\propto D^{2/3}$. Our M51 clusters show *no* relation between radius and galactocentric distance, meaning that they did not form in tidal equilibrium. It also means that the radius of the clusters is not related to the ambient pressure in their host galaxy, since this pressure decreases with galactocentric distance.

6 Conclusions

We measure the radii of 75 436 sources in M51 to select and study a large cluster sample covering the complete spiral disk. Some first preliminary results show a hint of enhanced cluster formation at the corotation radius and a preferred radius of $\sim$3 pc for YCs, which is similar to the preferred radius of much older GCs. However, in contrast to the GCs, the YCs in M51 do not show a relation between radius and galactocentric distance.

Acknowledgements. RS gratefully thanks Andrés Jordán for help in fitting our radius distribution with his function.

References

1. N. Bastian, M. Gieles, H. J. G. L. M. Lamers et al.: A&A **431**, 905 (2005a)
2. N. Bastian, M. Gieles, Y. N. Efremov & H. J. G. L. M. Lamers: A&A **443**, 79 (2005b)
3. J. Binney, S. Tremaine: *Galactic Dynamics* (Sprinceton University Press, 1987)
4. J. J. Feldmeier, R. Ciardullo & G. H. Jacoby: ApJ **479**, 231 (1997)
5. S. García-Burillo, F. Combes & M. Gerin: A&A **274**, 148 (1993)
6. M. Gieles, S. S. Larsen, R. A. Scheepmaker et al.: A&A **446**, L9 (2006)
7. W. E. Harris: AJ **112**, 1487 (1996)
8. A. Jordán, P. Coté, J. P. Blakeslee et al.: ApJ **634**, 1002 (2005)
9. S. S. Larsen: A&A Suppl. Ser. **139**, 393 (1999)
10. S. S. Larsen: MNRAS **319**, 893 (2000)
11. M. G. Lee, R. Chandar & B. C. Whitmore: AJ **130**, 2128 (2005)
12. A. F. J. Moffat: A&A **3**, 455 (1969)
13. M. Mutchler, S. V. W. Beckwith, H. E. Bond et al.: BAAS **37**, 452 (2005)
14. R. A. Scheepmaker, M. R. Haas, M. Gieles et al.: A&A **469**, 925 (2007)
15. R. B. Tully: ApJS **27**, 449 (1974)
16. S. Van den Bergh, C. Morbey & J. Pazder: ApJ **375**, 594 (1991)
17. P. Zimmer, R. J. Rand & J. T. McGraw: ApJ **607**, 285 (2004)

Extragalactic Star Clusters in Merging Galaxies

Gelys Trancho

[1] Gemini Observatory, 670 N. A'ohoku place, 96720, Hilo, Hawaii, USA;
[2] Universidad de La Laguna, Avenida Astrófisico Francisco Sánchez s/n, 38206, La Laguna, Tenerife, Canary Island, Spain `gtrancho@gemini.edu`

Abstract. The study of cluster populations as tracer of galaxy evolution is now quite possible with 8 m class telescopes and modern instrumentation. The cluster population can be used as a good tracer of the star forming episodes undergone by the merging system. We present two young galaxy mergers, NGC3256 and NGC4038, and the studies on the young cluster populations of those systems. We found that the cluster ages agree with the merger ages and their metallicities are consistent with being the progenitors of the old metal rich globulars in ellipticals.

1 Introduction

In the 70's [9] and [6] argued that a big fraction of ellipticals were formed by the merger of two spirals, rather than being ellipticals from the start. This argument was based on observational evidences, where remnants of mergers would present structures that were due to tidal interaction of disks. Looking for other clues to prove or disprove this argument, researchers eventually resourced to globular clusters. Globular clusters have always been very important for the determination of the evolutionary history of nearby galaxies. They are bright, numerous and provide an excellent chronometer and age metallicity indicator for the star formation burst from which they originated. In the early 90's, these studies led to the argument that, since globular clusters are more common in ellipticals than in spirals (higher specific frequency), if the number of globulars is conserved during a merger, ellipticals could no possibly come from a combination of spirals [11]. The question then becomes to determine if globulars clusters can be formed during and after a merger event. Evidence soon started to build up that the answer for that is yes [7,1,4]. Continuing along the lines of these works, we devised as the primary goal of my thesis to investigate the formation of elliptical galaxies as a result of the merger of two spirals, by using the star cluster populations present in the merger as tracers of the merger evolution through a study of their ages, metallicities, masses, sizes, etc. For this we selected a sample of merger systems of different ages, from very young (a few Myrs) to "bona fide" merger remnants (older than 700 Myr). In this work we will present the results of the two younger mergers in our sample: NGC4038/39 and NGC3256.

2 Observations

Globular clusters may be bright and numerous, but in the extragalactic domain, even for the closest mergers the light collecting power of 8m-class telescopes and the high spatial resolution of HST are needed to obtain the data required for a study like this. Starting with HST V and I images from the HST archive, we constructed colour-magnitude diagrams to select our cluster candidates:

- if the candidates are fainter than V=23, we have to use a combination of optical and near-IR photometry, comparing the data on colour–colour diagrams with Single Stellar Population (SSP) models.
- if the cluster candidates are brighter than V=23, spectroscopic followup is possible, and we then compare the properties of the cluster in age-metallicity spectral index diagrammes against the values from SSP models.
  Our spectroscopy data are mostly Gemini GMOS Multi-Object Spectroscopy (MOS), with one merger observed with the GMOS Integral Field Unit (IFU). Spectra were centered around 5000Å to include several Balmer lines, MgI at 5100Å, and a few iron features.

3 Data Reduction and Analysis

We followed the standard MOS data reduction procedures using the GEMINI GMOS IRAF package. For each galaxy, the clusters naturally separated in two samples:

- Clusters presenting only absorption lines (hydrogen, magnesium and iron are the most common).
  The radial velocity is accurately determined by using cross correlation of the absorption lines against a RV stars observed with the same instrumental setup. The age/metallicity indices are measured using INDEXF [2], which correctly takes into account the error propagation from random noise and from the uncertainty in the velocity determination. Once the indexes are measured, we plot them into a diagram such as Hγ × [Mgfe] or Hβ × [MgFe] [8], comparing with indexes from SSP models to obtain a first determination of the age and metallicity.
  We then compare our observed spectrum with the model spectrum for that age and metallicity.
- Clusters with the spectrum dominated by emission lines.
  Those are by definition young clusters, which still contain enough hot stars to ionize the surrounding gas. For these, we determine the radial velocity using the known emission lines in the spectrum. We then estimate the extinction from the gas in which the cluster is embedded, using the H$gamma$/H$beta$ ratio, assuming a Case B recombination and

the [3] extinction curve. Once the reddening is corrected, we calculate the gas metalicity as a surrogate for the embedded cluster, using the [O III]$\lambda\lambda$4363,4959+5007Å lines and the formulae from [10] and [5] for the oxygen abundance.

4 Results

- NGC4038:
  The Antennae galaxy. This system is at a distance of 19.2 Mpc, and is a young merger of 200 Myr, with the two merging galaxies still identifiable. From HST and GMOS imaging we selected 29 cluster candidates, of which 16 were confirmed, one being located in the tidal tail. Eight of them have a pure absorption spectrum, four have an emission spectrum and four are mixed. In summary, the clusters present the following properties:

  - Magnitudes $16 < V < 21.5$ (Reddening corrected)
  - The absorption line clusters span a range of ages between 70 to just over 300 Myr, with solar metallicity.
  - The emission line clusters are obviously younger, less than 10 Myr, and the metallicity obtained from the emission gas is a little under solar.
  - We find that the internal extinction can be quite large, up to $A_V \sim$ 2.5 mag, with the more reddened clusters located closer to the nucleus of the secondary galaxy (NGC4039).

- NGC3256
  This is a merger system twice as far as the Antennae and slightly older, but with the two nuclei still separated. The HST and GMOS imaging provided 109 cluster candidates, of which only 31 were spectroscopically confirmed (this galaxy is at lower galactic latitude so the field was quite contaminated by foreground stars). Three are located in the tidal tail.

  - They were still quite bright, despite being twice the distance of the Antennae ones, with $17.5 < V < 22.5$ (Reddening corrected).
  - The absorption line clusters span a range of ages quite similar to those in the Antennae, 80–300 Myr, but with solar or higher than solar metallicity.
  - The young cluster (those with emission lines) seem to present a slightly lower metallicity.
  - Again we measure quite a large reddening internal to the galaxy, up to $A_V \sim 3.5$ mag, with more reddened clusters being the ones closer to the center as expected.
  - We have also estimated the size of the clusters as between 1 and 10 pc, except the ones in the tidal tail, which are larger (10–18 pc).

5 Conclusion

The cluster population is a good tracer of the star forming episodes undergone by a merging system. As we initially proposed, all galaxies in our sample indicate that these new generations of clusters are formed at different epochs and with different metallicities. The cluster population is also a good indicator of the evolution of the merging system, being more uniformly older for the older mergers.

The new populations are systematically more metal rich, which is consistent with the ideas of merging events being what causes the bimodal distribution of clusters in ellipticals. The overall characteristics (sizes, masses) of the new clusters are consistent with them being the progenitors of the old metal rich globulars in ellipticals.

Clusters close to the centre of the merging system can be strongly reddened. This can have a very drastic effect in the resulting colours or in the line indexes and, if not corrected, will yield cluster ages much larger than the actual values.

The presence of clusters in the tidal tails of the merging systems, indicates that clusters formed in merger events can be ejected and end up forming part of the intragroup medium.

References

1. Burstein, D. 1987, Nearly Normal Galaxies, 47
2. Cenarro, A. J., Cardiel, N., Gorgas, J., Peletier, R. F., Vazdekis, A., & Prada, F. 2001, MNRAS, 326, 959
3. Edmunds, M. G., & Pagel, B. 1984, MNRAS, 211, 507
4. Kumai, Y., Hashi, Y., & Fujimoto, M. 1993, ApJ, 416, 576
5. Meyer, J.-P. 1985, ApJ Suppl., 57, 173
6. Schweizer, F. 1978, IAU Symp. 77, 77, 279
7. Schweizer, F., Ford, W. K. J., Jederzejewski, R., & Giovanelli, R. 1987, AJ, 320, 454
8. Schweizer, F., & Seitzer, P. 1998, AJ, 116, 2206
9. Toomre, A., & Toomre, J. 1972, ApJ, 178, 623
10. Vacca, W. D., & Conti, P. S. 1992, IAU Symp. 149: The Stellar Populations of Galaxies, 149, 497
11. van den Bergh, S. 1990, QJRAS, 31, 153

The Environment of Young Massive Clusters

Leonardo Vanzi

ESO, Alonso de Cordova 3107, Santiago, Chile lvanzi@eso.org

1 From Starburst Regions to Young Massive Clusters

Observations obtained in the last 10 years, from the ground and from space, have allowed to considerably improve our knowledge of nearby starburst galaxies, thanks to the increased sensitivity and resolution. In particular a few nearby extragalactic starburst regions were resolved in a number of compact sources. Figure 1 shows an image of the blue dwarf galaxy He 2-10 obtained combining HST optical images with a K band image from the VLT [1]. Some of the compact sources detected in the optical and infrared in this galaxy have radio counterparts that are mostly of characterized by a thermal radio spectrum [5]. These sources are interpreted as very young, few Myr old, massive, $10^5 M_\odot$ or more, clusters. Several nearby starburst galaxies show a similar situation. Cresci et al. [3], for instance, studied the cluster population of NGC 5253, deriving the optical-infrared color–color diagram of the hundreds of clusters detected in this galaxy, their luminosity and mass function. The youngest clusters can be extincted by several magnitudes in the optical, and be very bright in the IR. Calzetti et al. [2] found a young massive cluster (YMC) in NGC 5253 extincted by about 8 mag in the optical. Vanzi & Sauvage [8] found that this cluster emits in the IR as much energy as the entire galaxy in the optical-UV. The extremely low metallicity galaxy SBS 0335-052 is even more extreme in this sense hosting a YMC extincted by more than 15 mag in the optical [4].

We find therefore that, when observed with high enough spatial resolution, starburst regions show to host a large number of YMCs. The luminosity function of these clusters is a power law with index of about -2 [10]. The existence of YMCs opens a number of interesting questions. How common are they? Are YMCs the rule or the exception of the star formation process? Under which conditions do they form? What is the nature of IR bright clusters and how common are they? Addressing these questions is very important for understanding the star formation and galaxy formation process.

Fig. 1. HST–VLT combined image of He 2-10. The size of the image is $18'' \times 27''$, the resolution about $0.3''$.

2 New Infrared Observation

Observations in the IR can help answering some of the questions opened. Cresci et al. [3] showed in fact that in K band ($2.2\,\mu m$) both very young and exctincted clusters, and more evolved ones can be detected. Vanzi & Sauvage [9] observed a sample of nearby blue dwarf galaxies in K and L ($3.7\,\mu m$) with ISAAC at the VLT. The sample galaxies were selected to have firm detection by IRAS. They found that all galaxies observed are rich of clusters in K, but only 8% of the clusters detected in K are also detected in L. A careful comparison of the observations in L with radio observations available in the literature shows that most L sources have radio counterparts, with the remarkable exception of the optical bright cluster in NGC 1705. In Fig. 2 the radio image of IC 4662 from Johnson et al. [6] is displayed on the left side and the L image of the same object on the right side. Most radio sources can be associated to a faint source in L. We conclude that the clusters detected in L must be very young, and most likely embedde, their low detection rate indicates that such embedded phase must be very short in time.

3 YMC Versus Host Galaxy

Larsen & Richtler [7] compared the properties of the YMCs observed in U band with the properties of the host galaxies. They found a relation between the specific U luminosity of the cluster systems and the star formation rate of the host galaxy, measured from the FIR flux, and in particular with the star formation rate per unit area. They attributed this result to the possibility of a common controlling parameter, most likely the density of the ISM. A

The Environment of Young Massive Clusters   113

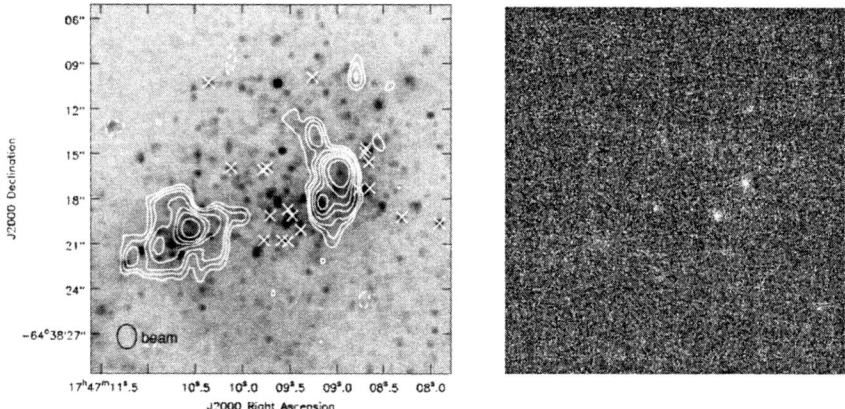

Fig. 2. IC 4662 radio contours from Johnson et al. [6] on the *left*, L image on the *right*.

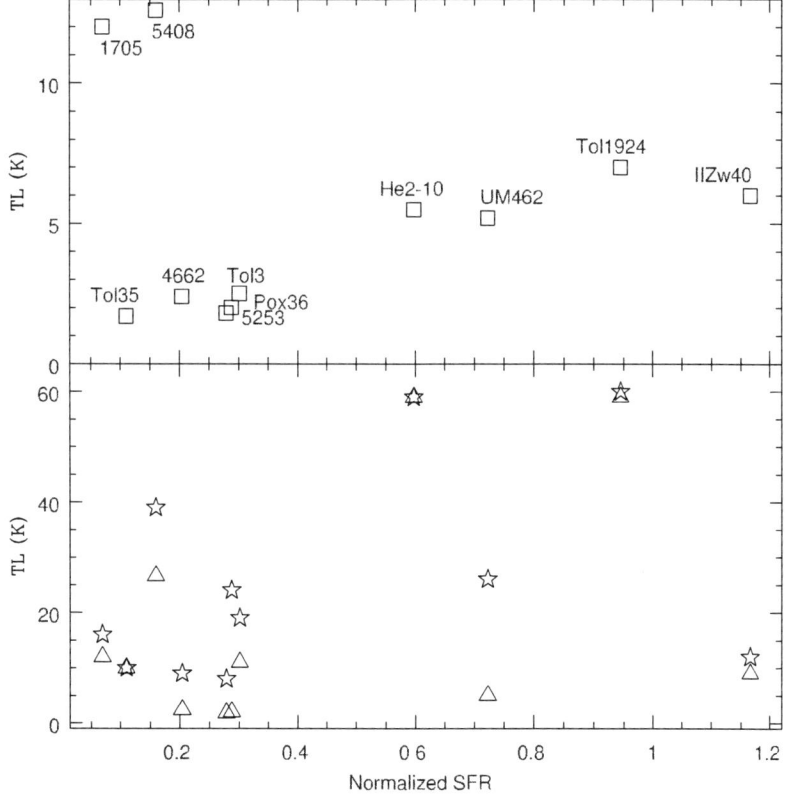

Fig. 3. K luminosity of the brightest cluster versus the host galaxy's SF rate – *upper panel*. Luminosity of the clusters brighter than M(K)=-12 (*stars*) and M(K)=-14 (*triangles*) – *lower panel*. Vanzi & Sauvage [9].

similar analysis based on IR data leads to different conclusions. A correlation is found in this case for the luminosity of the brightest cluster but not for the cluster system, see Fig. 3. The reasons for this discrepancy can be different. The U and K bands sample a different population of sources, in particular K could be more sensitive to the evolved, and to the young extincted, clusters. The correlation observed in U could be biased, or partially biased, by the contribution of few IR bright clusters to the SFR, as this is measured for the entire galaxy from the FIR IRAS fluxes. Extinction and age of the clusters could have an important effect too. Anyhow no correlation was found between the number of clusters and the properties of the host galaxies.

The presence of a correlation between the properties of the clusters and of the host galaxies, as for instance the SF rate, would indicate that YMCs require somehow special conditions to form and that they are not the simple result of statistically populating a luminosity function. However the evidences to support such conclusion at the moment seem not to be conclusive, and it is well possible that YMCs are commonly formed in any environment. Obviously further investigation is required to clarify this issue.

References

1. Cabanac R., Vanzi L., Sauvage M. 2005, ApJ 631, 252
2. Calzetti D., Meurer G. R., Bohlin R. C. et al. 1997, AJ 114, 1834
3. Cresci G., Vanzi L., Sauvage M. 2005, A&A 443, 447
4. Hunt L. K., Vanzi L., Thuan T. X. 2001, A&A 377, 66
5. Johnson K. E., Kobulnicky H. A. 2003, ApJ 597, 923
6. Johnson K. E., Indebetouw R., Pisano D. J. 2003, AJ 126, 101
7. Larsen S. S., Richtler T. 2000, A&A 354, 836
8. Vanzi L., Sauvage M. 2004, A&A 415, 509
9. Vanzi L., Sauvage M. 2006, A&A 448, 471
10. Whitmore B. C. et al. 1999, AJ 118, 1551

Star/Cluster Formation in Complexes: Insights from IFUs and HST

Nate Bastian

Department of Physics and Astronomy, University College London, London, UK
bastian@star.ucl.ac.uk

1 Cluster Complexes

It is now well established that star clusters tend not to form in isolation but rather in larger groupings, the so called cluster complexes (e.g. [6,1,2]). This observation has profound consequences for star cluster formation and evolution scenarios [1,5,3] (see Fig. 1), their impact on the surrounding ISM and progenitor GMC [2] (see Fig. 2) and even on their positions on the fundamental scaling relations of bound stellar systems [4]. In fact cluster complexes appear to be grouped themselves, on scales of ~ 800 pc in the Antennae galaxies [2]. This scale represented the largest scale of coherent star formation in these galaxies as well as that of the GMC associations [6].

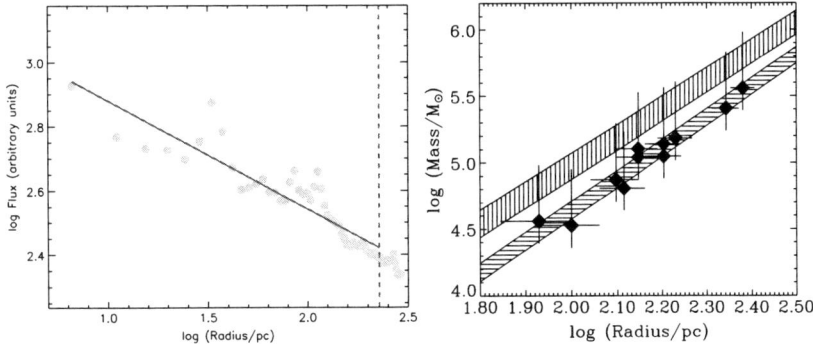

Fig. 1. Taken from [1]. *Left:* Surface brightness profile of Complex G2 in M51. The *solid red line* is the best fitting profile of the form $Flux \propto R^{-0.34}$ which corresponds to a 3D density distribution of $\rho \propto R^{-1.34}$ for similar to that observed for Galactic GMCs. *Right:* The mass-radius relation for GMCs in M51 (vertical filled area is the relation including errors) as well as that for the complexes in M51 (data points).

The full poster (pdf version) can be found at:
http://www.star.ucl.ac.uk/~bastian/Poster/concepcion.pdf

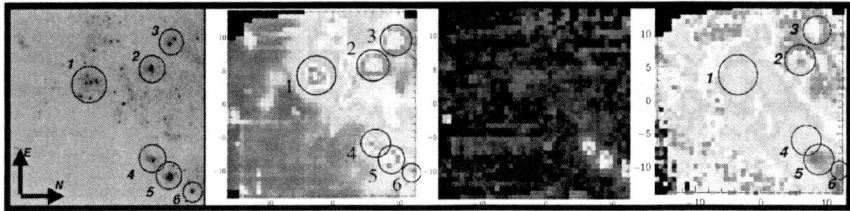

Fig. 2. Taken from [2]. Each panel has a width of ~ 2.5 kpc. *Left:* HST B-band image of part of the overlap region in the Antennae galaxies containing six complexes. *Second from left: VLT-VIMOS-IFU* intensity of the same region. *Second from right:* The reconstructed image of our *VIMOS-IFU* observations centred on the Wolf-Rayet feature at ~ 4650 Å, showing that complexes 4, 5 and 6 have similar ages (Δ(age) < 5 Myr). This, combined with their projected linear scale (~ 800 pc), suggests a large scale external triggering of their formation. *Right panel:* The velocity dispersion of the gas, the scale runs from blue (20 km/s) to red/white (100 km/s). Note that the complexes (except for 3) all have low velocities dispersions. This is consistent with outflows from the complexes (from the combined effects of supernovae and stellar winds) expanding as a bubble inside a relatively dense progenitor GMC.

Acknowledgements. NB gratefully thanks his collaborators Yuri Efremov, Eric Emsellem, Mark Gieles, Andrés Jordán, Markus Kissler-Patig, Henny Lamers, and Claudia Maraston on various studies of cluster complexes contained in this contribution.

References

1. Bastian, N., Gieles, M., Efremov, Yu.N., & Lamers, H.J.G.L.M. 2005, A&A, 443, 79
2. Bastian, N., Emsellem, E., Kissler-Patig, M., & Maraston, C. 2006, A&A, 445, 471
3. Fellhauer, M. & Kroupa, P. 2005, MNRAS, 359, 223
4. Kissler-Patig, M., Jordán, A., & Bastian, N. 2006, A&A, 448, 1031
5. Kroupa, P. 1998, MNRAS, 300, 200
6. Zhang, Q., Fall, M., & Whitmore, B.C. 2001, ApJ, 561, 727

Spectral Evolution of Blue Concentrated Star Clusters in the Large Magellanic Cloud

J.J. Clariá[1], J.F.C. Santos Jr.[2], A.V. Ahumada[1], E. Bica[3], A.E. Piatti[4], and M.C. Parisi[1]

[1] Observatorio Astronómico de Córdoba, Argentina claria@oac.uncor.edu
[2] Dpto. de Física, UFMG, Belo Horizonte, Brasil jsantos@fisica.ufmg.br
[3] Instituto de Física, UFRGS, Porto Alegre, Brasil bica@if.ufrgs.br
[4] Instituto de Astronomía y Física del Espacio, Bs. As., Argentina
andres@iafe.uba.ar

Integrated spectra of 17 LMC clusters were obtained in the (3600–6800) Å range using the CASLEO (Argentina) 2.15 m telescope. The typical resolution and dispersion were 12 Å and 3.5 Å/pixel, respectively. Cluster ages were derived by means of two methods: the *template matching*, in which the observed spectra are compared and matched to template spectra with well-determined properties, and the *Equivalent Width* (EW) method, in which diagnostic diagrams involving the sum of EWs of selected spectral lines were employed together with their calibrations with age and metallicity [8] (SP). The spectra were normalized to $F_\lambda = 1$ at approximately 5870 Å. The EWs of H Balmer, KCaII, G band (CH) and MgI were measured within the spectral windows defined by [2]. We then obtained the sum of EWs of the three metallic lines (S_m) and of the three Balmer lines H_β, H_γ and H_δ (S_h). As a first approach to get cluster ages, we used the diagnostic diagrams defined by SP. The clusters were then age-ranked according to the SP's calibrations. We used S_m to get a first age estimate using: $\log t(\text{Gyr}) = a_0 + a_1.S_m + a_2.(S_m)^2$, where $a_0 = -2.18 \pm 0.38$, $a_1 = 0.188 \pm 0.080$ and $a_2 = -0.0030 \pm 0.0032$. We then used S_h to get a second age estimate guided by the previous S_m estimate, since from S_h two solutions are possible: $\log t(\text{Gyr}) = \{-b \pm [b^2 - 4a(c - S_h)]^2\}/2a$, where $a = -6.35 \pm 0.18$, $b = -8.56 \pm 0.35$ and $c = 23.32 \pm 0.20$. The average of these two estimates is listed in column 5 of Table 1.

All 17 clusters in our sample are well represented by blue stellar populations, according to their spectral properties. Since the continuum distribution is affected by reddening, we firstly adopted a colour excess E(B–V) for each cluster, taking into account the [5] extinction maps. Secondly, we corrected the observed spectra accordingly and then we applied the *template matching* method. The resulting ages, together with estimates from the literature (whenever available), were used to get final averaged ages (Table 1). Within the expected uncertainties, the ages here derived agree with those given in the literature. Piatti et al. [6] observed in the Washington system 6 LMC clusters, which increased up to 37 the total sample with uniform estimates of age and metallicity. The general tendency is for the older clusters to lie in the outer disk regions of the galaxy while the younger ones tend to be located not far from or in the bar. This tendency is compatible with the findings of [9] who derived the LMC star formation history from HST observations of field stars. However, cluster formation does not seem to follow star formation in their detailed histories.

We used the integrated UBV colours of 624 LMC clusters and associations obtained by [3] to check the trend of (U–B) and (B–V) with age according to the present estimates. The colour gap seen in both (U–B) and (B–V) is a real feature first identified by [10]. The gap is probably a natural consequence of cluster evolution with increasing metallicities towards the present, and epochs of reduced cluster formation between $\approx$ 300 Myr and $\approx$ 1 Gyr. At least in the LMC bar, such a period of reduced cluster formation is not observed for the field stars [9].

Table 1. Cluster parameters

Cluster	E(B–V)	$t_{literatura}$ (Gyr)	Ref.	$t_{Sh,Sm}$ (Gyr)	$t_{template}$ (Gyr)	$t_{adopted}$ (Gyr)
NGC1804	0.08	0.08 ± 0.01	[7]	0.035 ± 0.004	0.05 ± 0.01	0.06 ± 0.02
NGC1839	0.06	0.10 ± 0.01	[7]	0.09 ± 0.02	0.06	0.09 ± 0.03
		0.125 ± 0.025	[6]			
SL237	0.07	0.038 ± 0.004	[7]	0.04 ± 0.02	0.05 ± 0.01	0.04 ± 0.02
		0.027 ± 0.009	[1]			
NGC1870	0.08	0.09 ± 0.01	[7]	0.033 ± 0.004	0.05 ± 0.01	0.06 ± 0.03
		0.07 ± 0.03	[1]			
NGC1894	0.09	0.071 ± 0.008	[7]	0.10 ± 0.08	0.13 ± 0.03	0.10 ± 0.03
NGC1902	0.04	–		0.07 ± 0.03	0.06	0.07 ± 0.03
NGC1913	0.09	0.024 ± 0.002	[7]	0.03 ± 0.02	0.06	0.04 ± 0.02
NGC1932	0.05	–		0.2 ± 0.2	0.4 ± 0.2	0.3 ± 0.2
NGC1943	0.08	0.14 ± 0.02	[7]	0.08 ± 0.06	0.28 ± 0.08	0.14 ± 0.06
		0.10 ± 0.01	[4]			
NGC1940	0.06	–		0.06 ± 0.02	0.06	0.06 ± 0.02
NGC1971	0.06	0.10 ± 0.01	[7]	0.05 ± 0.01	0.05 ± 0.01	0.06 ± 0.02
SL508	0.06	0.10 ± 0.01	[7]	0.07 ± 0.04	0.06	0.07 ± 0.04
NGC2038	0.06	0.13 ± 0.02	[7]	0.039 ± 0.008	0.06	0.08 ± 0.05
SL709	0.06	–		0.03 ± 0.02	0.13 ± 0.03	0.03 ± 0.02
NGC2118	0.07	–		0.05 ± 0.02	0.06	0.06 ± 0.02
NGC2130	0.05	–		0.03 ± 0.02	0.06	0.04 ± 0.02
NGC2135	0.05	–		0.085 ± 0.008	0.05 ± 0.01	0.07 ± 0.02

References

1. Alcaino, G., Liller, W., 1987, AJ, 94, 372
2. Bica, E., Alloin, D., 1986, A&A, 162, 21
3. Bica, E., Clariá, J.J., et al., 1996, ApJS, 102, 57
4. Bono, G., Marconi, M., Cassisi, S., et al., 2005, ApJ, 621, 966
5. Burstein, D., Heiles, C., 1982, AJ, 87, 1165
6. Piatti, A.E., Bica, E., Geisler, D., Clariá, J.J., 2003, MNRAS, 344, 965
7. Pietrzyńsky, G., Udalski, G., 2000, AcA, 50, 337
8. Santos Jr., J.F.C., Piatti, A.E., 2004, A&A, 428, 79 (SP)
9. Smecker-Hane, T.A., Cole, A., et al., 2002, ApJ, 566, 239
10. van den Bergh, S., 1981, A&AS, 46, 79

Young Star Clusters in the SMC

Katharina Glatt, Eva K. Grebel, and Andreas Koch

Astronomical Institute of the University of Basel, Department of Physics and Astronomy, Venusstrasse 7, CH-4102 Binningen, Switzerland
glatt@astro.unibas.ch, grebel@astro.unibas.ch, koch@astro.unibas.ch

Compact, long-lived star clusters can only form in very dense environements where star formation proceeds with high efficiency. Present-day globular cluster formation only seems to occur in starburst galaxies and/or in galaxies undergoing violent interactions. However, a less massive, compact and long-lived type of star clusters, the so-called populous clusters, are found in irregular galaxies like the Magellanic Clouds, where these clusters continue to form until the present day. The Small Magellanic Cloud (SMC), the Large Magellanic Cloud (LMC) and the Milky Way (MW) form an interacting triple system. The cluster age distribution of the LMC shows pronounced peaks that coincide with the times of past close encounters between the LMC, SMC and MW, indicative of interaction-triggered cluster formation. Curiously though, the field star formation history of the LMC does not show these variations. Moreover, the SMC appears to have formed clusters fairly continuously over the past 11 Gyr. In fact, the SMC is the only galaxy known so far to have experienced the seemingly continuous formation of long-lived, compact star clusters. Our target clusters were taken from the combined cluster catalogues of [1–3]. We determined ages for our clusters via isochrone fitting to colour-magnitude diagrams. Owing to the limiting magnitude of the MCPS photometric survey, only clusters with ages younger than 1 Gyr can be age-dated.

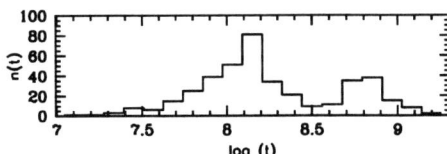

Fig. 1. Age distribution of our clusters, shown as *histogram* using logarithmic ages.

The resulting cluster age distribution is shown in Fig. 1 and may be viewed as evidence for episodic star formation. The periods of enhanced star formation roughly coincide with the periods of the closest approach of the SMC's neighbours. One maximum occurred 200 Gyr ago, in agreement with the predicted time of the closest approach of the LMC [4]. A second maximum is seen at 500 Myr, but here the true maximum may actually lie at higher ages where star clusters are more difficult to age-date due to the limited photometric depth of the MCPS. The age distribution does not only depend on the star formation rate, but also on the cluster dissolution. Due to the different structure of SMC and MW, the older clusters of the SMC are believed to have larger radii than the younger ones. The different structure and properties of the MW and the SMC is probably one of the main reasons for the longer lifetime of the SMC clusters and for the apparent absence of pronounced dissolution effects. The resulting spatial distribution of clusters in the SMC is shown in Fig. 2. Open black circles indicate clusters outside of the region for which [3] provide photometry.

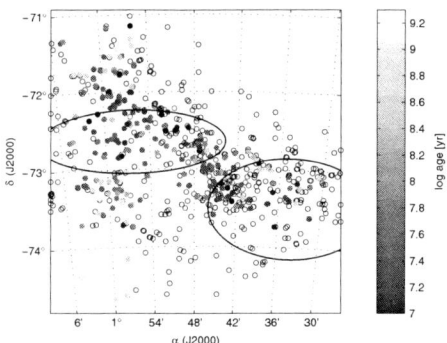

Fig. 2. The location of two of the supergiant shells of [5], namely 304A (*center*) and 37A (*lower right*), is indicated by *large ellipses*.

The most recently formed population in the SMC is traced by HII regions along the bar and the wing of the SMC, many of which coincide with the locus outlined by the HI supergiant shells (SGS). In fact, many of the young clusters in our sample are found near the rim of the two most prominent SGS in the SMC.

References

1. E. Bica et al. (1995) AJ 101, 41
2. G. Pietrzynski & A. Udalski (1999) AcA 49, 157
3. D. Zaritsky et al. (2002) AJ 123, 855
4. L.T. Gardiner & M. Noguchi (1996) MNRAS 278, 191
5. S. Stanimirović et al. (1999) MNRAS 302, 417

Molecular Clouds and Star Formation in the Magellanic System by NANTEN

A. Kawamura[1], T. Minamidani[1], Y. Mizuno[1], N. Mizuno[1], T. Onishi[1], A. Mizuno[1,2], Y. Fukui[1], and NANTEN team[1]

[1] Department of Astrophysics, Nagoya University, Furocho, Chikusa-ku, Nagoya 464-8602, Japan kawamura@a.phys.nagoya-u.ac.jp
[2] Solar-Terrestrial Environment Laboratory, Nagoya University, Furocho, Chikusa-ku, Nagoya 464-8601, Japan

1 GMC Surveys in the LMC and the SMC by NANTEN

Most massive stars and clusters are formed in giant molecular clouds (GMCs) so that it is of vital importance to understand the physical properties of the GMCs to elucidate cluster formation. The Magellanic system is the nearest neighbour to the Galaxy and is actively forming stars as populous clusters or OB associations [1]. It is the most suitable target for a detailed study of cluster formation, since it is located at ~ 50 kpc from the Sun, allowing us to identify young objects associated with GMCs unambiguously. We made CO $J = 1$–0 surveys toward the LMC and the SMC by NANTEN, a 4 m mm/sub-mm wave telescope located in Las Campanas, Chile. The sensitivity of the survey is equivalent to $N(\mathrm{H}_2) \sim 1 \times 10^{21}$ cm^{-2}, and the regions covered are $\sim$6 deg $\times$6 deg of the LMC and $\sim$1.5 deg^2 of the SMC, respectively, with $\sim$40 pc resolution. We have identified about 300 molecular clouds; the reliable estimates of physical quantities are available for the 164 GMCs among the 272 and 13 GMCs among the 23 in the LMC and the SMC. The mass of the GMCs are $\sim 10^3 < M_{\mathrm{vir}}[M_\odot] < 9 \times 10^6$ (e.g., [4, 5, 9]).

2 Star and Cluster Formation

We compared the GMC distributions with stellar clusters [1] and H II regions (e.g., [2, 3, 6]). The detection limit of H II regions is quite low, corresponding to those excited by a single O type stars [8]. The result indicates that the HII regions and the youngest clusters ($\tau < \sim 10$ Myr) exhibit strong spatial correlations, while the older clusters($\tau > 10$ Myr) show much weaker or no correlation [7, 12]. It is shown that $\sim$25% and $\sim$50% of the GMCs are associated with young clusters, and both young clusters and HII regions, respectively, while $\sim$25% show no signs of star formation, in the LMC. GMCs associated with young clusters tend to have a larger masses and sizes but there is no significant difference in frequency distribution of the CO line width among these three GMC groups. The completeness of the GMC sample enables us to infer the evolutionary timescales of the GMCs. The first stage corresponds to starless GMCs with ~ 7 Myr followed by a phase with small H II regions. The subsequent phase shows the most active formation of clusters followed by the dissipation of the GMCs. The total lifetime of a GMC is estimated as ~ 20 Myr. The rather lengthy timescale before star formation may allow the formation of proto-cluster molecular condensations as massive as $10^5 M_\odot$, which can lead to form the rich populous clusters [7].

3 CO Detections in the Magellanic Bridge

We have detected seven sites of ^{12}CO J =1–0 emission by NANTEN in addition to the one detected previously [11]. The integrated CO brightness ranges between 30 and 140 mK km s^{-1}; corresponding to an estimated molecular mass of 1–$7 \times 10^3 M_\odot$. The positions of the CO emission are co-incident with sites of bright 100 μm emission with $I_{100\mu m} > 2.6$ MJy sr^{-1}. The line widths of CO spectra are rather narrow, $< \sim 2$ km s^{-1}. These indicate the gas is in a cold and quiescent state. The velocity centroids of the CO and HI spectra are consistent. This may indicate the CO clouds are formed after the tidal encounter, rather than being extracted from the SMC. This is supported by the small lifetime, $\sim 10^7$ yr of the CO clouds, which is much less than the estimated 200 Myr age of the Bridge itself [10].

Acknowledgements. The NANTEN project is based on a mutual agreement between Nagoya University and the Carnegie Institution of Washington.

References

1. Bica, E., Claria, J. J., Dottori, H., Santos, J. F. C., Jr., & Piatti, A. E. ApJS, **102**, 57 (1996)
2. Davies, R. D. et al. MNRAS, **81**, 89 (1976)
3. Filipovic, M. D. et al. A&AS, **111**, 311 (1995)
4. Fukui, Y., Mizuno, N., Yamaguchi, R., Mizuno, A., Onishi, T., Ogawa, H., Yonekura, Y., Kawamura, A. et al. PASJ, **51**, 745 (1999)
5. Fukui, Y., Kawamura, A., Minamidani, T., Mizuno, Y., Mizuno, N., Onishi, T., Mizuno, A., Fukui, Y. in preparation (2006)
6. Henize, K. G. ApJS, **2**, 315 (1956)
7. Kawamura, A., Minamidani, T., Mizuno, Y., Mizuno, N., Onishi, T., Mizuno, A. in preparation (2006)
8. Kennicutt, R. C., Jr., Hodge, P. W. ApJ, **306**, 130 (1986)
9. Mizuno, N., Rubio, M., Mizuno, A., Yamaguchi, R., Onishi, T., Fukui, Y. PASJ, **53**, L45 (2001)
10. Mizuno, N., Muller, E., Maeda, H. Kawamura, A. Onishi, T. Mizuno, A. Fukui, Y. ApJ, submitted (2006)
11. Muller, E., Staveley-Smith, L., Zealey, W. J. MNRAS, **338**, 609 (2003)
12. Yamaguchi, R., Mizuno, N., Mizuno, A., Rubio, M., Abe, R., Saito, H., Moriguchi, Y., Matsunaga, K., Onishi, T. et al. PASJ, **53**, 985 (2001)

Two Star Cluster Populations in NGC 45

M.D. Mora[1], S.S. Larsen[1,2], and M. Kissler-Patig[1]

[1] European Southern Observatory, Karl-Schwarzschild-Strasse 2, 85748 Garching bei Munich Germany mmora@eso.org, mkissler@eso.org
[2] Astronomical Institute, University of Utrecht, Princetonplein 5, NL-3584 CC, Utrecht, The Netherlands larsen@astro.uu.nl

Star clusters are present in all galaxies (spirals, ellipticals, mergers, star burst, etc). Here we present the results of our study of star clusters in the low luminosity spiral galaxy NGC 45, that exhibits star cluster formation despite showing no sign of external perturbation.

NGC 45 ($M_B = -17.13$) is an outlying member of the Sculptor group. It is located at 5 Mpc from us. The images were acquired using the HST ACS and WFPC2 in U_{F336W}, B_{F435W}, V_{F555W}, I_{F814W}. Object detections were performed using SExtractor; for the photometry we used IRAF/PHOT and the cluster candidates selection was done using Baolab Ishape [6]. All "round", extended objects were treated as cluster candidates. Objects with B–V > 0.8 were selected as globular cluster candidates.

1 Globular Cluster Specific Frequency

The specific frequency [5] is defined as: $S = N_{GC} 10^{0.4*(M_V+15)}$ where N_{GC} is the total number of globular clusters that belong to the galaxy and M_V is the galaxy's absolute visual magnitude. Taking into account the detection completeness, we assumed that the globular cluster luminosity function is described by a Gaussian function. The total number of GCs was estimated by counting the number of GCs brighter than the turn-over magnitude (corrected for foreground reddening) multiplied by 2 plus the spatial incompleteness correction ($M_{TO} = -7.46 \pm 0.08$, $V = 20.89$ [4]). We derived a $S = 1.51^{+0.32}_{-0.48}$.

2 Cluster Ages and Masses

Photometry in 4 bands allows us to derive age, mass and extinction for each candidate using the 3D Fit method [3]. We minimized the difference between each observed star cluster color and the SSP model [1]. Independent of the metallicity, Fig. 1 right side, shows a concentration of young and not very massive clusters ($M < 10^3 M_\odot$) around $10^{6.8}$ yr. At older ages, the number of detected cluster candidates per age bin (note the logarithmic age scale) decreases rapidly. This is a consequence of fading due to stellar evolution (as indicated by the solid lines), as well as cluster disruption.

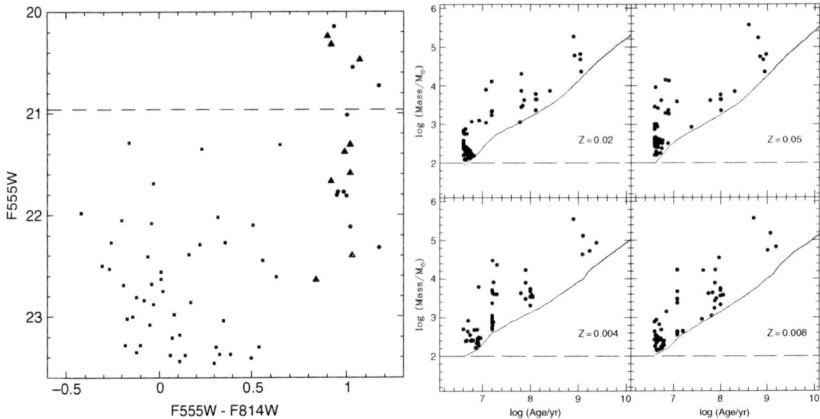

Fig. 1. *Left*: Color magnitude diagram for star clusters in NGC 45. *Squares* and *triangles* are objects with four band photometry, while *circles* are objects with only 3 bands (no F336W). *Circles* and *triangles* are globular clusters, while *squares* correspond to young star clusters. The *dashed line* is the TO of the old MW globular clusters system $M_{V,TO} \sim -7.4$. *Right*: Mass as a function of cluster age. All star cluster candidates with masses greater than 100 $M_\odot$ are plotted here. The *solid line* represents a cluster of $F435W = 23.5$ at different ages and masses.

3 Final Remarks

NGC 45 shows two main groups of star clusters: A number of relatively low-mass ($< 10^4 M_\odot$) objects similar to the open clusters observed in the Milky Way and a large population of old GCs for its luminosity with higher S in comparison to similar late type galaxies such as M33 (S = 0.5 ± 0.2) and LMC (S = 0.6 ± 0.2) [2].

References

1. Ander, P., & Fritze-v. Alvensleben, U. A&A **401**, 1063 (2003)
2. Ashman, K. M., & Zeprf, S. E. Globular Cluster Systems. Cambridge University press (1998)
3. Bik, A., Lamers, H.J.G.L.L., Bastian, N. et al. A&A **397**, 473 (2003)
4. Carney, B.W., & Harris, W.E. Saas-Fee Advanced Course 28. Lectures notes Swiss Society for Astrophysics and Astronomy Springer-Verlag (1998).
5. Harris, W.E., & van den Bergh, S., AJ **86**, 1627 (1981)
6. Larsen, S.S. A&A **354**, 836 (1999)

Characterization of Open Cluster Remnants

D.B. Pavani[1,2] and E. Bica[2]

[1] Instituto de Astronomia, Geofísica e Ciências Atmosféricas (IAG), Universidade de São Paulo - Rua do Matão 1226; 05508-900, São Paulo, SP, Brasil
daniela@astro.iag.usp.br
[2] Universidade Federal do Rio Grande do Sul, IF, CP 15051; 012501-970, Porto Alegre, RS, Brasil bica@if.ufrgs.br

The present work aims to better understand the relationship between open star clusters and their remnants. We selected 10 open cluster remnant candidates from Bica et al. [1] and searched for other poorly populated objects in open cluster catalogues. Bica et al. [1] referred to them as Possible Open Cluster Remnants (POCR). The candidates are analyzed by means of photometric and proper motion data. The data provide constraints for the objects and their fields, which we employed to establish criteria to characterize open cluster remnants, taking into account observational uncertainties.

1 Methods

From the observational point of view, an open cluster remnant (OCR) can be defined as a poorly populated concentration of stars resulting from the dynamical evolution of a more massive system [3]. Despite theoretical progress, and the growing number of candidates observed, open questions still remain, such as: (i) is there a preferential location in the Galaxy to find them? (ii) Among the remnants, do different evolutionary stages exist? (iii) Is it possible to define criteria for their characterization as OCRs?

The final sample contains 22 objects previously described as star clusters in open cluster catalogues and a new one (Object 1) which was found on a sky survey plate by E. Bica.

2MASS[3] J and H photometry is employed to (i) study the structural properties of the objects by means of radial stellar density profiles; (ii) test the similarity or not between objects and fields through a statistical method of comparison between distributions of stars in the CMD and, (iii) obtain ages, reddening values and distances using the CMD, together with the classification of the objects considering an index of isochrone fitting.

We conclude that the observed radial stellar density distribution in the sample objects is in general consistent with that expected for OCRs. They should be depleted as compared to open clusters, retaining their more central parts which determine their limiting radius. In the statistical method we test

---

[3]The Two Micron All Sky Survey, All Sky data release [4]

the hypothesis that the object can be reproduced by an equal area field fluctuation and derived the probability of the POCR to be such a fluctuation. This method showed one POCR to be a field fluctuation. We applied a numerical method to quantify the fit of an isochrone to the CMD of a POCR. Stars compatible with single and binary star isochrones taking into account error bars. We obtain their ratio to the total stars in the CMD. This is the fitting index I_f. In the sample, 13 POCRs with $I_f > 60\%$ are well described by the respective isochrones. Two POCRs are ranked with $60\% < I_f > 50\%$, five with $I_f < 50\%$, and for two others no fit was possible.

Individual POCRS show low statistics with respect to open clusters, and a composite CMD of POCRS might provide more insight on their nature. Proper motions extracted from UCAC2 allowed a comparison between their distribution in the POCRs area and their large offset fields. The objects were divided into two groups (A and B) according to their proper motion distributions. We transformed proper motions to linear velocities and considered their composite velocity histograms. For group A a well-defined low-velocity peak occurs, while for group B we find a higher-velocity distribution, which appear to be related to the distributions of single and unresolved binary stars as observed in open clusters [2]. After reddening and distance corrections we obtained absolute CMDs, which are in turn combined to build absolute composite CMDs for groups A and B, improving the statistics in the diagrams. Stars within the velocity range compatible with unresolved binaries are identified in CMD, in general occupying their expected loci.

2 Conclusions

The present methods allowed an objective analysis of POCRs and suggested the presence of 14 open cluster remnants among the 23 sample objects. Evidence of binarity was found, as expected from high binary fractions in dynamically evolved systems. We inferred on possible evolutionary stages among remnants from the structure, proper motion distribution and CMD properties of the objects.

References

1. Bica, E.; Santiago, B.X.; Dutra, C.M.; Dottori, H.; Oliveira, M.R.; Pavani, D. B.: A&A **366**, 827 (2001)
2. Bica, E.; Bonatto, C.: A&A **431**, 973 (2005)
3. Pavani, D.B.; Bica, E.; Dutra, C.M.; Dottori, H.; Santiago, B.X.; Carranza, G.; Díaz, R.J.: A&A **374**, 554 (2001)
4. Skrutskie, M.; Schneider, S.; Stiening, R.; et al.: *The Impact of Large Scale Near-IR Sky Surveys*, ed. Garzon et al. (The Netherlands: Kluwer 1997), **210**, p 187

HST Photometry of the Binary Globular Cluster Sersic 13N-S in NGC5128[1]

D. Villegas[1,2], D. Minniti,[1] and J.G. Funes[3]

[1] Department of Astronomy, P. Universidad Católica, Casilla 306, Santiago 22, Chile
[2] ESO, Karl-Schwarzschild-Str. 2, 85748 Garching, Germany
[3] Vatican Observatory Research Group, University of Arizona, Tucson, AZ 85721, USA

1 Introduction

Super star clusters (SSCs) are star clusters embedded in giant HII regions of merging and starbursting galaxies which features make us think they are recently-formed globular clusters [2]. R136 in 30 Doradus (LMC) is the best studied SSC [3], and its proximity allows the study of its star content directly showing that most of its components are type O3 stars, the most massive stars known (M $\sim$ 50–120 M$_\odot$). If SSCs are formed by such massive stars, we expect to define their individual components in nearby objects, out to a distance of a few Mpc, at least for the most luminous stars in each cluster.

With this goal we performed HST/WFPC2 photometry in Sersic 13, a giant HII region in the dust band of NGC5128 which ionizing central source is formed by a close pair of bright and blue SSCs [4]. We worked with images in the bands F336W, F555W and F814W (Johnson-Cousins U, V and I) in order to resolve individual stars in each cluster and estimate their ages.

2 Results

From the data we were able to resolve a total of nine "individual" stars in the clusters area. In the CMDs, these stars look bluer and more luminous than the other sources in the field. Their V magnitude is in the range $-7.7 < M_{F555W} < -5.9$, just comparable with the most luminous Galactic WR stars, which have magnitudes between $-8 < M_V < -2$ [5], with just the brighter 13% in the range of the resolved stars in the Sersic 13 region. These bright blue objects could be WR stars according to the characteristic wide emission features seen in the cluster spectra. Nevertheless, it is important to consider that this high luminosity could be an effect of blending due to the high crowding in the region.

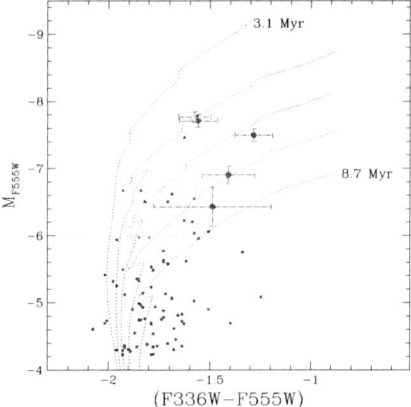

Fig. 1. Defined stars in each cluster plotted over Geneva evolutionary tracks between 3.1 and 8.7 Myr. *Yellow squares* are for Sersic 13-N and *red circles* for Sersic 13-S. *Small blue squares* correspond to the 65 O3 stars defined by Massey & Hunter (1998) in R136.

3 Conclusions

We estimated an age of $t \leq 3.9 \times 10^6$ yr for the clusters by comparison with Padova isochrones [6] and $t \leq 5.6 \times 10^6$ yr using Geneva isochrones [7], which agrees with the results from integrated spectroscopy [4]. The accuracy of this procedure is uncertain, especially for high mass stars and because of the unknown reddening in the region, but with the current data we are not able to obtain a better estimation.

Deeper HST photometry or the use of adaptive optics could improve the resolution, but certainly would not be able to obtain information on low and intermediate mass stars, which could help us to trace the formation history of the clusters. Additionally, further IR studies may allow to account for the reddening influence, and to obtain more information on the individual massive components of the clusters.

The age of $t \leq 6 \times 10^6$ years is compatible with the scenario where the clusters are gravitationally bound and will merge in the future. This is interesting in light of the recent study of Ferraro et al. [8] that suggests that ωCen is the product of a merger of two smaller star clusters.

References

1. Villegas, D., Minniti, D., Funes, J.G. 2005, A&A 442, 437
2. Whitmore, B.C. 2001, astro-ph/0012546
3. Selman, F., Melnick, J., Bosch, G., Terlevich, R. 1999, A&A 341, 98
4. Minniti, D., Rejkuba, M., Funes, J., Kennicutt, R. 2004, ApJ 612, 215

5. van der Hucht, K.A. 2001, New Astron. Rev. 45, 135
6. Girardi, L. et al. 2002, A&A 391, 195
7. Lejeune, T., Schaerer, D. 2001, A&A 366, 538
8. Ferraro, F.R. et al. 2004, ApJ 603, L81

Part IV

Globular Cluster Systems in Dwarf and Irregular Galaxies

LMC Cluster Abundances and Kinematics

Doug Geisler[1], Aaron Grocholski[2], Ata Sarajedini[2], Andrew Cole[3], and Verne Smith[4]

[1] Departments de Fisica, Universidad de Concepcion, Concepcion, Chile
   dgeisler@astro-udec.cl
[2] Department of Astronomy, University of Florida, Gainesville, FL, USA
   aaron@astro.ufl.edu, ata@astro.ufl.edu
[3] Department of Astronomy, University of Minnesota, Minneapolis, MN USA
   cole@physics.umn.edu
[4] US Gemini Project, La Serena, Chile vsmith@noao.edu

Abstract. We present results from a project aimed at better understanding the kinematics and metallicities of populous clusters in the LMC. We have utilized FORS2 on the VLT to obtain infrared spectra for more than 200 stars in 28 LMC clusters spanning a large range of ages (~ 1–$13\,\mathrm{Gyr}$) and metallicities ($-0.3 > [Fe/H] > -2.0$). The absorption lines of the calcium II triplet were then used to calculate radial velocities and [Fe/H]. We determine mean cluster velocities to typically 1.6 km s^{-1} and mean metallicities to 0.04 dex (random error). Similar to what was found by previous authors, this cluster sample has motions consistent with that of a single rotating disk system, with no indication of halo kinematics. However, in contrast to previous work, we find that the higher metallicity clusters in our sample show a very tight [Fe/H] distribution with no tail toward solar metallicities. The cluster distribution is similar to what has been found for red giant stars in the bar, which indicates that the bar and the intermediate age clusters have similar star formation histories. This is in good agreement with recent theoretical models that suggest the bar and intermediate age clusters formed as a result of a close encounter with the SMC. Our findings also confirm previous results which show that the LMC lacks the metallicity gradient typically seen in non-barred spiral galaxies, suggesting that the bar is driving the mixing of stellar populations in the LMC.

1 Introduction

In the current paradigm of galaxy formation, it is believed that the formation history of spiral galaxy spheroids, such as the Milky Way (MW) halo and bulge, may be dominated by the accretion/merger of smaller, satellite galaxies (e.g. [13, 16]). The Large Magellanic Cloud (LMC) and Small Magellanic Cloud (SMC) are two satellite galaxies currently interacting with the MW which may eventually be consumed into the MW halo. Their relative proximity allows us to easily resolve stellar populations down below their oldest main sequence turnoffs. Thus, the LMC and SMC offer us a golden opportunity to study the effects of dynamical interactions on the formation and evolution of satellite galaxies; this information plays an integral part in discovering the secrets of spiral galaxy formation.

LMC clusters are a particularly powerful tool in this study. Star clusters are Simple Stellar Populations with a unique age and metallicity and are thus excellent tracers of the star formation history of a galaxy, at least during the epochs when star and cluster formation were coupled. The LMC contains several thousand star clusters, many of which occupy regions of age-metallicity space uninhabited by MW clusters. In particular, there are a large number of massive young clusters [8] with subsolar metallicities.

We have begun a program to derive metallicities for a large number of LMC clusters. We report here on the technique and initial results. The details can be found in Grocholski et al. [7].

2 The Calcium Triplet Technique

Armandroff & Zinn [1] first applied the Calcium triplet technique to derive metallicities of red giant branch (RGB) stars in Galactic globular clusters. They obtained medium-resolution spectra of RGB stars at a wavelength of $\approx$ 8600 Å, centered on the very prominent triplet of calcium II (CaT) lines. This spectral feature is easily measured in distant targets and at medium resolution since the CaT lines are extremely strong and RGB stars are near their brightest in the near-infrared.

In a landmark study, Olszewski et al. ([9], hereafter OSSH) applied this technique to LMC clusters. They calculated metallicities and radial velocities for 72 clusters. To date, this has been the only large-scale spectroscopic metallicity determination for LMC clusters based on individual stars. They found a large range of metallicities, an age-metallicity relationship and no significant spatial gradient. Using radial velocities from the OSSH sample, Schommer et al. [12] found that the LMC cluster system rotates as a disk, with no indication that any of the clusters have kinematics consistent with that of a pressure supported halo.

However, despite their heroic effort, the results of OSSH were limited by the technology at the time. The combination of a single-slit spectrograph with a midsized telescope made it difficult for OSSH to build up the number of target stars necessary to differentiate between cluster members and field stars. Most of the resulting cluster values are based on only one or two stars; in some cases, there are metallicity or radial velocity discrepancies between the few stars measured, and it is unclear which of the values to rely on.

In addition to technology, subsequent advances have also ocurred both in the knowledge of the globular cluster metallicity scale to which the CaT strengths are referred [11] and in the standard procedure used to remove gravity and temperature dependencies from the CaT equivalent widths [10, 4] which allow much more reliable metallicities to be derived from the CaT technique. For example, Cole et al. [4] have shown that there is a very tight ($\sigma = 0.07$ dex) relationship between the gravity-corrected sum of the equivalent widths of the 3 CaT lines and metallicity for both MW open clusters (on the Friel et al. [6] metallicity scale) and globular clusters (on the Caretta & Gratton [3] scale), thus allowing accurate metallicities to be derived for clusters from 1–14 Gyr old. They also show that errors due to AGB contamination, variation of the Red Clump/Horizontal Branch magnitude with age or metallicity and [Ca/Fe] variations should all be < 0.1 dex. In addition, the technique eliminates reddening or distance errors.

3 Observations and Reductions

In an effort to produce a modern and reliable catalog of LMC cluster metallicities, we have obtained CaT spectra of a large number of stars in each of 28 LMC clusters. We have taken advantage of the multiplex capability and lightgathering power of the European Southern Observatory's 8.2 m Very Large Telescope, and of the great strides in the interpretation and calibration of Ca II triplet spectroscopy made in the past 15 years to provide accurate cluster velocities and abundances.

Our sample was intentionally biased towards those clusters with conflicting or uncertain OSSH abundance measurements, clusters thought to lie near the edge of the age gap, and those whose radial velocities might provide new insight into the dynamical history of the LMC-SMC system, based on their location. We selected targets lying near the cluster center and along the RGB, based on pre-images obtained at the VLT. We also selected field RGB stars to study their metallicity distribution as well.

The spectroscopic observations were carried out with FORS2 at the Antu (VLT-UT1) 8.2 m telescope at Paranal during December 2004. We used the FORS2 spectrograph in mask exchange unit (MXU) mode, with the 1028z+29 grism. We used slits that were 1 arcsec wide and 8 arcsec long. Typically 10 stars inside our estimated cluster radius were observed, with an additional ~ 20 stars outside of this radius. A total of ~ 850 stars were observed. The resulting spectra cover 1750 Å, with a central wavelength of 8440 Å and a dispersion of ~ 0.85 Å pixel^{-1} (resolution of 2–3 Å). The total exposure time in each setup was either 2×300 s, 2×500 s or 2×600 s.

Image processing was performed with a variety of tasks in IRAF. For the final spectra, S/N ratios are typically 25–50 pixel^{-1} with some stars as high as ~ 90 pixel^{-1} and, in only a few cases, as low as ~ 15 pixel^{-1}. Sample spectra showing the CaT region are presented in Fig. 1.

Radial velocities for all target stars were determined through crosscorrelation with 30 template stars using the IRAF task FXCOR. Typical total errors were roughly 7.5 km s^{-1}.

To measure the equivalent widths of the CaT lines, we follow the method of Armandroff & Zinn [1] and define continuum bandpasses on either side of each CaT feature. The 'pseudo-continuum' for each CaT line is easily defined by a linear fit to the mean value in each pair of continuum windows. The 'pseudo-equivalent' width is then calculated by fitting the sum of a Gaussian

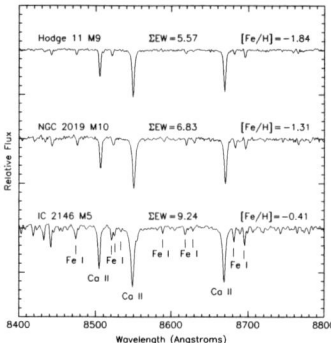

Fig. 1. A sample of spectra covering a range of metallicities. Summed equivalent widths and mean [Fe/H] values for each cluster are shown. The change in CaT line strength with [Fe/H] is readily visible.

and a Lorentzian, required to have a common line center, to each CaT line with respect to the 'pseudo-continuum'. We then summed the equivalent widths of the 3 CaT lines, corrected them for gravity effects using the reduced equivalent width, W', defined in Cole et al. [4] and finally used the Cole et al. [Fe/H]:W' relation to derive metal abundances. We estimate that individual metal abundances have errors of 0.15 dex.

4 Results

Cluster membership was determined by first deriving a cluster center and radius (by eye), based on the preimages. As noted above, stars lying along the cluster RGB sequence were our primary targets, followed by field RGB candidates to fill the spectrograph. Next, plots of velocity vs. radius and [Fe/H] vs. radius were used to further distinguish cluster members from field stars. Examples of each of these plots for Hodge 11 are given as Figs. 2 and 3. The likelihood of cluster membership of our final cluster candidates should be extremely high. This process resulted in obtaining velocities and metallicities for an average of 8 member candidates per cluster, as well as for about 20 field giants.

We then derive mean radial velocities and metallicities for each cluster, resulting in random mean errors of only 1.6 km s^{-1} and 0.04 dex, about 1/3 of the typical errors of OSSH. For the clusters in common between our studies, we find that our velocity and metallicity values agree to within the errors with the results of OSSH (after converting the OSSH metallicities to the same [Fe/H] scale). We determine the first spectroscopically derived metallicities and radial velocities for 6 clusters.

Positions on the sky for each cluster are shown in Fig. 4, along with the metallicity bin into which each cluster falls. In Fig. 5, we plot metallicity as

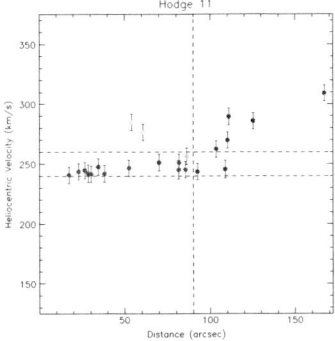

Fig. 2. Radial velocity vs. distance from the center of Hodge 11. The *horizontal lines* represent our velocity cut and have a width of $\pm 10\,\mathrm{km\,s^{-1}}$. The cluster radius is shown as the *vertical line*. Stars within the cluster radius falling within the velocity cut are considered cluster candidates.

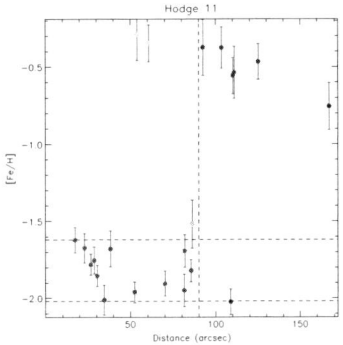

Fig. 3. Metallicity vs. distance from the center of Hodge 11. The *horizontal lines* represent our metallicity cut and have a width of ± 0.2 dex. Velocity members from Fig. 2 falling within the metallicity cut are our final cluster members.

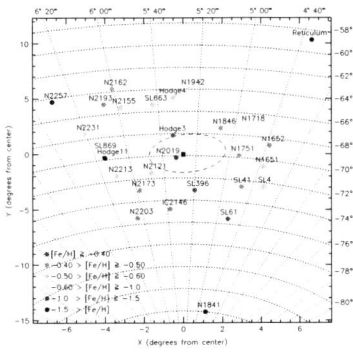

Fig. 4. Positions on the sky and derived mean metallicities for our target clusters. The adopted LMC center is marked with the *filled square* and the *dashed line* roughly outlines the bar.

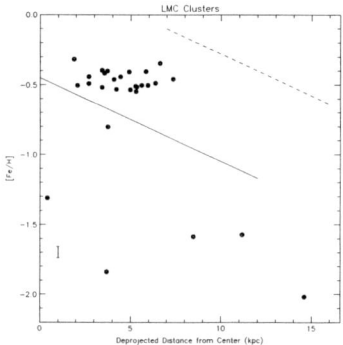

Fig. 5. Cluster metallicities are plotted as a function of deprojected distance (in kpc) from the center of the LMC. Overplotted are the metallicity gradients observed in the MW open clusters (*dashed line*; Friel et al. [6]) and M33 (*solid line*; Tiede et al. [14]).

a function of radial distance (in kpc). Combined, these two figures illustrate that, similar to what was found by OSSH, there is no [Fe/H] gradient for the higher metallicity clusters in our sample. In Fig. 5 we have over plotted both the MW open cluster [Fe/H] gradient and the M33 gradient. Neither of these disk abundance gradients resembles what we see among the LMC clusters. Zaritsky et al. [15] studied the HII region oxygen abundances in 39 disk galaxies. Their data suggest that disk abundance gradients are ubiquitous in spiral galaxies. However, the presence of a classical bar in the galaxy – one that extends a significant fraction of the disk length – tends to weaken the gradient. Thus, the flattened LMC distribution is likely caused by the presence of the central bar.

In Fig. 6, we compare the metallicity distributions of the OSSH sample and our clusters. From this figure it is clear that both the raw and converted OSSH samples show an extended distribution of intermediate metallicity clusters, whereas our sample exhibits a very tight distribution for these clusters. For the 20 clusters in common, we find a mean [Fe/H] = -0.47 with $\sigma = 0.06$, while the converted OSSH metallicities give [Fe/H] = -0.42 ± 0.14. We find only a single intermediate metallicity ($\sim$ intermediate age) cluster, NGC 1718, to have a metallicity significantly different from its counterparts in our sample. Otherwise, all our intermediate age clusters fall within the very narrow range from [Fe/H]= -0.32 to -0.55, with no more metal-rich clusters. Clearly, our superior dataset allows us to show that the metallicity spread given in the OSSH dataset is due in large part to greater errors and that intermediate aged clusters, which formed between 1 and 3 Gyr ago (e.g. OSSH) over a large part of the LMC, formed out of material which was very well mixed. Additionally, we find that our intermediate age clusters have a mean metallicity and distribution similar to that of the metal rich component of the bar field studied by Cole et al. [5]. The similarity between these two

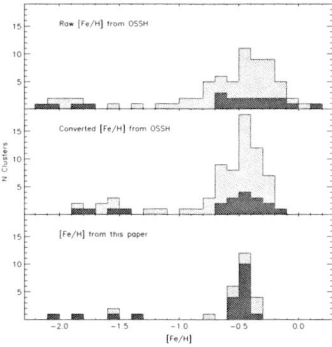

Fig. 6. The metallicity distribution of LMC clusters as determined by OSSH and this work. Published values from OSSH are given in the *top panel* while the *middle panel* shows their values converted onto our metallicity scale; in the *bottom panel* we show our results. In all three panels, the *dark shaded region* shows the distribution for the 20 clusters in common between OSSH and our work, while the *light shaded region* shows the entire cluster sample from each study.

populations is in good agreement with the models of Bekki et al. [2], in which the formation of the LMC bar and the restart of cluster formation (end of the 'age-gap') are both a result of the same very close encounter with the SMC.

Finally, we plot in Fig. 7 Galactocentric radial velocity versus position angle on the sky for our sample, along with velocity data for all clusters listed

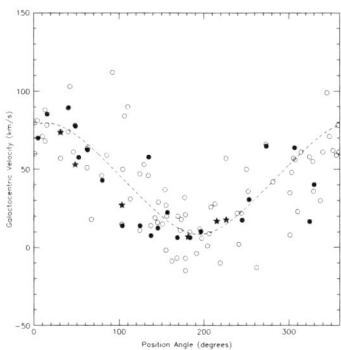

Fig. 7. Galactocentric radial velocities as a function of position angle on the sky are plotted for the clusters in our sample (*filled symbols*) as well as those from Schommer et al. ([12]; *open circles*); the six clusters in our sample with no previous velocity determinations are plotted as *filled stars* and all others in our sample are *filled circles*. Rotation curve solution number 3 from Schommer et al. [12] is overplotted as the *dashed line*, showing that both data sets are consistent with circular rotation. We note that we have not plotted a representative error bar since our plotting symbols are roughly the same size as the average random velocity error.

in Schommer et al. [12], who used the OSSH data to analyze the cluster kinematics. We adopted the same procedures and parameters as Schommer et al. [12] to ensure both datasets were analyzed identically. For the clusters in common between these two data sets, we find excellent agreement. Additionally, the derived velocities for the six 'new' clusters show that their motions are consistent with the findings of Schommer et al. [12] in that the LMC cluster system exhibits disk-like rotation that is very similar to the HI disk, and has no obvious signature of a stellar halo.

Acknowledgements. AJG was supported by NSF CAREER grant AST-0094048 to AS. AAC was supported by a fellowship from the Netherlands Research School for Astronomy (NOVA). DG gratefully acknowledges support from the Chilean *Centro de Astrofísica* FONDAP No. 15010003. VVS has been supported by the NSF through grant AST03-07534.

References

1. Armandroff, T.E. & Zinn, R. 1988, AJ, 96, 92
2. Bekki, K., Couch, W.J., Beasley, M.A., Forbes, D.A., Chiba, M., & DaCosta, G.S. 2004, ApJL, 610, 93
3. Carretta, E. & Gratton, R.G. 1997, A&A, 121, 95
4. Cole, A.A., Smecker-Hane, T.A., Tolstoy, E., Bosler, T.L., & Gallagher, J.S. 2004, MNRAS, 347, 367
5. Cole, A.A., Tolstoy, E., Gallagher, J.S., & Smecker-Hane, T.A. 2005, AJ, 129, 1465
6. Friel, E.D., Janes, K.A., Tavarez, M., Scott, J., Katsanis, R., Lotz, J., Hong, L., & Miller, N. 2002, AJ, 124, 2693
7. Grocholski, A.J., Cole, A.A., Sarajedini, A., Geisler, D., & Smith, V.V. 2006, AJ, 132, 1630
8. Hodge, P.W. 1960, ApJ, 131, 351
9. Olszewski, E.W., Schommer, R.A., Suntzeff, N.B., & Harris, H.C. 1991, AJ, 101, 515 (OSSH)
10. Rutledge, G.A., Hesser, J.E., Stetston, P.B., Mateo, M., Simard, L., Bolte, M., Friel, E.D., & Copin, Y. 1997a, PASP, 109, 883
11. Rutledge, G.A., Hesser, J.E., & Stetson, P.B. 1997b, PASP, 109, 907
12. Schommer, R.A., Olszewski, E.W., Suntzeff, N.B., & Harris, H.C. 1992, AJ, 103, 447
13. Searle, L. & Zinn, R. 1978, ApJ, 225, 357
14. Tiede, G.P., Sarajedini, A., & Barker, M.K. 2004, AJ, 128, 224
15. Zaritsky, D., Kennicutt, R.C., Jr., & Huchra, J. P. 1994, ApJ, 420, 87
16. Zentner, A.R. & Bullock, J.S. 2003, ApJ, 598, 49

Globular Clusters in Dwarf Galaxies

Bryan W. Miller

Gemini Observatory, Casilla 603, La Serena, Chile `bmiller@gemini.edu`

Abstract. Recent work on globular cluster systems in dwarf galaxies outside the Local Group is reviewed. Recent large imaging surveys with the *Hubble Space Telescope* and follow-up spectroscopy with 8-m class telescopes now allow us to compare the properties of massive star clusters in a wide range of galaxy types and environments. This body of work provides important constraints for theories of galaxy and star cluster formation and evolution.

1 Introduction

Studies of globular clusters (GCs) in dwarf galaxies provide very important insights into galaxy formation, the formation and evolution of GCs, and the relationship between GCs and nuclei. Comparisons of the properties of star clusters in different types of galaxies can test the theories of galaxy formation. In hierarchical scenarios of galaxy formation dwarf-size galaxies form first and then merge into larger systems. If star cluster formation coincided with galaxy formation, then a significant fraction of the star clusters in massive galaxies should have been formed in dwarfs. In this case the star clusters in dwarf galaxies in dense environments should be at least as old and metal-poor as the oldest star clusters in giant galaxies. However, recently evidence has mounted that stellar populations in surviving low mass galaxies are younger than in giant ellipticals [27]. In this "downsizing" view the dwarf galaxies formed after the giants or at least had their star formation rates suppressed at early times. A signature of downsizing would be that the star clusters in dwarfs are younger than those in giant galaxies.

Another question that star clusters can help answer is the relationship between dwarf irregular (dI) and dwarf elliptical (dE) galaxies. All dwarf galaxies must have formed with substantial gas fractions like today's dI galaxies. However, in massive local galaxies clusters the majority of the dwarfs are gas-free, smooth-isophote dEs. The differences may be due to environment or dIs may get transformed into dEs by gas stripping, supernovae winds, or galaxy interactions. A comparison of the star clusters in the two types of dwarfs provides in-site into the processes that shaped these galaxies and into why some dEs form nuclei.

In addition, the shape of the initial mass function of star clusters and how it evolves is not well understood. There are still debates about whether the form of initial mass function is a single or broken power-law (resulting in a log-normal distribution in magnitudes) and about the effects of various destruction processes [1,7,29]. By comparing the present-day mass functions in dwarf galaxies with those in giant galaxies it may be to disentangle the destructive processes and therefore determine the shape of the initial star cluster mass function.

This paper reviews the properties of star cluster systems in dwarf galaxies outside of the Local Group. Large imaging surveys with the *Hubble Space Telescope* are now starting to provide us with statistically significant samples of GCs in dEs and dIs in different environments. Follow-up spectroscopy with 8-m class telescopes are now providing complementary results on the ages, metallicities, abundance ratios, and kinematics of GCs and nuclei in dwarf galaxies.

2 Radial Distributions

A common problem when studying the globular cluster systems (GCSs) of dwarf galaxies is that any given galaxy generally has too few clusters to draw broad conclusions. Therefore, the standard approach is to combine the clusters from a large number of galaxies into a "master" dE GCS. Various studies have found that the radial distribution of GCs in dEs follows that of the background light and that it has a power-law form with a slope ranging between -1.6 and -3.5 [10,6,20,24]. Figure 1 show the background-subtracted radial distribution of GCs from the WFPC2 dE Snapshot Survey. The distribution is a power-law with $\alpha = -3.5 \pm 0.2$.

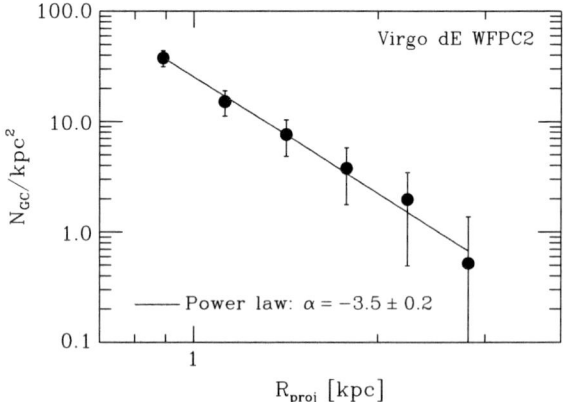

Fig. 1. The projected radial distribution of globular clusters in the Virgo Cluster sample of the WFPC2 dE Snapshot Survey. The data is well-fit by a single power-law with a slope $\alpha = -3.5 \pm 0.2$.

An alternative way of characterizing the radial distribution is to scale the projected radius of each cluster by the scale length of the host galaxy. This allows a direct comparison between the GC and background light distributions for a combined sample with galaxies of varying sizes. Data from the WFPC2 dE Snapshot Survey have shown that the radial distribution of the complete sample of GCs follows the background light extremely well [15]. However, the distribution of the GCs with $M_V < -8$ shows a deficit at small radii that may be the result of dynamical friction. The dynamical friction timescales in dEs are short enough that the merging of GCs via this process is one avenue of producing nuclei [11, 21, 15, 3]. However, simple dynamical friction calculations over-predict the luminosities of the nuclei so other processes may be counteracting it [15].

3 Luminosity Functions

The observed GC luminosity function (GCLF) gives the present-day GC mass function if the ages and metallicities (M/L ratios) of the clusters are known. Modeling the processes that can destroy GCs in dwarfs, mainly two-body relaxation with stellar evolution will hopefully allow us to determine the initial GCLF.

The GCLF in dEs has been measured recently for galaxies in nearby groups and the field [24] and in the Virgo and Fornax Clusters [19, 12]. In the Virgo Cluster the combined GCLF from WFPC2 data plotted as a function of magnitude is fit by a t_5 distribution with a peak at $M_V^0 = -7.3 \pm 0.1$. This is consistent with a GCLF peak of $M_V^0 \approx -7.5 \pm 0.3$ in VCC 1087 [2] and in the nearby group sample there is a peak at $M_V^0 = -7.4$ but after a small decline the numbers continue to rise at fainter magnitudes [24].

A key issue is whether the GCLF peak in dEs is the same as the peak seen in old. metal-poor GCs in giant galaxies. Di Criscienzo et al. have recently compared the GCLFs for the Milky Way, M31, and several giant ellipticals in Virgo using consistent selection criteria and distance scale [5]. Fits to a t_5 distribution give very consistent peaks with an average value of $M_V^0 = -7.66 \pm 0.2$. The GCLF peak for the dEs is consistent with this value, suggesting that the GCLF peak for old, metal-poor GC populations is nearly universal. However, there is a suggestion that the peak in dEs is ~ 0.3 mag fainter than in giant galaxies, perhaps as result of less efficient disk shocking [18].

The GCLF can also be plotted as a function of luminosity rather than magnitude. In this representation the peaks discussed above correspond to breaks in a power-law distribution. The bright-end GCLF in dwarfs is consistent with $\phi(L)/L \propto L^\alpha$ with $\alpha \sim -1.9$, similar to the slopes of the mass functions of Galactic molecular clouds and the luminosity functions of very young star clusters in starburst galaxies [18].

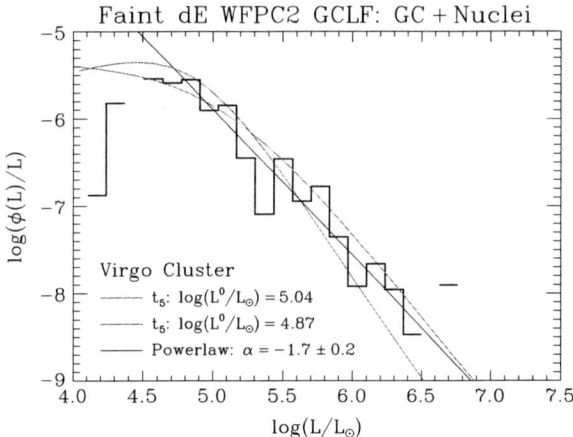

Fig. 2. Background-subtracted luminosity function for GCs and nuclei for Virgo galaxies in the WFPC2 Virgo sample with $M_V > -15.75$. For $\log(L/L_\odot) > 4.8$ the data is well-fit by a power-law with $\alpha = -1.7 \pm 0.2$. The data is also reasonably consistent with the t_5 fit to the entire WFPC2 sample which has a peak at $\log(L/L_\odot) = 4.87$ (*blue line*). The *red line* is the best fitting t_5 function to the faint-galaxy GCLF and it has a brighter and broader peak than the standard GCLF.

Recently, van den Bergh has proposed that the GCLF for galaxies with $M_V > -16$ is a single power-law, without a break [28]. The WFPC2 dE Snapshot data is also consistent with this but low number statistics make it difficult to distinguish various models (Fig. 2).

4 Colors, Ages, and Metallicities

The mean (V–I) color of dE GS is (V–I) ~ 0.9, similar to the colors of Galactic halo GCs and to the GCs in the "blue peak" in giant elliptical galaxies [16]. However, several studies have found that the mean color becomes slowly redder with increasing galaxy luminosity [25, 16, 22].

Recently work has been proceeding to use 8-m telescope to measure the ages and metallicities of GCs in dEs outside the Local Group [23, 2, 4, 18]. The metallicities fall in the range $-1.0 < $ [Fe/H] < -1.5 and the ages are all greater than 10 Gyr. The [α/Fe] ratio is more difficult to measure but it is important since it indicates whether the clusters formed after a significant starburst or after a period of quiescent star formation. Current measurements indicate that [α/Fe] is either solar or slightly enhanced.

Figure 3 shows preliminary results of GMOS spectroscopy of GCs and nuclei in three Virgo dEs and one Fornax dE [18]. The [α/Fe] ratios are between 0.0 and 0.3. Using solar [α/Fe] models we find that the ages are > 10 Gyr and the metallicities are [Fe/H] ~ -1.5. Interestingly, the bright

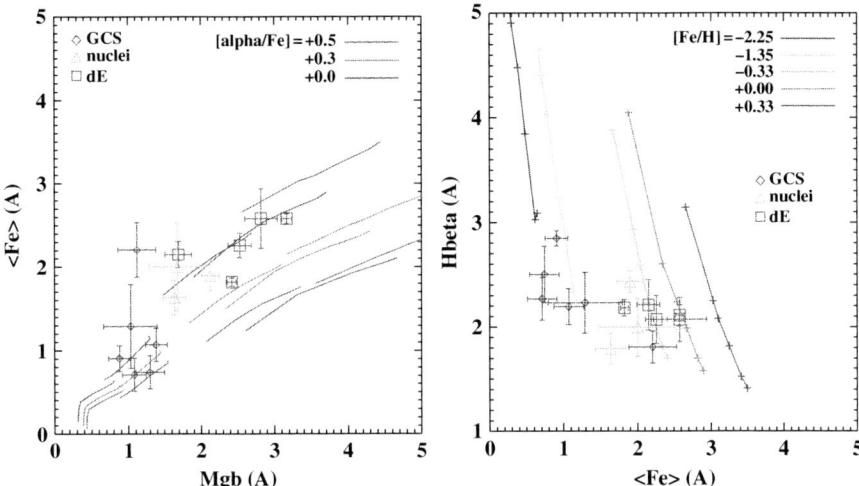

Fig. 3. Line indices for GCs and nuclei from GMOS spectroscopy [19] are compared with dE background light [9] and stellar evolutionary models [26]. The *left plot* shows <Fe> versus Mgb with models for different [α/Fe]. The *right plot* uses models with [α/Fe] = 0.0 and shows that the GCs are old and metal-poor while the nuclei and dE are somewhat younger and more metal-rich.

nuclei are more metal rich ([Fe/H] ∼ −0.5) and somewhat younger than the typical GC. As found from photometry, the ages and metallicities of the nuclei are intermediate between the properties of the GCs and the background stellar light [16].

With the ages known the $(V-I)$ colors can be converted to metallicities using stellar models. The GC color–galaxy luminosity relation then gives that $Z_{\rm MP,GC} \propto L_B^{0.2}$. However, the metallicity–luminosity, or mass, relation for all GCs including the red GCs that are more common in brighter galaxies is $Z_{\rm GC} \propto M^{0.4}$ (Fig. 4) [22]. This is the same as the dependence for the underlying field stars and suggests that GCs and field stars follow a similar chemical enrichment history.

5 Specific Frequency and GC Mass Fraction

The specific globular cluster frequency is useful for comparing globular cluster systems and it is related to the efficiency of globular cluster formation. In order to compare GC populations in different type of galaxies one can calculate the T parameter

$$T = \frac{N_{\rm GC}({\rm tot})}{M_G/10^9 M_\odot} \qquad (1)$$

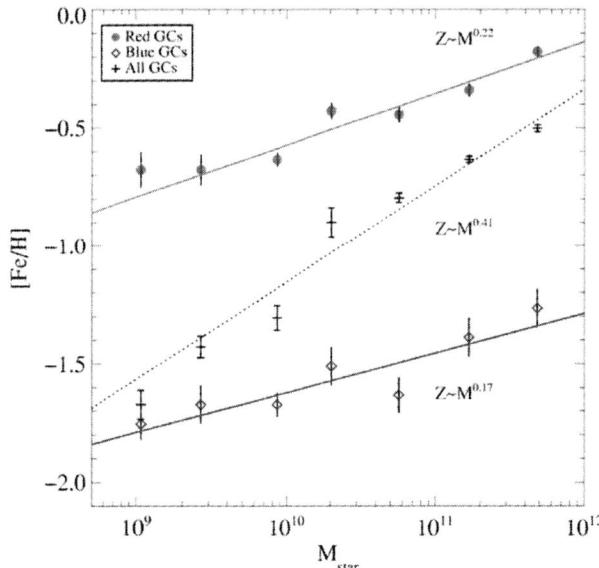

Fig. 4. Trends of [Fe/H] with galaxy stellar mass for *red, blue*, and all GC candidates from the ACS Virgo Survey [22]. The fraction of *blue* clusters increases with lower galaxy mass and the slope of the relation for the *blue* clusters is consistent with the results of [25] and [16].

which is corrected for the differences in galaxy M/L [30]. Assuming a universal GCLF (Sect. 3) then the GC mass fraction is $F = 0.0433T$. Figure 5 shows how T and F for metal-poor GCs correlate with galaxy mass [18]. The increase in T with mass for $\log(M_{\rm gal}) > 10.5$ can be explained by hierarchical galaxy formation models (solid line [14]). Below $\log(M_{\rm gal}) = 10.5$ T also increases with decreasing mass. This can be explained by models of GC formation that include the suppression of star formation in low-mass halos from supernovae winds (dashed lines, [17]).

It is also found that nucleated dEs have a mean value of T about a factor of two higher than that for non-nucleated dEs [18]. In addition, dE,N galaxies are more centrally concentrated within galaxy clusters and they have lower velocity dispersions than dE,noN galaxies. Therefore, the differences in T values and the presence of nuclei may be explained if dE,Ns experienced higher star formation rates due to "hot-mode" gas accretion in the high density cluster environment [13, 8]. Conversely, dE,noNs would have formed in lower density environments where star formation is lower due to "cold-mode" gas accretion.

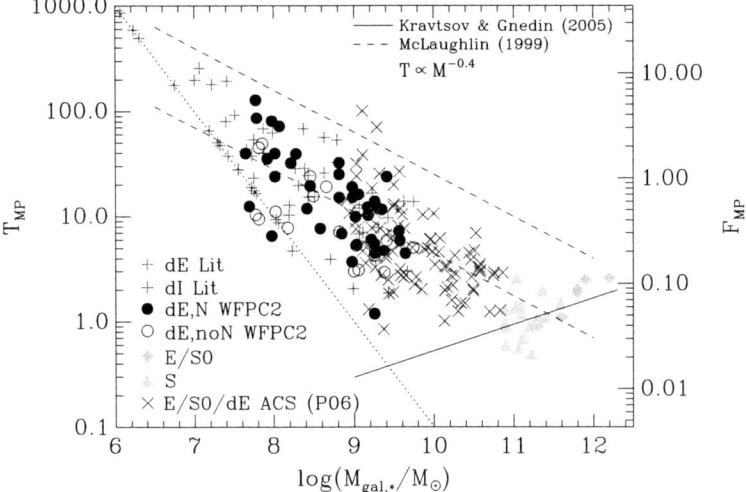

Fig. 5. The T parameter for metal-poor (MP) GC populations vs. galaxy stellar mass for dE galaxies from the WFPC2 dE Snapshot Survey and from the literature (see [19]). The equivalent mass fraction, F_{MP}, is shown on the right axis assuming a universal GC mass function. The *dotted line* on the *left* is the line of constant $N_{GC} = 1$. The *solid line* is the prediction of F_{MP} with galaxy mass for $\log(M_{G,*}) > 10.5$ from [14]. The *dashed lines* have a slope of -0.4, from the SNe-driven wind models of [17]. The *lower dashed line* is the prediction from [17] for $M/L_V = 5$ for the galaxies. The *upper dashed line* is an approximation of the upper envelope to the points.

6 Conclusions

Substantial progress in understanding the GCSs of dwarf galaxies has been made in recent years due to large imaging surveys in different environments with *HST*, new spectroscopic work using 8-m class telescope, and the inclusion of globular clusters in cosmological galaxy formation models.

More work is still needed on photometry and spectroscopy of GCs in dIs in order to improve the comparisons with the results on dEs. Also, there is much to be learned from the kinematics of GCs that could not be discussed here. GCs will continue to be a fundamental tool for understanding the formation of dwarf galaxies and testing theories of galaxy formation in general.

Acknowledgements. This work was supported by the Gemini Observatory, which is operated by the Association of Universities for Research in Astronomy, Inc., on behalf of the international Gemini partnership of Argentina, Australia, Brazil, Canada, Chile, the United Kingdom, and the United States of America.

References

1. H. Baumgardt: A&A, **330**, 480 (1998)
2. M. A. Beasley, et al.: AJ, **131**, 814 (2006)
3. K. Bekki, et al.: ApJL, **642**, L133 (2006)
4. C. J. Conselice: ApJ, **639**, 120 (2006)
5. M. Di Criscienzo, et al.: MNRAS, **365**, 1357 (2006)
6. P. R. Durrell, et al.: AJ, **112**, 972 (1996)
7. S. M. Fall & Q. Zhang: ApJ, **561**, 751 (2001)
8. D. A. Forbes: ApJL, **635**, L137 (2005)
9. M. Geha, et al.: AJ, **126**, 1794 (2003)
10. W. E. Harris: AJ, **91**, 822 (1986)
11. X. Hernandez & G. Gilmore: MNRAS, **297**, 517 (1998)
12. A. Jordán, et al.: in these proceedings (2006)
13. D. Kereš, et al.: MNRAS, **363**, 2 (2005)
14. A. V. Kravtsov & O. Y. Gnedin: ApJ, **623**, 650 (2005)
15. J. M. Lotz, et al.: ApJ, **552**, 572 (2001)
16. J. M. Lotz, B. W. Miller, & H. C. Ferguson: ApJ, **613**, 262 (2004)
17. D. E. McLaughlin: AJ, **117**, 2398 (1999)
18. B. W. Miller, et al.: in preparation (2006)
19. B. W. Miller & J. M. Lotz: AJ, **670**, 1074 (2006)
20. D. Minniti, et al.: A&A, **312**, 49 (1996)
21. K. S. Oh & D. N. C. Lin: ApJ, **543**, 620 (2000)
22. E. W. Peng, et al.: ApJ, **639**, 95 (2006)
23. T. H. Puzia, T. H., et al.: AJ, **120**, 777 (2000)
24. M. E. Sharina, et al.: A&A, **442**, 85 (2005)
25. J. Strader, et al.: AJ, **127**, 3431 (2004)
26. D. Thomas, et al.: MNRAS, **339**, 897 (2003)
27. T. Treu, et al.: ApJ, **622**, L5 (2005)
28. S. van den Bergh: AJ, **131**, 304 (2006)
29. E. Vesperini: MNRAS, **322**, 247 (2001)
30. S. E. Zepf & K. M. Ashman: MNRAS, **264**, 611 (1993)

Globular Clusters in Dwarf and Giant Galaxies

Sidney van den Bergh

Dominion Astrophysical Observatory, Herzberg Institute of Astrophysics,
National Research Council of Canada, 5071 West Saanich Road, Victoria, BC,
V9E 2E7, Canada sidney.vandenbergh@nrc.gc.ca

Abstract. The luminosity distribution of globular clusters shows a dramatic dependence on parent galaxy luminosity. Dwarf galaxies contain far more faint globulars than do luminous galaxies. This difference is significant at the 99.7% level. On the other hand the luminosity distribution of globular clusters in dwarf galaxies does not appear to depend strongly on their host's morphological type. The dichotomy of globular cluster masses occurs at a host galaxy mass of $\sim 4 \times 10^8$ $M_\odot$, which is almost two orders of magnitude lower than the onset of the dichotomy in globular color characteristics at $\sim 3 \times 10^{10}$ $M_\odot$ that was recently noted by Forbes.

1 Introduction

According to current ideas on galaxy formation dwarf galaxies and giant galaxies follow quite different evolutionary tracks. One might therefore expect these two types of galaxies to also have experienced very different histories of star and globular cluster formation. The first indication for such differences [18] was provided by the observation that the globulars associated with giant galaxies are systematically more metal- rich than those hosted by dwarf galaxies. It is the purpose of the present investigation to see if there are also systematic differences between the luminosities (masses) of the globular clusters associated with giant and dwarf galaxies. The main difficulty with such an investigation is that it is not easy to find galaxy samples for which the globular cluster data are complete down to sufficiently faint magnitude limits.

2 Data on Globular Clusters

A recent Hubble Space Telescope snapshot survey of nearby dwarf galaxies with $D < 4.0$ Mpc by Sharina, Puzia and Marakov [16] [hereinafter referred to as SPM] has provided a compilation of information on clusters which is essentially complete down to $M_v \sim -5.0$. Dwarf galaxies are mostly low in metals. One therefore expects the clusters within them to generally exhibit low reddening values. Most of the red clusters with $(V - I) > 0.70$ in the SPM catalog are therefore probably globular clusters, rather than reddened young blue clusters.

A listing of such red clusters in galaxies with $D < 4.0$ Mpc is given in Table 1. For more distant host galaxies the SPM data suffer from increasing incompleteness for faint clusters. Since most of the galaxies contained in the

Table 1. Globular clusters in galaxies with $D < 4.0$ Mpc that are fainter than $M_v = -18.9$. a: Probably a galaxy nucleus

Galaxy	T	M_V	Cluster	M_V
LMC	9	-18.5	N 1466	-7.26
			N 1754	-7.09
			N 1786	-7.70
			N 1835	-8.30
			N 1841	-6.82
			N 1898	-7.49
			N 1916	-8.24
			N 1928	-6.06:
			N 1939	-6.85:
			N 2005	-7.40
			N 2019	-7.75
			N 2210	-7.51
			N 2257	-7.25
			Hodge 11	-7.45
			Retic	-5.22
			ESO121	-4.37
SMC	9	-17.1	N 121	-7.89
NGC 205*	-5	-16.4	Hu I	-7.62
			Hu II	-7.82
			Hu IV	-6.02
			Hu VI	-6.62
			Hu VII	-6.52
			Hu VIII	-7.92
NGC 6822*	10	-16.0	Hu VII	-8.5
			SC 1	-7.26
			SC 2	-5.6
NGC 185*	-5	-15.6	I	-6.31
			II	-4.98
			III	-7.91
			IV	...
			V	-7.97
			VI	...
NGC 147*	-5	-15.1	No. 1	-6.97
			No. 2	-6.25
			No. 3	-7.65
			No. 4	-3.53
WLM*	10	-14.4	...	-8.33
Ho. IX	10	-13.8	3-1565	-5.31
			3-1932	-6.61
			3-2373	-6.04

Table 1. continued

Galaxy	T	M_V	Cluster	M_V
Sagit.	-5	-13.8	N 4147	-6.16
			N 6715	-10.01^a
			Ter. 7	-5.05
			Arp 2	-5.29
			Ter. 8	-5.05
			Pal. 12	-4.48
DDO 53	10	-13.74	3-1120	-5.88
KDG 61	-1	-13.58	3-1325	-7.55
Fornax	-5	-13.1	No. 1	-5.32
			No. 2	-7.03
			NGC 1049	-7.66
			No. 4	-6.83
			No. 5	-6.82
KDG 63	-3	-12.82	3-1168	-7.09
DDO 78	-3	-12.75	1-167	-7.23
			3-1082	-8.81
DDO 113	10	-12.67	2-579	-5.60
			4-690	-5.27
KK 211	-5	-12.58	3-917	-6.86
			3-149	-7.82
KK 27	-3	-12.32	4-721	-6.36
KK 77	-3	-12.21	4-939	-5.01
			4-1162	-5.37
			4-1165	-5.69
KK221	-3	-11.96	2-608	-8.04
			2-883	-7.07
			2-966	-9.80
			2-1090	-7.77
			3-1062	-6.10
UA 438	10	-11.94	3-2004	-8.67
			3-3325	-5.96
BK 6N	-3	-11.93	2-524	-5.40
			4-789	-5.60
E 540	-3	-11.84	4-1183	-5.37
E 294	-3	-11.40	3-1104	-5.32
KDG 73	10	-11.31	2-378	-5.75

SPM catalog are situated in small clusters they are located in environments similar to that of our own Local Group [14,19]. It therefore seems reasonable to combine the cluster data collected by SPM with those on globular clusters in the Local Group. For those globular clusters in Local Group dwarf galaxies that are situated within 150 kpc the data on the globular clusters luminosity distribution are probably also complete down to $M_v \sim -5.0$. The data [15] for the globular clusters in the LMC, the SMC, the Fornax dwarf and the Sagittarius dwarf are listed in Table 1. Additional information, which may not be quite as complete at faint magnitudes, is available for the globular clusters in the more distant Local Group dwarfs NGC 205 ($D = 760$ kpc), NGC 6822 ($D = 500$ kpc), NGC 185 ($D = 660$ kpc), NGC 147 ($D = 660$ kpc) and the Wolf-Lundmark-Melotte system ($D = 925$ kpc). Data on the luminosities of the globular clusters in these galaxies was taken from Harris (1991), [11, 12, 1, 10, 4, 22, 17, 3]. The uncertainties and ambiguities in the identifications of the faintest globular ! clusters in distant Local Group galaxies are well summarized in the appendix to the paper by Ford, Jacoby and Jenner [8]. Additional evidence for the possible incompleteness of the data sample on faint globular clusters in the most distant Local Group galaxies is provided by the very recent discovery of two additional globular cluster suspects in the outer regions of NGC 6822 [13]. To remind the reader of Table 1 of this possible incompleteness entries for the parent galaxies of Local Group galaxies with $D > 150$ kpc have been marked by an asterisk.

3 Comparison of the Globular Clusters in Dwarf and in Giant Galaxies

Available data [6,9] suggest that most giant galaxies have similar log-normal cluster luminosity distributions that peak at $M_v \sim -7.6$. The question that we wish to ask is: Do the globular cluster systems associated with dwarf galaxies also have a similar luminosity distribution? In attempting to answer this question the luminosity distribution of Galactic globulars associated with the main body of the Galaxy, i.e. those having $R_{gc} < 15$ kpc [21, 15], for which data are most complete and reliable, will be used as the template for the cluster systems hosted by giant galaxies. This template can then be compared to the luminosity distributions of the globular clusters hosted by various classes of fainter parent galaxies.

Using the listing of van den Bergh and Mackey [21] for the combined data on the 16 globulars in the LMC (M_v = -18.5) and the single globular cluster in the SMC ($M_v = -17.1$) one finds that these objects have a mean absolute magnitude $< Mv >= -7.10 \pm 0.25$. A Kolmogorov-Smirnov test shows no significant difference between the luminosity distributions of the 17 globular clusters in the Magellanic Clouds and the 110 globulars with $R_{gc} < 15$ kpc that are hosted by the Galaxy. It is therefore tentatively concluded that the luminosity distribution of the globulars in the Magellanic Clouds is similar to that in typical giant galaxies.

Table 2 and Fig. 1 show a comparison between the luminosity distributions of the globular clusters associated with the main body of the Galaxy and that of the nearby globulars associated with dwarf galaxies having $M_v > -17.0$. The Figure shows that the Galactic globular clusters have an approximately log-normal luminosity distribution with a maximum between $M_v = -7.0$ and $M_v = -7.5$, whereas the globulars associated with dwarf galaxies appear to have a luminosity distribution that exhibits a monotonic increase down to

Table 2. Luminosity distribution of globular clusters. a: dwarf galaxy core?

M_V	Galaxy $R_{gc} < 15$ kpc	Dwarf galaxies $M_V > -17.0$	Dwarf galaxies $M_V > -14.5$
-10.25	1[a]	1[a]	1[a]
-9.75	1	1	1
-9.25	7	0	0
-8.75	8	3	2
-8.25	13	2	2
-7.75	18	10	4
-7.25	19	5	4
-6.75	19	7	4
-6.25	9	7	4
-5.75	5	7	6
-5.25	2	11	11
-4.75	3	1	0
-4.25	2	1	1
-3.75	3	1	0
total	110	57	40

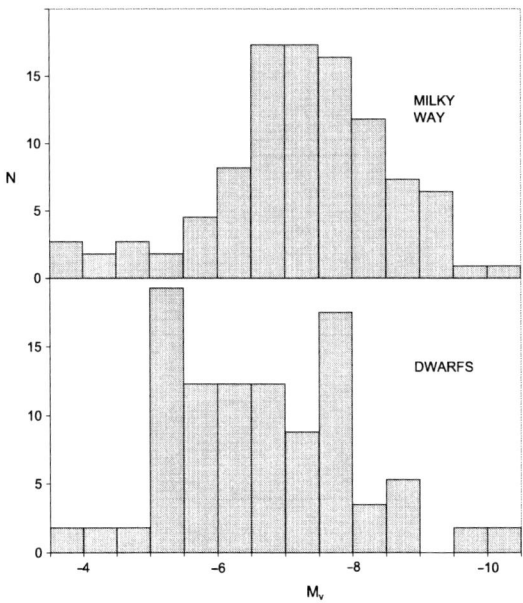

Fig. 1. Luminosity distributions of globular clusters. The figure shows that the globular clusters associated with giant Galaxy, which is a giant system, have an approximately log-normal distribution, whereas those hosted by dwarf galaxies fainter than $M_v = -17.0$ appear to have a luminosity distribution that rises up to the completeness limit of the data at $M_v \sim -5$.

the completeness limit of the data at $M_v \sim -5$. A Kolmogorov-Smirnov test on the data in Table 2 shows that there is only a 0.9% probability that the data for the Galactic clusters and for the globulars associated with all nearby dwarf galaxies were drawn from the same parent population. A similar test shows that the luminosity distributions of the globulars hosted by "bright dwarfs" with $-17.0 > M_v > -14.5$ differs from that of the "faint dwarfs" with $-14.5 > M_v > -11.0$ at the 5% significance level. The observed difference is in the sense that the luminosity distribution of globular clusters in faint dwarfs is more heavily weighted towards faint clusters than is that of the globular clusters hosted by giant galaxies. The data in Table 1 show that this effect is particularly pronounced for the cluster systems hosted by very faint parent systems having $M_v > -14.5$. A K-S test shows that there is only a 0.3% probability that the 40 globulars in such faint dwarfs were drawn from the same parent population as are the 110 globular clusters associated with the main body of the Galaxy. The luminosity distribution of the clusters hosted by dwarfs fainter than $M_v = -14.5$ peaks at $M_v \sim -5.25$, i. e. just above the completeness limit of the data at $M_v \sim -5.0$. This strongly suggests that the difference between the luminosity distributions of globular cluster systems hosted by dwarf and giant galaxies would be even more striking if cluster data could be extended to fainter completeness limits.

4 Discussion

It is interesting to ask if the characteristics of the globular cluster systems hosted by dwarf galaxies depend not only on host galaxy luminosity, but also on the morphological types of these hosts. In first approximation the dwarfs in Table 1 may be divided into two broad morphological types: (1) Early-type galaxies with de Vaucouleurs type $T < 0$, and (2) late-type galaxies with $T > 0$. A K-S test shows no significant difference between the luminosity distributions of the 44 globular clusters in early type galaxies and that of the 13 globulars hosted by late-type galaxies. It should, however, be emphasized that this conclusion is based on a relatively small data sample. It would clearly be of interest to extend this data base by carrying out HST snapshot surveys of larger samples of galaxies with distances smaller than 4 Mpc. Nevertheless the data that are already in hand clearly suggest that: (1) The luminosity distribution of globular clusters in dwarf galaxies differs dramatically from that in giants, and (2) there is no evidence that the luminosity distribution of the globular clusters in dwarf galaxies depends on the morphological types of their hosts. Observations by Chandar, Whitmore and Lee [2] suggest that M101, which appears to have a power-law luminosity distribution of globular clusters down to $M_v \sim -6$, may be an exception to the rule that giant galaxies contain few faint globulars.

The observation that faint globular clusters are common in dwarf galaxies, but rare in giants, might be accounted for by the fact that disk and bulge shocks will destroy clusters more efficiently in giants than in dwarfs. However, it is not yet possible to exclude the alternative possibility that real differences in star and cluster forming conditions in dwarf and giant galaxies were responsible for the observed differences between the luminosity distributions of globular clusters hosted by dwarf and giant galaxies. For example Dekel and Birnboim [5] have recently suggested that halos below a certain shock-heating mass will build star and cluster forming disks from cold streams, whereas the clusters in more massive halos might form from in hot flows. For an alternative suggestion about the early history of globular cluster formation the reader is referred to van den Bergh [20].

Recently Forbes [7] has drawn attention to the fact that globular cluster systems with host masses greater than 3×10^{10} $M_\odot$ tend to have globular clusters with bimodal color distributions, whereas the clusters surrounding less massive galaxies are embedded in unimodal blue globular cluster systems. The dichotomy in the globular cluster luminosities occurs at host galaxy masses of $\sim 4 \times 10^8$ $M_\odot$. This difference of almost two orders of magnitude in host galaxy masses shows that the effects discussed in the present paper, and those to which Forbes has drawn attention, are probably due to different physical causes. In particular it is noted that the unimodel blue LMC globular cluster system belongs to Forbes' low mass parent class, whereas the luminosity distribution of the LMC's globulars places it among the subset of galaxy hosts that have high galaxy luminosities and masses.

Acknowledgements. I am indebted to Narae Hwang for permission to include the results on two new globular clusters in NGC 6822 in the present paper. I also thank Avishai Dekel, Oleg Gnedin, and Dougal Mackey for helpful exchanges of e-mail.

References

1. Ables, H. D., & Ables, P. G. 1977, ApJS, 34, 245
2. Chandar, R., Whitmore, B., & Lee, M. G. 2004, ApJ, 611, 220
3. Da Costa, G. S. 2003, in New Horizons in Globular Cluster Astronomy (=ASP Conference Series No. 296), Eds. G. Piotto, G. Meylan, S. G. Djorgovski, and M. Riello (San Francisco: ASP), p. 545
4. Da Costa, G. S. & Mould, J. R. 1988, ApJ, 334, 159
5. Dekel, A. & Birnboim, Y. 2004, Astro-ph/0412300
6. Di Criscienzo, M., Caputo, F., Marconi, M. & Musella. I. 2006,MNRAS (in press = astro-ph/0511128)
7. Forbes, D. A. 2005, astro-ph/0511291
8. Ford, H. C., Jacoby, G., Jenner, D. C. 1977, ApJ 213, 18
9. Harris, W. E. 1991, ARA&A 29, 543
10. Harris, W. E. & Racine, R. 1979, ARAA, 17, 241
11. Hodge, P. W. 1974, PASP, 86, 289
12. Hodge, P. W. 1976, AJ, 81, 25

13. Huang, N. et al. 2005, in Near Field Cosmology with Dwarf Elliptical Galaxies = IAU Colloquium No. 198, Eds. H. Jergen, and B. Bingelli (=astro-ph/0510802)
14. Hubble, E. 1936, The Realm of the Nebulae (New Haven: Yale University Press), p. 124
15. Mackey, A. D. & van den Bergh, S. 2005, MNRAS, 360, 631
16. Sharina, M. E., Puzia, T. H. & Makarov, D. I. 2005, A&A (in press = astro-ph/0505624)
17. Strader, J., Brodie, J. P. & Huchra, J. P. 2003, MNRAS 339, 707
18. van den Bergh, S. 1975, ARAA, 13, 217
19. van den Bergh, S. 2000, The Galaxies of the Local Group (Cambridge: Cambridge University Press)
20. van den Bergh, S. 2001, ApJ, 559, L113
21. van den Bergh, S. & Mackey, A. D. 2004, MNRAS, 354, 713
22. Wyder, T. K., Hodge, P. W. & Zucker, D. B. 2000, PASP, 112, 1162

The Age-Metallicity Relation of the SMC

Andrea Kayser[1], Eva K. Grebel[1], Daniel R. Harbeck[2], Andrew A. Cole[3], Andreas Koch[1], John S. Gallagher[2], and Gary S. Da Costa[4]

[1] Astronomical Institut, U. Basel, Venusstr. 7, 4102 Binningen, Switzerland
    akayser@astro.unibas.ch, grebel@astro.unibas.ch
[2] Department of Astronomy, U. Wisconsin, 475 N. Carter St. Madison, WI 53706, USA
[3] Department of Astronomy, U. Minnesota, 116 Church St. SE, MN 55455, USA
[4] ANU, Mt Stromlo Observatory, Cotter Rd, Weston ACT 2611, Australia

1 Introduction

The Small Magellanic Cloud (SMC), as one of our nearest galactic neighbours, provides a very important laboratory for the study of galaxy evolution. Its proximity enables us to resolve stellar populations well below the oldest main sequence turn-offs (MSTO) and thus allows reliable age dating. Moreover the star cluster (SC) system of the SMC shows no sign of any substantial age gap, as found, e.g., in the LMC. In fact it is the only dwarf galaxy in the Local Group known to have formed and preserved populous SCs continuously over the past ∼12 Gyr. This provides a unique, closely spaced set of single-age, single-metallicity tracers.

In a former study, spectroscopic metallicities for 6 populous SMC clusters with ages from 3 to 12 Gyr were obtained by [4]. They found that the chemical evolution of the SMC was generally consistent with a simple "closed-box" model and a smoothly varying star-formation rate. However, this conclusion has been challenged by studies based on additional photometric data [7] suggesting bursty evolution, or even a bimodal age distribution [9].

All existing age-metallicity relation studies suffer from uncertainties due to either small number statistics or photometry based abundances. Moreover, deep high resolution HST imaging data are available for only a subset of SCs in the SMC. Thus all present age determinations are based on different studies of varying quality.

2 Data and Analysis

2.1 Data

In order to eliminate as many uncertainties as possible, we obtained homogeneous spectroscopy (VLT) and deep photometry (HST/ACS) for a large sample of SMC clusters. In this proceedings we will present the first results of the spectroscopic part of this project. Our spectroscopic sample comprises 12 SMC clusters, 2 of which are in common with the sample of [4] (see Fig. 1).

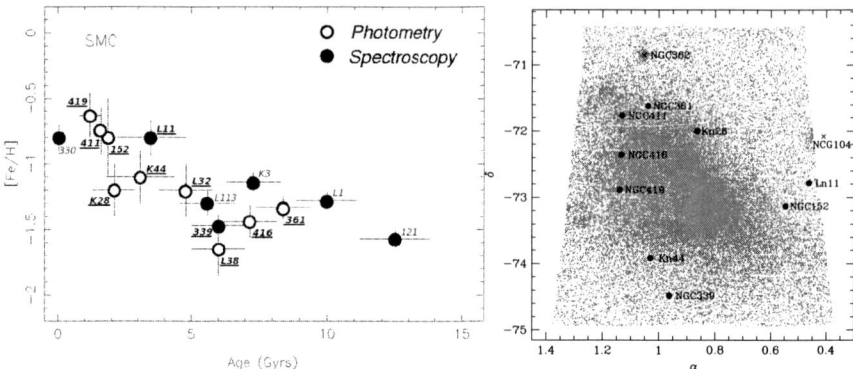

Fig. 1. (*left*) The age-metallicity relation of the SMC composed of star clusters with spectroscopic (*fill circles*) and photometric (*open circles*) metallicities. The names of the clusters of our sample are underlined (figure adopted from [5]). (*right*) Position of the clusters in the SMC with respect to the photometry by [10].

Additionally, we observed 3 Galactic globular clusters for the abundance calibration. The observations were carried out in October 2005 at the VLT at ESO/Paranal using the FORS2 spectrograph and the multi-object facility MXU. With the 1028z+29 (R≈ 3400) grism this configuration yields a spectral coverage of ≈ 1700 Å and a dispersion of ≈ 0.85 Å pixel^{-1} in the region of the near-infrared Ca II Triplet (CaT) The $\lambda\lambda$8498, 8542, 8662 Å lines were observed for 30–50 red giants (RGs) in each of the 15 clusters.

2.2 Ca Triplet Method

The use of the CaT index (ΣW) in terms of magnitude difference from the horizontal branch (HB) is one of the most widely applied techniques for the derivation of abundances in individual RGs [1,3]. The CaT index is formed by the linear combination of the pseudo-equivalent width of the 3 individual lines. As this index is strongly dependent on stellar T_{eff} and $\log g$, one corrects for these effects by using the linear relation between the magnitude difference from the HB, ($V_{\text{HB}}-V$) and ΣW for stars of the same cluster (i.e. metallicity). The resulting iso-metallicity lines in the ($V_{\text{HB}}-V$)-ΣW -plane are then used to derive absolute [Fe/H] abundances based on a reference [Fe/H]-scale [2]. The typical accuracy of this method is of the order of 0.2 dex [3] have shown that the CaT method is valid over an [Fe/H] range from -2.2 dex to -0.2 dex and is essentially age-independent for ages ranging from 13 to 2.5 Gyr. As our HST photometry is not fully available yet, we preliminarily use the photometric catalogue by [10] to estimate the ($V_{\text{HB}} - V$) values. This catalogue provides homogeneous stellar UBVI photometry of the central 18 deg^2 of the SMC. Unfortunately, some of our target clusters lie outside of this area.

3 Results

From their radial velocities and position in the CMD, we selected the candidate member RGs for each cluster. For all clusters we plotted the $(V_{HB} - V)$-ΣW -plane. The majority of the datapoints for the SMC clusters follow the slope of the iso-metallicity lines. As SCs can be presumed to be single metallicity objects without any substantial abundance spread in Fe and Ca, we assume that the very few outliers in this diagram are most likely non-cluster-members. They were ignored in the further analysis. As an example we present the results for 4 clusters of our sample in Fig. 2.

Using the [Fe/H] calibration by [3] and adopted ages from different sources, we can update the age-metallicity relation by [4]. In Fig. 3 (left)

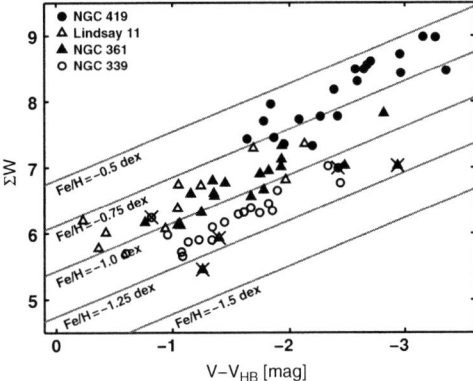

Fig. 2. CaT index vs. $(V_{HB} - V)$ for individual RGs in four of our SMC clusters. Stars of different clusters are marked by different symbols, probable non-cluster-members are crossed out. Calibration lines in this plot have been taken from [3].

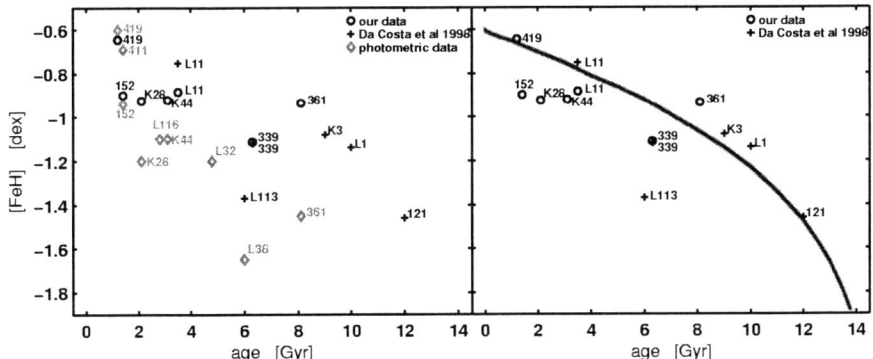

Fig. 3. (*left*) The age-metallicity data of our study are compared with those of the former study by [4] and photometric results. (*right*) Comparison of the spectroscopic data with the closed box model of continuous star formation (*solid line*) by [4].

we compare our results with the spectroscopic abundances by [4] and recent photometric results [6–8]. We see excellent agreement with the results of [4] for NGC339. The results for L11 differ by ≈ 0.13 dex, which is within the error range given by [4]. The comparison with the photometric metallicities gives a good agreement only for the 2 youngest clusters in our sample.

We compare the age-metallicity data of our study and of [4] with a simple closed box model of star formation. Fig. 3 shows that the general trend is reproduced well by this model. Nevertheless, the flat plateau in age seen in our data between ~ 2 and 4 Gyr, combined with the steep rise towards the younger end suggests that the star formation history of the SMC is probably more complicated. The inflow of unenriched gas may have played an important role. We are awaiting the results for NGC411, L116 and L32, which lie in this region of the diagram, to see whether this pattern in the age-metallicity relation is confirmed.

The next step will be the measurement of [Fe/H] based on CaT ΣW determinations for the remaining 5 clusters in our sample. In total, we will have reliable spectroscopic metallicities for 10 new SMC clusters. Furthermore, HST photometry will be available very shortly. This will enormously diminish the errors due to uncertainties and inhomogeneities in the age determination. With these results we will obtain a well-sampled, well-defined age-metallicity relation and quantify the abundance dispersion at a given age. The comparison of the derived age-metallicity relation with different theoretical models (closed box, leaky box, infall) will greatly improve our knowledge about the star formation history and chemical evolution of the SMC.

Acknowledgements. A.K., E.K.G. and A.K. were supported by the Swiss National Science Foundation through the grant 200020-105260. J.S.G. acknowledges partial research support from NSF AST-9803018 to the University of Wisconsin.

References

1. Armandroff, T. E. and Da Costa, G. S.: AJ **101**, 1329 (1991)
2. Carretta, E. and Gratton, R. G.: A&AS **121**, 95 (1997)
3. Cole et al.: MNRAS **347**, 367 (2004)
4. Da Costa, G. S. and Hatzidimitriou, D.: AJ **115**, 1934 (1998)
5. Da Costa, G. S.: IAU Symp. **207**, 83 (2002)
6. de Freitas Pacheco, J. A., Barbuy, B. and Idiart, T.: A&A **332**, 19 (1998)
7. Mighell, K. J., Sarajedini, A. and French, R. S.: AJ **116**, 2395 (1998)
8. Piatti, A. E. et al.: MNRAS **325**, 792 (2001)
9. Rich, R. M., Shara, M., Fall, S. M. and Zurek, D.: AJ **119**, 197 (2000)
10. Zaritsky, D. et al.: AJ **123**, 855 (2002)

Integrated Spectroscopic Analysis of Galactic and Small Magellanic Cloud Clusters

A.V. Ahumada[1], J.J. Clariá[1], and E. Bica[2]

[1] Observatorio Astronómico de Córdoba, Argentina andrea@oac.uncor.edu
[2] Instituto de Física, UFRGS, Porto Alegre, Brasil bica@if.ufrgs.br

Integrated spectra of 42 Galactic open clusters and 24 SMC star clusters were obtained in the (3600–7000) Å range using the CASLEO (Argentina) 2.15 m and CTIO (Chile) 1.5 m telescopes. The method used to determine ages and reddening of the clusters consists of the following steps: (1) Estimation of the cluster age from equivalent widths of the Balmer lines. This age is practically independent of the reddening. (2) Choice of the template whose spectral features better resemble those of the observed spectrum. This choice was made by using the libraries of template spectra which were available at the moment of making use of this methodology. In a first approach, the age inferred by the previous method was adopted. (3) Variation of the reddening of the observed spectrum until obtaining the best match to the chosen template. The reddening corrections were done using the normal reddening law.

Except ESO429-SC2, the remaining Galactic open clusters are located within two 90° sectors centered at $l = 347°$ and $l = 257°$, respectively. Thirty three out of the 42 studied open clusters have not been previously studied so that their fundamental parameters here determined turn out to be the first of their kind. Three SMC clusters (L114, NGC 416 and K3) are found to be quite old, with ages equal or larger than 4.5 Gyr, while NGC 121 is the only genuine globular cluster known in the SMC with an age of about 12 Gyr. These four clusters as well as NGC 411 and K5 are all well-known objects so that they can be used as templates corresponding to their respective ages. The resulting parameters for the Galactic and SMC clusters together with a detailed analysis and discussion of the individual spectra are given in [1]. A comparison of the properties of the Galactic open clusters here studied to those of well-known clusters located in the above mentioned sectors, shows that, unless major star forming events had occurred in the Galactic disk in the last 100 Myr or so, the present results would favour an important dissolution rate of star clusters in the above mentioned Galactic sectors.

Twenty two new Galactic templates were created, characterized by a high S/N ratio and a good temporal resolution. Twenty out of these 22 templates were published by [2] (PBCSA). The two remaining new templates correspond to the age groups of (4–5) Myr (Fig. 1, left panel) and 30 Myr (Fig. 1, right panel), respectively. Since the quality of the PBCSA's 20 Myr template contrasts with that of the remaining PBCSA templates, it was redefined by

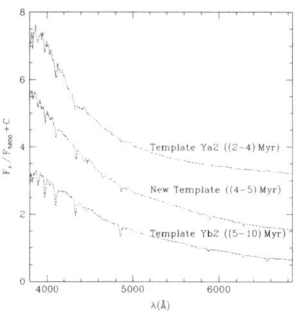 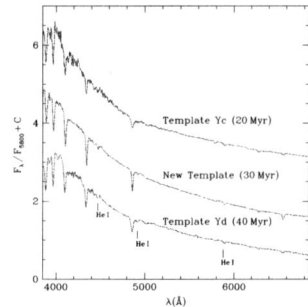

Fig. 1. *Left panel*: Galactic template spectra for the age groups of (5–10) Myr (*bottom*), (4–5) Myr (*middle*) and (2-4) Myr (*up*). *Right panel*: Galactic template spectra for the age groups of 40 Myr (*bottom*), 30 Myr (*middle*) and 20 Myr (*up*).

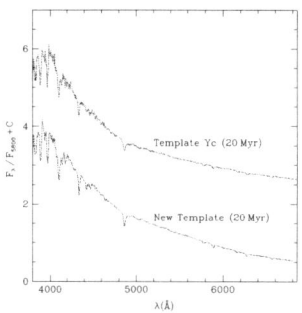 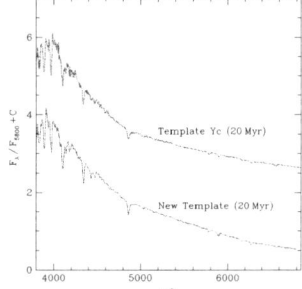

Fig. 2. *Left panel*: Old 20 Myr galactic template of PBCSA (*up*) and new 20 Myr (*bottom*). *Right panel*: Old (3–4) Gyr galactic template of PBCSA (*up*) and new (3-4) Gyr template (*bottom*).

averaging only the spectra of Hogg 15 and BH 217 (Fig. 2, left panel), instead of averaging also the spectra of BH 245, NGC 6318 (low S/N ratio) and Ruprecht 119 (H_α emission). Finally, since the PBCSA's (3-4) Gyr template has a low S/N ratio, it was redefined by averaging the spectra of ESO93-SC8, NGC 6253 and Ruprecht 2, previously corrected by reddening (Fig. 2, right panel).

References

1. Ahumada, A.V., 2004, Tesis Doctoral, FaMAF, Univ. Nacional de Córdoba.
2. Piatti, A.E., Bica et al., 2002, MNRAS, 335, 233 (PBCSA).

Variable Stars in the Globular Clusters and in the Field of the Fornax dSph Galaxy

C. Greco[1], G. Clementini[1], E.V. Held[2], E. Poretti[3], M. Catelan[4], L. Dell'Arciprete[3], M. Gullieuszik[2], M. Maio[1], L. Rizzi[5], H.A. Smith[6], B.J. Pritzl[7], A. Rest[8], and N. De Lee[6]

[1] INAF- O.A.Bologna, Via Ranzani, 1, 40127, Bologna, Italy
claudia.greco@bo.astro.it
[2] INAF- O.A.Padova, Vicolo dell'Osservatorio 5, 35122, Padova, Italy
[3] INAF- O.A.Brera, Via E. Bianchi 46, 23807 Merate, Italy
[4] Departamento de Astronomía y Astrofísica, Pontificia Universidad Católica de Chile, Avenida Vicuña Mackenna 4860, 782-0436 Macul, Santiago, Chile
[5] Institute for Astronomy, University of Hawaii, 2680 Woodlawn Drive, Honolulu, HI, USA
[6] Department of Physics and Astronomy, Michigan State University, East Lansing, MI 48824-2320, USA
[7] Macalester College, 1600 Grand Avenue, Saint Paul, MN 55105, USA
[8] Cerro Tololo Inter-American Observatory, Casilla 603, La Serena, Chile

1 Fornax Globular Cluster System

Fornax is a dwarf spheroidal galaxy (dSph) placed at about 140 kpc from the Milky Way (MW). Along with Sagittarius, it is the only dSph known to contain globular clusters (GCs). The Fornax GC system is made up of 5 GCs [3–5].

Galactic GCs seem to follow the Oosterhoff dichotomy [7], a sharp division into two distinct types according to the mean periods of the *ab*-type RR Lyrae stars (OoI: $\langle P_{ab} \rangle \sim 0.55$ d, OoII: $\langle P_{ab} \rangle \sim 0.65$ d). Do variable stars in Fornax GCs and in the galaxy field conform to the Oosterhoff dichotomy as observed in the MW? We present the first variability study of For 4, the cluster in the central region of the Fornax dSph, and results from well-sampled light curves of the variable stars in 3 other clusters, namely: For 2, 3 and 5. Our data allow us to firmly place the Fornax GCs in the Oosterhoff plane, and supersede a previous study of For 2, 3 and 5, based on HST archival data, by Mackey and Gilmore [6].

2 Observations and Results

Time-series B,V photometry of Fornax GCs 2, 3, 4 and 5 was collected with the 2.2 m ESO-MPI and the 6.5 m Magellan/Clay telescopes and combined with HST archival data. Variable stars were identified with the Image Subtraction Technique [1]. Photometry was performed with DAOPHOT-ALLSTAR-ALLFRAME [8]. We obtained light curves and derived periods

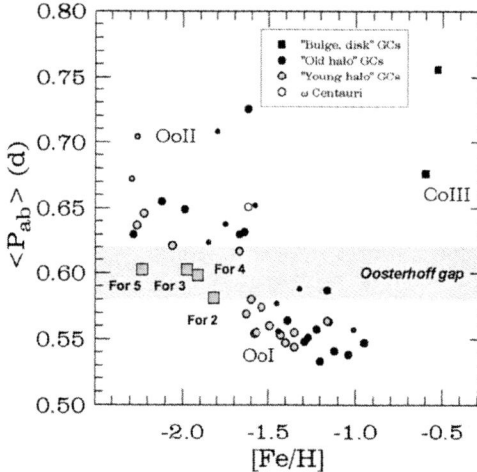

Fig. 1. Positions occupied in the Oosterhoff plane by different GC systems, as discussed in [2]. Squares represent Fornax GCs. Unlike most Galactic GCs, they seem to be either Oosterhoff-intermediate or lie at the edges of the Oosterhoff gap.

Table 1. For each cluster, number of RR Lyrae stars, Dwarf Cepheids and Anomalous Cepheids, mean period of the ab- and c-type RR Lyrae stars (in days), metallicity and number of observations

Cluster	RR Lyrae	DCs	ACs	$\langle P_{ab} \rangle$	$\langle P_c \rangle$	[Fe/H]	B	V	I
For 2	32 (13ab, 18c, 1d)	–	–	0.578	0.356	−1.79	6	32	16
For 3	29 (14ab, 5c, 10d)	2	2	0.606	0.358	−1.96	61	38	16
For 4	18 (14ab, 3c, 1d)	–	–	0.599	0.358	−1.90	19	59	–
For 5	13 (7ab, 5c, 1d)	–	–	0.602	0.354	−2.20	10	55	16
Fornax field (Chip 6)	82 (58ab, 14c, 10d)	9	21	0.595	0.361	−1.78	61	16	–

for the variables in all 4 GCs. A similar study was conducted in the galaxy field. Preliminary results are summarized in Table 1. In Fig. 1 we compare the Oosterhoff types determined for Fornax GCs with those of the MW GCs. Final results, both for Fornax clusters and field, will be published soon.

References

1. C. Alard: A&AS **144**, 363 (2000)
2. M. Catelan: Horizontal Branch Stars: Observations, Theory and Insights into the formation of the Galaxy In: *Resolved Stellar Populations*, in press, ed. by D. Valls-Gabaud, M. Chavez (ASP Conf. Ser. 2006) (astro-ph/0507464)
3. P. W. Hodge: AJ **66**, 83 (1961)
4. P. W. Hodge: ApJ **141**, 308 (1965)
5. P. W. Hodge: AJ **720**, 249 (1969)
6. A. D. Mackey & G. F. Gilmore: MNRAS **345**, 747 (2003)
7. P. Th. Oosterhoff: Observatory **62**, 104 (1939)
8. P. B. Stetson: PASP **106**, 250 (1994)

Physical Parameters of Intermediate-Age LMC Clusters from Modelling of HST CMDs

L.O. Kerber[1], B. Barbuy[1], and E. Brocato[2]

[1] IAG-Universidade de São Paulo, Rua do Matão 1226, Cidade Universitária, São Paulo, 05508-900 SP, Brazil. kerber@astro.iag.usp.br
[2] Osservatorio Astronomico di Collurania, Via M. Maggini, 64100 Teramo, Italy

Abstract. We analyzed HST/WFPC2 colour-magnitude diagrams (CMDs) from 12 populous Large Magellanic Cloud (LMC) stellar clusters with ages between ~ 1 Gyr and ~ 3 Gyr. These (V, B-V) CMDs, provided by Brocato et al. [1], are photometrically homogeneous and reach typically V$\sim$22. Accurate physical parameters (age, metallicity, distance modulus and reddening) were extracted for each cluster by comparing the observed CMDs with synthetic ones, applying statistical tools.

The LMC is a unique nearby case of a gas rich, star forming, irregular galaxy containing thousands of clusters with varying masses, ages and metallicities [2]. Therefore, this cluster system is a useful record of the history of star formation, chemical enrichment and dynamics of a quite distinct type of galaxy relative to the Galaxy. Although global features of this cluster system were revealed in the 80's and 90's, like age and metallicity gap [3,4], some important improvements are still in progress (Geisler et al., this conference). Besides, the physical parameters of each cluster (specially the age) can be better determined using sophisticated analysis of CMDs obtained with HST data, capable to resolve stars even in the cluster centre.

In the present work we analyzed 12 populous LMC clusters: NGC 1651, 1718, 1777, 1868, 2155, 2162, 2173, 2209, 2213, 2249, SL506, 663. We used (V, B-V) CMDs presented by Brocato et al. [1] and kindly provided by these authors. Figure 1a illustrates the CMD in a central region ($R < 2\ R_{core}$) of NGC 2213, revealing two clear features that are common in all clusters in this sample: the main-sequence (MS), reaching $V \sim 22$, and the red clump.

Our CMD modelling process [5] assumes that the cluster is a simple stellar population (SSP) where we introduce as model inputs the information about age (τ), metallicity (Z) (given by a Padova isochrone [7]), intrinsic distance modulus ($(m-M)_0$), reddening value ($E(B-V)$), Present Day Mass Function (PDMF) slope (α) and fraction of unresolved binaries (f_{bin}). The models explored a regular grid in the parameter space consistent with previous results found in the literature (see Table 2 from [6] for a good summary). We keep fixed $\alpha = 2.00$ (where Salpeter is 2.35) and $f_{bin} = 0.30$.

By means of model vs. data comparisons we determined the following physical parameters for each cluster: τ, Z, $(m-M)_0$ and $E(B-V)$. In these comparisons we confront both the MS fiducial line and the median red clump position using statistical tools, offering objective and robust criteria to find the best models. Figure 1b illustrates an artificial CMD for a best model. Comparing Fig. 1a and Fig. 1b one can see the agreement between the data and model MS fiducial lines and red clump positions.

Our preliminary results show that the best models show a satisfactory fit to the data. Besides, these models constrain well the physical parameters of each cluster, resulting in an age-metallicity relation for this cluster sample with lower spread than found in the literature (Fig. 1c).

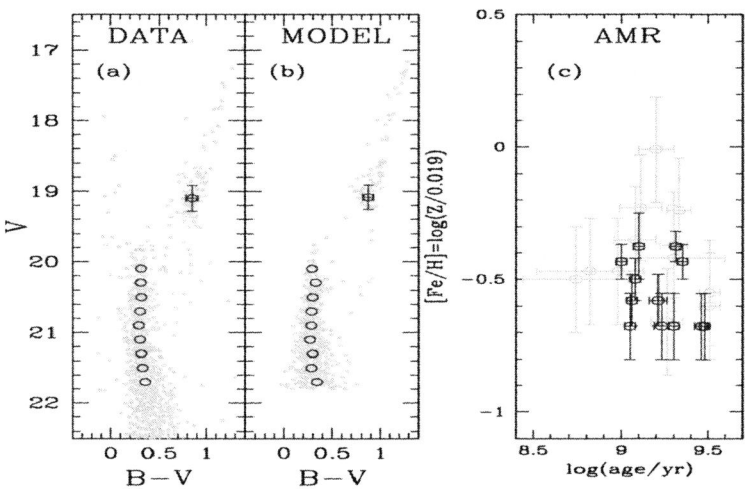

Fig. 1. Panel **a**: observed CMD for NGC 2213 and its MS fiducial line and red clump position; Panel **b**: the same for an artificial CMD; Panel **c**: age-metallicity relation derived from our results (black) and the one (grey) summarized by [6].

References

1. E. Brocato, E. Di Carlo, G. Menna: A&A **374**, 523 (2001)
2. E.W. Olszewski, N.B. Suntzeff, M. Mateo: ARA&A **34**, 511 (1996)
3. E.W. Olszewski, R.A. Schommer, N.B. Suntzeff, H. Harris: AJ **101**, 515 (1991)
4. D. Geisler, E. Bica, H. Dottori et al: AJ **114**, 1920 (1997)
5. L. Kerber, B. Santiago: A&A **435**, 77 (2005)
6. A. Mackey, G. Gilmore: MNRAS **338**, 85 (2003)
7. L. Girardi, G. Bertelli, A. Bressan et al.: A&A **391**, 195 (2002)

RGB Properties of the LMC/SMC Clusters in the Infrared

Radostin Kurtev[1], Valentin Ivanov[2], Jura Borissova[2], and Márcio Catelan[3], and Douglas Geisler[4]

[1] University of Valparaiso, Chile `radostin.kurtev@uv.cl`
[2] ESO, Chile `vivanov@eso.org, jborisso@eso.org`
[3] Catholic University, Santiago, Chile `mcatelan@astro.puc.cl`
[4] University of Concepcion, Chile `dgeisler@astro-udec.cl`

1 Introduction and Motivation

Red giant branch (RGB) stars are among the brightest red stars in stellar systems that are older than a few Gyrs. Globular clusters, with their single age and metallicity, are the ideal sites for calibrating the RGB parameters and the infrared regime has many advantages to do this. It can also help to break the age-metallicity degeneracy which plagues the optical. It has been demonstrated both empirically and theoretically that the slope of the RGB in a K vs. $(J-K)$ color-magnitude diagram is sensitive to the metallicity of the population. All these investigations concern the globular clusters in the Milky Way galaxy.

The slope of the RGB at a constant metallicity is little affected by age for ages between 10 and 14 Gyr. Between 1 and 10 Gyr, however, the RGB slope at fixed metallicity changes significantly and the RGB slope is comparably sensitive to both age and metallicity. To derive the general relation between slope of the RGB, [Fe/H] and age we have to extend the existing calibrations to the clusters in external galaxies, because there are no Galactic counterparts of the young yet metal-poor and massive Magellanic Cloud clusters. The globular clusters there span a wide range of metallicity and ages. We are observing both young and old MC clusters in order to study this relation. Observations list contains: ESO 121-3, NGC 1754, 1786, 1795, 1835, 1846, 1866, 1898, 1900, 2005, 2019, 2155, Hodge 60, Lindsay 113, NGC 339, and 361. Observations have been made with the Wide Field Infrared Camera (WFIRC) attached to the 2.5-m du Pont telescope at the Las Campanas observatory. Stellar photometry was carried out with IRAF DAOPHOT II. The final photometry list contains J and K_S magnitudes of 17665 stars with "formal" DAOPHOT errors less than 0.15 mag. 2MASS stars were used for the transformation to the standard JHK_S system.

2 The [Fe/H] vs. RGB Slope Diagram

The first step was to remove the fore- and background contamination, which constituted in some cases ∼30–40% of the stars in the designated cluster area.

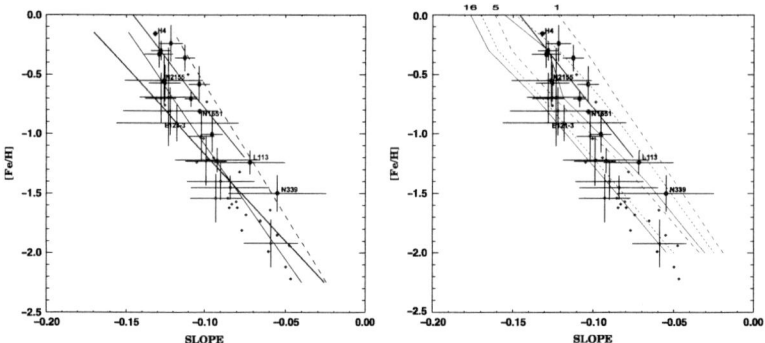

Fig. 1. The [Fe/H] vs. RGB slope relation for the LMC/SMC cluster sample.

We used a special Monte-Carlo technique, which simultaneously removes the background contamination and derives the RGB slope and the zero point on the corrected color-magnitude diagram. The RGB locus was defined after inspecting each CMD. Figure 1a represents the [Fe/H] vs. slope relation for the LMC/SMC clusters' sample. The clusters are shown as filled diamonds. The data for the clusters NGC 1651 and Hodge 4 are taken from Grocholski et al. [1]. In the figure are also overplotted the Milky Way (MW) globular clusters from the sample of Ivanov and Borissova [2] with open small diamonds and the sample of metal rich MW globular clusters of Tiede et al. [3] with big dots. The dashed line shows the relation for the sample of MW globular clusters of Valenti et al. [4], the thin line is the relation of Ivanov and Borissova [2], the short line is the relation of Tiede et al. [3], and the thick line is the relation of this sample of LMC/SMC clusters. As can be easily seen, the locus of the old LMC/SMC clusters on the [Fe/H] vs. slope diagram is the same as that of the MW old clusters. On Fig. 1b observational data are compared with the [Fe/H] vs. RGB slope relation for the clusters of different ages, predicted by theoretical isochrones. The youngest clusters are presented with big diamonds (Hodge 4, NGC 339, 1651, 2155, and Lindsay 113). There is a clear separation between young and old clusters on this diagram, in agreement with the theoretical predictions.

Acknowledgements. Support for MC is provided by Proyecto Fondecyt Regular #1030954.

References

1. Grocholski, A., Sarajedini, A., Olsen, K., Tiede, G. 2005, astro-ph/0506760
2. Ivanov, V., Borissova, J. 2002, A&A, 390, 937
3. Tiede, G., Martini, P., Frogel, J. 1997, AJ, 114, 694
4. Valenti, E., Ferraro, F., Origlia, L. 2004, MNRAS, 351, 1204

WLM-1: A Non-Rotating, Gravitationally Unperturbed, Highly Elliptical Extragalactic Globular Cluster

Andrew W. Stephens[1], Márcio Catelan[2], and Roxana P. Contreras[3]

[1] Gemini Observatory, 670 N. A'ohoku Place, Hilo, HI 96720
   astephens@gemini.edu
[2] Pontificia Universidad Católica de Chile, Departamento de Astronomía y Astrofísica Av. Vicuña Mackenna 4860, Santiago, Chile mcatelan@astro.puc.cl
[3] University of Missouri, Department of Physics & Astronomy, 503J Benton Hall, 8001 Natural Bridge Road, St. Louis, MO 63121 rpcontre@astro.puc.cl

Globular clusters have long been known to show (at times significant) deviations from perfect spherical symmetry, however, the origin of this nonsphericity has yet to be conclusively determined [7]. Several possible causes have been suggested, including stresses due to the presence of strong galactic tidal fields [6,3], gravothermal shocks during the passage of a cluster through the disk [5], velocity anisotropies, and internal rotation. While rotation has been the main proposed explanation, complicating factors such as interaction with the strong gravitational potential of the host galaxy have made it difficult for a consensus to be reached.

An ideal test, from an empirical viewpoint, of the impact of rotation on the shapes of globular clusters, is to measure the rotation velocity of a highly flattened, isolated extragalactic globular cluster, where there is little possibility that tidal effects and/or gravothermal shocks may be responsible for its flattening. The closest such example is the old (14.8 ± 0.6 Gyr [4]) globular cluster in the Wolf-Lundmark-Melotte (WLM) dwarf irregular galaxy (DDO 221). HST WFPC2 images show it to be highly elliptical [9], with a mean ellipticity of 0.17 ± 0.04 in the region $0.5'' < r < 5''$, which is comparable to what is found in our Galaxy for the most elliptical globular clusters [8].

We have obtained high-resolution VLT+UVES long-slit spectra of WLM-1 with two slit positions, one aligned with the major axis (PA= $6.8°$), and another aligned along the minor axis (PA= $96.8°$), with 3060 seconds of integration in each. The wavelength coverage is 3275–6650 Å, and with a $0.5''$ slit we obtain a spectral resolution of $\sim 60,000$.

We extracted spectra from different cluster annuli, and used Fourier cross-correlation [10] to look for evidence of rotation. By cross-correlating all possible permutations of the various extractions, and assuming that we should find equal but opposite velocities at the same radius on either side of the cluster, we are able to solve for the rotation profile which is most consistent with all extractions and cross correlations. This results in **no** statistically significant rotation along the line of sight at any distance along either axis of WLM-1.

We estimate the velocity dispersion from the width of the cross-correlation peak, using two high-resolution F8V stellar spectra obtained from the UVES Paranal Observatory Project [1] as templates. Cross-correlating the WLM-1 spectra with the templates we find that all spectral extractions are consistent with zero difference in the dispersions between the major and minor axes, and the integrated cluster spectra measured along the major and minor axes yields a velocity dispersion difference of only -0.4 ± 2.6 km/s.

In order to estimate how much rotation and/or velocity anisotropy is required to produce the observed ellipticity, we employ the tensor virial theorem [2] and assume that a globular cluster is an oblate spheroid, rotating around the z axis, and is seen edge-on. The resulting relation [9] illustrates the kinematical combinations available to explain the ellipticity of WLM-1.

At one extreme, the ellipticity of WLM-1 could be produced entirely by rotation. In this scenario there is no velocity anisotropy, and taking a conservative estimate of 10 km/s for the central velocity dispersion of the cluster the required rotation would be 5.7 km/s. However, as we find no measurable rotation ($v_{\rm rot} < \sim 0.1$ km/s), we exclude this possibility.

At the other extreme, the ellipticity of WLM-1 is due entirely to velocity anisotropy. It is possible to produce the observed ellipticity if $\sigma_z \approx 0.93\,\sigma_x$. Again taking the nominal velocity dispersion to be 10 km/s, this velocity anisotropy would only produce a difference of 0.7 km/s in velocity dispersion between the major and minor axes. Such a difference would be too small to be detected on the basis of our data.

Thus neither cluster rotation nor galactic tides can be responsible for the flattened morphology of WLM-1. We argue that the required velocity dispersion anisotropy between the semi-major and semi-minor axes that would be required to account for the observed flattening is relatively small, of order 1 km/s. Even though our errors preclude us from conclusively establishing that such a difference indeed exists, velocity anisotropy remains at present the most plausible explanation for the shape of this cluster.

References

1. Bagnulo, S., Jehin, E., Ledoux, C., Cabanac, R., Melo, C., & Gilmozzi, R. Messenger, **114**, 10 (2003)
2. Binney, J., & Tremaine, S. Galactic Dynamics (Princeton Univ. Press 1987)
3. Combes, F., Leon, S., & Meylan, G. A&A, **352**, 149 (1999)
4. Hodge, P. W., Dolphin, A. E., Smith, T. R., & Mateo, M. ApJ, **521**, 577 (1999)
5. Kontizas, E., Kontizas, M., Sedmak, G., & Smareglia, R. AJ, **98**, 590 (1989)
6. Longaretti, P.-Y., & Lagoute, C. A&A, **319**, 839 (1997)
7. Meylan, G., & Heggie, D. C. A&AR, **8**, 1 (1997)
8. Navarro, C., Catelan, M., & Stephens, A. W. in preparation (2006)
9. Stephens, A. W., Catelan, M., & Contreras, R. P. AJ, **131**, 1426 (2006)
10. Tonry, J., & Davis, M. AJ, **84**, 1511 (1979)

Part V

Globular Cluster Systems in Spiral Galaxies

Star Clusters in M33 – Clues to Galaxy Formation and Evolution

Rupali Chandar

Johns Hopkins University, 3400 N. Charles St., Baltimore, MD 21218, USA
rupali@pha.jhu.edu

1 Introduction

The properties of ancient star cluster systems (e.g., velocities, ages, and chemical abundances) provide important constraints on the assembly history of galaxies. For example, the globular cluster (GC) system of the Galaxy has sub-components associated with the bulge, thick disk, and halo (e.g., [12]). Measured properties of these sub-systems paint a picture where the halo of our Galaxy formed through chaotic merging and accretion. The thick disk appears to have resulted from a major merger $\sim$ 9–10 Gyr ago which heated the pre-existing thin disk (see review in [21]). Given that clusters residing in the thin disk all appear to be younger than those associated with the halo, bulge, and thick disk (e.g., [1]), it appears that the existing thin disk formed after these components. Thus, the properties of ancient star clusters are fundamental markers for establishing the order in which various galaxian components assembled.

While on-going studies of the GC system in M31 will provide additional, much needed information for understanding the formation of massive spiral galaxies, the third and final spiral in the Local Group, M33, is a lower mass system which has received far less attention. At a distance of only $\sim$ 800 kpc, this late-type (Scd) system provides an important link between the earlier-type spirals in the Local Group and numerous later-type dwarf galaxies. M33 also lacks a bulge (e.g., [15]), leaving fewer structural components to study than earlier-type spirals. This simplified morphology and proximity make M33 an ideal target for understanding how a spiral galaxy is put together.

The main issues which are addressed in this contribution are:

- What are the overall properties of clusters in M33?
- What do the known properties of the ancient clusters in M33 tell us about how M33 assembled?
- Do the field stars and ancient clusters in M33 tell a consistent story?

2 What We Currently Know About the M33 Cluster System

2.1 Cluster Surveys

A number of ground-based surveys of the cluster system of M33 have been undertaken over the last half-century. For nearly 25 years, the work of Christian and Schommer [8,9] has provided the broad picture of the M33 cluster system, and their catalogs and photometry are still used regularly today. They established that the M33 cluster population has a large range of ages, and that these objects have properties similar to clusters in the Magellanic Clouds. Most relevant to this proceedings, they discovered 25 objects which have red BVI colors similar to those of Galactic globular clusters (GCs). While there have been more recent ground-based surveys of the M33 cluster system

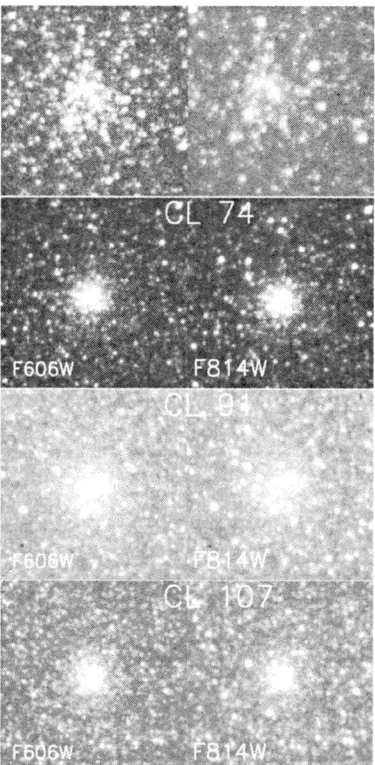

Fig. 1. Four clusters discovered by Chandar et al. [5] in HST/WFPC2 images which have colors typical of Galactic GCs, and which were not previously catalogued in ground-based surveys. The left panel shows the images in the F606W filter (broad-V band) and the right hand panel shows the clusters in the F814W (I band). Each cluster image is roughly 12 arcseconds on a side.

(e.g., [14]), to date no published cluster catalog covering the entire extent of M33 with high quality ground based observations exists.

M33 cluster studies conducted with the *Hubble Space Telescope* (*HST*) have shown the power of high resolution for detecting additional, previously unknown clusters (e.g., [4, 2]). From 55 *HST*/WPFC2 pointings, Chandar et al. [4, 5] discovered $\sim$ 115 new compact clusters which were not in the Christian & Schommer catalogs. From this sample, they found 25 new clusters with integrated colors similar to those of Galactic GCs (see Fig. 1 for a few examples), bringing the total number of ancient cluster candidates in this low mass spiral galaxy to $\sim$ 50. *HST* observations currently cover $\sim$ 30–40% of the optically luminous galaxy, so there is still a need for broad *HST* coverage of M33.

In total, the current catalogs have identified roughly 350 clusters in M33. Chandar et al. [4] estimated that there should be $\sim$ 800 clusters brighter than V=19. Of the 350 known, roughly 50 clusters have colors similar to ancient Galactic GCs.

While this proceedings focuses on the ancient clusters and what they tell us about the formation of M33, the more numerous young clusters also provide important information on the stellar content of spiral galaxies. Previous photometric and spectroscopic cluster age estimates suggest that M33 contains a population of intermediate age clusters, with ages roughly 0.5 to several Gyr; although the existence of clusters with ages $\sim$ 2–8 Gyrs has yet to be unambiguously confirmed. The properties of the clusters (for e.g., their ages, masses, and structural parameters) can be used to study the rate at which the clusters in M33 are born, and the rate at which they are destroyed. This type of study requires a more complete cluster sample than is currently available, but would be very interesting to compare with the similar results in more massive spirals (e.g., M101, M51, etc.).

2.2 General Properties of Ancient Clusters

Here, we focus on the known properties of the ancient cluster candidates. It should be noted that these have been selected to have integrated colors similar to ancient GCs in the Milky Way, and there are still many things that we need to learn about them. Chandar, Whitmore, and Lee [7] presented an updated luminosity function for the 50 ancient cluster candidates in M33. Although ancient cluster systems in many massive galaxies have been found to have GC luminosity functions which have very similar turnovers (e.g., [12]), the candidate M33 GCs appear to have an "excess" of fainter clusters, relative to a fiducial peak around $M_V \sim -7.5$. Making corrections for additional, currently undetected ancient cluster candidates which could be detected with future *HST* observations, Chandar et al. [7] estimate that M33 contains 75 ± 14 clusters with integrated colors similar to Galactic GCs. This would give a specific frequency of S_N of 1.3 ± 0.3. However, there are a number of explanations which can easily account for this difference in the number and luminosity function of the ancient clusters in M33 relative to what is thought to be "normal" based on massive spirals such as the Milky Way. Many of the ancient cluster candidates may actually be younger than 10 Gyr, and represent a mix of ages. Alternatively, if the specific frequency of M33 GCs is calculated by correcting for the excess faint clusters (i.e., assuming a gaussian distribution with a peak at $M_V = -7.4$ where the counts are essentially double the number of clusters brighter than this value), the S_N value is only 0.6 [7], very similar to that found for other spirals. This excess of faint clusters suggests that the survival chances may be better in M33 than in our Galaxy and other massive systems (e.g., since two-body evaporation dominates cluster destruction on longer timescales, and becomes even more efficient in the presence of an external gravitational potential, the weaker potential provided by M33 may improve a cluster's long-term ability to survive).

Ten ancient M33 cluster candidates were imaged by Sarajedini et al. [16,17] with HST/WFPC2, in order to investigate the metallicity and horizontal branch (HB) morphologies. They found that 8/10 clusters possess HB morphologies which are too red for their metallicity, when making a direct comparison with the properties of the GC system in the Milky Way (see Fig. 2), and suggested that one explanation is that the M33 clusters are younger (by several Gyrs) than their MW counterparts.

A possible counter-argument for younger ages was presented by Larsen et al. [13], who obtained high resolution spectroscopy for four of the Sarajedini et al. objects. They found that the M/L ratios are indistinguishable from those of MW GCs, which argues against an age difference. However, given the small number of clusters, it is possible that the objects in their sample all happen to be relatively ancient.

A. Moretti in her thesis has studied the metallicity ([Fe/H]) of the clusters in M33 from integrated spectroscopy. By comparing with SSP model predictions, she finds that most of the clusters are relatively metal-poor, with a peak near [Fe/H] of -1.5. This is similar to the peak [Fe/H] for the metal-poor system of Galactic GCs (e.g., see Fig. 8 in [10]).

There have been several studies of the kinematics of the M33 cluster system (e.g., [18,6]). These have revealed that blue, young clusters rotate nicely with the HI disk. However, the ancient clusters have a large velocity dispersion around this HI disk, and have no measurable rotation (although a weak rotational signature could be washed out due to the velocity uncertainties). Is this a thick disk or a halo component? Although not yet definitive, the lack of rotation and large velocity dispersion ~ 70 km/s are consistent with a pressure supported, dynamically hot halo component. Chandar et al. [6] also compared the velocities of ancient cluster candidates with different galaxy models, and found evidence that these objects reside in two different galactic components – a disk and halo, with the halo component dominating (Fig. 3).

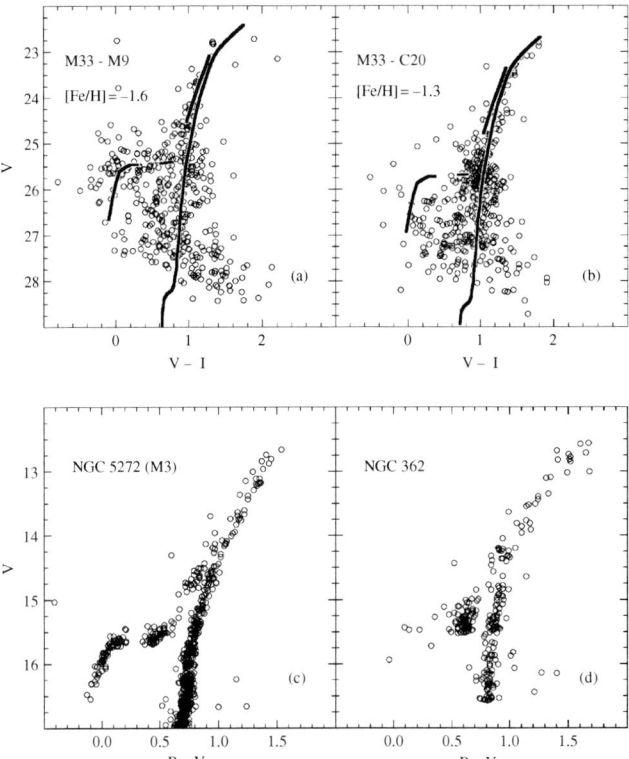

Fig. 2. From Sarajedini et al. [16]. An example of resolved CMDs for M33 clusters M9 and C20. The solid lines show the fiducial sequence of the Galactic GC M5. For comparison, the lower panels show CMDs of Galactic globular clusters M3, which is similar to M33-M9 in metal abundance, and NGC 362, which is similar to M33-C20 in metal abundance. While M33-M9 has a blue horizontal branch component (similar to that of M3), M33-C20 has a significantly redder horizontal branch than seen in Milky Way globular clusters at a similar metallicity.

What do these features of the M33 cluster system tell us about the early formation of M33? The fact that the GC candidates have large velocity dispersion and generally low metallicities is consistent with a halo component. If previous suggestions for a large age spread among these halo clusters can be confirmed, it would suggest that the formation of the M33 halo continued for significantly longer timescale than in the Milky Way. The best way to assess this possibility would be via very deep *HST* imaging, in order to directly derive the age from the main sequence turnoff in a representative sample of clusters A second possibility is to obtain high S/N, integrated spectroscopy of the GC candidates, particularly the ones which already have known HB morphologies. One ambiguity which exists in deriving ages from integrated spectroscopy is the possible existence of blue HB stars in more metal-rich clusters,

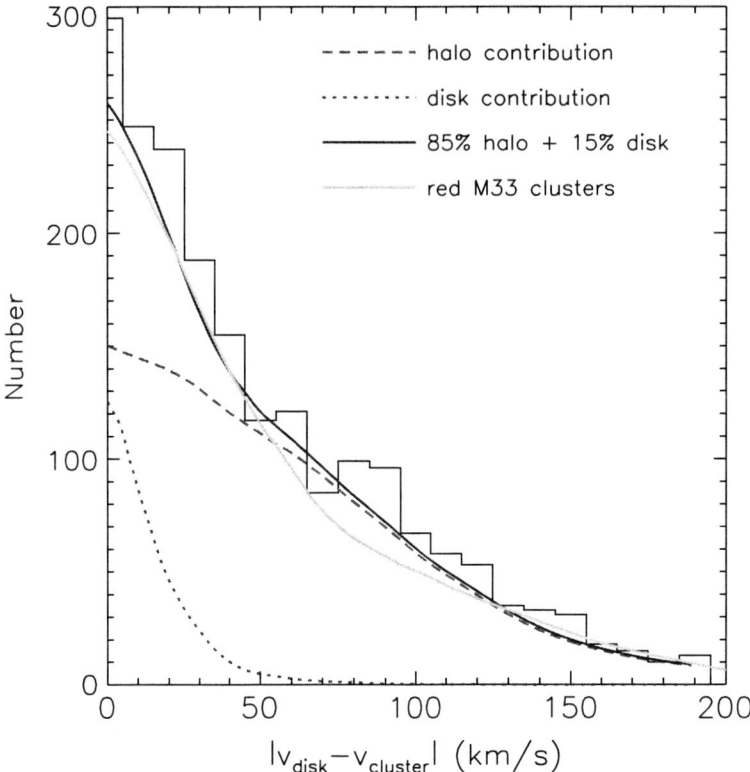

Fig. 3. This figure shows one representation of a simulation of M33's cluster system. The model includes both a halo and a disk contribution, and compares the summed components for the best fit M33 disk+halo model with a scaled version of the actual ancient cluster velocities from Chandar et al. [6]. Clearly, neither a pure disk nor a pure halo population fit the observations.

which can mimic younger ages due to the increase Balmer line strength. However, in M33 clusters we already know that many are relatively metal-poor, and we also have the HB morphologies which rule out any significant blue HB component. Therefore, high quality integrated spectroscopy should be able to establish the existence or absence of an age spread among the ancient clusters.

Knowing whether there is a large age spread among the GC candidates which deviate significantly from local HI disk motions will also help in deciphering whether the clusters reside in a halo or a thick disk. If there is a large age spread, this would boost the halo argument, since it is believed that thick disks typically result from the ingestion of a relatively massive satellite galaxy, during a single "event". In the Milky Way, the GCs associated with the thick disk have a very small age spread.

3 Reconciling Results from Clusters and Field Stars

In addition to GCs, ancient field stars provide important constraints on galaxy formation. In the Milky Way, the field stars and GCs paint a consistent picture, where both ancient stellar populations show little evidence for major merging in the Galactic halo more recently than about ~ 9 Gyr ago (e.g., [20, 22]), and both are consistent with the thick disk forming in a single accretion event that heated the pre-existing thin disk.

What do the field stars in M33 tell us about the formation of this galaxy? Given the faintness of individual stars, studies of the M33 field star population have lagged behind those of the GCs. However, there have recently been several works which studied the distributions of red giant stars in this galaxy.

Two recent studies [3, 19] find a mean [Fe/H]~ -1 for M33 field stars, based on the colors of RGB stars. Tiede et al. also find that out to a projected distance of 10 kpc, there is a gradient in [Fe/H], which is consistent with what is seen in the inner M33 disk. This suggests that M33 is disk-dominated out to this radius.

Surface photometry from wide-area photometry of the red giants show a very smooth and regular stellar density, with no apparent evidence for substructure, as is found in M31 (Ferguson et al. 2006). The RGB counts show a luminosity profile which has an exponential decline out to about 8 kpc; it steepens significantly beyond this radius. The steep profile continues out to ~ 14 kpc. Ferguson et al. [11] suggest that there is little contribution from a power-law stellar halo. The lack of substructure also implies that there has been very little recent accretion in this galaxy.

At first glance, the scenarios painted by the ancient clusters and field stars in M33 appear to be at odds, with the GCs having kinematics which deviate significantly from local HI motion, and the field stars having a profile which is most consistent with a disk component. How can these results be reconciled? There are at least two possibilities:

First, that M33 possesses a thick disk, which has a similar scale length to that of the thin disk. This would make it difficult to differentiate the two components based on star counts alone, and would also imply that the ancient star clusters belong to the thick disk component.

Alternatively, M33 may have a stellar halo which has a low surface brightness, and therefore does not significantly influence the field star counts from the disk component. In this situation, the GCs would also reside in a halo component.

4 Where Do We Go from Here

One important clue to telling difference between the two possibilities raised above is the age distribution of M33 clusters which have discrepant velocities with respect to the HI disk. If these clusters have a relatively large age spread, this would favor a halo component, since it is difficult to understand how a thick disk could form with such a large range of ages. Therefore, one of the most important pieces of information that we still need is direct constraints on the ages of ancient clusters in M33. At the same time, it is also very important to investigate the kinematics of individual red supergiant stars (currently being investigated by T. Smecker-Hane and collaborators), and compare the results with those for the GC system. With the ages, velocities, and chemical compositions of field stars and GCs in M33, significant progress can be made in understanding the formation and evolution of this low mass spiral galaxy.

References

1. K. Ashman & S. Zepf *Globular Cluster Systems* (Cambridge University Press, Cambridge)
2. L. R. Bedin et al.: A & A **444** 831 (2005)
3. R. S. Brooks, C. Wilson, & W. Harris: AJ **128** 237 (2004)
4. R. Chandar, L. Bianchi, & H. Ford: ApJS **122** 431 (1999)
5. R. Chandar, L. Bianchi, & H. Ford: A&A **366** 498 (2001)
6. R. Chandar, L. Bianchi, H. Ford, & A. Sarajedini: ApJ **564** 712 (2002)
7. R. Chandar, B. Whitmore, & M. G. Lee: ApJ **611** 220 (2004)
8. C. Christian & R. Schommer: ApJS **49** 405 (1982)
9. C. Christian & R. Schommer: AJ **95** 704 (1988)
10. P. Cote: AJ **118** 409 (1999)
11. A, Ferguson et al.: Resolving the Stellar Outskirts of M31 and M33: *Island Universes – Structure and Evolution of Disk Galaxies*, ed by R. de Jong (Springer: Dordrecht) (astro-ph/0601121)
12. W. E. Harris: ARA&A **29** 543 (1991)
13. S. Larsen, J. Brodie, A. Sarajedini, & J. Huchra: AJ **124** 2615 (2002)
14. B. J. Mochejska, J. Kaluzny, M. Krockenberger, D. D. Sasselov, & K. Z. Stanek: AcA **48**, 455 (1998)
15. M. Regan, & S. Vogel: ApJ **434**, 536 (1994)
16. A. Sarajedini, D. Geisler, P. Harding, & R. Schommer: ApJ **508** 37L (1998)
17. A. Sarajedini, D. Geisler, P. Harding, & R. Schommer: AJ **120** 2437 (2000)
18. R. A. Schommer, C. A. Christian, N. Caldwell, G. D. Bothun, & J. Huchra: AJ **101**, 873 (1991)
19. G. Tiede, A. Sarajedini, & M. Barker: AJ **128**, 224 (2004)
20. M. Unavane, R. Wyse, & G. Gilmore : MNRAS **278**, 727 (1996)
21. R. Wyse: The Merging History of the Milky Way Disk. In: *Galaxy Disks and Disk Galaxies*, vol 230, ed by J. Funes & E. Corsini (ASP Conference Series, San Francisco 2001) pp. 71–80
22. R. Wyse et al.: ApJ **639**, L13 (2006)

M31 and its Globular Clusters

Kathy Perrett

University of Toronto, 60 St. George Street, Toronto, Ontario, Canada M5S 3H8

1 Introduction and Early Studies

As our nearest bright spiral galaxy neighbor and the most prominent member of the Local Group, M31 has long provided astronomers with an excellent target for the investigation of the structure, dynamics, and stellar populations of external galaxies. Due to its proximity ($D \sim 770$ kpc), orientation ($i = 77.7°$) and relatively large globular cluster (GC) population size, the M31 globular cluster system (GCS) in particular provides an ideal target for probing the dynamical and chemical processes in galaxies.

In 1932, Hubble [23] identified 140 objects near M31 that had the appearance of "nebulous stars" on photographic plates obtained using the 100-inch telescope at Mount Wilson. On the basis of their number and spatial distribution, Hubble proposed that these globular nebulae were associated with M31 itself. Since Hubble's pioneering observations, many studies have contributed to the difficult task of taking inventory of the M31 GCS (see review by Hodge [22] and references therein). Subsequent searches have revealed an M31 GCS population that is more than three times the size of the Milky Way GCS.

As part of a major photographic survey of cluster candidates around M31, the Bologna Group published a compendium that included 353 "probable" and 150 "plausible" M31 GC candidates having $14 \lesssim V \lesssim 19.5$ mag within $3°$ of the M31 center. This *Bologna Catalogue* [20] and the recently updated and expanded *Revised Bologna Catalogue* – now containing nearly 700 known and candidate clusters with additional near-IR photometry – have become widely used by later studies that have continued the endeavour of adding positive identifications, photometry and spectroscopy for M31 GCs.

Combining new observations with existing data, Barmby et al. [4] produced a comprehensive photometric database of 435 M31 GCs and cluster candidates. In an effort to quantify the incompleteness of published catalogues, Barmby and Huchra [5] undertook a search for M31 GCs in archival HST images and found that the existing sample was nearly complete for $V < 18$ mag at galactocentric radii of $R_{gc} > 5'$. Based on these results, Barmby and Huchra [5] estimated a total population size of 459±69 M31 GCs.

The first major spectroscopic survey of the M31 GCS was performed by van den Bergh [41], who measured radial velocities for 43 clusters and discovered that they formed a rotating system. Kinematics and metallicities for ~ 150 M31 GCs were published based on the works of Huchra, Brodie and collaborators, who confirmed the rotation of the GCS. Huchra, Brodie and Kent [24] found that, in broad terms, the properties of the M31 GCS were "remarkably similar" to those of the Milky Way's GCS. They observed that the metal-rich clusters within $R \sim 7'$ (2 kpc) of the galaxy center form a rapidly-rotating $(100 - 200$ km s$^{-1})$ disk-like system, while the metal-poor clusters within this radius demonstrate no net rotation.

Perrett et al. [34] presented a sample of more than 200 M31 GC radial velocities good to ± 12 km s^{-1}, along with spectroscopic metallicity estimates. They showed that the metallicities are well-represented by a bimodal distribution with peaks at [Fe/H] ~ -1.4 and -0.5. Perrett et al. [34] also found that both the metal-rich and metal-poor cluster populations exhibit significant rotation. Overall, the correlation between kinematics and metallicity in the M31 GCS seems much less significant than in the Milky Way. Separating the various stellar populations within M31 is complicated by the inclination of the galaxy (77°) as the bulge, disk and halo components all overlap in both position and velocity parameter space. Photometric catalogues can suffer from contamination due to foreground stars; kinematics alone cannot uniquely identify the clusters, since there is overlap between the velocity distribution of the Milky Way's stellar disk and halo with that expected for the M31 GCS.

2 Thin Disk

Most of the GCs of the Milky Way are associated with the metal-poor, pressure-supported halo, with none known to belong to the thin disk. The only Milky Way clusters with thin-disk kinematics are the younger open clusters that are a few orders of magnitude less luminous than the M31 GCs. The majority of the M31 GCs have been shown to have significant rotational support [34]. Morrison et al. [33] find that the M31 GCS kinematics can be well explained by a two-component model: a thin, rapidly-rotating disk with a low velocity dispersion, plus a higher velocity dispersion rotating component with kinematic properties similar to the bulge. Within 2 kpc (projected) from the major axis, about half of the GCs seem to belong to the thin disk and extend across the full radius of M31's disk. While there may be clusters with kinematics appropriate to a non-rotating halo in M31, they do not seem to dominate the sample at least out to these radii. This is in stark contrast with the Milky Way where there is a prominent halo system of GCs with little or no mean rotation.

The presence of this distinct disk system with old- or low-metallicity clusters (down to -2.0 dex) would imply that M31 formed this component early in its history. Unfortunately, the majority of the current kinematic sample lies within a projected distance of 5 kpc of the major axis. A larger sample of high-precision velocities to greater radii above and below the major axis is required along with age estimates for the GCS members.

3 GC Ages

The majority of M31 GCs appear to be old, although recent studies have found evidence for sub-populations of intermediate-age and young clusters. Using optical and near-IR colors in combination with population synthesis models, Barmby and Huchra [4] suggested that the metal-rich M31 GCs in their sample may be as much as 4 Gyr younger than the metal-poor GCs (see also Jiang et al., [28]). As more detailed spectroscopy is obtained, evidence for multiple age populations in the M31 GCS are now emerging: very young (< 1 Gyr), young (1–2 Gyr), intermediate (3–6 Gyr), as well as the complement of old GCs.

The intermediate-age globular clusters (IAGCs) found by Beasley et al. [8] have ages in the range $\sim$ 2–6 Gyr, are of intermediate metallicity, and show no sign of net overall rotation. Puzia et al. [36] find that although their sample spans a range of ages, the majority are consistent with an old population of 10–12 Gyr (60% are older than 9 Gyr). They also find evidence for spatially-clumped GC sub-populations in M31, including a population of intermediate-age GCs between $\sim$ 5–8 Gyr, in agreement with Beasley et al. [8]. The M31 IAGCs in the Puzia et al. [36] sample are typically within $R_{gal} \sim 5$ kpc, implying an inner-disk/bulge population that formed somewhat later than the original epoch of galaxy formation.

Barmby et al. [3] identified $\sim$ 50 M31 GCs in their catalogue as having strong Balmer lines or comparatively very blue colors and flagged these as "young" clusters. Several recent studies have confirmed the presence of young M31 clusters that are $\lesssim 1$ Gyr old, younger than the any GCs known within the Milky Way [7,13,36,30,31].

A study of 67 very blue and likely young massive clusters in M31 from the Revised Bologna Catalogue was performed by Fusi Pecci et al. [21] focusing on objects similar in luminosity and shape to Milky Way GCs but having integrated colors that are significantly bluer. These blue luminous compact clusters (BLCCs) in M31 have estimated ages $\sim$ 200 Myr–2 Gyr and have kinematics consistent with the thin disk. Fusi Pecci et al. [21] go on to propose that at least 15% of the entire sample of confirmed M31 GCs are instead young BLCCs located in the outskirts of the thin disk. No corresponding population has been found in the Milky Way, although similar objects have been found in other local spiral galaxies [29] and in the LMC [13]. The exact age, spatial distribution, and true nature of these objects is still uncertain. Four of the BLCCs from Fusi Pecci et al. [21] have since been shown to be asterisms by Cohen et al. [15]; it seems that the M31 GCS catalogues still contain a significant number of non-GC interlopers which must be culled.

Detailed color-magnitude diagrams (CMDs) are needed to help disentangle the effects of metallicity and age on integrated cluster spectra. Direct age and metallicity estimates of individual star members are currently only available for one M31 GC. Brown et al. [11] present a deep CMD of the M31 globular cluster G312 with photometry reaching more than one mag below the main-sequence turnoff. They determine an age of $\approx 10^{+2.5}_{-1}$ Gyr for this cluster, 2–3 Gyr younger than Galactic GCs at similar metallicities.

4 Streams and Substructure

Recent discoveries of large-scale stellar substructure in the M31 halo lend weight to the satellite-accretion scenario as a viable galaxy formation mechanism for M31 (e.g. [26,18,46]). Ferguson et al. [19] also present evidence for large-scale stellar population inhomogeneities in the outer regions of M31. Contradictory evidence is presented by Brown et al. [12], who compare deep CMDs for stream and halo fields and find that although the regions are kinematically distinct, the fields are very similar in age and metallicity distribution. The stream characteristics observed by Brown et al. [12] are difficult to explain within the context of the dwarf-accretion scenario.

M31 is host to approximately 15 satellites of which the most luminous are the dwarf ellipticals NGC 205 and M32. M32 lacks GCs, although from its luminosity one would expect ~ 20 [42]. It is possible that M32's GCs may have already been stripped and become incorporated into the M31 GCS. N-body simulations by Perrett et al. [35] reveal that it is possible to identify some fraction of a disrupted satellite's initial GCS based on grouping in projected position and radial velocity, even for timescales of $\gtrsim 1$ Gyr after the initial encounter. In addition to the observed stellar substructure, tentative evidence has also been found for kinematic and chemical sub-clustering within the GC population of the Andromeda Galaxy [2,39,35].

5 A Milky Way Comparison

M31 and the Milky Way are of similar Hubble types and masses, and likely evolved in similar environments. Until fairly recently, the M31 GCS was generally viewed as a nearly identical analogue of the Milky Way GCS apart from minor differences in mean metallicity, size, and CN enhancement. As studies of the stellar populations in M31 have improved in both quality and quantity in recent years, it is becoming increasingly clear that these two galaxies may differ considerably.

M31's GCs are significantly enhanced in CN compared with their Milky Way counterparts at similar metallicities (e.g. [13,36]). It is possible that the M31 proto-globular clouds were enriched in nitrogen at some early epoch, whereas those in the Milky Way were not.

The M31 and Milky Way GC luminosity functions appear to differ as a function of metallicity and galactocentric radius [6]. Both GCSs have bimodal color and metallicity distributions. Studies of the Milky Way GCS have demonstrated that its metal-poor clusters belong to the Galactic halo [45] and its metal-rich clusters to either the bulge/bar [32,16] or thick disk [45,1]. In M31, however, things don't seem quite so straightforward.

Both the Milky Way and M31 are dominated by objects reaching oldest ages at $\sim$ 11–13 Gyr [36] although M31 hosts GCs with a wider range of ages and chemical compositions. As more detailed spectroscopic observations are obtained, evidence is mounting for the presence of multiple age populations in the M31 GCS: very young ($<$ 1Gyr), young (1–2 Gyr), intermediate (3–6 Gyr) as well as the complement of old GCs [3, 13, 7, 8, 36, 21]. Deep imaging and CMDs from HST and Keck are now uncovering the properties of individual stars in M31 GCs. No counterparts for the intermediate-age or young clusters found in the M31 sample have yet been identified in the Milky Way, although ongoing spectroscopic studies of Local Group GCs may yet reveal other intermediate-aged GC systems.

M31 appears to have a thin disk of GCs [33] although no such kinematic population is seen in the Milky Way GCS. The detection of several old metal-poor GCs having thin-disk kinematics implies that this disk component has survived since the early stages of galaxy evolution [36]. In such a case, no significant merger event could have occurred in the intervening time, since the disk would have been disrupted (or at least heated into a thick disk). More detailed studies of ages and chemical abundances are required for the disk members.

Irwin et al. [27] show that the minor-axis surface-brightness profile changes from a de Vaucouleurs $R^{1/4}$ law to a more flattened $R^{-2.3}$ power law beyond 1.4° or $\sim$ 20 kpc. For comparison, the density profile of the outer M31 GCS is $\propto R^{-3}$ [37]. The M31 spheroid appears to be a homogeneous and moderately metal-rich subsystem of the galaxy out to $R_{\rm gc} \sim$ 20–35 kpc, although the distribution of M31 halo stars covers a wide range in metallicity ($-2 <$ [Fe/H] < 0) (e.g. [9,17,43]). On average, the Milky Way halo is comparatively more metal-poor by nearly an order of magnitude. Stellar populations with a range of ages have been found at different positions throughout the M31 halo, although the abundance distributions of widely-dispersed fields in the halo are remarkably similar despite sampling different environments and galactocentric radii [9, 17]. The presence of an intermediate-age (6–8 Gyr) metal-rich component found by Brown et al. [10] conflicts with the usual picture of stellar halos as being old and metal-poor, although it is possible that this region may include some contamination from a disk population [38], some type of "thick-disk" component [44,40], or stream debris [19].

The Milky Way's halo of predominantly old, metal-poor stars becomes its most significant component a few kpc from the Galactic center and above its disk. In M31, by contrast, Hurley-Keller et al. [25] find that the kinematics of PNe even well above the disk ($|Y| > 6$ kpc) are clearly dominated by rotation, consistent with a large-bulge population, in agreement with the GC kinematics [34]. It has therefore been suggested that in M31 the bulge is the dominant component even at very large distances from the galaxy center (also see [43,14]). Direct comparisons of the Milky Way halo with comparable positions in M31 could be very misleading as M31's halo may not dominate at such radii.

6 M31 Formation Scenarios

Despite their proximity and observed similarities, the formation history of the Milky Way may still have been dramatically different from that of M31. The age distribution of M31 GCs rules out any scenario wherein the bulk of the GCS forms before or during the early stages of the formation of the galaxy. The formation of intermediate-age GCs could be possible in an extended collapse if the collapse of the gas clumps occurs over several Gyr (i.e. through hierarchical assembly). In our present understanding of galaxy formation, it seems clear that all galaxies must have undergone some degree of merging and accretion.

A major merger scenario would help to explain the bimodality in M31's GC color and metallicity distributions. Such a merger might also explain the flattened rotating spheroid observed in M31 if one traces this component to a disrupted early disk: pre-existing metal-rich bulge or thick-disk stars would be heated to larger radii, potentially explaining the intermediate-age stellar halo population found by Brown et al. [10]. M31's current disk would then have formed after the merger, and any GCs in the original thin disk would have been heated into a spheroid or thick disk. Detailed age and metallicity measurements of the present-day thin-disk GC population would help to either support or rule out the proposed model of an equal-mass merger [33]. If all of the thin-disk GCs in M31 are younger than ~ 6–8 Gyr and have metallicities consistent with some level of enrichment since early epochs, then this represents a viable scenario. Based on the ages of 6 GCs consistent with disk membership, Beasley et al. [7] find that the disk may be rather young (~ 1 Gyr) and metal-rich; in this case the presence of this disk is no constraint as the current thin disk would have had time to reform. Puzia et al. [36], however, find that the 18 disk GCs in their sample exhibit a wide range in ages (1–11.9 Gyr) and metallicities (-1.8 dex to solar). The fact that there are old thin-disk members suggests that this component formed early and continued to form clusters. A more extensive and uniform study of the ages and abundances of putative thin-disk population is clearly in order.

The relatively low metallicity of the intermediate-age GCs from Beasley et al. [8] does not fit well with a major merger origin, unless the merger progenitors were significantly metal-poor. The fact that the IAGCs are enhanced in α-elements also implies that the progenitor clouds were preferentially enriched by SNe II. The group of six IAGCs from Beasley et al. [8] also cannot be directly (spatially and kinematically) associated with known stellar streams. Puzia et al. [36] find that the IAGCs in their sample are relatively metal-rich (-0.6 dex) with $[\alpha/Fe]$ values nearly solar. A more detailed investigation of this age population is currently underway (see the contribution by J. Strader in these proceedings).

Its large spread in halo metallicity, the presence of stellar substructure and chemical inhomogeneities in the halo, and evidence for kinematically-coupled groups of GCs in M31 are factors which point to an active merger history for M31. However, the apparent uniformity in age and metallicity between stream and halo populations found by Brown et al. [12] is rather perplexing. If the stars in the halo substructure originated from merging/accreting systems, each with their own independent formation and enrichment histories, one would certainly expect to see differences. Most of the young, metal-rich GCs reside in the disk and are kinematically cold, making an accretion-type origin for this component unlikely unless the trajectory of the incident satellite was aligned with the galactic plane [13,36].

The chemical compositions of intermediate-age and young M31 GCs seem consistent with an extended timescale of formation. The intermediate-age population could have formed from previously-enriched material a few Gyr after the old halo population formed, and the young GCs may have arisen from star-formation resulting from recent infall of material and/or the accretion of dwarf satellites. Our current formation model for M31 must support the presence of multiple age and metallicity sub-populations within the M31 GCS. A more uniform and complete distribution of *confirmed* globular clusters with well-determined velocities, abundances, and ages is essential to improving our understanding of the thin disk and the various age populations that have recently been discovered in M31.

References

1. Armandroff, T. E. 1989, AJ, 97, 375
2. Ashman, K. M., & Bird, C. M. 1993, AJ, 106, 2281
3. Barmby, P., et al. 2000, AJ, 119, 727
4. Barmby, P., & Huchra, J. P. 2000, ApJ, 531, 29
5. Barmby, P., & Huchra, J. P. 2001, AJ, 122, 2458
6. Barmby, P., Huchra, J. P., & Brodie, J. P. 2001, AJ, 121, 1482.
7. Beasley, M. A., et al. 2004, AJ, 128, 1623
8. Beasley, M. A., et al. 2005, AJ, 129, 1412
9. Bellazzini, M., et al. 2003, A&A, 405, 867
10. Brown, T. M., et al. 2003, ApJ, 592, L17
11. Brown, T. M., et al. 2004, ApJ, 613, L125
12. Brown, T. M., et al. 2006, ApJ, 636, L89
13. Burstein, D., et al. 2004, ApJ, 614, 158

14. Chapman, S. C., et al. 2006, astro-ph/0602604
15. Cohen, J. G., Matthews, K., & Cameron, P. B. 2005, ApJ, 634, 45.
16. Côté, P. 1999, AJ, 118, 406
17. Durrell, P. R., Harris, W. E., & Pritchet, C. J. 2004, AJ, 128, 260
18. Ferguson, A. M. N., et al. 2002, AJ, 124, 1452
19. Ferguson, A. M. N., et al. 2005, ApJ, 622, L109
20. Fusi Pecci, F., et al. 1993, ASP Conf. Ser. 48, 410
21. Fusi Pecci, F., et al. 2005, AJ, 130, 554
22. Hodge, P. 1992, *The Andromeda Galaxy* (Dordrecht: Kluwer)
23. Hubble, E. 1932, ApJ, 76, 44
24. Huchra, J. P., Brodie, J. P., & Kent, S. M. 1991, ApJ, 370, 495
25. Hurley-Keller, D., et al. 2004, ApJ, 616, 804
26. Ibata, R., et al. 2001, Nature, 412, 49
27. Irwin, M. J., et al. 2005, ApJ, 628, L105
28. Jiang, L., et al. 2003, AJ, 125, 727
29. Larsen, S. S., & Richtler, T. 2000, A&A, 354, 836
30. Lee, H., & Worthey, G. 2005, ApJs, 160, 176
31. Ma, J., et al. 2006, A&A, 449, 143
32. Minniti, D. 1995, AJ, 109, 1663
33. Morrison, H. L., et al. 2004, ApJ, 603, 87
34. Perrett, K. M., et al. 2002, AJ, 123, 2490
35. Perrett, K., Stiff, D. A., Hanes, D. A., & Bridges, T. J. 2003, ApJ, 589, 790
36. Puzia, T. H., Perrett, K. M., & Bridges, T. J. 2005, A&A, 434, 909
37. Racine, R. 1991, AJ, 101, 865.
38. Rich, R. M., et al. 2004, AJ, 127, 2139
39. Saito, Y., & Iye, M. 2000, ApJ, 535, L95
40. Sarajedini, A., & Van Duyne, J. 2001, AJ, 122, 2444
41. van den Bergh, S. 1969, ApJS, 171, 145
42. van den Bergh, S. 2000, PASP, 112, 932
43. Worthey, G., et al. 2005, ApJ, 631, 820
44. Wyse, R. F. G., & Gilmore, G. 1988, AJ, 95, 1404
45. Zinn, R. 1985, ApJ, 293, 424
46. Zucker, D. B., et al. 2004, ApJ, 612, L117

IR Integrated Light Colors For Galactic GCs and An Update on Young M31 Globular Clusters

Judith Cohen

California Institute of Technology, Mail Code 105-24, Pasadena, Ca. 91125 USA
jlc@astro.caltech.edu

1 On the Reality of Young Globular Clusters in M31

In our recent publication, Cohen, Matthews and Cameron [3], we presented observations made with the newly commissioned Keck laser-guide star adaptive optics system of 6 objects in M31 that are alleged in multiple recent studies to be young globular clusters (GCs); all are supposed to have ages ≤5 Gyr. The resulting FWHM of the PSF core in our images was ~70 mas. The four youngest of these objects are asterisms; they are with certainty not young GCs in M31. Based on their morphology, the two oldest are GCs in M31. While the M31 GCs with ages 5–8 Gyr appear to be mostly genuine, it appears that many of the alleged very young GCs in M31 are spurious identifications. This problem will be even more severe in studies of the GC systems of more distant spiral galaxies now underway, for which imaging at the spatial resolution of our observations in M31 may not be adequate to detect sample contamination by asterisms.

Here we provide a brief update for this work. First, we have searched the HST Archive, and found additional cases of M31 GCs which appear to be asterisms, particularly among the bluest purported young M31 GCs. Second, we have determined the zero point of our K' mags via wide field IR images from the 5-m Hale Telescope combined with the 2MASS Point Source Catalog. Our deepest images reach $K' \sim 23.3$ mag for isolated point sources not in the crowded cluster regions. Finally, with this calibration in hand, we have created a color-magnitude diagram for the M31 GC B232, combining our LGS/AO images taken with the K' filter with a short optical WFPC2 exposure from the HST Archive. We find that the RGB tip lies at the K' mag level expected for a GC with an age comparable to that of the Galactic GCs and a metallicity between that of M3 and 47 Tuc. There is no evidence for the presence of stars more luminous than the RGB tip. This absence of bright AGB stars is consistent with an old age for the M31 GC B232. It is clear that such CMDs can be created with either optical HST or LGS/AO J images in concert with K' LGS/AO images, boding well for future ground-based LGS/AO imaging studies.

2 IR Integrated Light Colors for Galactic GCs

The only existing large dataset for IR photometry of Galactic GCs was obtained by M. Aaronson, M. Malkan and S. Kleinmann in the late 1970s and was used by Aaronson et al. [1], but was never published in detail due to Marc Aaronson's untimely death. Now with the release of 2MASS (see [6,4]) it is possible to determine accurate IR integrated light photometry for a much larger sample of Galactic GCs. In collaboration with Scott Hsieh, Stan Metchev and George Djorgovski, we have done this, producing mosaic images of the fields of Galactic GCs and fitting King profiles. Imposing a lower limit to the SNR, we successfully carried this out for a sample of 102 Galactic GCs.

Figure 1 shows $V - K_s$ colors (upper panels) and $J - K_s$ (lower panels) for the integrated light of Galactic GCs for a fixed circular aperture size of 50"

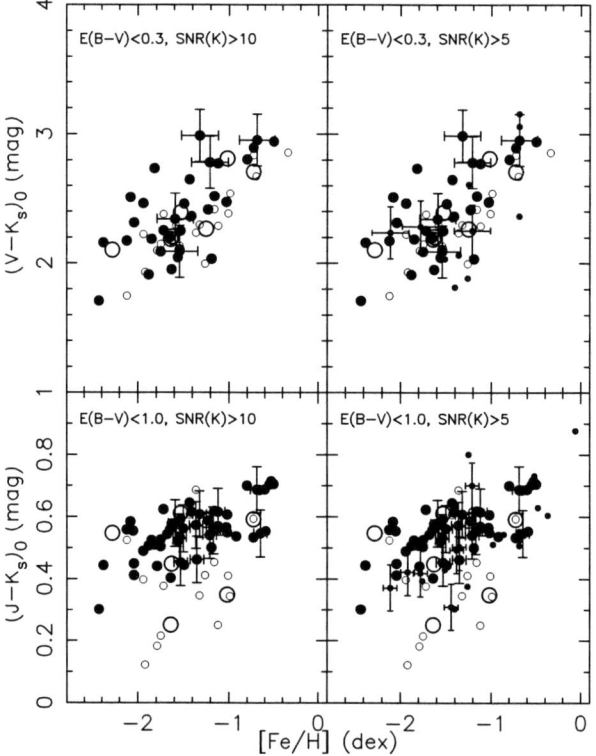

Fig. 1. The top panels show $V - K_s$ versus [Fe/H] from our BEST (*left, large filled circles*) and FAIR (*right, small filled circles*) samples; the bottom panels show the same for $J - K_s$. A 50" radius circular aperture is used. Probable core collapsed GCs are marked with crosses. The 1978 data of Aaronson, Malkan & Kleinmann, updated and transformed as required, is superposed (*large open circles* are their calibrating clusters, which they believed to have accurate metallicities and reddenings; smaller open circles are other GCs in their sample).

radius. The colors have been calculated by integrating the King profile fits in the IR that we have carried out, combined in the case of $V - K_s$ with integrations of similar fits from the literature to optical surface brightness measurements. The left panels contains those GCs with the highest SNR and with low $E(B-V)$ ("low" being defined as less than 0.3 mag for $V - K_s$ and less than 1.0 mag for $J - K_s$), while the right panels relaxes the SNR requirement somewhat. Superposed on this figure are the 1978 measurements of Aaronson, Malkan & Kleinmann. All colors have been corrected for reddening, and the 1978 colors have been transformed from the Johnson system (to which it is assumed they were calibrated) to the 2MASS system. There is good agreement between the two datasets for $V - K_s$, but a hint of a possible systematic color-dependent transformation at the level of ∼0.15 mag for $J - K_s$.

A comparison with several recent models for the integrated light of old simple stellar populations is given in Fig. 2. The same transformation issue

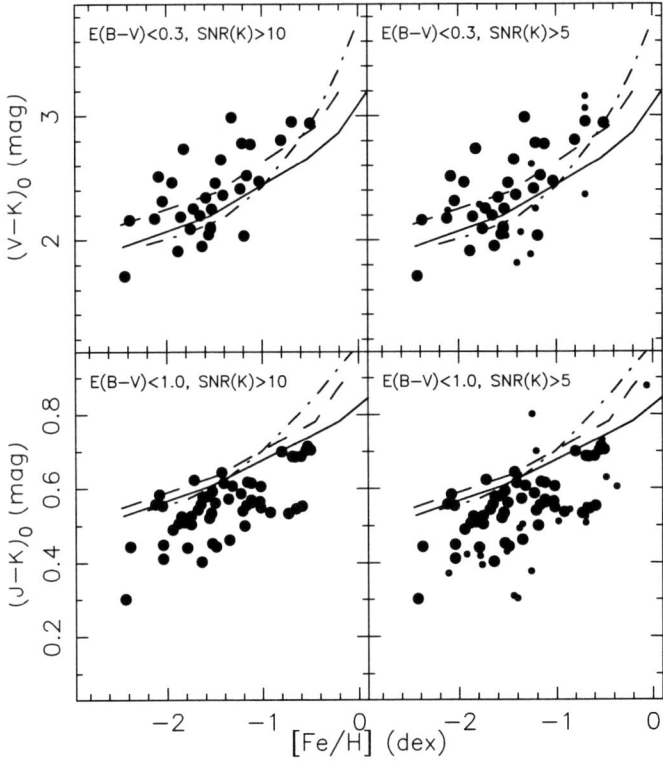

Fig. 2. Our derived 2MASS integrated light colors $V - K_s$ and $J - K_s$ are shown as a function of [Fe/H] for the "best" and "fair" Galactic GC samples, again for a circular aperture with radius 50". Predicted SSP colors of a 10 Gyr model from Maraston [5], a 12 Gyr model from Worthey [7] and a 12.5 Gyr model from Buzzoni [2] are superposed. These have been transformed from the Johnson into the 2MASS system and [Fe/H] values for each of the model curves have been adjusted for the α-enhancement characteristic of Galactic GCs.

persists in $J - K$, with the models being redder than the our derived colors, particularly at the lowest metallicities.

Full details of this work will be submitted for publication soon.

References

1. Aaronson, M., Cohen, J. G., Mould, J. R. & Malkan, M., ApJ, **223**, 824 (1978)
2. Buzzoni, A., ApJS, **71**, 817 (1989)
3. Cohen, J. G., Matthews, K. & Cameron, P. B., ApJL, **634**, L45 (2005)
4. Cutri, R. M. et al., 2003, "Explanatory Supplement to the 2MASS All-Sky Data Release", http://www.ipac.caltech.edu/2mass/releases/allsky/doc/explsup.html
5. Maraston, C., MNRAS, **362**, 799 (2005)
6. Skrutskie, M. F. et al., AJ, **131**, 1163 (2006)
7. Worthey, G., ApJS, **95**, 107 (1994)

Nuclear Star Clusters in Edge-on Galaxies

Anil C. Seth[1], Julianne J. Dalcanton[1,2], Paul W. Hodge[1], and Victor P. Debattista[1,3]

[1] University of Washington, seth@astro.washington.edu
[2] Alfred P. Sloan Research Fellow
[3] Brooks Prize Fellow

From observations of edge-on, late-type galaxies, we present morphological evidence that some nuclear star clusters have experienced *in situ* star formation. We find three nuclear clusters that, viewed from the edge-on perspective, have both a compact disk-like component and a spheroidal component. In each cluster, the disk components are closely aligned with the major axis of the host galaxy and have bluer colors than the spheroidal components. We spectroscopically verify that one of the observed multiple component clusters has multiple generations of stars. These observations lead us to suggest a formation mechanism for nuclear star clusters, in which stars episodically form in compact nuclear disks, and then lose angular momentum, eventually forming an older spheroid. The full results of this study can be found in a forthcoming paper.

1 Background

Nuclear star clusters are a common feature of dwarf elliptical and spiral galaxies. Surveys of both face-on bulgeless spirals (type Scd and later) and dwarf ellipticals find that roughly 75% have a single bright star cluster as their nuclei [2,5]. The sizes of these nuclear clusters are similar to Galactic globular clusters ($r_{eff} \sim 3$ pc), but they are significantly brighter, with absolute I-band magnitudes of -8 to -16 [2,3]. This luminosity is due both to their high masses (typically a few $\times 10^6$ $M_\odot$) and to the presence of younger stellar populations [14, 15].

Nuclear clusters are interesting objects in a galaxy evolution context. A number of groups have recently shown that the masses of nuclear clusters correlate with their host galaxy masses along the same relation found for supermassive black holes, and thus appear to be directly connected to the process of galaxy formation [6, 10, 16]. Furthermore, observations of nuclear clusters may provide clues to the formation of unseen supermassive black holes. Lastly, nuclear clusters are possible progenitors to massive globular clusters such as ω Cen and ultracompact dwarfs [1, 8].

Two scenarios have been suggested to explain the formation of nuclear star clusters: (1) nuclear clusters form from multiple globular clusters accreted via dynamical friction [13], and (2) nuclear clusters form *in situ* from gas channeled into the center of galaxies [9]. In this proceeding we present evidence for the latter scenario.

2 Results

Our sample of galaxies consists of 14 nearby (2–20 Mpc), late-type (Sbc+), edge-on spiral galaxies observed with HST/ACS as part of a Cycle 12 snapshot program (sample details can be found in [12]). In these 14 galaxies, we identified 9 nuclear cluster candidates.

Nuclear Cluster Morphologies and Luminosities

All nine of the detected cluster candidates were at least partially resolved in the HST/ACS images. Fits of convolved King profiles gave effective (half-light) radii ranging from 1 to 20 pc, with most of the clusters having effective radii between 1 and 4 pc. This size range is similar to what has been found previously for nuclear star clusters in face-on, late-type galaxies [3]. Furthermore, the absolute I-band magnitudes are also similar to previously observed nuclear clusters, ranging from –8 to –15 [2].

Three of the cluster candidates (in IC 5052, NGC 4206, and NGC 4244) have unusual morphologies. As shown by the contours in Fig. 1, these three candidates are elongated and appear to have both a disk-like and spheroidal component, much like miniature S0 galaxies. Fitting these clusters with both an exponential disk component and an elliptical King profile component gave much smaller residuals than single component fits.

The elongations and disk components of the three multi-component clusters are aligned to within 10° of the major axis of the edge-on galaxy disks (see Fig. 1). Previous studies of nuclear clusters have focused on face-on galaxies, making detection of similar multi-component clusters difficult.

Nuclear Cluster Stellar Populations

The color maps in Fig. 1 show that the multiple morphological components have clearly different F606W-F814W colors. In each cluster, the disk components are bluer than the spheroid, with a color difference >0.3 magnitudes. This color difference is most simply interpreted as a difference in age, with the disk being made of younger stars than the spheroid. Although the reddening is unknown, based on Padova single-stellar population models in the ACS filters [7], the observed color difference implies that the stellar ages of the disk are younger than ~1 Gyr.

For the nuclear cluster in NGC 4244, the nearest in our sample (D = 4.4 Mpc), we obtained a long-slit spectrum of the cluster using the DIS spectrograph on the Apache Point Observatory 3.5 m telescope (Fig. 2). This

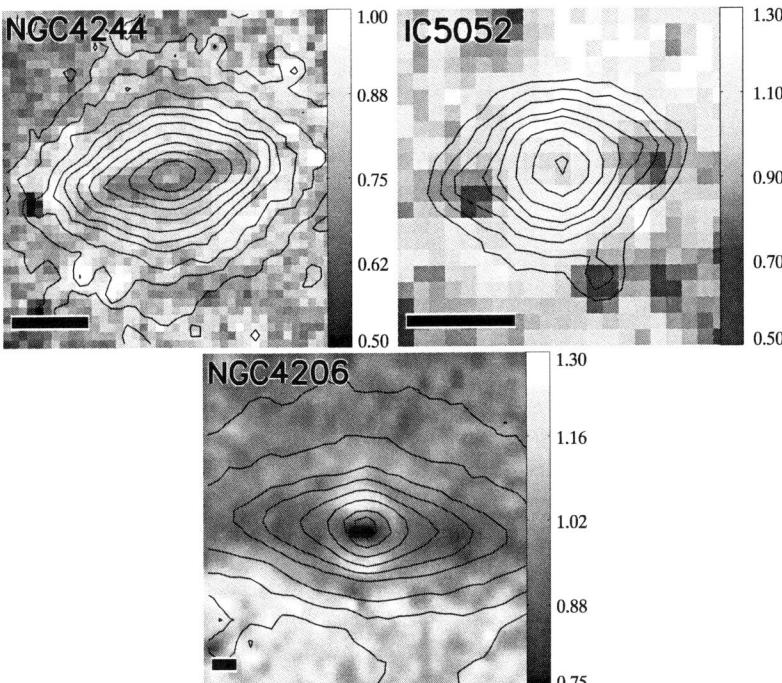

Fig. 1. Color maps of the three multi-component nuclear clusters overlaid with contours showing the F606W brightness. The colorbars indicate the F606W-F814W in each cluster; dark colors indicate blue regions. Each image has been rotated so that the x-axis is parallel to the major axis of the galaxy disk. The black bar in the bottom left corner indicates a length of 10 pc.

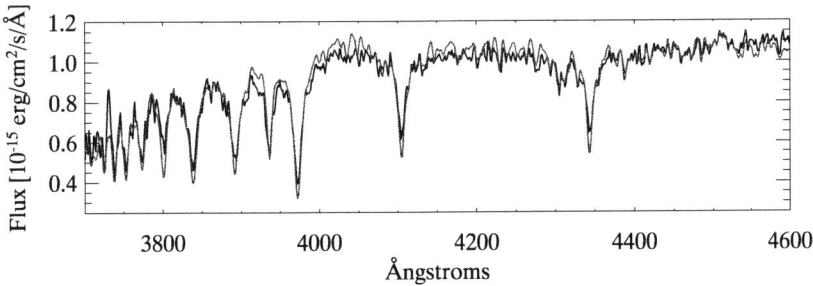

Fig. 2. Spectrum of the NGC 4244 nuclear cluster obtained with the Apache Point Observatory 3.5 m telescope using the DIS spectograph (*black line*). The gray line shows the best-fitting two-age spectrum, with stellar populations of 0.1 and 1 Gyr and a total mass of 3.5×10^6 $M_\odot$.

spectrum verifies that multiple stellar populations are present in the cluster, with the youngest component having an age of ∼0.1 Gyr. We fit the spectrum using combinations of Bruzual & Charlot models [4] assuming Z = 0.008.

As would be expected from the color maps, the spectrum is much better fit by multiple stellar populations than any single stellar population. In Fig. 2 we show the best fitting two-age fit with ages of 0.1 and 1 Gyr, and a mass of 3.3×10^6 $M_\odot$, 5% of which is in the younger 0.1 Gyr component. The luminosity of this young component matches the disk luminosity of the best morphological fit. However, many different combinations of masses and ages fit the data well, including a three-age model with a significant old (10 Gyr) component, and a constant star formation rate model.

3 Summary and Discussion

Three of the nine nuclear cluster candidates in our sample have young disk components (<1 Gyr) in addition to an older spheroidal component. These disks cannot be formed by accretion of globular clusters and must instead be formed *in situ* from gas accreted into the nuclear regions. The multiple stellar populations observed in many nuclear clusters [15], and the direct detection of a molecular gas disk coincident with the nuclear cluster in IC 342 [11] provide additional evidence that nuclear cluster formation is an ongoing process. We propose a model for nuclear cluster formation in which the stars in the cluster form episodically in nuclear disks. Such episodic star formation would naturally result from stochastic accretion events and/or feedback from star formation. Then, over time, the stars in the disk lose angular momentum and end up in a more spheroidal component. We are currently investigating the mechanism by which the stellar disks could lose angular momentum.

References

1. K. Bekki, M. Chiba: A&A, **417**, 437 (2004)
2. T. Böker, S. Laine, R. van der Marel et al.: AJ, **123**, 1389 (2002)
3. T. Böker, M. Sarzi, D. McLaughlin et al.: AJ, **127**, 105 (2004)
4. G. Bruzual, S. Charlot: MNRAS, **344**, 1000 (2003)
5. P. Côté, S. Piatek, L. Ferrarese et al.: astro-ph/0603252 (2006)
6. L. Ferrarese, P. Côté, E. Dalla Bonta et al.: astro-ph/0603252 (2006)
7. L. Girardi: http://pleiadi.pd.astro.it (2006)
8. O. Gnedin, H. Zhao, J. Pringle et al.: ApJL, **568**, L23 (2002)
9. M. Milosavljević: ApJL, **605**, L13 (2004)
10. J. Rossa, J. van der Marel, T. Böker et al.: astro-ph/0604140 (2006)
11. E. Schinnerer, T. Böker, D. Meier et al.: ApJL, **591**, L115 (2003)
12. A. Seth, J. Dalcanton, R. de Jong: AJ, **129**, 1331 (2005)
13. S. Tremaine, J. Ostriker, L. Spitzer: ApJ, **196**, 407 (1975)
14. C.-J. Walcher, R. van der Marel, D. McLaughlin et al.: ApJ, **618**, 237 (2005)
15. C.-J. Walcher, T. Böker, S. Charlot et al.: astro-ph/0604138 (2006)
16. E. Wehner, W. Harris: astro-ph/0603801 (2006)

HST ACS Wide-Field Photometry of the Sombrero Galaxy Globular Cluster System

Lee Spitler

Centre for Astrophysics & Supercomputing, Swinburne University
Hawthorn, VIC 3122, Australia lspitler@astro.swin.edu.au

The Hubble Heritage Team acquired a six pointing mosaic of the Sombrero Galaxy with the HST Advanced Camera for Surveys (ACS). The unusually long exposure times and the extended spatial coverage make this dataset perhaps the best demonstration of the ACS instrument's scientific merit for studying a GC system. The Sombrero Galaxy (NGC 4594) is a massive ($M_B = -21.4$) Sa-type spiral located in the field. The distance to the Sombrero is taken to be m−M = 29.77 or 9.0 Mpc. A detailed description of the data reduction process, contamination estimates and the data analysis are published in [1].

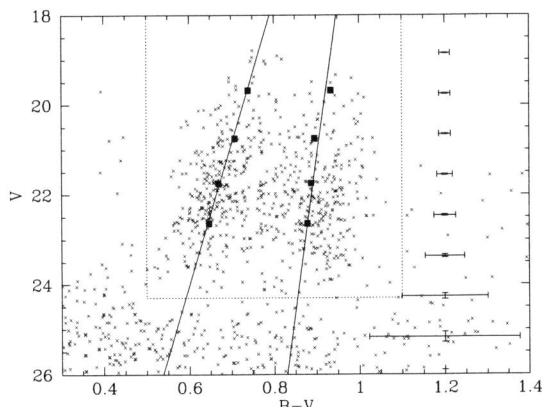

Fig. 1. Color-magnitude diagram of the objects selected on size, shape and a visual inspection. Objects falling with the region bounded by the dotted lines are the Sombrero GC candidates. A statistically significant color-magnitude trend is detected among the metal-poor GCs but not the metal-rich GCs.

Figure 1 shows that the bulk of the candidate GCs are found above the luminosity where contamination from faint background galaxies begins to increase significantly. Thus by avoiding much of this contamination, the Sombrero ACS sample is largely free of spurious objects over the relevant magnitude range. Figures 2 and 3 show the GC luminosity functions (GCLFs) in

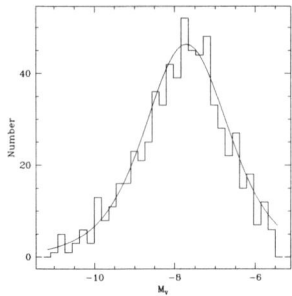
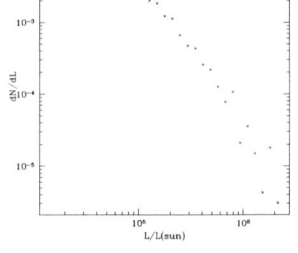

Fig. 2. Absolute V-band GC luminosity function with a Student-$t5$ distribution ($M_V^{TOM} = -7.71$, $\sigma_{t5} = 1.08$).

Fig. 3. GC luminosity function in solar luminosity units.

units of magnitudes and luminosity, respectively. The faint end of the GCLF can now be examined directly and is found to be a continuation of the brighter half of the GCLF.

1 Blue Tilt

Among the metal-poor GCs, a color-magnitude trend was discovered that is interpreted as a mass-metallicity trend. When a line is fitted to the brightest four of five equally-spaced magnitude intervals (see Fig. 1) all color and magnitude combinations of the three available bands (B, V and R) yield a statistically significant trend in the metal-poor and not the metal-rich sub-population. This is the first such example of a "blue tilt" in a spiral galaxy and first example of a blue tilt found in a low-density galaxy environment. Blue tilts have been found in other GC systems [2–5], as presented in Table 1. GC systems in galaxies of all morphological and environmental types show this trend. One exception is the M49 GC system, which has been reported to show no blue tilt [2]. If confirmed, this now atypical GC system will help constrain any ideas that attempts to explain the blue tilt. An examination of

Table 1. Blue Tilt Comparison

Location	Galaxy Type	Environment	Magnitude Extent	Metallicity$\propto$Mass
8 BC/BGGs [3]	Es	High	TOM−1.0	Z$\propto$M$^{0.55}$
M87 [2]	E0	High	TOM	Z$\propto$M$^{0.48}$
NGC 4649 [2]	E2	High	TOM	Z$\propto$M$^{0.42}$
Sombrero [1]	Sa	Low	TOM+0.5	Z$\propto$M$^{0.27}$
NGC 5170 [4]	Sc	Low	TOM+0.5	Z$\propto$M$^{0.34}$
Virgo Cluster [5]	Various	High	−	0.3 − 0.5

the tilt slopes when converted to a metallicity-mass proportionality, suggests that the spiral galaxies found in low-density environments have a shallower trend. This fact and observations of M49 likely indicate the existence and specific properties of the blue tilt are influenced by its host galaxy and therefore is not solely an intrinsic feature of metal-poor GC formation.

2 GC Size-Magnitude Trend

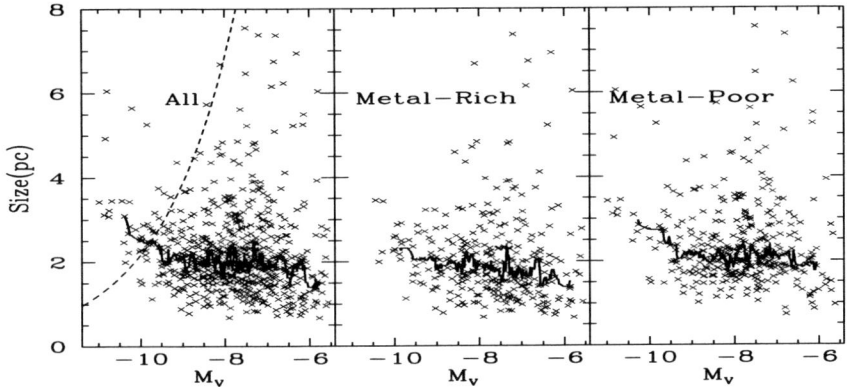

Fig. 4. Size-magnitude plots of the entire sample, metal-rich, and metal-poor GCs are shown in the left, center and right panels, respectively. Sizes tend to increase as a function of magnitude for objects brighter than $M_V = -9.5$, as demonstrated by the thick solid line moving median estimates. The dashed line is a division between typical MW GCs and those suspected to be accreted dwarf nuclei [8].

Figure 4 contains size-magnitude diagrams which show that the sizes of the brighter Sombrero GCs ($M_V < -9.5$) have increasing sizes with increasing luminosity. Fainter GCs show the expected magnitude-independent sizes, although a shallow trend might be present among all metal-rich GCs. The separation between typical Milky Way (MW) GCs and the massive and extended MW objects, suspected to be accreted dwarf elliptical nuclei, coincides with the location where the Sombrero trend size-magnitude begins to emerge. This observation and the established similarities between metal-poor GCs and dwarf nuclei in terms of color [6,7,3] size [7], luminosity [6] and the color-magnitude trends [3,2], suggests these stellar systems have a similar structure and supports the idea that an unknown fraction of the Sombrero GC system contains accreted nuclei. Similar observations of large and massive objects in other GC systems may indicate this Sombrero trend is not unique [8,9,3]. If this is confirmed, it suggests there exists smooth transition between GCs and more massive stellar systems such as ultra compact dwarf galaxies.

3 Subpopulation Size Difference: A Projection Effect

Two theories have been proposed to explain the observed mean GC subpopulation size difference [10, 11]. The first study pointed out that projections effects are more important closer to the center of a galaxy, which could lead to an observed subpopulation mean size difference [10]. An alternative interpretation suggested that mass-segregation and a metallicity-dependent GC main-sequence lifetime could cause an intrinsic difference between the subpopulation sizes. To differentiate between the two theories, GCs from both subpopulations must be observed at large galactocentric radii where projection effects are less important. The Sombrero mean subpopulation size difference (14%) is noticeably smaller than past observations (~20–30%), but is consistent with previous results if just the inner GCs are considered. Thus as illustrated in Figure 5, the Sombrero observations are best explained by a projection effect.

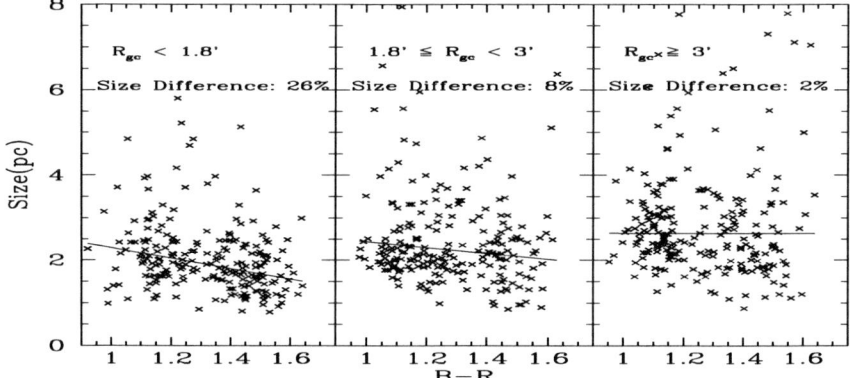

Fig. 5. Size-color plots of GCs in three galactocentric radius intervals. Lines are linear fits to each distribution and the mean size difference between the subpopulation is shown. The sizes of the metallicity subpopulations become more alike at larger radii, which is consistent with the idea that projection effects are largely responsible for the Sombrero subpopulation mean size difference.

References

1. Spitler, L., Larsen, S., Strader, J., et al.: 2006, AJ, **132**, 1593
2. Strader, J., Brodie, J. P., Spitler, L., Beasley, M.: 2006, AJ, **132**, 2333
3. Harris, W. E., Whitmore, B. C., Karakla, D.: 2006, ApJ, **636**, 90
4. Forde, K., et al.: in prep
5. Mieske, S., et al.: Jordán, A., Côté, P. et al.: 2006, ApJ, **653**, 193
6. Lotz, J. M., Miller, B. W., Ferguson, H. C.: 2004, ApJ, **613**, 262

7. De Propris, R., Phillipps, S., Drinkwater, M. J., et al.: 2005, ApJL, **623**, 105
8. Mackey, A. D., van den Bergh, S.: 2006,MNRAS, **360**, 631
9. Martini, P., Ho, L. C.: 2004, ApJ, **610**, 233
10. Larsen, S. S., Brodie, J. P.: 2003, ApJ, **593**, 340
11. Jordán, A.: 2004, ApJL, **613**, L117

Intermediate-Age Globular Clusters in M31

Jay Strader

UCO/Lick Observatory, 1156 High St., Santa Cruz, CA 95060, USA
strader@ucolick.org

Abstract. Several groups have claimed the discovery of intermediate-age globular clusters in M31 on the basis of low-resolution spectroscopy. We use high-resolution Keck spectra and HST imaging to show that three such objects have mass-to-light ratios like those of old clusters, and thus are unlikely to be of intermediate age.

1 Introduction

A number of recent papers [3, 4, 7, 11] have presented evidence, mostly from low-resolution spectroscopy, that M31 has a significant population of intermediate-age (1–6 Gyr) globular clusters (GCs). The Milky Way has no such clusters, and these results lead one to wonder whether the GC formation history (and thus perhaps the violent star formation history) of M31 was very different than the Milky Way. Such an interpretation could be consistent with the extensive accretion history of M31, as indicated by the presence of large stellar streams and plumes in the halo (e.g., [9]).

The standard approach to deriving GC ages from low-resolution spectroscopy is to use a combination of Balmer and metal lines to break the age-metallicity degeneracy inherent in any single line. At fixed metallicity, stronger Balmer lines indicate a hotter main sequence turnoff and younger ages. However, in old GCs a variety of hot star populations—most notably blue horizontal branch stars—can lead to "artificially" strong Balmer lines and thus to an underestimation of GC ages.

There are several ways to address this concern, including color-magnitude diagrams reaching below the horizontal branch and the direct detection of the integrated light of such stars using UV photometry. We have chosen a different approach, deriving mass-to-light (M/L) ratios from high-resolution integrated spectroscopy and Hubble Space Telescope (HST) imaging. The spectra yield velocity dispersions, and when combined with a radius measured from the HST imaging an application of the virial theorem gives the GC mass. Maraston [10] models using a Kroupa initial mass function predict that V-band M/L increases by approximately a factor of two from 5 to 12 Gyr. This is easily detectable with the ~ 20–25% relative errors in M/L_V we achieve in our analysis (see below).

2 Data and Analysis

Spectra of three candidate intermediate-age GCs and one old GC were taken with Keck/HIRES on 8 Oct 2005, with a resolution of ~ 46000, corresponding to ~ 5 km/s FWHM. The slit width was 0.85"; M31 GCs have typical half-light diameters of $\sim 1 - 1.5$". The wavelength coverage was ~ 3800–6500 Å, and exposure times were 30 minutes per GC. The properties of the GCs are given in Table 1, including age estimates from Burstein et al. [4] or Beasley et al. [3]. The GCs chosen all have HST/WFPC2 imaging, and for three GCs their King model structural parameters were given in Barmby, Holland, and Huchra [2]. For GC 126-184 we derived structural parameters using the same method as Barmby et al.

Velocity dispersions were derived through order-by-order cross-correlation with 13 late F-K template stars. The relationship between the FWHM of the cross-correlation peak and velocity dispersion was established by convolving the template stars with Gaussians in steps of 0.5 km/s, and then cross-correlating the template stars with themselves and the other templates. No systematic trends were identified with wavelength or with the spectral type of the template star. The final error on the velocity dispersion of an individual GC was estimated as the RMS of the velocity dispersions derived from each combination of template stars. This may of course be an underestimate of unknown systematic errors, but should be accurate in a relative sense for intercomparing values among GCs.

The velocity dispersions thus estimated are implicitly integrated over the finite width of the slit. We integrated over the respective King models to derive central (σ_0) and global (σ_∞) velocity dispersions for each GC. Only a single GC (358-219) is in common with previous work, but for this GC our σ measurement agrees to within a surprising 0.1 km/s with the previous estimate [6].

We used the radii and σ_∞ values of the GCs to estimate virial masses. These values are quite similar to King model masses except for large values of the cluster concentration. In any case, we are making only a relative comparison of M/L among GCs, and neither mass measure is amenable to direct comparison with values predicted by stellar population models (due, for example, to the effects of mass segregation). GC luminosities were derived

Table 1. Properties of GCs

ID	age (Gyr)	M_V (mag)	σ_∞ (km/s)	Mass ($10^5 M_\odot$)	M/L_V
232-286	2-5	-9.0	11.3 ± 1.2	6.9	2.04 ± 0.53
126-184	3-6	-7.8	6.8 ± 0.5	1.8	1.53 ± 0.32
311-033	5	-9.7	11.1 ± 0.6	6.9	1.07 ± 0.19
358-219	11	-9.3	7.8 ± 0.5	5.7	1.18 ± 0.23

from V magnitudes in the catalog of Barmby et al. [1]. A distance modulus of 24.47 was assumed [8]. Reddenings were taken from the unpublished work of P. Barmby, which utilized photometry over a wide wavelength range and in some cases low-resolution spectroscopy. The final estimates of M/L_V are given in Table 1. Propagation of errors in the values of the half-light radius, luminosity, and σ_∞ gave a final error estimate for the M/L_V; the total error is dominated by the uncertainty in σ.

Djorgovski et al. [5] provide velocity dispersions for a sample of 21 old M31 GCs. We selected those with structural parameters in Barmby et al. [1] and calculated M/L_V using the same method as above, and used these GCs as a comparison sample.

3 Results and Discussion

The median virial M/L_V for the Djorgovski et al. [5] M31 GCs is 1.38. As is clear from Table 1, the M/L_V of the three candidate intermediate-age GCs are consistent with this value. None have $M/L_V < 1$, while a relative comparison to Maraston [10] models (using the Djorgovski sample as a reference) predict a 5 Gyr GC would have $M/L_V \sim 0.7$, with even lower values for younger GCs. There is no evidence from our high-resolution spectroscopy that the GCs studied have intermediate ages.

The GC 232–286 has $M/L_V = 2.04 \pm 0.53$. While only slightly more than 1σ deviant from the median value of old M31 GCs, it is worthwhile to consider whether the discrepancy might be real. The GC has a relatively high ellipticity (0.18), and similarly flattened Galactic GCs are generally rotating. Such rotation could artificially inflate the virial mass estimate. Anisotropy in the velocity tensor could also lead to an overestimate of the mass, but a *different* anisotropy would be needed to account for the flattening of the GC. Stephens, Catelan, and Contreras [12] have found that the single GC in the dIrr WLM is flattened solely by anisotropy.

Since we have observed only three candidate intermediate-age GCs in M31, we cannot make broad statements about the true global fraction of younger GCs. However, our results do indicate the need for ongoing vigilance in considering systematic uncertainties in age estimates of GCs with low-resolution optical spectroscopy.

Acknowledgements. I would like to thank my collaborators Jean Brodie, Soeren Larsen, and Graeme Smith for their ongoing advice and assistance, and to Tom Richtler for organizing a beautiful and stimulating conference. The pudús were especially cute. This work was supported by an NSF Graduate Research Fellowship and NSF Grant AST-0507729.

References

1. Barmby, P., Huchra, J.P., Brodie, J.P., Forbes, D.A., Schroder, L.L., Grillmair, C.J. 2000, AJ, 119, 727
2. Barmby, P., Holland, S., Huchra, J.P. 2002, AJ, 123, 1937
3. Beasley, M.A., Brodie, J.P., Strader, J., Forbes, D.A., Proctor, R.N., Barmby, P., Huchra, J.P. 2005, AJ, 129, 1412
4. Burstein, D., et al. 2004, ApJ, 614, 158
5. Djorgovski, S.G., Gal, R.R., McCarthy, J.K., Cohen, J.G., de Carvalho, R.R., Meylan, G., Bendinelli, O., Parmeggiani, G. 1997, ApJ, 474, 19
6. Dubath, P., Grillmair, C.J. 1997, A&A, 321, 379
7. Fusi Pecci, F., Bellazzini, M., Buzzoni, A., De Simone, E., Federici, L., Galleti, S. 2005, AJ, 130, 554
8. Holland, S. 1998, AJ, 115, 1916
9. Ibata, R., Irwin, M., Lewis, G., Ferguson, A.M.N., Tanvir, N. 2001, Nature, 412, 49
10. Maraston, C. 2005, MNRAS, 362, 799
11. Puzia, T.H., Perrett, K.M., Bridges, T.J. 2005, A&A, 434, 909
12. Stephens, A.W., Catelan, M., Contreras, R.P. 2006, AJ, 131, 1426

Metal-Poor Globular Clusters of the Galactic Bulge

B. Barbuy[1], M. Zoccali[2], S. Ortolani[3], Y. Momany[3], D. Minniti[2], V. Hill[4], A. Renzini[3], E. Bica[5], A. Alves-Brito[1], A. Goméz[4], L. Pasquini[6], and R.M. Rich[7]

[1] U. São Paulo, Brazil `barbuy,abrito@astro.iag.usp.br`
[2] PUC, Chile `mzoccali,dante@astro.puc.cl`
[3] U. Padova, Italy `ortolani,momany,arenzini@pd.astro.it`
[4] Obs. Paris-Meudon, Chile `vanessa.hill,anita.gomez@obspm.fr`
[5] UFRGS, Brazil `bica@if.ufrgs.br`
[6] ESO, Germany `lpasquin@eso.org`
[7] UCLA, USA `rmr@astro.ucla.edu`,

1 Introduction

Metal-poor bulge field stars and clusters represent a crucial piece in the puzzle of the Milky Way formation. In fact, if the Galaxy formed via dissipational collapse, then the metal-poor objects in the central region would be the first to have formed (e.g. [5]).

1.1 HP 1 and NGC 6558

The spectroscopic analysis provided a metallicity [Fe/H] = -1.00 ± 0.2 for HP 1 and [Fe/H] = -0.90 ± 0.2 for NGC 6558. They are therefore intermediate metallicity clusters, showing a blue extended Horizontal Branch (HB) and a post-core-collapse structure, that may characterize a distinct class of clusters in the bulge.

Together with NGC 6388 and NGC 6441 of [Fe/H]$\sim$-0.6 they would be another two ones with such characteristics, but differing from them, since these two other clusters have also a populous Red HB, and a normal slope of the RGB for their metallicity, which is not the case of HP-1 and NGC 6558.

It is more difficult to reconcile our iron abundance with the steep slope of the Red Giant Branch (see diagram of HP 1 in Ortolani et al. [4]); the slopes of HP 1 and NGC 6558 are consistent with [Fe/H]≈ -1.5 like NGC 6752. The RGB slope is reflection of line blanketing in the bluer bands. We find both HP1 and NGC 6558 to be enhanced in the α elements O and Si, but Ti is Solar. We speculate that the TiO blanketing in HP1 and NGC 6558 might be lower than in other clusters of similar metallicity.

In Table 1 are given the abundance ratios [X/Fe] for the following metal-poor bulge clusters: preliminary values for NGC 6558 [6], HP 1 [1], UKS 1 [3] and Terzan 4 [2]. The α-elements Oxygen and Silicon showed to be

Table 1. Abundances for metal-poor bulge clusters

Species	$\epsilon(X)_\odot$	NGC 6558	HP 1	UKS 1	Terzan 4
[Fe/H]	7.50	−0.90	−0.99	−0.78	−1.60
[OI/Fe]	8.77	+0.40	+0.40	+0.27	+0.55
[NaI/Fe]	6.33	+0.00	+0.00	—	—
[MgI/Fe]	7.58	+0.20	+0.10	+0.32	+0.41
[SiI/Fe]	7.55	+0.20	+0.30	+0.28	+0.55
[CaI/Fe]	6.36	+0.15	+0.03	+0.38	+0.53
[TiI/Fe]	5.02	+0.0	+0.02	+0.32	+0.43
[TiII/Fe]		0.00	+0.1	—	
[BaII/Fe]	2.13	+0.20	+0.15	—	
[LaII/Fe]	1.22	0.00	+0.00	—	
[EuII/Fe]	0.51	+0.40	+0.15	—	

[α/Fe] ≈ +0.3, Calcium and Titanium showed solar ratios. Oxygen and Silicon enhancements, together with that of the r-process element Eu may be indicative of a fast early enrichment by Supernovae type II. On the other hand, the solar ratios Ca and Ti may be indicative of important contribution from SN Ia. The s-element La shows a solar ratio [La/Fe]=0.0 and Ba is enhanced with [Ba/Fe]=+0.2. Finally, a comparison with the results by Origlia and Rich [2] for the metal-poor cluster Terzan 4, also located in the inner bulge, of [Fe/H]=-1.6, for which significant enhancements of α-elements were found, may indicate a halo origin, whereas HP-1 and NGC 6558 showing a more peculiar pattern might be revealing characteristics of the bulge chemical enrichment.

References

1. Barbuy, B. et al. 2006, A&A, 449, 349
2. Origlia, L., Rich, R.M. 2004, AJ, 127, 3422
3. Origlia, L. et al. 2005, MNRAS, 363, 897
4. Ortolani, S. et al. 1997, MNRAS, 284, 692
5. van den Bergh, S. 1993, ApJ, 411, 178
6. Zoccali, M. et al 2006, in preparation

Globular Cluster System and Milky Way Properties Revisited

C. Bonatto[1], E. Bica[1] B. Barbuy[2], and S. Ortolani[3]

[1] Universidade Federal do Rio Grande do Sul, Instituto de Física, CP 15051, Porto Alegre 91501-970, RS, Brazil charles@if.ufrgs.br, bica@if.ufrgs.br
[2] Universidade de São Paulo, Dept. de Astronomia, Rua do Matão 1226, São Paulo 05508-090, Brazil barbuy@astro.iag.usp.br
[3] Università di Padova, Dipartimento di Astronomia, Vicolo dell'Osservatorio 2, I-35122 Padova, Italy ortolani@pd.astro.it

Updated data of the 153 Galactic globular clusters are used to readdress fundamental parameters of the Milky Way, such as the distance of the Sun to the Galactic centre, bulge and halo structural parameters, and cluster destruction rates. For this analysis we build a reduced sample that has been decontaminated of all the clusters younger than 10 Gyr and of those with retrograde orbits and/or evidence of relation to dwarf galaxies. The reduced sample contains 116 globular clusters that are tested for whether they were formed in the primordial collapse.

The 33 metal-rich globular clusters ([Fe/H] ≥ -0.75) of the reduced sample basically extend to the Solar circle and are distributed over a region with the projected axial-ratios typical of an oblate spheroidal, $\Delta x : \Delta y : \Delta z \approx 1.0 : 0.9 : 0.4$ (left panel of Fig. 1). Those outside this region appear to be related to accretion. The 81 metal-poor globular clusters span a nearly spherical region of axial-ratios $\approx 1.0 : 1.0 : 0.8$ extending from the central parts to the outer halo (right panel of Fig. 1), although several clusters in the external region still require detailed studies to unravel their origin as accretion or collapse. A new estimate of the Sun's distance to the Galactic centre, based on the symmetries of the spatial distribution of 116 globular clusters, is provided with a considerably smaller uncertainty than in previous determinations using globular clusters, $R_O = 7.2 \pm 0.3$ kpc. The metal-rich and metal-poor radial-density distributions flatten for $R_{GC} \leq 2$ kpc and are represented well over the full Galactocentric distance range both by a power-law with a core-like term and Sérsic's law; at large distances they fall off as $\sim R^{-3.9}$ (Fig. 2).

Both metallicity components appear to have a common origin that is different from that of the dark matter halo. Structural similarities between the metal-rich and metal-poor radial distributions and the stellar halo are consistent with a scenario where part of the reduced sample was formed in the primordial collapse and part was accreted in an early period of merging. This applies to the bulge as well, suggesting an early merger affecting the central parts of the Galaxy. The present decontamination procedure is not sensitive to all accretions (especially prograde) during the first Gyr, since the observed radial density profiles still preserve traces of the earliest merger(s). We estimate that the present globular cluster population corresponds to $\leq 23 \pm 6\%$ of the original one. The fact that the volume-density radial distributions of the metal-rich and metal-poor globular clusters of the reduced sample follow both a core-like power-law, and Sérsic's law indicates that we are dealing with spheroidal subsystems at all scales.[4]

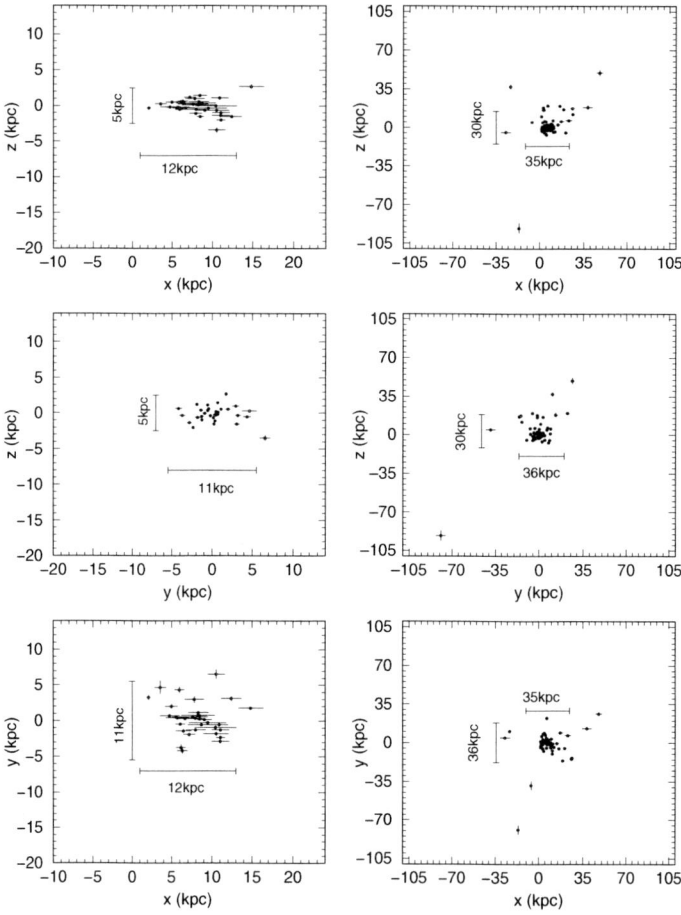

Fig. 1. *Left panels*: Spatial projections of the heliocentric positions of the metal-rich GCs of the reduced sample. Right panels: metal-poor GCs.

---

[4]Bica, E. et al. 2006, A&A 450, 105

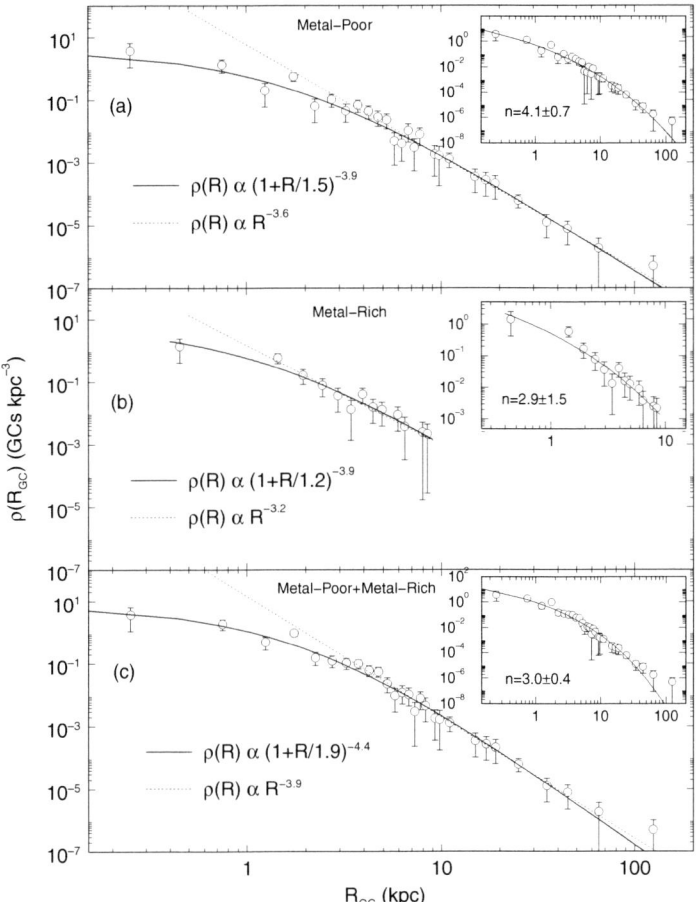

Fig. 2. Radial density profiles of the GCs in the reduced sample as a function of Galactocentric distance. Panel (**a**) metal-poor GCs; Panel (**b**) metal-rich; Panel (**c**) all GCs. Dashed line: single power-law fit for large Galactocentric distances. Solid line: fit of $\rho(R) = \rho_0/(1 + R/R_C)^\alpha$. Insets: fit of Sérsic's law $\rho(R) = a\, e^{-b\left[(R/R_C)^{(1/n)} - 1\right]}$.

RR Lyrae-Based Calibration of the Globular Cluster Luminosity Function

M. Di Criscienzo[1,3], F. Caputo[2], M. Marconi[1], and I. Musella[1]

[1] INAF-OAC, Via Moiarello 16, I-80131, Naples, Italy
[2] INAF-OAR, Via di Frascati 33, I-00040, Monte Porzio Catone, Italy
[3] Universitá "Tor Vergata", via della Ricerca Scientifica, 1,I-00040, Rome, Italy

1 Introduction

In several fields of modern astronomy, the determination of extragalactic distances is based on a ladder which is firmly anchored to Classical Cepheids and RR Lyrae stars, the "primary" standard candles for Pop. I and Pop. II stellar systems, respectively, with the properties of these variables used to calibrate "secondary" indicators which step-by-step lead us through the Local Group up to cosmologically significant distances. In this context, the Globular Cluster Luminosity Function (GCLF) is playing an ever increasing role to estimate the distance to galaxies within $\sim$ 30 Mpc, as witnessed by the huge amount of relevant papers published in the last decade. In the past, its use was hampered by the lack of observations of Globular Clusters (GC) beyond the Local Group but with the advent of modern telescopes, above all the Hubble Space Telescope (HST), it is now possible to resolve stellar populations in faraway galaxies, identify the GC candidates, measure their integrated magnitude and finally build the related luminosity function.

The GCLF method is based on the assumption that within each galaxy hosting statistically significant numbers of GCs, the frequency of the cluster integrated magnitude $V(GC)$ exhibits a universal shape which can be fitted with a Gaussian distribution

$$\frac{dN}{dV} = A e^{-\frac{[V(GC)-V(TO)]^2}{2\sigma^2}}$$

where dN is the number of clusters in the magnitude bin dV, $V(TO)$ is the magnitude of the peak or turnover, σ is the Gaussian dispersion and A the normalization factor. Once the turnover absolute value $M_V(TO)$ is known to be constant or varying in a predictable way, the distance to the parent galaxy follows immediately from the apparent (reddening corrected) magnitude of the GCLF peak.

2 Results

In Di Criscienzo et al. [2] we have tested whether the peak absolute magnitude $M_V(TO)$ of the Globular Cluster Luminosity Function (GCLF) can be used for reliable extragalactic distance determinations. Starting with the luminosity function of the Galactic Globular Clusters listed in Harris [1] catalog, in this paper we determine $M_V(TO)$ either using current calibrations of the absolute magnitude $M_V(RR)$ of RR Lyrae stars as a function of the cluster metal content [Fe/H] and adopting selected cluster samples. We show that the peak magnitude is slightly affected by the adopted $M_V(RR)$-[Fe/H] relation, with the exception of that based on the revised Baade-Wesselink method, while it depends on the criteria to select the cluster sample. Moreover, grouping the Galactic Globular Clusters by metallicity, we find that the metal-poor ([Fe/H] < -1.0, $\langle$[Fe/H]$\rangle \sim -1.6$) sample shows peak magnitudes systematically brighter by about 0.36 mag than those of the metal-rich ([Fe/H] > -1.0, $\langle$[Fe/H]$\rangle \sim -0.6$) one, in substantial agreement with the theoretical metallicity effect suggested by synthetic Globular Cluster populations with constant age and mass-function. Moving outside the Milky Way, we show that the peak magnitude of the metal-poor clusters in M31 appears to be consistent with that of Galactic clusters with similar metallicity, once the same $M_V(RR)$-[Fe/H] relation is used for distance determinations. As for the GCLFs in other external galaxies, using Surface Brightness Fluctuations (SBF) measurements we give evidence that the luminosity functions of the blue (metal-poor) Globular Clusters peak at the same luminosity within ~ 0.2 mag, whereas for the red (metal-rich) samples the agreement is within ~ 0.5 mag even accounting for the theoretical metallicity correction expected for clusters with similar ages and mass distributions. Then, using the SBF absolute magnitudes provided by a Cepheid distance scale calibrated on a fiducial distance to LMC, we show that the $M_V(TO)$ value of the metal-poor clusters in external galaxies is in excellent agreement with the value of both Galactic and M31 ones, *as inferred by a RR Lyrae distance scale referenced to the same LMC fiducial distance.* Eventually, adopting μ_0(LMC)=18.50 mag, we derive that the luminosity function of metal-poor clusters in the Milky Way, M31, and external galaxies peak at $M_V(TO)$=-7.66 ± 0.11 mag, -7.65 ± 0.19 mag and -7.67 ± 0.23 mag, respectively. This would suggest a value of -7.66 ± 0.09 mag (weighted mean), with any modification of the LMC distance modulus producing a similar variation of the GCLF peak luminosity.

References

1. W. Harris, 1996, AJ, 112 (update 2003 available on http://physun.physics.mcmaster.ca)
2. M. Di Criscienzo, F. Caputo, M. Marconi, I. Musella, 2006, MNRAS, 365, 1357

Globular Cluster Systems in Spiral Galaxies Using ACS Imaging

Kieran Forde

University of California, Santa Cruz forde@ucolick.org

1 Introduction

Extensive studies have been carried out on the populations of extragalactic GC in *early-type* (E/S0) galaxies using HST wide-field camera (see [2] for a review). However relatively very little information exists on the GCs of the *late-type* (Sa/Sb/Sc) galaxies. This can, in part, be attributed to the fact that GCs are more difficult to detect on the patchy star forming background typical of a spiral galaxy and also that E/S0 galaxies exhibit a much richer population of GCs than observed in Spirals.

We present a preliminary investigation into the globular cluster systems of three *late-type* galaxies (see Table 1). The images were taken using the *Hubble Space Telescope* Advanced Camera for Surveys Wide Field Channel (ACS/WFC) with the F435W and F814W filters.

Table 1. Properties of the observed galaxies. Morphological types taken from NASA/IPAC Extragalactic database; Magnitudes taken from LEDA

Galaxy ID	Type	RA	DEC	D (Mpc)	B	A_B
NGC 2683	Sb	08 52 41.68	+33 25 10.4	7.73[a]	10.64	0.089[d]
NGC 3957	Sa	11 54 01.0	−19 34 05	22.3[b]	13.08	0.200[d]
NGC 5170	Sc	13 29 48.8	−17 57 57	19.1[c]	12.35	0.343[d]

[a] Kregel and Kruit 2005; [b] from Pohlen et al. 2004; [c] from Karachentsev et al. 2004; [d] taken from Schlegel, Finkbeiner and Davis 1998

2 Photometry

Photometric reductions were carried out using DAOPHOT package within IRAF. The star finding was carried out on a median-subtracted image with sky level restored to that of the original frame. Aperture photometry was carried out using the PHOT task using an aperture of 4 pixels (see [3] for more details on optimal aperture choice). The photometry was then adjusted to a 0.5" aperture. These magnitudes were then corrected to an infinite aperture using the values from Sirianni et al. [3]. Photometric calibration and transformations to BVRI magnitudes from ACS/WFC magnitudes was achieved via the standard procedure outlined in Sirianni et al. [3].

Figure 1 shows both the color distribution and the 'CMD for NGC5170 with a best-fit line superposed.

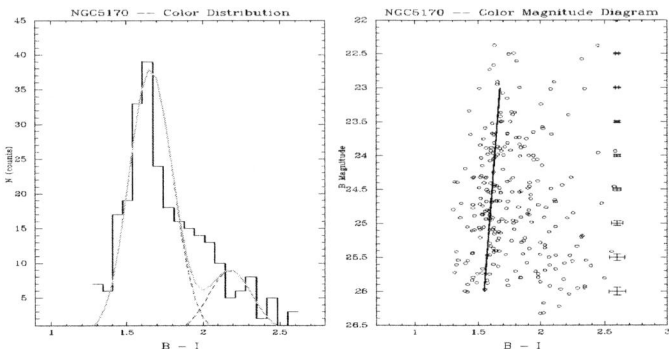

Fig. 1. The color distribution for the NGC5170 GCs in the ranges $22.0 < B < 26.5$. A KMM [1] test was applied to the data yeilding peaks at $(B - I)_0 blue = 1.66$ and $(B - I)_0 red = 2.18$. Over-plotted are Gaussians corresponding to the two color peaks detected and their sum. (right panel) $[(B - I)_0, B]$ CMD for NGC5170. Error bars indicate the mean color and B magnitude errors for the data.

3 Conclusions and Future Work

The *tilt* of the color distribution which we observe in NGC5170 indicates that the blue GCs seem to become redder at higher luminosities. This trend has been noted in ellipticals and S0 galaxies but never before in a *late-type* spiral. This implies that the reddening of the blue population might be a *general* property of galaxies. A deficiency of red GCs in both CMDs (*above*) is observed. This is what is expected from an Sc galaxy (NGC5170) however NGC3957 (Sa-type) also exhibits this red GC deficiency. The analysis of NGC2683 reveals very few detected GC candidates. A more detailed analysis is required to fully investigate the GC systems of NGC5170 and NGC3957.

References

1. Ashman, K. M., Bird, C, M. & Zepf, S. E., 1994, *AJ*, 108, 2348
2. Brodie, J. P., & Strader, J. 2006, *ARA&A* 44, 193
3. Sirianni, M., et al. 2005, *PASP*, 117, 1049–1112

Laser Guide Star Imaging of M31 Globulars

Michael D. Gregg,[1,2] Arna Karick,[1,2] and Bruce Macintosh[2]

[1] University of California, Davis
[2] IGPP/LLNL gregg,akarick,bmac@igpp.ucllnl.org

The respective fundamental planes of globular clusters and elliptical galaxies have very different slopes and locations in parameter space [1]. This is not understood, though it must be a key to the formation of both types of stellar systems. The recent discovery of large populations of "ultra-compact dwarf" galaxies (UCDs) in the Fornax and Virgo clusters has renewed interest in understanding the globular/galaxy connections and differences. The UCDs overlap the globulars at luminosities of $M_V \sim 10$, and distinguishing the two classes requires high dispersion spectroscopy and high resolution imaging to measure their small velocity dispersions (< 30 km/s) and sizes (< 20 pc). The plentiful UCDs may account for a sizable portion of the populous "globular" cluster systems around giant ellipticals; if so, they provide constraints on the earliest periods of formation of giant ellipticals.

UCDs may be masquerading as globular clusters right here in the Local Group. Meylan et al. [2] make an excellent case for the M31 globular Mayall-II/G1 being the remnant nucleus of a dwarf elliptical. It has $\sigma = 28$ km/s, M/L $\sim 2-3$ times higher than typical globulars, a large ellipticity, and an abundance spread, all consistent with it being a UCD galaxy, not a globular. There are at least three other M31 "globulars" with $\sigma > 25$ km/s, and many more comparably bright and not yet characterized.

We have begun a program using Keck NIRC2 Laser Guide Star Adaptive Optics imaging to investigate the structure and stellar populations of the brightest globular clusters in M31 (Fig. 1) We will add HIRES echelle spectra next year. These data will refine the globular cluster fundamental plane, and will help to understand why galaxies and star cluster dynamical relations differ and will reveal which M31 globulars may be UCDs rather than simple star clusters, yielding insights into spiral galaxy formation. Excellent conditions during our first run enabled us to produce some $J - K$ vs. K CMDs (Fig. 2); when fully calibrated, these will yield metallicity estimates.

Keck LGSAO observations are now routine, making it feasible to image > 100 M31 globulars in just a few nights of observing. HIRES observations will be more time consuming, but they can be carried out for the brightest 50 clusters in just 5–10 nights. The resulting data set will characterize the M31 globular cluster fundamental plane and reveal the presence of any UCDs in the M31 halo.

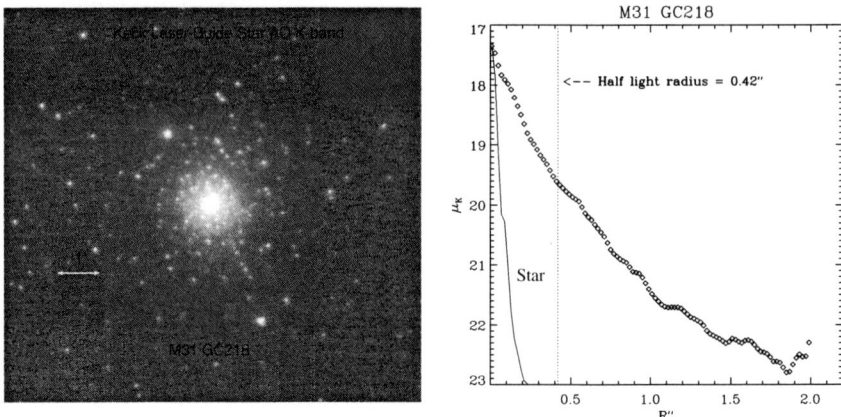

Fig. 1. *Left:* An example LGSAO K-band image of a typical M31 globular, GC218. Resolution is comparable to NICMOS on HST. *Right:* The surface brightness profile shows that GC218 has a half-light radius of 0.42″ and the ease with which the LGS system resolves the cluster cores, enabling measurement of cluster core sizes down to 0.1″ in size (twice the stellar psf).

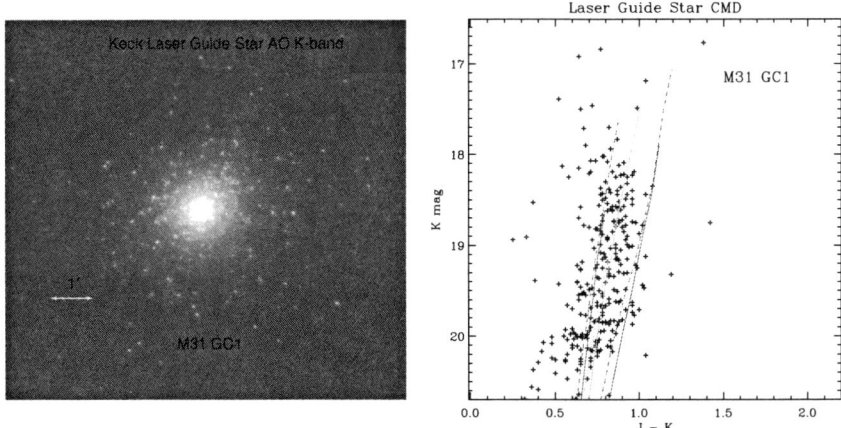

Fig. 2. *Left:* LGSAO K-band image of the brightest M31 globular, GC1, obtained in September 2005 under good conditions, strehl ∼0.35. A J-band image of comparable quality, though not as deep, was also obtained. *Right:* Resulting CMD, calibrated by shifting zeropoints to bring the cluster into rough agreement with the appropriate Padova isochrone (shown are [Fe/H] = -1.68, -1.28, -0.67, age 13 Gyr).

Acknowledgements. We acknowledge support from NSF grant No. 0407445 and the Institute of Geophysics and Planetary Physics, under the auspices of the U.S. Dept. of Energy, Lawrence Livermore National Laboratory, contract W-7405-Eng-48.

References

1. Djorgovski, S.G., Gal, R.R., McCarthy, J.K. et al. 1997, ApJ 474, L19
2. Meylan, G., Sarajedini, A., Jablonka, P. 2001, AJ 122, 830

GALEX UV Observations of M31 Globular Clusters

Soo-Chang Rey[1], Sangmo T. Sohn[2], R. Michael Rich[3], Suk-Jin Yoon[4], Chul Chung[4], Sukyoung K. Yi[4], and Young-Wook Lee[4]

[1] Department of Astronomy and Space Science, Chungnam National University, Daejeon 305-764, Korea screy@cnu.ac.kr
[2] Korea Astronomy and Space Science Institute, Daejeon 61-1, Korea
[3] Department of Physics and Astronomy, University of California at Los Angeles, Box 951547, Knudsen Hall, Los Angeles, CA 90095
[4] Center for Space Astrophysics, Yonsei University, Seoul 120-749, Korea

1 Age Distribution of M31 Globular Clusters

The hot He-burning horizontal-branch (HB) stars and their progeny are most likely dominant ultraviolet (UV) sources in the old stellar population systems such as globular clusters (GCs). The integrated FUV flux can be an age indicator of GCs and allows us to investigate age distributions of GCs within a given galaxy or between galaxies [1,5]. The unprecedented set of UV photometry for M31 by *Galaxy Evolution Explorer* (*GALEX*), coupled with most recent detailed population models enable to study detailed global UV properties of M31 GCs.

The M31 images were obtained by *GALEX* in two UV bands: FUV (1350–1750Å) and NUV (1750–2750Å). Sources in our *GALEX* photometry were cross-matched with the M31 GCs and GC candidates in the Revised Bologna Catalog [2]. Details on the observations and analysis can be found in Rey et al. [4], (2006 in preparation).

Figure 1 shows the distribution of $(FUV-V)_0$ vs. $(B-V)_0$ for M31 (open circles) and Galactic (crosses) GCs, which is a representative plot for the age distribution of GCs. We also superpose our model isochrones for different ages. Old GCs of M31 and Milky Way are well located in the age range (± 2 Gyr) of model predictions (dotted lines) and show no significantly different age distribution between two GC systems.

We detect three candidates UV bright metal-rich ([Fe/H]>-1) clusters (filled circles) which have $FUV - V$ colors similar to or smaller than those of NGC 6388 and NGC 6441, peculiar metal-rich Galactic GCs with prominent blue HB populations. The inference is that these anomalous UV bright metal-rich GCs may be usual in more or most galaxies.

Intermediate-age GCs (IAGCs) are relatively faint in FUV since they contain only warm main-sequence turnoff stars [3]. If the IAGCs are truly intermediate-age, then they should not be detected from our *GALEX* FUV photometry within the current detection limit. Among the IAGC candidates proposed by recent spectroscopic observations, more than 50% GCs are

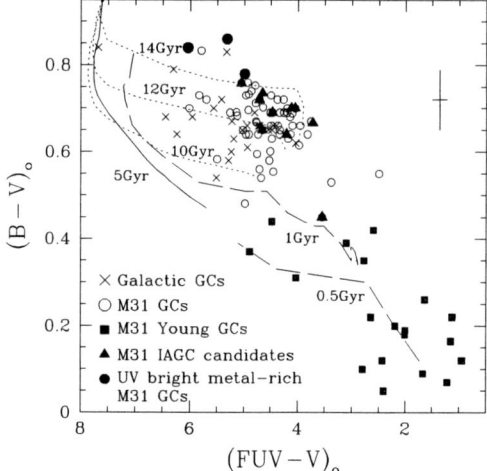

Fig. 1. $(FUV-V)_0$ vs. $(B-V)_0$ in M31 and Galactic GCs. Our model isochrones for different ages are also superposed. Three metal-rich ([Fe/H]>-1) M31 GCs (*filled circles*) show significant FUV flux comparable to those of NGC 6388 and NGC 6441, peculiar metal-rich Galactic GCs with anomalous hot HB stars. Among IAGC candidates proposed by recent spectroscopic observations, many GCs are detected in *GALEX* observations (*filled triangles*). We confirm that there is a significant fraction of young GCs (*filled boxes*) in M31.

detected in *GALEX* observations (filled triangles). This suggests that some of the spectroscopically suggested IAGCs may not be truly intermediate-age, but rather older GCs with developed HB stars.

We confirm that there is a significant fraction of young (< 2 Gyr) GCs (filled boxes) in the whole GC system of M31 which show UV and optical properties distinct from GCs in the Milky Way. A large fraction of young GCs show kinematical properties of a thin, rapidly rotating disk component, which indicates that significant cluster formation has been triggered recently in the disk of M31.

References

1. Catelan, M. 2005, astro-ph/0507464
2. Galleti, S., Federici, L., Bellazzini, M., Fusi Pecci, F., & Macrina, S. 2004, A&A, 416, 917
3. Lee, H.-C., Lee, Y.-W., & Gibbson, B. K. 2003, in Extragalactic Globular Cluster Systems, ed. M. Kissler-Patig (Berlin: Springer), 261
4. Rey, S.-C., Rich, R. M., et al. 2005, ApJ, 619, L119
5. Yi. S. 2003, ApJ, 582, 202

Integrated Spectroscopy of Galactic Globular Clusters

Ray M. Sharples[1] and Jaeil Cho[2]

[1] Department of Physics, University of Durham `r.m.sharples@durham.ac.uk`
[2] Department of Physics, University of Durham `jaeil.cho@durham.ac.uk`

1 Introduction

The Milky Way globular cluster system provides a benchmark against which the analysis of extragalactic globular cluster systems should be compared. Integrated spectroscopy of Galactic globular clusters is made difficult by their large angular extent on the sky and the fact that many of the most interesting high metallicity clusters are viewed towards the Galactic centre, where the contamination by foreground/background field stars is high and there are additional complications due to high and variable extinction by dust. This paper presents the results from a new intermediate spectral resolution survey of integrated spectra and compares the derived line indices with previous comparable surveys [1, 2] and spectral synthesis models [3–5].

2 Observations

Integrated spectra for 24 Galactic globular clusters were obtained using the Intermediate Dispersion Spectrograph on the 2.5 m Isaac Newton Telescope. For each globular cluster, the slit was scanned in two orthogonal directions and integrated spectra were extracted from a square aperture with sides of length equal to twice the core radius of the cluster. The spectra have a wavelength range from 4000–5400Å with an instrumental resolution of FWHM=2.0Å. Spectral reductions were performed using standard IRAF routines. The orthogonal scans have been used to estimate the systematic uncertainties in line indices for diffuse clusters due to background corrections and foreground contamination. A large number of Lick standard stars were also observed using the same instrumental setup for calibration.

2.1 Results

As an initial approach we have used the smoothed spectra to derive Lick line indices and compare with published models in the literature [3–5]. In the future we intend to use the spectra to refine the line index definitions and exploit the higher spectral resolution available. Our data are well calibrated onto the Lick line index system and agree well with published data for common clusters from [1] and [2]. Figure 1 shows a comparison of Balmer line index measurements with the models from [5] and [3]. A flattening in the trend of Balmer line strength with decreasing metallicity at low metallicites is best matched by models with a significant horizontal branch component (cf. [3]).

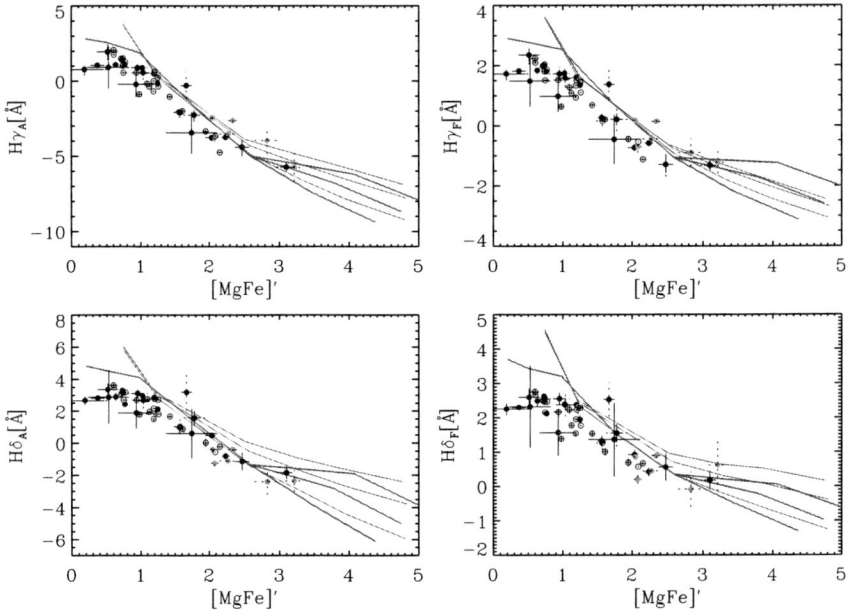

Fig. 1. Age-metallicity diagnostic plots with higher order Balmer line indices. The thin grey lines are SSP models from [5]; thick red lines are SSP models from [3]. Solid circles are this Galactic GC sample, the shaded circles are from [1], and open circles from [2]. The combined sample gives a total of 54 clusters with high quality line index data.

References

1. Puzia, T. H. et al. 2002, A&A, 395, 45
2. Schiavon, R. P. et al. 2005, ApJS, 160, 163
3. Lee, H.-C. & Worthey, G. 2005, ApJS, 160, 176
4. Thomas, D., Maraston, C. & Bender, R. 2003, MNRAS, 339, 897
5. Thomas, D., Maraston, C. & Korn, A. 2004, MNRAS, 351, L19

Part VI

Globular Cluster Systems in Early-Type Galaxies

Globular Cluster Systems: Do They Really Trace Star Formation? (Or Rather: What Mode of Star Formation Do They Trace?)

Michael A. Beasley

Instituto de Astrofísica de Canarias, La Laguna, Tenerife, Spain beasley@iac.es

1 Globular Clusters and Their Host Galaxies

A good reason to believe that globular clusters (GCs) are representative of the star formation in galaxies is that the mean colours of GC systems are correlated with the luminosity of their host galaxies [9,35]. The GC systems of massive galaxies appear bimodal in their colour distributions [38], and it has now been demonstrated that both the metal-rich (red) and metal-poor (blue) GC peak positions are a function of galaxy luminosity [20, 24, 32]. Although there is increasing evidence for the existence of (an undetermined fraction of) young GCs in elliptical galaxies [23, 28](T Bridges, these proceedings), the colour distributions of GC systems are generally thought to reflect underlying metallicity rather than age distributions. Thus, more massive galaxies have, on average, more metal-rich GC systems. However, if GC formation were to mirror the formation of the galaxy field stars at a fixed efficiency (by field stars, we refer to all the stars now in the spheroid), the metallicity distributions of GCs and field stars should be very similar. They do not appear to be [17] (Sect. 5). Understanding this difference between the GCs and galaxy stars is essential if we are to use GCs to unravel the detailed formation histories of galaxies. In the following some general ideas about the formation of GCs in the context of galaxy formation are presented. We then discuss ongoing work on the comparison of the GCs and field stars of the nearest elliptical galaxy NGC 5128.

2 The Importance of Being Merged

Stars and star clusters form efficiently in regions where the local gas density is high [29], possibly coupled with pressure-induced shock fronts [2]. One environment where these criteria are met is in galaxy mergers (Fig. 1). Indeed, there is ample evidence that massive, metal-rich star clusters form in mergers [36]. Moreover, mergers are appealing in that they may unify high- and low-redshift galaxy populations, such as Lyman-Break galaxies which would need to grow by factors of up to ∼100 in stellar mass to be associated with massive galaxies at the present day [11, 30]. Models in which the red peak in the GC systems of massive galaxies is formed through successive dissipative mergers

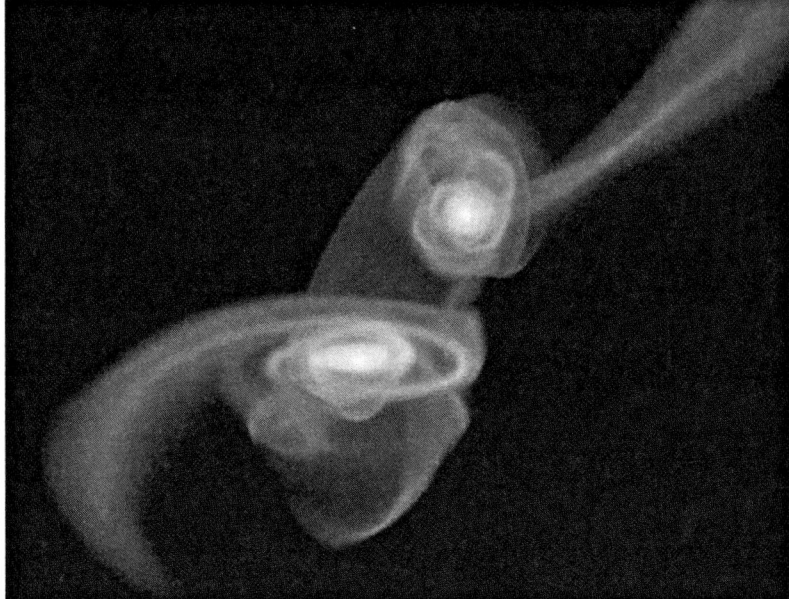

Fig. 1. 3×10^8 particle N-body simulation of the merger of the M31-Milky Way system. Image courtesy of J. Dubinski.

have had some success in reproducing the observations [1,5,6]. In this picture, the low-mass galaxy regime provides interesting constraints. The discovery of red populations of GCs in dwarf galaxies [24] would require the dissipative merging of dwarf-dwarf systems. Cosmological simulations show that some 30% of cluster substructure undergoes a gravitational encounter per orbit [18]. However the high ratio of the relative galaxy velocity to internal velocity dispersion in dwarfs makes their mutual mergerging a rare event in galaxy clusters. Of course, all such mergers could occur prior to virialisation (while the dwarfs retain gas to form GCs). Old ages for the red GCs in dwarfs would support this notion [3]. Correlating the incidence of kinematically decoupled cores in dwarf galaxies [12] (which are generally interpreted as signatures of mergers in massive galaxies) with the existence of red GCs in dwarfs would provide a test of the dwarf-merger hypothesis.

Interestingly, the existence of red GCs in ellipticals with very high [α/Fe] ratios has been reported (T. Puzia, these proceedings). Such high α-abundances–higher than the bulk of the galaxy star – suggest very short star formation timescales. Chemical evolution models require either very high initial [α/Fe] ratios in the proto-Globular gas, or some level of self-enrichment in the nascent GCs to achieve these abundance ratios [10]. In the former scenario, the question then becomes why aren't the galaxy stars enriched with α-products similar to the GCs? Are these red GCs some of the first metal-rich stars to form in the galaxy? On the other hand, the possibility of self-enrichment is intriguing, and has already been cited as the possible origin of the "blue-tilt" increasingly seen in blue GC subpopulations [15, 31] (L. Spitler, these proceedings).

3 Then Destroyed

However, in terms of GC formation, for some time the key issues have been what fraction of clusters will survive the effects of internal and external dynamical processes, and whether the power-law mass distribution of young clusters can dynamically evolve to resemble the log-normal distribution of GCs seen in old systems. Recent work on intermediate-age cluster systems does indeed show a turn-over in their mass distributions [14] (Fig. 2), which are broadly consistent with analytical destruction models of GC systems [13]. These results seem to suggest that merger-formed clusters can evolve to resemble the old, red GCs in elliptical galaxies. However, it has been argued that the intermediate-aged clusters in [14] may be *too* faint for existing models to comfortably evolve the clusters to look like "classical" GCs over a Hubble time [7]. The dynamical evolution of GCs in a "live" halo (i.e., non-

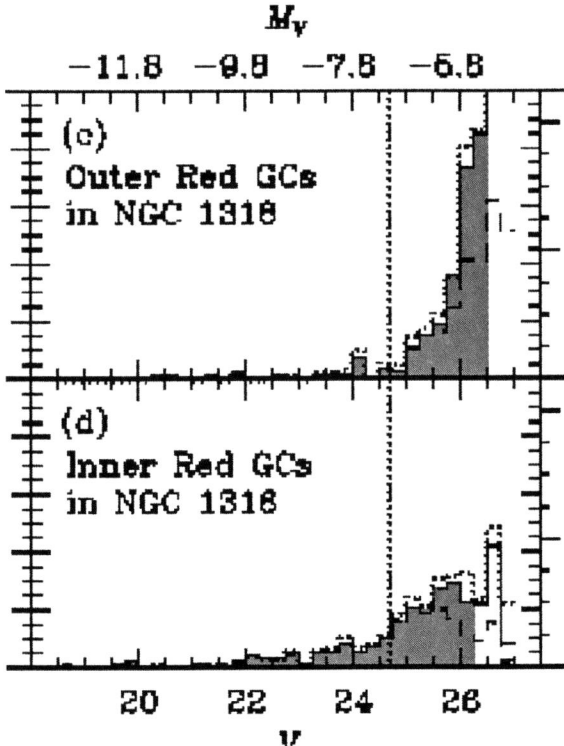

Fig. 2. Luminosity functions of inner and outer red GCs in the merger remnant candidate NGC 1316 [14]. The inner clusters show a turnover at $M_V \sim -6$.

230   M.A. Beasley

static, non-spherical potential) has yet to be simulated, and would prove very valuable in this regard.

4 GC Formation in Disks?

It has long been recognised that massive clusters can also form in a variety of other environments, such as in interacting irregulars (e.g., the LMC-SMC system) and in quiescent spiral disks [21]. Indeed the inference is that the density threshold, not the process itself, is that which is most important. High resolution numerical simulations [19] show that at high redshifts, significant cluster formation may be achievable in regions of high local density (such as in the spiral arms) of disks (Fig. 3). Indeed, it is a natural expectation that in the absence of heating, gas will cool into rotationally supported disks, as implemented in semi-analytical treatments [30]. It is worth noting that in the simulations of [19], the star formation in disks may hardly be called "quiescent" since they continue to undergo merger and accretion events up until the end of the simulations at $z = 3.3$. Subsequent large-scale interactions (mergers) may reconfigure the extant cluster systems into a spheroid, with the occurrence of further star formation dependent upon the cold gas fraction in the disk. Alternatively, the disk may remain relatively unperturbed which could perhaps lead to an old disk system of clusters [22].

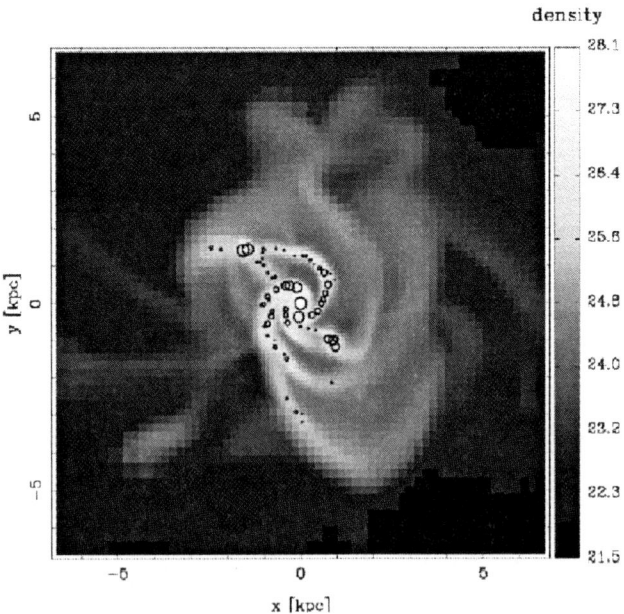

Fig. 3. Density map showing the sites of possible GC formation from ART SPH simulations of a disk at high redshift [19].

The term "clusters" is deliberately used in the above discussion. Do we expect classical GCs to form in disks? In this context, the existence of "faint-fuzzies" is relevant [8]. These objects, which typically exhibit an effective half-light radius of $\sim 10\,\mathrm{pc}$, rather then the more typical $\sim 3\,\mathrm{pc}$ for GCs, generally show disk-like spatial distributions, and where available, disk-like kinematics [8, 25]. We speculate that the size difference between GCs and faint-fuzzies may originate in the presence or absence of external pressure differences. Specifically, a shock induced pressure increase, which may aid in triggering cluster formation in mergers, could plausibly lead to compact GCs. The absence of such external pressure may result in the formation of more spatially extended objects. However, the sizes of young clusters in the M51 spiral galaxy (Scheepmaker et al., these proceedings) may rule out this naïve interpretation. We know mergers are not a *requirement* for massive cluster formation, but they may be for the majority of "true" GCs.

5 NGC 5128 – A Test Case

A key to understanding the connection between the GCs and field stars in galaxies is by directly comparing their metallicity and (ideally) age distribution functions (here abbreviated to MDF and ADF respectively). In principal, an MDF and ADF can be constructed for an extragalactic GC system using low-resolution spectroscopy coupled with modern stellar population models. However, the necessity of obtaining photometry of individual stars in a galaxy to derive its stellar MDF limits us to the nearest galaxies. The nearest giant elliptical in which this has been performed is NGC 5128, lead by the efforts of B. Harris and G. Harris and co-workers. A secure estimate of the age (let alone the full ADF) for this galaxy requires reaching the main-sequence turn-off, which at the distance of NGC 5128 corresponds to $V_{\rm TO} \sim 31$, is beyond the capabilities of current instrumentation. Here we will focus on ongoing work on the MDFs of the GCs and field stars in this galaxy. We [4] have undertaken a spectroscopic survey of the GC system of NGC 5128 galaxy for the purpose of studying its MDF, ADF and kinematics (K. Woodley, these proceedings) Candidate lists were constructed from wide-field photometric studies [16, 26]. Spectroscopy was performed using the 2-degree field (2dF) instrument at the Anglo-Australian telescope between the nights of May 6th and May 11th 2003.

Metallicities for the GCs were derived using two complementary methods: multivariate fits to stellar population models (PFB04) [27], and a principal component analysis (PCA) [33]. The PFB04 approach allows for the estimate of metallicity, age and α-abundance. The PCA approach, based upon Galactic GCs and the metallicity scale of [39], can be used to derive accurate metallicities from relatively lower S/N spectra, assuming old ages. In what follows, we use the (empirical) PCA metallicities, after having satisfied ourselves that the majority ($\sim 90\%$) of the GCs in NGC 5128 are old: using the PFB04 approach

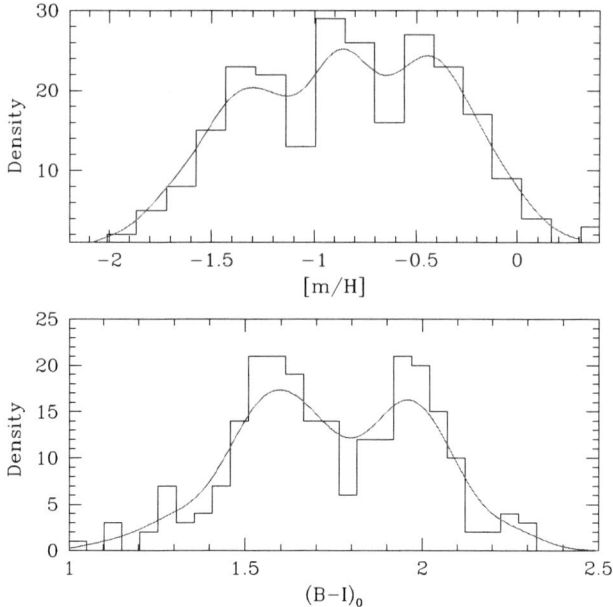

Fig. 4. (*upper panel*) Metallicity distribution of 250 GCs in NGC 5128. Metallicities derived using principle components analysis [33]. The smooth distribution is a Guassian kernel density estimate from the unbinned data. (*lower panel*) corresponding $B - I$ colours of the GCs.

does not yield significantly different MDFs. The spectroscopic MDF for 250 GCs in NGC 5128, and their corresponding $B-I$ colours, are shown in Fig. 4. The MDF exhibits three peaks, rather than the expected two. The "trimodality" persists when a kernel density estimate is used, and is not an artefact of our analysis methods since it is evident in individual line-strength indices (e.g. Mg b). However, the *colour* distribution appears bimodal, as is typically observed in the GC systems of massive ellipticals. This clearly demonstrates that the colour-metallicity relation is non-linear; the intermediate metallicity peak in the MDF becomes incorporated into the *blue* peak of the colour distribution. On the basis of stellar population modelling, it has been proposed that the colour-metallicity relation is non-linear with a quasi-inflection point, which can make an intrinsically unimodal metallicity distribution appear bimodal in colour [37]. This clearly would have profound implications for the study of extragalactic GC systems if correct, and would simplify enormously models for the formation of GC systems. Our data certainly support the notion that the conversion between colour and metallcity is nonlinear, however it does not appear consistent with a unimodal distribution. Rather, the MDF shows fairly complex substructure.

A comparison of the MDFs of the Milky Way GC system, the NGC 5128 GC system and the halo field stars in NGC 5128 is shown in Fig. 5. The MDF

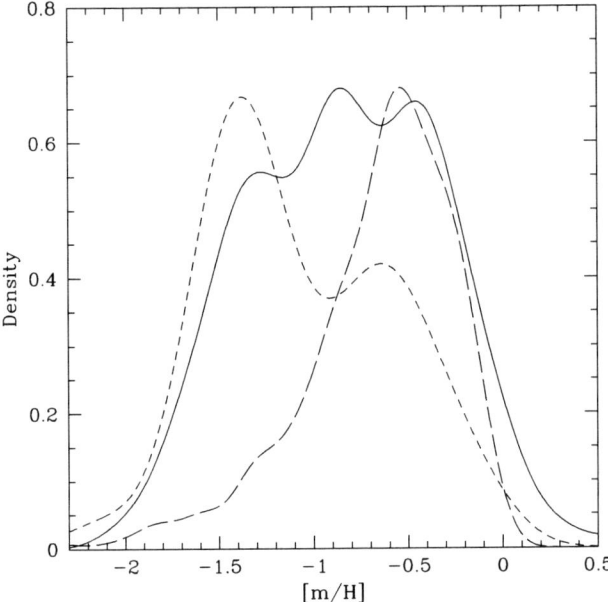

Fig. 5. Comparison of the metallicity distributions of Milky Way GCs (short-dashed line) NGC 5128 GCs (long-dashed lines) and the NGC 5128 field-star distribution at 20 kpc [17].

of the field stars was derived from photometry rather than spectroscopy [17], however, all three metallicity scales are that of [39], which broadly reflects total metallicity [Z/H] [34]. One obvious issue is at what radius the clusters and field stars should be compared. We have opted to compare with an outer (20 kpc) stellar field (~ 4 effective radii), since the stellar metallicity gradient appears flat at that radius. The three MDFs are clearly very different. The positions of the blue peaks in the Milky Way and NGC 5128 are consistent with the galaxy metallicity-GC metallicity relation (e.g., [32]). The red peak of the NGC 5128 GCs and the field stars are roughly coincident at [m/H]$\sim$ -0.5. The position of the red peak of the Milky Way GCs is more metal-poor than in NGC 5128, but is also consistent with the relation in [32]. The intermediate-metallicity peak of the NGC 5128 GCs appears to have no obvious counterpart in either the field stars or the Milky Way GCs. These results emphasize the importance of deriving spectroscopic metallicities for extragalactic GC systems.

Acknowledgements. MB would like to thank the organisers for hosting an excellent conference. MB would like to acknowledge his collaborators on the NGC 5128 project: Terry Bridges, Duncan Forbes, Bill Harris, Gretchen Harris, Eric Peng and Glen Mackie.

References

1. Ashman, K.M. & Zepf, S.E.: ApJ **384**, 50 (1992)
2. Barnes, J.E.: MNRAS **350**, 798 (2004)
3. Beasley, M.A., Strader, J., Brodie, J.P., Cenarro, A.J., & Geha, M.: AJ **131**, 814 (2006)
4. Beasley, M.A., Bridges, T.J., Peng, E.W., Forbes, D.A., Harris, W.E., Harris, G.L.H., & Mackie, G.: MNRAS in prep. (2006)
5. Beasley, M.A., Baugh, C.M., Forbes, D.A., Sharples, R.M., & Frenk, C.S.: MNRAS **333**, 383 (2002)
6. Bekki, K., Forbes, D.A., Beasley, M.A., & Couch, W.J.: MNRAS **335**, 1176 (2002)
7. Brodie, J.P. & Strader, J.: ARA&A in press (2006)
8. Brodie, J.P. & Larsen, S.S.: AJ **124**, 1410 (2002)
9. Brodie, J.P. & Huchra, J.P.: AJ **379**, 157 (1991)
10. Fritze-v. Alvensleben, U. & Burkert, A.: A&A **300**, 58 (1995)
11. Conselice, C.J.: ApJ **638**, 686 (2006)
12. De Rijcke, S., Dejonghe, H., Zeilenger, W.W., & Hau, G.K.T.: A&A **426**, 53 (2004)
13. Fall, M.S. & Zhang, Q.: ApJ **561**, 751 (2001)
14. Goudfrooij, P., Gilmore, D., Whitmore, B., & Schweizer, F.: ApJ **613** 121 (2004)
15. Harris, W.E., Whitmore, B.C., Karakla, D., Okon, W., Baum, W.A., Hanes, D.A., & Kavelaars, J.J.: ApJ **636**, 90 (2006)
16. Harris, G.L.H. et al.: AJ **128**, 712 (2004)
17. Harris, G.L.H. & Harris, W.E.: AJ **120**, 2423 (2000)
18. Knebe, A., Gill, S.P.D., & Gibson, B.K.: PASA **21**, 216 (2004)
19. Kravtsov, A.V & Gnedin, O.Y.: ApJ **623** 650 (2005)
20. Larsen, S.S., Brodie, J.P., Huchra, J.P., Forbes, D.A., & Grillmair, K.: AJ **121**, 2974 (2001)
21. Larsen, S.S. & Richtler, T. : A&A **345**, 59 (1999)
22. Olsen, K.A.G., Miller, B.W., Suntzeff, N.B., Schommer, R.A., & Bright, J.B.: AJ **127**, 2679 (2004)
23. Pierce M.: MNRAS **368**, 325 (2006)
24. Peng, E.W. et al.: ApJ **631**, 95 (2006)
25. Peng, E.W. et al.: ApJ **639**, 838 (2006)
26. Peng, E.W., Ford, H.C., & Freeman, K.C.: ApJS **150**, 367 (2004)
27. Proctor, R.N., Forbes, D.A., & Beasley, M.A.: MNRAS **355**, 1327 (2004)
28. Puzia, T.H. et al. : A&A **439**, 997 (2005)
29. Schmidt, M.: ApJ **129**, 243 (1959)
30. Somerville, R.S., Primack, J.R., & Faber, S.M.: MNRAS **320** 504 (2001)
31. Strader, J., Brodie, J.P., Spitler, L., & Beasley, M.A.: AJ in press (2006)
32. Strader, J., Brodie, J.P., & Forbes, D.A.: AJ **127** 3431 (2004)
33. Strader, J. & Brodie, J.P.: AJ **128**, 1671 (2004)
34. Thomas, D., Maraston, C., & Bender, R.: MNRAS **339**, 897 (2003)
35. van den Bergh, S. : ARA&A **13**, 217 (1975)
36. Whitmore, B.C. & Schweizer, F.: AJ **109**, 960 (1995)
37. Yoon, S.-K., Yi, S.K., & Lee, Y.W.: Science **311**, 1129 (2006)
38. Zepf, S.E. & Ashman, K.M.: MNRAS **264**, 611 (1993)
39. Zinn, R. & West, M.J.: ApJS **55**, 45 (1984)

Globular Clusters in Early Type Galaxies

Jean P. Brodie

UCO/Lick Observatory, University of California, Santa Cruz, CA 95064, USA
brodie@ucolick.org

1 Introduction

Over the last 15 years or so, attempts to understand the origin of bimodality in globular cluster (GC) color distributions have been organized around three main scenarios linking GC formation with the formation of their host galaxies. The "major merger" model [23,1], the "in situ multiphase collapse" model [9], and the "accretion/stripping" model [6,7] make radically different predictions about the relative ages of the red and blue GCs. This fact focussed significant research effort over this period into determining accurate ages for GCs in galaxies spanning a range of luminosities, morphological types and environments. While elements of each of these scenarios has survived to the present, our current ideas on GC formation are closely tied to hierarchical galaxy formation in a ΛCDM cosmology. This short article will highlight recent progress in establishing GC ages, metallicities, specific frequencies and spatial distributions, as well as scaling relations for GCs, and between GC and galaxies. An intimate connection between GC and field star formation is implicit in this work and provides the foundation for our conclusions. The potential for insight into the epoch and homoegeneity of reionization and the link between baryons and dark matter are also discussed in the context of blue GC formation early in the history of the universe. These and related topics are discussed in more detail in the recent review of extragalactic globular clusters and galaxy formation by Brodie and Strader [5].

2 Spirals vs. Ellipticals

It has been known for more than twenty years that the GC system of the Milky Way (MW) is bimodal with red and blue peak metallicities of [Fe/H] $\sim$ –1.5 and –0.6 respectively. High quality imaging surveys of large numbers of early type galaxies, (e.g. [16, 14, 19, 25]) reveal peak colors for the blue and red subpopulations consistently occurring for massive ellipticals near the values $(V - I)$ =0.95 or $(g - z)$ =0.95 for the blue GCs, and $(V - I)$ =1.18 or $(g - z)$ =1.40 for the red ones. While there are some uncertainties in the details of the color-metallicity transformations, the inferred metallicities for the peaks in early type galaxies correspond quite closely to those measured directly for the MW GCs, with appropriate adjustment for galaxy luminosity (see Sect. 5). More recent work on the GC systems of other spiral galaxies confirms the notion that the GC systems of spirals and ellipticals are remarkably similar (e.g. [15, 24, 11]). This suggests that we should be looking for a universal formation mechanism for GCs in luminous galaxies, independent of morphological type.

3 GC Ages

We know too, of course, that MW GCs are essentially all extremely old. The results of more than a decade's worth of integrated light spectroscopy of individual extragalactic GC, carried out on 8–10 m class telescopes, confirms that the vast majority of both blue *and* red GCs are extremely old. These results are summarized in Fig. 1, taken from Strader et al. [26]. This meta-analysis of the highest quality Keck data directly compares age and metallicity-sensitive indices measured in MW GCs to those measured in blue and red GCs in a variety of galaxies, including spirals and recent mergers. At all metallicities the extragalactic subpopulations appear coeval and at least as old as the MW GCs. This suggests mean ages >10–13 Gyr for the extragalactic GCs. The paper by J. Cenarro in this volume illustrates the superb data quality

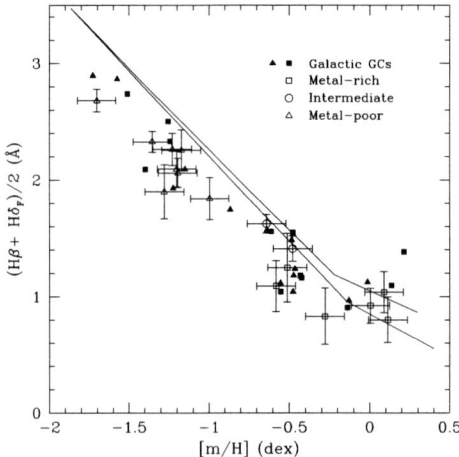

Fig. 1. Combined Balmer index vs. metallicity for extragalactic GC subpopulations and Galactic GCs [26]. The extragalactic subpopulations: *open triangles* (metal-poor), *open circles* (intermediate-metallicity), and *open squares* (metal-rich). Galactic GCs: *filled triangles* [12] and *filled squares* [20]. At all metallicities, the extragalactic subpopulations appear coeval with (or older than) the comparison Galactic GCs. The data are not fully calibrated to the models (*solid lines*), so cannot be directly compared, and the offset between the data and the model lines is not significant.

that can be achieved with today's largest telescopes, and the high level of confidence that can be placed the estimates of old ages.

4 Specific Frequencies

Recent mergers should, of course, only affect the metal-rich peak; independent of the details of new star/GC formation, the specific frequency S_N will not increase. Rhode, Zepf, and Santos [21] measured T (similar to S_N but normalized to host galaxy mass rather than light) for metal-poor GCs in 13 massive nearby early and late type galaxies. Figure 3, taken from Brodie and Strader [5], shows both T_{blue} and T_{red} vs. stellar galaxy mass. Rhode et al. found an overall correlation between T_{blue} and galaxy mass. The spirals are all consistent with $T_{blue} \sim 1$, while cluster Es lie higher at $T_{blue} \sim 2$–2.5. Field/group Es, have values similar to those of the spirals. Based solely on the metal-poor GCs, these comparisons seem to rule out the formation of cluster gEs (and some massive Es in lower-density environments) by major mergers of disk galaxies. However, see section 6 on biasing and the paper by K. Rhode (this volume) for a more complete picture.

The T_{red} data show a similar correlation with galaxy mass, although with a smaller dynamic range. These data are consistent with the hypothesis of near-constant formation efficiency for metal-rich GCs in both spirals and field Es with respect to bulge mass (e.g., [10]).

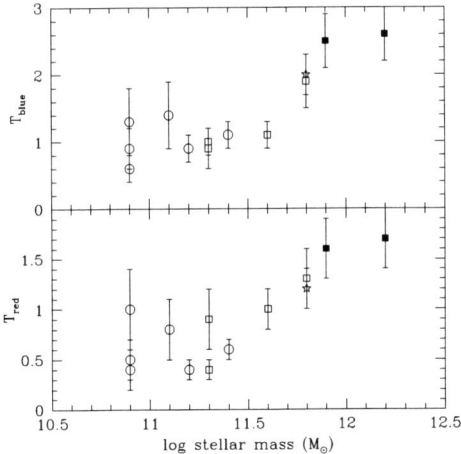

Fig. 2. T_{blue} and T_{red} (*top and bottom* panels, respectively) vs. galaxy mass for a range of spirals and Es [21]. Filled squares are cluster Es, open squares are field/group Es and S0s, open circles are field/group spirals, and the open star is the Sa/S0 galaxy NGC 4594. There is a general trend of increasing T_{blue} and T_{red} with galaxy mass. (Data courtesy K. Rhode).

5 Correlations with Host Galaxy Mass

The correlation between red GC peak color and host galaxy luminosity described in Larsen et al. [16], Strader et al. [27], Peng et al. [19], and Strader et al. [25], whose slope has consistently been found to be similar to that of the color-magnitude relation for early-type galaxies ($V - I \propto -0.018 M_V$), revealed the connection between the formation of metal-rich GCs and the bulk of the field stars in their parent galaxies. Such a connection was expected under all three of the "classical" GC formation scenarios. When first reported [16], the trend between GCs color and host galaxy luminosity for metal-poor GCs was much more controversial, since such a relation implies that these GCs, forming very early in the history of the universe, already "knew" about the final galaxy to which they would ultimately belong. Lotz et al. [17] suggested a correlation between GC color and galaxy luminosity for dwarf galaxies. Strader et al. [28] analyzed the GC data for galaxies spanning > 10 magnitudes in luminosity and found a relation with > 5σ statistical significance. Strader et al. pointed out that, at face value, this relation argues against both major mergers and accretion as the mechanism by which galaxies assemble; it is not possible to build a more massive galaxy from any combination of lower mass galaxies, since the resulting blue GCs will have the "wrong" color – too blue for the host's luminosity. The metal-poor relation does indeed rule out mergers and accretion, but only in the local universe for structure forming at the present day. How these results can be incorporated into our understanding of hierarchical galaxy assembly is discussed in

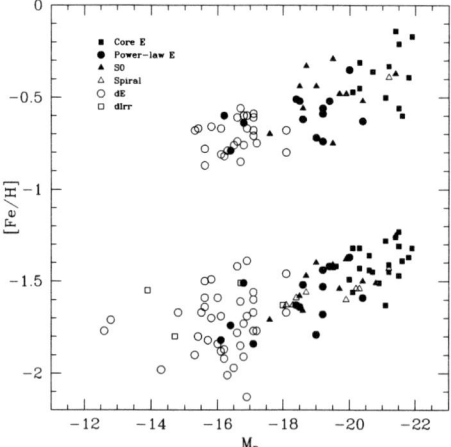

Fig. 3. Peak GC metallicity vs. galaxy luminosity (M_B) for metal-poor and metal-rich GCs in a range of galaxies, from Brodie and Strader [5]. The points have been converted from $V-I$ and $g-z$ to [Fe/H] using the relations of Barmby et al. [2] and Peng et al. [19], respectively. Linear relations exist for both subpopulations down to the limit of available data.

the following section. In the relations connecting GC color with host galaxy luminosity it was noted that the red relation was significantly steeper than the blue one. Interestingly, when the GC color-galaxy luminosity relations are transformed to GG metallicity- galaxy luminosity space (Fig. 3, taken from Brodie and Strader [5], and see also Peng et al. [19]) the metal-poor and metal-rich relations are found to have very similar slopes. With newer data, we can also see that the sequences continue down to very low galaxy luminosities [19,25]. We see clearly that even dwarf galaxies can often have red GCs and their colors are consistent with an extrapolation of the relation for more massive galaxies down to lower luminosities. This is a revelation with significant implications for galaxy formation. If dwarf galaxies exist that have not suffered mergers then their red GCs cannot have a merger origin. Note, though, that because of the large errors on the peak colors for the dwarfs (few GCs per galaxy), the slope is very poorly constrained at low galaxy luminosities.

6 The Biasing Scenario

In order to accommodate the constraints imposed by the GC metallicity-host galaxy luminosity relations, as well as the evidence that most luminous galaxies have suffered some degree of merging since $z \sim 2$, Brodie and Strader [5] explored the idea that galaxy assembly is biased in the context of GC formation. Their suggestion that the metal poor GC – galaxy L relation was different at high z, and that merging was biased, offers a strong end constraint on hierarchical structure formation. N-body simulations of hierarchical merging in a lambda CDM cosmology [8,18] suggest that dark matter (DM) halos "sitting" on top of large overdensities collapse (and form stars) first, at very high z (>10), and that these first-forming halos are the ones that combine to form massive galaxies at z=0. Halos more distant from such overdensities survive independently to become dwarf galaxies. The in situ/multiphase collapse scenario of Forbes, Brodie and Grillmair [9] and the subsequent analysis by Beasley et al. [4], who traced the formation of GCs in merging DM halos, both required GC formation to be halted temporarily at high z in order to reproduce the observed bimodality in GC color distributions. Santos [22] first suggested that blue GC formation could plausibly have been truncated by cosmic reionization. This idea has interesting consequences. First, it naturally explains the blue GC metallicity – host galaxy luminosity relation because halos that collapse first can reasonably be supposed to have formed GCs of higher metallicity, either because they have had more time for enrichment or because of closer proximity massive protogalaxies and their outflows of enriched material [26]. Second, it naturally explains the correlation between specific frequency (or T value) and host galaxy luminosity (Sect. 4), since first-forming halos will have a larger number of metal-poor GCs than similar mass halos in lower density environments (which got started later) by the

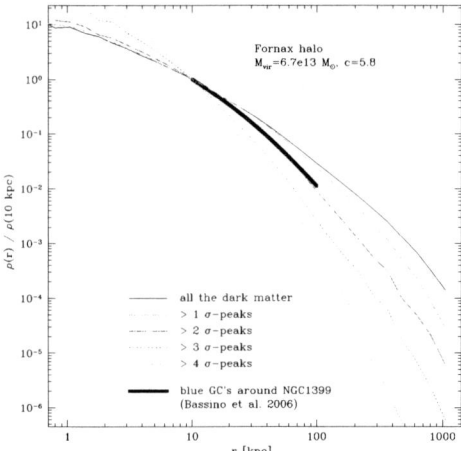

Fig. 4. Unprojected 3D Radial density profiles from N-body simulations of the formation of the Fornax cluster showing the distribution of DM halos corresponding to various overdensity biases. The solid line represents the observed surface density distribution of metal poor GCs in NGC 1399 from Bassino et al. [3]. Figure courtesy of Juerg Diemand.

time reionization (or some other mechanism) truncates their formation [21]. Third, it offers the possibility that the surface density distribution of blue GCs in galaxies of different masses and environments can trace the epoch and homogeneity of reionization. In Fig. 4, courtesy of Juerg Diemand, we show results from his simulation of the formation of the Fornax cluster of galaxies. The different radial density distribution curves reflect different peak heights in the DM distributions and these peaks collapse at different redshifts depending on the details of the adopted cosmology. In this set of simulations variable minimum masses were used (8×10^8 to 3×10^9 $M_\odot$) corresponding to the peak heights given by the labels. The Fornax halo hosts the entire cluster and is taken to have a virial mass of 6.7×10^{13} $M_\odot$, a virial radius of 1050 kpc and a concentration of 5.8. Bassino et al. ([3], and this volume) fitted the surface brightness of blue GC with a projected NFW with a scale radius of 35 kpc. This profile is plotted with a thick black line in Fig. 4. It corresponds to the 2 σ curve in the simulation. The value for the metal-poor GC distribution in the MW is ∼2.5 σ. For a minimum halo mass of 2×10^8 $M_\odot$ in the MW this corresponds to a redshift of formation z>11.3 (see Moore et al. [18] for details).

7 The Blue Tilt

A new discovery, made possible by the enhanced imaging capabilities of the ACS on HST, is a correlation between color and luminosity for *individual* metal-poor GCs in some giant Es (the "blue tilt"; see Fig. 5). The "blue tilt"

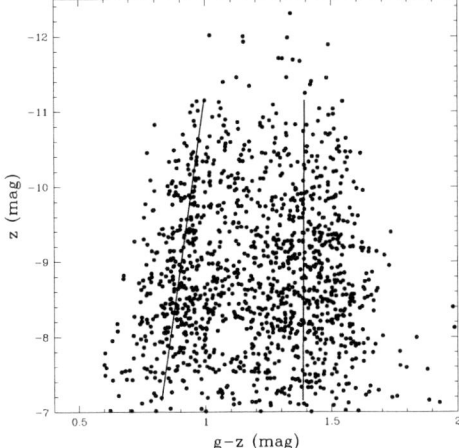

Fig. 5. z vs. $g-z$ color-magnitude diagram for M87 GCs from Strader et al. [25]. A correlation between color and luminosity for the bright metal-poor GCs is apparent (the "blue tilt"). The solid lines are fitted linear relations.

was found by Strader et al. [25] in the Virgo giant Es (gEs) M87 and NGC 4649 and by Harris et al. ([13], and this volume) in their sample of Brightest Cluster Galaxies (BCGs). Strader et al. [25] suggested that the mass-metallicity relation for individual metal-poor GCs argued for self-enrichment and speculated that these metal-poor GCs were able to self-enrich because they once possessed dark matter halos that were subsequently stripped. Harris et al. interpretated the relation in terms of a mass threshold and mass-dependent ability of star/cluster forming clumps to retain enriched SN ejecta. We now know that the blue tilt phenomenon is not confined to galaxies in high density environments or even just to E galaxies. It has recently been reported for NGC 4594 ([24], and this volume), a luminous Sa galaxy that lies in a loose group, as well as spiral galaxies of later type ([11], and this volume). Curiously, the Virgo gE NGC 4472 (also studied by Strader et al.) shows no evidence for the blue tilt. If this lack of a tilt is confirmed with better data, it will be a strong constraint on any potential "universal" model for explaining the phenomenon in massive galaxies. The Milky Way itself does not show evidence for the tilt, but this could be due to the small number of metal-poor GCs (~ 100) compared to massive galaxies or to the inhomogeneity of metallicities and integrated photometry in current catalogs. The CMD in Figure 5 also demonstrates that the color dispersion of the metal rich and metal poor GCs is significantly different. In fact, the spread is almost a factor of 2 more for the red GCs than the blue ones in $g-z$ (0.7 vs. 0.4). However, Peng et al. [19] find that this translates into a larger spread in metallicity for the *blue* GCs based on their color-metallicity transformation.

Summary and Conclusions

Newly observed relations between mass and luminosity for individual GCs, as well as between GC color and galaxy luminosity, point to strong connections between GCs and dark matter, at least for metal-poor GCs forming during the earliest phases of hierarchical galaxy assembly. In fact, GCs may be one of the best observable links between baryons and dark matter. The old ages of the vast majority of both red and blue GCs strongly suggests that GC formation, and by extrapolation, the bulk of star formation in spheroids, took place at early times ($z > 2$). The GC systems of spirals and ellipticals, even dwarf ellipticals, are remarkably similar, which argues for a universal formation mechanism for GCs and by implication, their host galaxies. We argue that the red GCs trace the buildup of bulges while the blue GCs may trace DM halos at the earliest times. In particular the relation between GC metallicity and host galaxy luminosity is a strong end constraint on hierarchical merging, understandable in terms of biases merging; structure formation is not self-similar. The surface density distributions of GCs around galaxies of different masses and in different environments may offer measures of the epoch and (in)homogeneity of reionization.

Acknowledgements. I would like to thank my many SAGES collaborators and especially Jay Strader. This work was supported by NSF grant AST-0507729.

References

1. Ashman K.M., Zepf S.E. 1992, *Ap. J.* 384:50
2. Barmby P., Huchra J.P., Brodie J.P., et al. 2000, *Astron. J.* 119:727
3. Bassino L., Richtler T., Dirsch B., 2006, *MNRAS* 367:156
4. Beasley M.A., Baugh C.M., Forbes D.A., et al. 2002, *MNRAS* 333:383
5. Brodie J.P., Strader J. 2006, *Annu. Rev. Astron. Astrophys.*
6. Côté P., Marzke R.O., West M.J. 1998, *Ap. J.* 501:554
7. Côté P., McLaughlin D.E., Hanes D.A., et al. 2001. *Ap. J.* 559:828
8. Diemand J., Madau P., Moore B. 2005, *MNRAS* 364:367
9. Forbes D.A., Brodie J.P., Grillmair C.J. 1997, *Astron. J.* 113:1652
10. Forbes D.A., Brodie J.P., Larsen S.S. 2001, *Ap. J. Lett.* 556:83
11. Forde, K., et al. 2006. *Astron. J.* in preparation
12. Gregg M.D. 1994, *Astron. J.* 108:2164
13. Harris W.E, Whitmore B.C., Karakla D., et al. 2006, *Ap. J.* 636:90
14. Kundu A., Whitmore B.C. 2001, *Astron. J.* 121:2950
15. Larsen S.S., Brodie J.P., Beasley M.A., Forbes D.A. 2002, *Astron. J.* 124:828
16. Larsen S.S., Brodie J.P., Huchra J.P., et al. 2001, *Astron. J.* 121:2974
17. Lotz J.M., Miller B.W., Ferguson H.C. 2004, *Ap. J.* 613:262
18. Moore, B. et al. 2006, astro-ph/0510370
19. Peng E.W., Jordán A., Côté P., et al. 2006, astro-ph/0509654
20. Puzia T.H., Saglia R.P., Kissler-Patig M., et al. 2002, *A & A 391:453*

21. Rhode K.L., Zepf S.E., Santos M.R. 2005, *Ap. J. Lett.* 630:21
22. Santos M.R. 2003, Extragalactic Globular Cluster Systems, ESO Astrophysics Symposia, ed. M. Kissler-Patig, Springer-Verlag 2003, p. 348
23. Schweizer F. 1987, "Nearly Normal Galaxies", Eighth Santa Cruz Summer Workshop in Astronomy and Astrophysics, ed. S. Faber, Springer-Verlag 1987, p. 18
24. Spitler L., Larsen S., Strader J., Brodie J.P., et al. 2006, *AJ* 132:1593
25. Strader J., Brodie J., Beasley M., Spitler L., 2006, astro-ph/0508001
26. Strader J., Brodie J.P., Cenarro A.J. 2005, *Astron. J.* 130:1315
27. Strader J., Brodie J.P., Forbes D.A. 2004, *Astron. J.* 127:295
28. Strader J., Brodie J.P., Forbes D.A. 2004, *Astron. J.* 127:3431

Globular Clusters and Galaxy Formation

Duncan A. Forbes

Centre for Astrophysics and Supercomputing, Swinburne University, Hawthorn VIC 3122, Australia dforbes@swin.edu.au

Abstract. We first discuss recent progress in using the Milky Way globular cluster (GC) system as a 'test-bed' for properties derived from integrated spectra and stellar population models. Standard techniques may give rise to spuriously high alpha-element ratios at low metallicities. We then discuss evidence for early epoch ($z \geq 2$) formation for most GCs in galaxies today. Recent accretions of GCs (and their host galaxy) make a small contribution but recent mergers form few if any new GCs in today's elliptical galaxies. The early formation of metal-poor GCs and the bimodality seen in GC specific frequency requires a 'truncation' which may be due to reionization.

1 Milky Way Globular Cluster System

The globular cluster (GC) system of our own Galaxy provides an obvious test-bed for the integrated spectral properties that we can measure for extragalactic GCs, such as age, metallicity and alpha-element abundance. It also allows us to test the single stellar population (SSPs) models that we rely on. Limitations on this approach include the small number of Milky Way GCs (∼150), strong extinction to some (particularly metal-rich bulge GCs), a limited age range (most GCs are > 10 Gyrs old) and the difficulty of obtaining well calibrated, high S/N spectra. Some inroads on the latter have been made by Puzia et al. [24], Cohen et al. [7] and Schiavon et al. [26]. From this combined dataset we have formed a database of 48 distinct Milky Way GCs on the Lick system with S/N ∼ 100.

In Proctor, Forbes and Beasley [23] we showed that more reliable results are obtained from using all available Lick lines than Hβ or Hγ alone. Recent fits to the combined dataset by J. Mendel shows a good agreement with the CMD-derived ages and metallicities of de Anglei et al. [8]. However, the trend for alpha-elements with metallicity rises to unrealistically high values at low metallicities (see Fig. 1). Given that high resolution spectra of Carney and coworkers indicates a near constant [E/Fe] = +0.3 for Milky Way GCs, we suspect this is an SSP modelling effect. We have modified the base SSP model of Lee & Worthey [18] to include the Houdashelt et al. [16] formulation for alpha-elements. The refit results are shown in Fig. 2 – they now reveal the expected trend. So although preliminary, this approach shows promise.

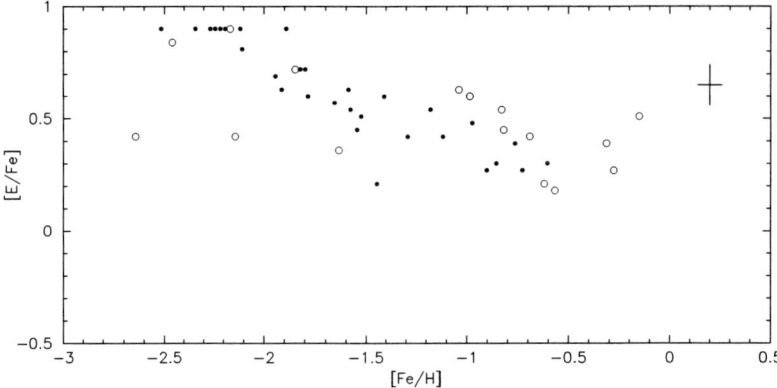

Fig. 1. Alpha-element ratio vs metallicity for Milky Way GCs. Symbols represent different datasets. Fits were carried out using Korn et al. [17] SSP models. The rise to large alpha ratios at low metallicities is not seen in high resolution spectra.

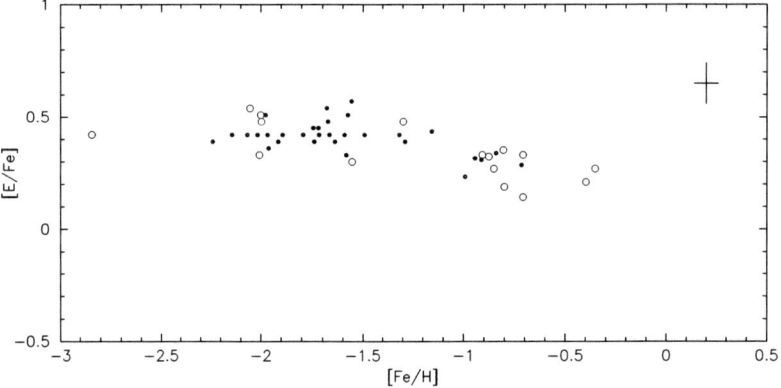

Fig. 2. Alpha-element ratio vs metallicity for Milky Way GCs. Same as Fig. 1 except we used the base SSP model of Lee and Worthey [18] modified by Houdashelt et al. [16]. Here the expected alpha-element trend is found.

2 Early Globular Cluster Formation

Forbes, Brodie and Grillmair [11] and Beasley et al. [1] favoured a formation scenario in which GCs were formed in a dissipative process in the early Universe. Both required a mechanism to cutoff or truncate the formation of metal-poor GCs (which formed first) before the second phase of metal-rich GC formation occurred. Reionization now seems to be the 'best bet' process for this truncation (e.g. [25]). If correct, a fossil record of the epoch of reionization can be found by examining the surface density profile of metal-poor GCs [19,2]. This prospect is particularly exciting given the current constraints of $z_{reion} \sim 6$ from QSOs and $z_{reion} \sim 11$ from the WMAP satellite.

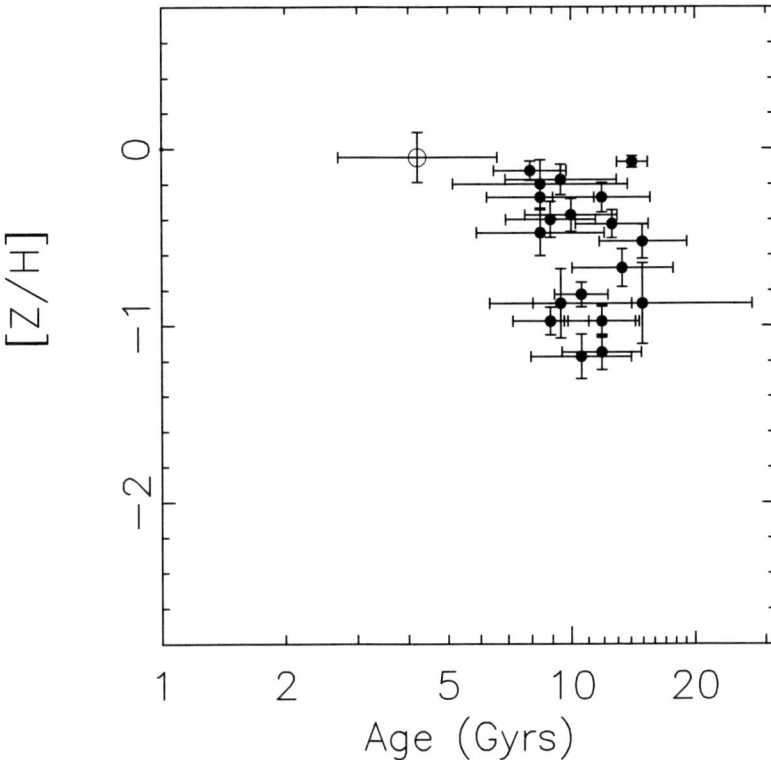

Fig. 3. Age-metallicity relation for NGC 4365 GCs. All sampled GCs appear to be old. The one possible exception (*open symbol*) has a blue horizontal branch that causes us to underestimate its age.

The combination of truncation of metal-poor GC formation and a galaxy mass-to-light ratio that varies with mass, can explain the U-shaped distribution in GC specific frequency with host galaxy luminosity [3, 10]. Such bimodality in GC properties with galaxy mass may be explained by the transition from hot to cold flows [9].

The most direct way of probing the formation epoch of GCs is to determine GC ages from high S/N spectra. The SAGES team have been carrying out such work for several years with the Keck telescope. It appears that the vast bulk of GCs in elliptical galaxies are old, i.e. ages >10 Gyrs. An interesting, and controversial case, is NGC 4365 in the Virgo cluster. In Fig. 3 we show the age-metallicity relation (AMR) for GCs in this galaxy. Ages and metallicities have been fit using the multi-line method of Proctor, Forbes and Beasley [23] by M. Pierce with indices from Brodie et al. [4]. The AMR shows that our sample GCs are all old, i.e. forming at redshifts $z \geq 2$. One possible exception (open symbol) has a clear Calcium H+K 'inversion' in its spectrum indicative of a blue horizontal branch; so in this case the age is underestimated and the metallicity overestimated.

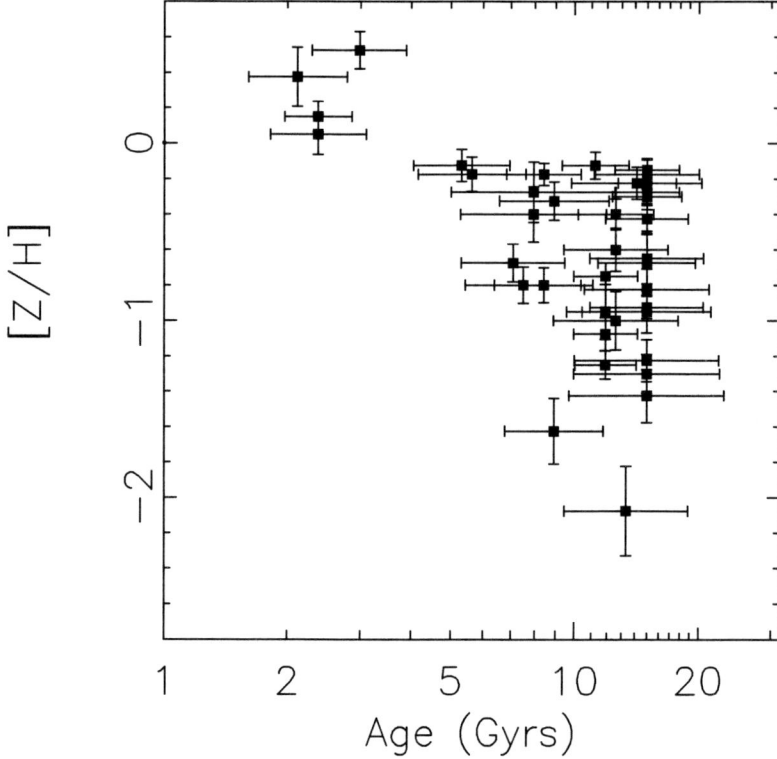

Fig. 4. Age-metallicity relation for NGC 4649 GCs. Most GCs are old, however 4 GCs appear to be quite young (2–3 Gyrs). These GCs are not all centrally located and we speculate that they may have been accreted from a dwarf galaxy.

3 Recent Accretions

In Forbes et al. [14] we presented AMRs for the GCs associated with the accreted Sgr and Canis Major dwarf galaxies. We further suggested that dwarf galaxy accretion contributed to the increased age dispersion at intermediate metallicities of the Milky Way GC system, but overall the contribution of accreted GCs was small. A rare example of a dwarf galaxy (and presumably its GC system) in the process of being accreted by a larger spiral galaxy was presented in *Science* by Forbes et al. [13].

A possible example of accreted GCs in a nearby elliptical galaxy is NGC 4649 [22]. The AMR shown in Fig. 4 reveals that the bulk of the GCs are old, however there are 4 GCs which appear to be young (∼2–3 Gyrs) ages. Interestingly, these GCs are not all centrally located or on the side of the galaxy towards its nearest neighbour NGC 4647, but lie in different parts of the galaxy. We speculate that they may have been accreted from a gaseous dwarf galaxy, of which there is no current sign.

4 Recent Mergers

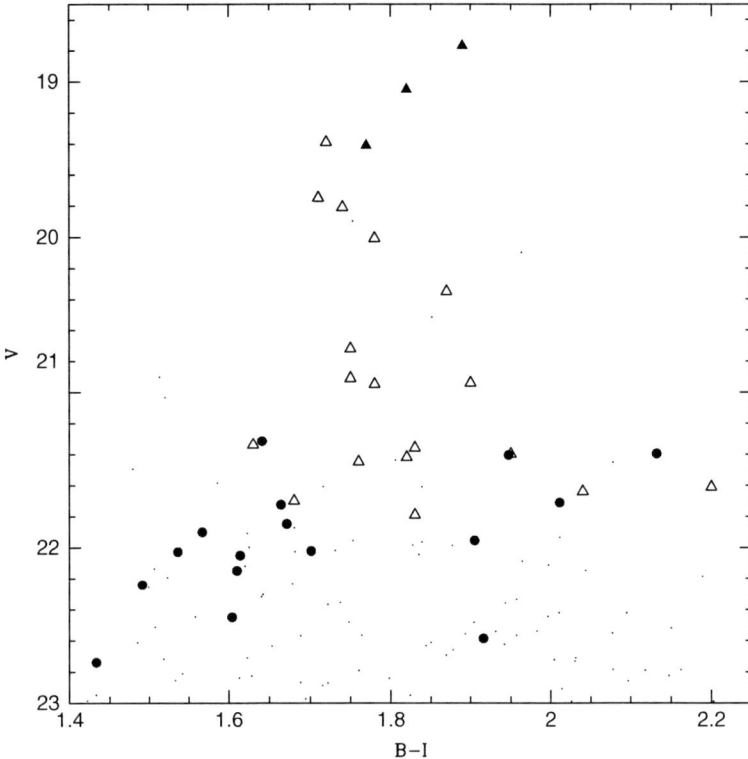

Fig. 5. Colour-magnitude diagram for the NGC 1052 and NGC 1316 GC systems. The NGC 1316 GCs are shown by triangles (*filled symbols* when spectral age estimates are available). The NGC 1052 GCs are shown by circles (*large symbols* when spectral ages are available). The NGC 1052 GC system reveals few if any young, bright GCs associated with the central starburst of $\sim$2 Gyrs, unlike NGC 1316.

Proto-globular clusters certainly appear to be forming in gas-rich mergers today such as the Antennae, but the question remains 'what contribution do GCs formed in recent mergers make to the overall GC system of today's elliptical galaxies?' An interesting comparison is the cases of NGC 1316 and NGC 1052. Both are thought to be 1–3 Gyr old merger remnants. NGC 1316 has a very disturbed morphology with extensive gas and dust. NGC 1052 has an elliptical appearance with some fine structure [27], and evidence for HI gas infalling onto the nucleus and in extended tails [28]. The CMDs for the GC systems on these two galaxies, placed at the same distance, are shown in Fig. 5. The NGC 1316 GC system reveals several GCs at intermediate colour and bright magnitudes of $18.5 < V < 21.5$. The brighter three are confirmed by spectra to be $\sim$ 3 Gyrs old from Goudfrooij et al. [15]. NGC

1052 has no equivalent counterpart GCs [12]; here the CMD distribution of GCs suggests they are predominately old. Subsequent Keck spectra of 16 NGC 1052 GCs confirm they are old despite the galaxy having a young central stellar population of $\sim$ 2 Gyrs [21].

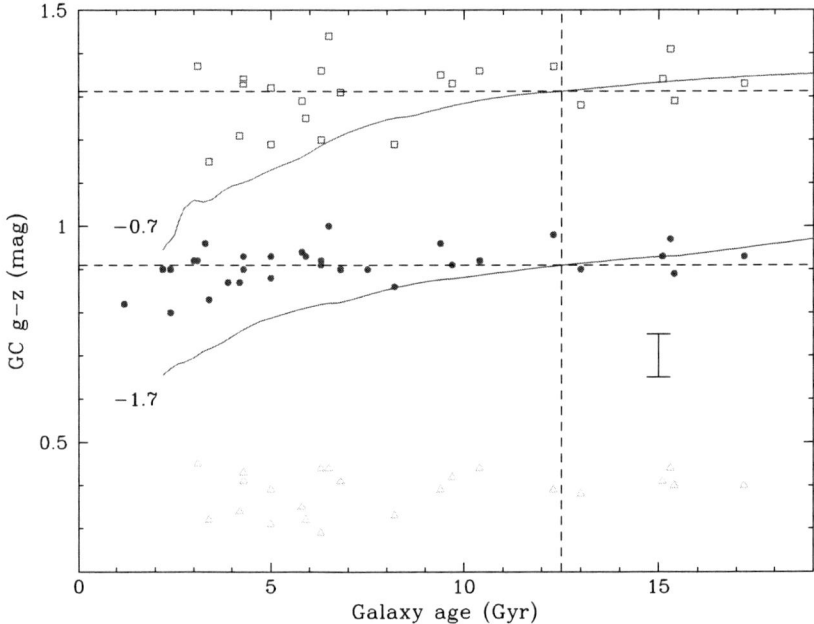

Fig. 6. Mean colour of GCs vs galaxy age for early-type Virgo galaxies. The dotted vertical line indicates an age of 12.5 Gyrs (typical of Milky Way GCs). The solid lines show the change in colour with age for two fixed metallicities, i.e. [Fe/H] = –0.7 and –1.7. The horizontal lines show the colour of a 12.5 Gyr population at these two metallicities. The mean GC colours for the red, blue and difference do not follow an evolutionary trend as expected if they formed at the time of the galaxy starburst. Rather, they have nearly constant colours consistent with them all having an age of $\sim$12.5 Gyr. A typical error bar is shown.

Many galaxies are found to have young central stellar populations, indicative of an interaction or merger which induced some star formation. If a significant number of GCs are formed during this star formation episode we might expect the mean colours of GCs to correlate with the time since the starburst. We investigate this issue using early-type galaxies in the Virgo cluster. Figure 6 shows the mean colour of the red and blue GC subpopulations, and their difference, with galaxy 'age'. Colours come from Peng et al. [20] and galaxy ages from Caldwell et al. [6]. The dotted vertical line indicates an age of 12.5 Gyrs (typical of Milky Way GCs). The solid lines show the change in colour with age for two fixed metallicities, i.e. [Fe/H] = –0.7

and −1.7 using models from Bruzual and Charlot [5]. Where these intersect the 12.5 Gyr line we draw a horizontal line to guide the eye. The mean GC colours for the red/blue subpopulations and difference do not follow an evolutionary trend as expected if they formed at the time of the galaxy starburst. Rather, they have nearly constant colours consistent with them all having an age of ∼12.5 Gyr. Thus any GCs formed in a recent merger associated with the galaxy starburst appear to make little, if any, contribution to the overall GC system.

5 Conclusions

- The ages, metallicities and alpha-element ratios of Milky Way GCs measured with integrated spectra, and modified SSP models, are in good agreement with CMD and higher resolution spectral results.
- Early formation (i.e. $z \geq 2$, age older than 10 Gyrs) is favoured for most GCs.
- Recent accretion is responsible for some GCs in large galaxies today.
- Recent mergers contribute few, if any, GCs to today's elliptical galaxies.
- Reionization is the current best-bet for truncating the formation of metal-poor GCs.
- The bimodality seen in GC specific frequency mirrors the mass-to-light variations of the general galaxy population.

References

1. Beasley, M., et al. 2002, MNRAS, 333, 383
2. Bekki, K., 2005, ApJ, 626, L93
3. Bekki, K., Yahagi, H., Forbes, D.A., 2006, ApJ, 645, L29
4. Brodie, J., et al. 2005, AJ, 129, 2643
5. Bruzual, G., Charlot, S., 2003, MNRAS, 344 1000
6. Caldwell, N., et al. 2003, AJ, 125, 2891
7. Cohen, J., Blakeslee, J., Ryzhov, A., 1998, ApJ, 496, 808
8. de Angeli, F., et al. 2005, AJ, 130, 116
9. Dekel, A., Birnboim, Y., 2006, MNRAS, 368, 2
10. Forbes, D.A. 2005, ApJ 635, L 137
11. Forbes, D. Brodie, J., Grillmair, C., 1997, AJ, 113, 1652
12. Forbes, D., Georgakakis, A., Brodie, J., 2001, MNRAS, 325, 1431
13. Forbes, D., Beasley, M., Bekki, K., Strader, J., Brodie, J., 2003, Science, 301, 1217
14. Forbes, D., Strader, J., Brodie, J., 2004, AJ, 127, 3349
15. Goudfrooij, P., et al. 2001, 322, 643
16. Houdashelt, M., Trager, S., Worthey, G., Bell, R., 2002, BAAS, 34, 1118
17. Korn, A., Maraston, C., Thomas, D., 2005, A&A, 438, 685
18. Lee, H., Worthey, G., 2005, ApJS, 160, 176

19. Moore, B., Diemand, J., Madau, P. 2006, MNRAS, 368, 563
20. Peng, E., et al. 2005, astro-ph/0509654
21. Pierce, M., et al. 2005, MNRAS, 358, 419
22. Pierce, M., et al. 2006, astro-ph/0601531
23. Proctor, R., Forbes, D., Beasley, M., 2004, MNRAS, 355, 1327
24. Puzia, T.H., Saglia, R.P., Kissler-Patig, M. et al. 2002, A&A, 395, 45
25. Santos, M., 2003, Extragalactic globular clusters, ed. M. Kissler-Patig, Springer, Heidelberg, p. 348
26. Schiavon, R., Rose, J., Courteau, S., MacArthur, L., 2005, ApJS, 160, 163
27. Schweizer, F., Seitzer, P., 1992, AJ, 104, 1039
28. van Gorkom, J., et al. 1986, AJ, 91, 791

Globular Cluster Systems in Giant Ellipticals: New and Old Patterns

William E. Harris

McMaster University, Department of Physics and Astronomy, Hamilton, Canada
harris@physics.mcmaster.ca

1 Introduction

In this paper I summarize new findings about the GC populations in giant E galaxies and how they connect to current issues of their metallicity and age distributions. The discussion concentrates on (a) color and metallicity distributions of GCs in Brightest Cluster Galaxies and the newly discovered Mass-Metallicity Relation; (b) an intriguing possible connection between the massive, GC-like nuclei in dE galaxies and the supermassive black holes in larger galaxies, both of which together appear to define a new class of "Central Massive Objects"; and (c) new optical/IR photometry relevant to the age distribution of GCs in giant ellipticals.

2 Supergiant Ellipticals and the GC Mass-Metallicity Relation

The very biggest globular cluster populations of all are found in supergiant (often cD-type) Brightest Cluster Galaxies. A single BCG can contain upwards of 10^4 globular clusters extending over a region of space 100+ kpc in radius (for a striking example, see the work on the M87 system described by Tamura at this conference). By studying them, we probe a uniquely rich GC-forming environment deep in the biggest potential wells that galaxies have to offer; and because of their sheer size, we are free of statistical uncertainties due only to size-of-sample effects. Of course, BCGs are rare, and to gain a broader picture of the GC patterns that BCGs have to offer, we must search outward to larger redshift than just the closest few (NGC 5128 in Centaurus, M87 in Virgo, NGC 1399 in Fornax). In addition, learning more about them than just the first-order information on total GC populations (specific frequencies) and spatial distributions requires deep and precise photometry.

Recently we have carried out the first deep, comprehensive survey of BCGs deliberately aimed at their globular cluster systems, using the ACS/WFC camera on HST. Photometry for 8 galaxies is presented in [9], with 4 more to appear in [10]. The ACS/WFC camera with its larger field and much higher blue sensitivity provides a huge gain over the older WFPC2 photometry. In

Fig. 1, the composite data in our first ACS imaging study are shown. Two features of the color-magnitude distribution immediately leap out: first, the classic "bimodality" is obvious, with roughly equal numbers of GCs within each of the blue ($\langle B-I \rangle \simeq 1.64$) and red ($\langle B-I \rangle \simeq 2.06$) sequences. Second, the internal color spread (that is, the intrinsic dispersion in metallicity) within each sequence is well resolved (see [9]) – characteristic of an extended period of formation in *both* modes with large amounts of gas being progressively turned into stars. On the red (metal-rich) side, the BCGs have clusters over a wider metallicity range than we find in the Milky Way, extending up to roughly Solar abundance and perhaps higher. Here, we run into the generic problem affecting all broadband color indices that we do not yet have calibrations of GC color against metallicity that extend reliably above $[Fe/H] \simeq -0.2$, the highest metallicities for Milky Way clusters. The nearest giant E galaxies such as NGC 5128, M87, and M49 in fact provide our best hope for constructing such calibrations over the full range of GC metallicities that nature provides (see also the paper by Beasley at this conference).

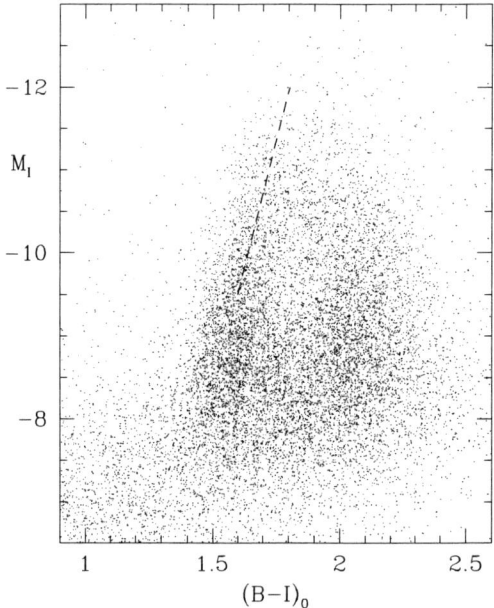

Fig. 1. Globular cluster luminosity M_I against color index $(B - I)$ for globular clusters in 8 Brightest Cluster Galaxies, from HST/ACS photometry [9]. The dashed line shows the mass-metallicity relation on the blue GC sequence.

An entirely new feature of the color distribution has, however, come to light from these data. Starting at a GC luminosity level of about $M_I \simeq -9.5$, the blue sequence by itself starts to slant redward with increasing cluster luminosity, all the way up to the brightest clusters known at $M_I < -12$. In contrast, the red sequence maintains a uniform mean color with luminosity. The trend, combined with the intrinsic color dispersion in each mode, means that for the uppermost range $M_I < -10.5$, the two sequences overlap and appear more like a single broad, unimodal distribution. This latter pseudo-unimodal effect was first noticed for NGC 1399 by [17] and [3], and in M87 itself the same blue-sequence color trend has been pointed out [19,24]. The effect is generally most obvious in the most GC-rich galaxies. In other words, the general phenomenon of the color trend along the blue sequence (which is essentially a *mass/metallicity relation* or MMR) could *only* have been discovered in the BCGs with their huge GC samples. We were not aware of it in earlier years not just because most of the data were taken in rather metallicity-insensitive indices like $(V - I)$, but also because of the smallness of the available samples.

The mean colors/metallicities of the sequences are replotted in Fig. 2, adapted from [9]. Now that we know what to look for, two immediate questions come up: (a) Is the MMR a *universal* phenomenon? and (b) What does it mean? What does it tell us about the formation history of the oldest, metal-poor cluster population, and about galaxy formation in general?

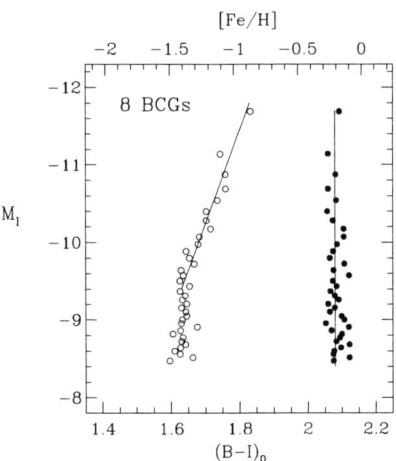

Fig. 2. Mass-Metallicity Relation (MMR) for the blue and red GC sequences in the eight BCGs studied by Harris et al. [9]. For the blue sequence, the straight line drawn in for $M_I < -9.5$ is equivalent to a heavy-element abundance scaling with cluster mass of $Z \sim M^{0.55}$.

At this conference we have already seen fascinating replies emerging to question (a). The same basic MMR, albeit at lower statistical confidence, can be seen in other large elliptical and disk galaxies. The most impressive evidence is from the Virgo Cluster Survey, as described by Mieske at this meeting: the combined data from their sample of GCs in 100 Virgo galaxies shows that the "amplitude" of the MMR, as measured by the slope of the upper blue sequence $\Delta(\text{color})/\Delta(\text{magnitude})$, increases with host galaxy luminosity. In short, the MMR is more dramatic in bigger galaxies regardless of type. At least one puzzling anomaly may, however, already exist: M49 (NGC 4472), the second supergiant Virgo E after M87, shows no apparent trace of the effect. But at this point, enough evidence is in front of us to conclude that the MMR represents an important new trend that formation models will have to account for.

Answering the underlying question (b) regarding physical interpretation will be more difficult, but it already seems clear that it will take us directly into understanding more about cluster formation and chemical enrichment processes that were going on in the pregalactic clouds, or at least in their earliest stages of hierarchical merging. We suggest [9] that the MMR arises because the clusters more massive than $\sim 10^6 M_\odot$ formed within dwarf-galaxy-sized clouds of $10^8 M_\odot$ or more, which were massive enough to hold on to the metal-enriched gas from the first rounds of SNe. Smaller clusters, *on average*, formed within less massive clouds that could not self-enrich in this way. By contrast, the red-sequence GCs are generally thought to have formed in later epochs during which the larger potential well of the merged galaxy had already developed, so that the degree of enrichment would be more nearly independent of GC mass.

An explanation of this type is no more than an idea at present, and must be followed up with detailed simulations. But if it is in the right direction, then the increase in amplitude of the MMR with galaxy size may indicate that larger galaxies were formed from systematically more massive populations of progenitor clouds. A variation of this view might be that the blue GCs formed late enough in the progression of hierarchical merging *in large galaxies* that they were already self-enriched to a small degree. We can look forward to intriguing developments in learning about this new phenomenon.

An interesting side effect of the MMR is on the classic 'bimodality' paradigm itself (Fig. 3). Bimodality was originally based on the assumption that each sequence lies at a constant mean color. But if we take the total metallicity distribution function (MDF) to mean the total *mass* in GC stars at a given abundance, the highest-luminosity clusters become quite important since they take up a high fraction of the total mass in the system. When constructed this way (Fig. 3b), the split between the two sequences is much less prominent than before.

3 Ultra-Massive GCs, dE Nuclei, Black Holes

The mass distribution function for globular clusters continues smoothly upward to more than $10^7 M_\odot$ (corresponding to $M_I < -12$). These ultra-massive GCs are quite rare in the Local Group (ω Cen, M31-G1, M31-B327), but the larger databases from the BCGs show that exist in significant numbers elsewhere.

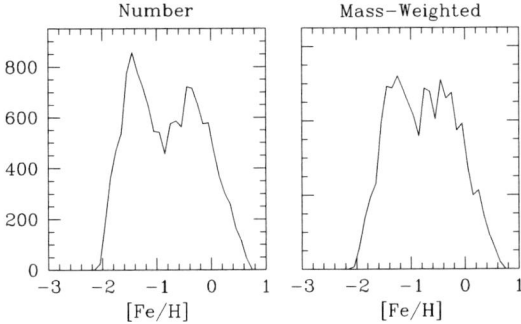

Fig. 3. Two different views of the "bimodal" metallicity distribution function for globular clusters. The left panel shows the number of clusters in the BCGs versus metallicity. The right panel shows the same data, now weighted by the *mass* of each cluster, assuming the same M/L for all of them. The top-end merging of the blue and red sequences caused by the MMR blurs out the bimodality effect.

New work on the Virgo and Fornax galaxies especially has told us a great deal about another type of object that resembles these massive globular clusters: the nuclei sitting at the centers of dwarf elliptical galaxies. The Virgo Cluster Survey data in particular suggest that half or more of all dE's may contain these small, distinct nuclei [2], many of which were not detectable from ground-based imaging. The bulges of spiral galaxies are also sites for nuclear star clusters (see, e.g., [23] and the paper by Seth at this conference).

The HST-based photometry shows unequivocally that the dE,N nuclei structurally resemble massive GCs like ω Cen quite closely. So, too, do the newly discovered UCDs (ultra-compact dwarf galaxies) that are also discussed extensively at this conference. Many UCDs and dE nuclei fall on a region of the fundamental plane of structural parameters that is very near the intersection of the E-galaxy line and the normal GC line, so they become quite hard to distinguish clearly from classic GCs on that basis alone. In addition, the Virgo dE nuclei fall on very much the same color-magnitude trend and MMR defined by the blue GCs in Figs. 1 and 2 [9]. It seems possible to claim that at least *some* of the biggest objects that we have long thought to be part of the normal GC sequence have had their origin this way [6,28], and that the uppermost part of the GCLF might actually be a mixed bag of objects. To put this another way, nature has clearly provided more than one channel to form massive, compact stellar systems, and that 10^{10} years later, it may be hard to decide which channel any given object originated from.

At masses higher than $10^7 M_\odot$, there is an intriguing changeover in the type of object at the center of E galaxies and bulges. Rather than a compact, GC-like stellar system we instead find a supermassive black hole (SBH). Still more intriguingly (Fig. 4), there is a smooth mass continuum between these two types of Central Massive Objects (CMO's). Perhaps nuclear star clusters and SBHs both start out in the same way, by early accumulation of gas at

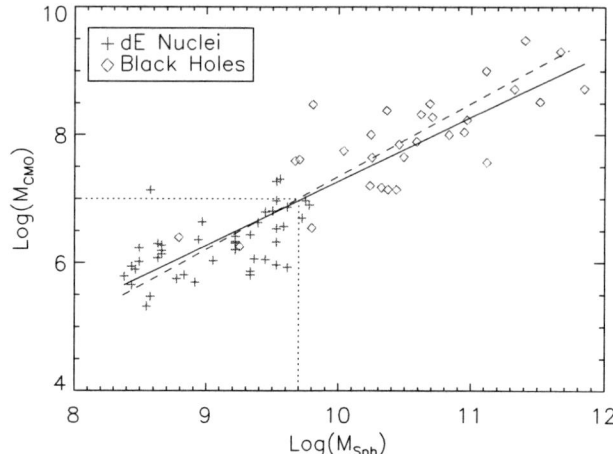

Fig. 4. Mass of the Central Massive Object (CMO) in a galaxy, plotted against the total spheroid mass of the galaxy (see [26]). Nuclei of dE galaxies (data from [15]) are shown as crosses, while supermassive black holes in larger galaxies (data from [25]) are shown as circles. The two least-squares fitted lines shown are both closely equivalent to $M_{CMO} \sim M_{sph}^{1.0}$.

the deepest points of their host-galaxy potential wells; but that for masses $M_{CMO} > 10^7 M_\odot$ (corresponding roughly to bulge masses $M_{sph} > 10^{10} M_\odot$), an SBH results rather than a compact stellar system. The CMO correlation – presented at this conference for the first time – is discussed more completely by [26] and [4]. In [26] we speculate that the key factor determining which type of CMO is formed may be the rate at which gas is accumulated at the center of the spheroid at early times; since the mass deposition rate is roughly proportional to the spheroid mass (e.g. [7]), a SBH may result if it accumulates too quickly for star formation to convert most of it, and that this critical transition rate may occur near $10^7 M_\odot$. Detailed hydrodynamic simulations – and a much more detailed understanding of the process of cluster formation – will clearly be needed to understand this link further.

4 NGC 5128 and the Case of the Metal-Poor Clusters

The nearest BCG of all is NGC 5128, just 4 Mpc distant. An enormous advantage it gives us is the chance to compare the metallicities and ages of the clusters and field-halo stars *directly in the same galaxy* – so far, a unique possibility for any gE galaxy. Assessment of the age distributions is still in development, but the field stars and GCs appear to have similar mean ages of at least 8 Gy [22, 18]. Figure 5 shows the metallicity comparison based on photometric indices (see the paper by Beasley for a spectroscopically based MDF for the clusters). This Figure confronts us with what I view as the most

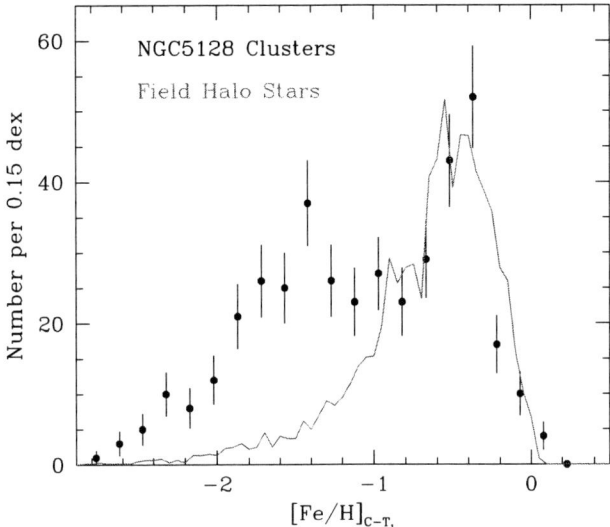

Fig. 5. Metallicity distribution for globular clusters and field halo stars in NGC 5128. The points with error bars show the number of GCs at a given [Fe/H], determined from $\simeq 420$ GCs with measured $(C - T_1)$ indices [27]. The solid line gives the relative number of outer-halo stars at a given [Fe/H] [8].

important GC "specific frequency problem" of all, something which cuts right to the basic issues of cluster formation efficiency at different early epochs.

The issue is essentially that there are far more blue, metal-poor GCs *per unit halo star* than metal-rich ones. The red GCs can plausibly be viewed as the "normal" population for the gE, since they match closely with the halo stars that define the MDF for the galaxy as a whole. Then if we divide the sample in half roughly at [Fe/H] $\simeq -1$, we find for NGC 5128 in particular that $S_N(red) = 0.85 \pm 0.12$ for the metal-richer GCs, and $S_N(blue) = 4.2 \pm 0.6$ for the metal-poor ones – a difference of a factor of 5. This phenomenon is likely to prove a generic one among all giant E galaxies, evidence for which is already accumulating (e.g. [5]). Did the blue GCs form at much higher efficiency? Was their accompanying field-star formation truncated, perhaps by the reionization epoch? Was the IMF for the first, most metal-poor field stars much flatter than normal (*without* the same being true for cluster stars)? We have no clear answers as yet.

5 Ages and Optical/IR Colors

To reconstruct the formation history of a GCS and its host galaxy, we need to build up their age distribution function (ADF). This task is difficult. The best route at present is to use the various line-index strengths from integrated

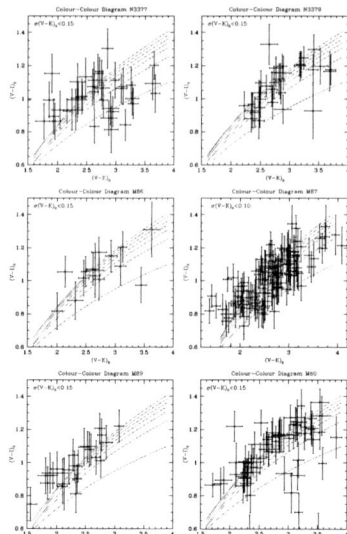

Fig. 6. Two-color diagrams $(V - I)$ vs. $(V - K)$ for globular clusters in 6 giant E galaxies, from [16]. Starting at *top* and reading from *left* to *right*, the galaxies shown are NGC 3377, NGC 3379, M86, M87, M89, and M60. The isochrone lines (*bottom to top*) are for 1, 2, 3, 5, 7, 10, 13 Gy.

spectra (e.g. [20] among others), but to do this reliably, many indices, many clusters, and spectra with $S/N > 30$–40 are needed.

Limited progress towards understanding the ADF has been made with the use of combined optical and near-infrared color indices such as $(V-I), (V-K)$ (e.g. [21, 12, 14, 13, 11]). In Fig. 6, new VIK data for six Leo and Virgo ellipticals obtained by [16] are shown. Comparison with isochrone lines (adapted from [1]) shows on average that the majority of the GCs in these galaxies are classically "old" (8 Gy or more); simulations show that most of the scatter is due simply to photometric measurement scatter. In two of the gEs (M87 and M60), perhaps $\sim 20\%$ of the GC population may be noticeably younger; although even here it is difficult to draw very general conclusions since these data (like all the near-IR photometry so far) sample only the inner few kiloparsecs of the galaxy where the widest age mixture might be expected. The Virgo gE in which strongest claims for a significant intermediate-age GC population have been made is NGC 4365, but even here the results are still inconclusive (cf. the references cited).

Quite a bit more precise photometry, both optical and near-IR, will be needed to explore the full potential of this technique in the future. Glimpses of the ADF are beginning to emerge, but we have a long way to go.

References

1. G. Bruzual & S. Charlot, MNRAS, **344**, 1000 (2003)
2. P. Côté et al., astro-ph/0603252 (2006)
3. B. Dirsch et al., AJ, **125**, 1908 (2003)
4. L. Ferrarese et al., astro-ph/0603840 (2006)
5. J.C. Forte, F. Faifer, & D. Geisler, MNRAS, **357**, 56 (2005)
6. K.C. Freeman, in ASP Conf. Ser. **48**, ed. G.H. Smith & J.P. Brodie (San Francisco: ASP), 608 (1993)
7. M.G. Haehnelt, P. Natarajan, & M.J. Rees, MNRAS, **300**, 817 (1998)
8. W.E. Harris & G.L.H. Harris, AJ, **123**, 3108 (2002)
9. W.E. Harris et al., ApJ, **636**, 90 (2006a)
10. W.E. Harris et al., in preparation (2006b)
11. M. Hempel, D. Geisler, D.W. Hoard, & W.E. Harris, AAp, **439**, 59 (2005)
12. M. Kissler-Patig, J.P. Brodie, & D. Minitti, AAP, **391**, 441 (2002)
13. A. Kundu et al., ApJ, **634**, L41 (2005)
14. S.S. Larsen, J.P. Brodie, & J. Strader, AAp, **443**, 413 (2005)
15. J.M Lotz, B.W. Miller, & H.C. Ferguson, ApJ, **613**, 262 (2004)
16. W.M. Okoń & W.E. Harris, AJ, submitted (2006)
17. P.G. Ostrov, J.C. Forte, & D. Geisler, AJ, **116**, 2854, 2854 (1998)
18. E.W. Peng, H.C. Ford, & K.C. Freeman, ApJ, **602**, 705 (2004)
19. E.W. Peng et al., ApJ, **639**, 95 (2006)
20. T.H. Puzia et al., AAp, **439**, 997 (2005)
21. T.H. Puzia et al., AAp, **391**, 453 (2002)
22. M. Rejkuba et al., ApJ, **631**, 262 (2005)
23. J. Rossa et al., astro-ph/0604140 (2006)
24. J. Strader, J.P. Brodie, L. Spitler, & M.A. Beasley, astro-ph/0508001 (2005)
25. S. Tremaine et al., ApJ, **574**, 740 (2002)
26. E.H. Wehner & W.E. Harris, astro-ph/0603801 (2006)
27. K.A. Woodley, in preparation (2006)
28. H. Zinnecker et al., in IAU Symposium **126**, ed. J.E. Grindlay & A.G.D. Philip (Dordrecht: Kluwer), 603 (1988)

The ACS Virgo Cluster Survey

Andrés Jordán

European Southern Observatory, Karl-Schwarzschild-Straße 2, 85748 Garching bei München, Germany `ajordan@eso.org`

1 Introduction

Much of our understanding of galaxy formation and evolution is based on observations of galaxies in cluster environments. The Virgo cluster is probably the most thoroughly studied cluster of galaxies by virtue of being the cluster nearest to our galaxy and observations of it have had an important role in the study of several astrophysical problems. In Cycle 11 of the Hubble Space Telescope we initiated the ACS Virgo Cluster Survey (ACSVCS; [1]),[1] an imaging survey of 100 galaxies in the Virgo cluster carried out with the Advanced Camera for Surveys (ACS; [2]) on board the Hubble Space Telescope (HST). In this contribution we describe the ACSVCS. One of its main scientific objectives is the study of the globular cluster (GC) systems of the target galaxies.

2 The Sample and Observations

We selected a sample of 100 early-type galaxies from the Virgo Cluster Catalog of [3]. The sample includes morphological types E, S0, dE, dE,N and dS0. It is complete down to $B = 12$ mag and then has a varying degree of completeness down to the survey limit of $B \approx 16$ mag.

Each galaxy is observed in a single orbit in the F475W ($\approx$ Sloan g) and F850LP ($\approx$ Sloan z) bands. This allows us to obtain a census of the GC system of each galaxy that is $\gtrsim 90\%$ complete, a testimony to the efficiency of ACS. The filter combination was chosen for two main reasons. First, it gives a large color baseline, and thus good metallicity sensitivity. Indeed, the combination of ACS/gz gives twice as much sensitivity to age and metallicity changes as the most widely used combination used to study GC systems before the installation of the ACS, namely WFPC2/VI. Secondly, the z-band is ideal

---

[1] The team members of the ACS Virgo Cluster Survey are: J.P. Blakeslee, P. Côté (PI), L. Ferrarese, A. Jordán, S. Mei, D. Merritt, E.W. Peng, J.L. Tonry and M.J. West. Several additional persons have contributed to various aspects of the scientific analysis as reflected in the author lists of the Virgo Cluster Survey series of papers published in the *Astrophysical Journal* and listed in the references.

to measure surface brightness fluctuations (SBF; [4]) in the target galaxies, one of the main scientific objectives of the survey.

In addition to the ACS images of the central regions of the target galaxies, the ACSVCS includes a significant parallel component, using WFPC2 to target 100 "blank" fields that are $5.8' \sim 29$ kpc from the galaxies. These observations are carried out using the F606W and F814W filters. We will not address the parallels in this contribution. They are discussed in the contribution by M. Takamiya in this volume.

The data reduction procedures for the survey are detailed in [5]. The data reduction pipeline includes modeling of the underlying galaxy, selection of GC candidates, fitting their light distribution with PSF-convolved King [6] models in order to perform photometry and estimate their half-light radii r_h [7] and selection of a clean sample of GCs by use of a statistical clustering method (Jordán et al. 2006, in preparation). In [1] we present a detailed account of the sample selection, scientific motivations and observational setup. The reader is directed to the papers referred to in this paragraph for detailed information about the observations and the reduction procedures.

3 Results

In this section we summarize the results of the ACSVCS to date (as of April 2006). An up-to-date list of publications resulting from the survey can be found at the project's website: http://www.cadc.hia.nrc.gc.ca/community/ACSVCS/

3.1 Surface Brightness Fluctuations and Distance Moduli

In [8,9] we have described the data reduction procedures appropriate for SBF measurements with the ACS. In those works we have demonstrated the feasibility of these measurements, a non-trivial issue due to the strong geometric distortion of the ACS which requires drizzling the data to an undistorted frame. The interpolations necessary in the last step can affect the power-spectrum of the fluctuations, something that will depend on the kernel used for the drizzling, but as shown in [8] the measurements are feasible in spite of this caveat.

In order to measure distances with the SBF method a calibration of the dependence of the fluctuation magnitude on the stellar population content of the galaxy is needed. The stellar population dependence of the fluctuation magnitude is parametrized with the galaxy color. We have calibrated the SBF distance measurement method for the survey bands gz [9]. We find that a single linear relation between the fluctuation magnitude and galaxy color was not satisfactory and propose instead a calibration based on two linear relations for different color regimes.

Using our calibration we measured distance moduli for 89 galaxies of our sample. This allows us to study the three dimensional structure of the Virgo cluster. We find that Virgo has a slightly elongated structure along the line of sight, that the distribution of the dwarfs is consistent with that of the giants and that a group of galaxies associated with NGC 4365 is infalling from behind. The results of this study will be presented in full in an upcoming paper [10].

3.2 The Structure of the Surface Brightness Profiles of Early-Type Galaxies

We measured surface brightness profiles and did a detailed study of dust morphology and nuclear properties for all sample galaxies [11]. Each surface brightness profile $I(r)$ is modelled as $I(r) = I_g(r) + I_k(r)$, where $I_g(r)$ is the model of the galaxy and $I_k(r)$ is a King model and is added when there is a nucleus present. We fitted parametric models to $I(r)$, and prior to fitting to the observed surface brightness profiles the model was convolved with the point-spread function. For I_g we choose for each galaxy the best description between a Sersic and a core-Seric profile. The latter is made up of a Sersic profile outside a break radius r_b and by a power-law with exponent γ within r_b. For galaxies that are better described by a pure Sersic profile, we estimate γ by calculating the logarithmic slope of the Sersic profile at $0''.1$.

We find no evidence for a strong bimodality in γ as claimed by previous works. Luminous galaxies in our sample are better described by core-Sersic profiles, while lower luminosity galaxies are better described by pure Sersic profiles. Ignoring luminous galaxies (termed core galaxies due to the fact that they are better described by core-Sersic profiles), the remaining galaxies in our sample form a continuous family in morphological properties as a function of galaxy mass. Detailed results are presented in [11].

3.3 The Properties of Nuclei of Early-Type Galaxies

We have studied the properties of the nuclei for our sample galaxies [12]. This topic is also addressed in another contribution in this volume by P. Côté. An interesting finding that was unanticipated at the outset of our survey has been the realization that previous ground-based studies of early-type galaxies greatly underestimated the number of galaxies which contain compact stellar nuclei near their photocenters. We find $\approx 70\text{--}80\%$ of our program galaxies to contain such nuclei, roughly three times higher than previously believed.

The nuclei in our sample galaxies show a number of remarkable properties. They appear with roughly equal frequency in both giant and dwarf galaxies. They are almost always located at the precise photocenters of their host galaxies. They span a large range in size (from 62 pc down to ≤ 2 pc, with a median of 4 pc). Their half-light radii depend on luminosity according to

$r_h \propto \sqrt{\mathcal{L}}$. The colors of the nuclei in galaxies fainter than $M_B \approx -17.6$ are tightly correlated with their luminosities, and less so with the luminosities of their host galaxies, suggesting that their chemical enrichment histories were governed by local or internal factors.

Additionally, we have found that nuclei contain a mean fraction, $\eta \approx 0.3\%$, of the total galaxy luminosity. This fraction is, to within the errors, identical to the fractional mass contributed by the central supermassive black holes (SBHs) in massive early-type galaxies. The latter class of galaxies are the only objects in our survey which, as a class, do *not* appear to contain stellar nuclei. These findings suggest that the compact stellar nuclei found in many low- and intermediate-luminosity galaxies may be the low-mass counterparts of the SBHs detected in the bright galaxies. If this interpretation is correct, then one should think in terms of *Central Massive Objects (CMOs)*—either SBHs or compact stellar nuclei—that accompany the formation of almost all early-type galaxies and contain a mean fraction $\approx 0.3\%$ of the total bulge mass. In this view, SBHs would be the dominant formation mode of CMOs above $M_B \approx -20.5$.

Using long-slit spectra for several dozen Virgo galaxies, we have measured velocity dispersion profiles for a number of the host galaxies from the ACSVCS. Using these data we infer the existence of a common $\mathcal{M}_{\rm CMO}$-$\mathcal{M}_{\rm gal}$ relation that leads smoothly from SBHs to nuclei as one moves down the mass function for early-type galaxies strongly suggests that a single mechanism is responsible for the growth—and perhaps even the formation—of both nuclei and SBHs [13]. It also points to galaxy mass as the primary (though not necessarily only) parameter regulating such growth.

3.4 The Properties of Ultracompact Dwarfs and Diffuse Star Clusters

Recent years have seen the discovery of stellar systems that fill regions of star cluster parameter space that had not been well explored. Determining the nature of these systems has been an active area of research in the last few years.

One such set of systems are the so-called ultra-compact dwarfs (UCDs), which are characterized by masses $M \gtrsim 5 \times 10^6 M_\odot$ and by sizes $\lesssim 50$ pc. We performed a study of the properties of these objects – which we termed Dwarf Globular Transition Objects (DGTOs)–around M87 and determined integrated velocity dispersions for 6 of the objects, which allowed us to estimate virial masses [14]. We find that some of the objects show larger sizes ($r_h \sim 20$ pc) and V-band mass-to-light ratios in the range 6–9 and have properties similar to the nuclei of nucleated dwarfs, while some objects show properties which are consistent with those of bright GCs. Thus, DGTOs seem to be a mixture of GCs and bona-fide UCDs, with the latter showing evidence for high mass-to-ligh ratios. The results of this study are presented in detail in [14] and a larger set of mass-to-light measurements for additional UCD candidates in the ACSVCS will be presented in upcoming work (Haşegan et al. 2006, in preparation).

Another set of stellar systems is characterized by its diffuse nature and are thus termed Diffuse Star Clusters (DSCs; [15]). They have g-band surface brightness $\mu_g \gtrsim 20$ mag arcsec^{-2}. These star clusters include the "faint fuzzy" clusters found in nearby lenticular galaxies by [16]. We find that 12 galaxies in our sample contain a significant population of DSCs, with 9 of them morphologically classified as S0s. The mean colors of DSCs are red, $1.1 < (g - z) < 1.6$, redder than metal-rich GCs and often as red as the galaxy light. DSCs are also discussed in a contribution in this volume by E.W. Peng and detailed results are presented in [15].

3.5 The (g–z) Color Distributions of GCs

We studied the (g–z) color distributions of the 100 galaxies of the ACSVCS. We find that galaxies have bimodal or asymmetric color distributions, with almost all galaxies possessing a component of metal-poor GCs and a varying fraction of metal-rich GCs that increases systematically as the galaxy mass increases. We find that the mean colors of both subpopulations correlate with host galaxy luminosity. We derived a color-metallicity transformation based on Galactic, M49 and M87 GCs. Using this transformation, which is not linear over its whole domain, we show for the first time that the mean *metallicities* of metal-rich and metal-poor GCs vary in a similar fashion with respect to galaxy mass, namely as [Fe/H] $\sim M^{0.2}$, where M is the galaxy stellar mass. Using our newly derived transformation we determined that while the metal-rich GCs show a wider dispersion in color, it is the metal-poor GCs that have an equal or larger dispersion in *metallicity*. Color distributions are also discussed in a contribution in this volume by E.W. Peng and detailed results are presented in [17].

3.6 Half-Light Radii of Globular Clusters in Early-Type Galaxies

The study of GC sizes beyond the Local Group has only been possible with HST: at the distance of Virgo, the typical half-light radius of a GC is $\approx 0\rlap{.}''03$. In our Galaxy, r_h is found to scale with galactocentric distance R_g as $r_h \propto R_g^{0.5}$ [18]. Once this dependence is taken into account, the half-light radii of Galactic GCs do not show a dependence on any other observable [19]. The lack of variation of $\langle r_h \rangle$ is roughly equivalent to the fundamental plane for GCs [19].

We have measured r_h for thousands of GCs belonging to the ACSVCS sample galaxies and the elliptical galaxy NGC 4697. An analysis of the dependencies of the measured half-light radii on both the properties of the GCs themselves and their host galaxies reveals that, in analogy with GCs in the Galaxy but in a milder fashion, the average half-light radius increases with

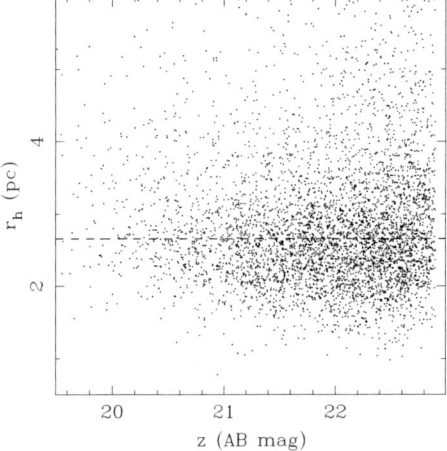

Fig. 1. Half-light radius, r_h, plotted as a function of z-band magnitude for the combined GCs of our sample galaxies. The data are consistent with $\langle r_h \rangle$ being independent of magnitude in the range $20 \lesssim z \lesssim 23$. The dashed line marks the median value of $\langle r_h \rangle = 2.66$ pc.

increasing galactocentric distance or, alternatively, with decreasing galaxy surface brightness. For the first time, we find that the average half-light radius decreases as the host galaxy color increases. We also show that there is no evidence for a variation of r_h with GC luminosity for GCs with masses in the range $2 \times 10^5 \lesssim (M/M_\odot) < 3 \times 10^6$ (Fig. 1).

In agreement with previous observations we find that the average r_h depends on the color of GCs, with red GCs being $\sim 17\%$ smaller than their blue counterparts. We show that this difference is probably a consequence of an intrinsic mechanism, rather than projection effects, and that it is in good agreement with the mechanism proposed in [20].

After accounting for the dependencies on galaxy color, galactocentric radius and underlying surface brightness, we show that the average GC half-light radii $\langle r_h \rangle$ can be successfully used as a standard ruler for distance estimation (see Fig. 2). We outline the methodology and provide a calibration for its use. We find $\langle r_h \rangle = 2.7 \pm 0.35$ pc for GCs with $(g-z) = 1.2$ mag in a galaxy with color $(g-z)_{\rm gal} = 1.5$ mag and at an underlying surface z-band brightness of $\mu_z = 21$ mag arcsec^{-2}. Detailed results are presented in [7].

3.7 The GC—Low Mass X-Ray Binary Connection

The exceptional angular resolution of *Chandra* has opened the study of Low Mass X-ray Binaries (LMXBs) to nearby galaxies. In analogy with our Galaxy, it has been found that GCs are very efficient at producing LMXBs, something which is thought to arise due the increased efficiency of dynamical formation processes for LMXBs in the dense cores of GCs (e.g. [21] and references therein). We used a deep (154 ks) *Chandra* exposure of M87 to detect

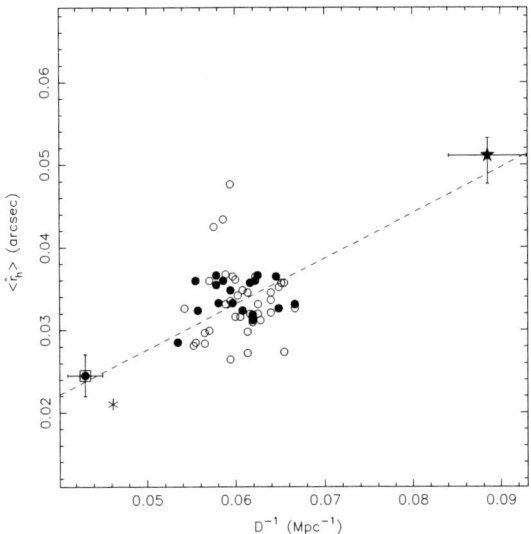

Fig. 2. Corrected half-light radii $\langle \hat{r}_h \rangle$ versus inverse distance D^{-1}. The corrected half-light radii $\hat{r}_h$ are constructed from the observed ones by the relation $\hat{r}_h \equiv r_h 10^{-0.016(\mu_z-21)+0.17[(g-z)_{\rm gal}-1.5]+0.17[(g-z)-1.2]}$, where μ_z is the z-band surface brightness, $(g-z)_{\rm gal}$ is the color of the galaxy and $(g-z)$ is the GC color (see [7]). *Filled symbols* are galaxies with $M_B < -19$ mag, while *open symbols* are those with $M_B \geq -19$ mag. The star is NGC 4697 and the *filled circle* surrounded by a square is VCC 731. The *dashed line* represents the relation $\langle r_h \rangle \propto D^{-1}$ normalized to be $0\farcs033$ at $D = 16.5$ Mpc, and it is *not* a fit. Error bars are included only for NGC 4697 and VCC 731 for clarity. The *asterisk* is VCC 575. We can see from this figure that $\langle \hat{r}_h \rangle$ can be used as a distance indicator.

174 X-ray point sources of which ≈ 150 are likely LMXBs. Combining this catalog with the ACSVCS catalog we determined the properties of GCs that determine the presence of an LMXB in a GC. We found that massive and redder GCs are more likely to harbour an LMXB and used our measured GC structural parameters to probe for the first time the relevance of dynamical processes in the formation of GC LMXBs in early-type galaxies [21]. We have now analyzed all galaxies in the ACSVCS with available *Chandra* data in order to perform a comprehensive study of the connection of GCs and LMXBs in early-type galaxies. Results will be presented in an upcoming paper [22] and are described in another contribution in this volume by Jordán et al.

4 Conclusion

We have briefly described the ACS Virgo Cluster Survey, a program to image in the Sloan g and z bands 100 galaxies in the Virgo cluster with ACS on board HST. We have summarized some of the results of this survey to date which have been presented fully in a series of publications and which include:

measurement of surface brightness fluctuations, a study of the structure of the surface brightness profiles of early-type galaxies and of their nuclei, a study of the properties of new families of hot stellar systems (UCDs, DSCs) and a comprehensive study of the properties of GC systems of the target galaxies (colors, luminosities, sizes, X-ray binary content).

In Cycle 13 of the Hubble Space Telescope we were awarded 43 orbits to extend our study to 43 galaxies in the Fornax cluster of galaxies. The ACS Fornax Cluster Survey [23, 24] will let us extend the studies performed in Virgo to a different cluster environment.

References

1. Côté, P. et al. 2004, ApJS, 153, 223 (Paper I)
2. Ford, H.C. et al. 1998, Proc. SPIE, 3356, 234
3. Binggeli, B., Sandage, A., & Tammann, G.A., 1985, AJ, 90, 1681
4. Tonry, J.L., & Schneider, D.P. 1988, AJ, 96, 807
5. Jordán, A. et al. 2004a, ApJS, 154, 509 (Paper II)
6. King, I.R. 1966, AJ, 71, 64
7. Jordán, A. et al. 2005a, ApJ, 634, 1002 (Paper X)
8. Mei, S. et al. 2005a, ApJS, 156, 113 (Paper IV)
9. Mei, S. et al. 2005b, ApJ, 625, 121 (Paper V)
10. Mei, S. et al. 2007, ApJ, 655, 144
11. Ferrarese, L. et al. 2006, ApJS, 164, 334
12. Côté, P. et al. 2006, ApJS, 165, 57
13. Ferrarese, L. et al. 2006, ApJL, 644, L21
14. Haşegan, M. et al. 2005, ApJ, 627, 203 (Paper VII)
15. Peng, E.W. et al. 2006b, ApJ, 639, 838 (Paper XI)
16. Brodie, J.P., & Larsen, S.S. 2002, AJ, 124, 1410
17. Peng, E.W. et al. 2006a, ApJ, 639, 95 (Paper IX)
18. van den Bergh, S., Morbey, C., & Pazder, J. 1991, ApJ, 375, 594
19. McLaughlin, D.E. 2000, ApJ, 539, 618
20. Jordán, A. 2004, ApJ, 613, L117
21. Jordán, A. et al. 2004b, ApJ, 613, 279 (Paper III)
22. Sivakoff, G.R. 2007, ApJ, 660, 1246
23. Jordán, A. et al. 2005b, in IAU Colloq. 198, "Near-Field Cosmology with Dwarf Elliptical Galaxies", ed. H. Jerjen & B. Binggeli (Cambridge: CUP), p. 368
24. Jordán, A. et al. 2007, ApJS, 169, 213

Globular Clusters at the Centre of the Fornax Cluster: Tracing Interactions Between Galaxies

Lilia P. Bassino[1,2], Tom Richtler[3], Favio R. Faifer[1,2], Juan C. Forte [1], Boris Dirsch[3], Doug Geisler[3], and Ylva Schuberth[4]

[1] Facultad de Ciencias Astronómicas y Geofísicas, Universidad Nacional de La Plata, Paseo del Bosque S/N, 1900-La Plata, Argentina
lbassino@fcaglp.unlp.edu.ar
[2] IALP – CONICET, Argentina
[3] Universidad de Concepción, Departamento de Física, Casilla 160-C, Concepción, Chile
[4] Argelander-Institut für Astronomie, Auf dem Hügel 71, D-53121 Bonn, Germany

Abstract. We present the combined results of two investigations: a large-scale study of the globular cluster system (GCS) around NGC 1399, the central galaxy of the Fornax cluster, and a study of the GCSs around NGC 1374, NGC 1379 and NGC 1387, three low-luminosity early-type galaxies located close to the centre of the same cluster. In both cases, the data consist of images from the wide-field MOSAIC Imager of the CTIO 4-m telescope, obtained with Washington C and Kron–Cousins R filters, which provide good metallicity resolution.

The colour distributions and radial projected densities of the GCSs are analyzed. We focus on the properties of the GCSs that trace possible interaction processes between the galaxies, such as tidal stripping of globular clusters (GCs). For the blue GCs, we find tails between NGC 1399 and neighbouring galaxies in the azimuthal projected distribution, and the three low-luminosity galaxies show low specific frequencies and a low proportion of blue GCs.

1 Introduction

It is widely known that GCs are a useful tool for studying the origin and evolution of galaxies, in the case of isolated galaxies as well as for those within groups or clusters.

In our first wide-field CCD study of the GCS around NGC 1399, at the centre of the Fornax cluster [4], we found that there are GCs out to the very limits of the studied field (Field 3 in Fig. 1), corresponding to a projected galactocentric distance of 100 kpc. In a later run, we obtained images of three adjoining fields, using the same observational set up, which is described in Sect. 2. We added one field to the West (Field 4 in Fig. 1) and two fields to the East of NGC 1399 (Fields 1 and 2 in Fig. 1). As there were several low-luminosity early-type galaxies with their own GCSs in the western field (NGC 1374, NGC 1379 and NGC 1387), we decided to keep this field to study them [1], and we used the eastern fields, where there were no conspicuous galaxies, to study the NGC 1399 GCS over a larger field [2].

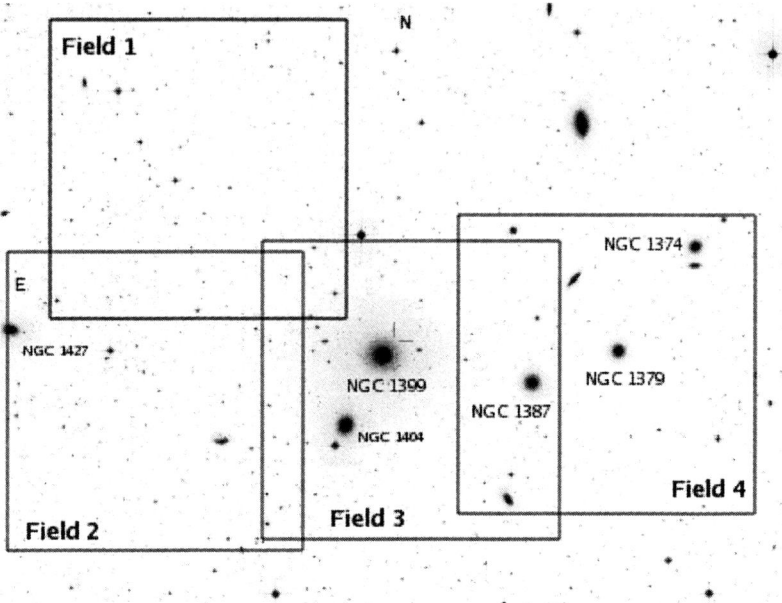

Fig. 1. MOSAIC fields overlaid on a DSS image of the Fornax cluster. North is up and East to the left.

2 Observations and Data Reduction

The observations were performed with the MOSAIC camera and 4–m Blanco telescope at the Cerro Tololo Inter–American Observatory (CTIO). The MOSAIC wide-field camera has a field of view of 36 × 36 arcmin (200 × 200 kpc at the Fornax distance). For more information on the MOSAIC camera we refer to the homepage *http://www.noao.edu/kpno/mosaic/mosaic.html*.

Kron–Cousins R and Washington C filters were used. We remind the reader that R and Washington T_1 magnitudes are very similar, with a very small colour term and zero-point difference [8]. For more details on the point source selection, photometric calibration, and the identification of GC candidates we refer the reader to [1] and [2]. Statistical subtraction of the contamination by the background was performed in all cases.

3 GCSs Around NGC 1374, NGC 1379 and NGC 1387

It is shown, for the first time, that the colour distributions of these three low-luminosity galaxies are clearly bimodal, with very similar colours for the blue GC peaks [1]. The red peak in NGC 1387, the galaxy located closer to NGC 1399 (see Fig. 1), is redder than the others and its whole colour distribution is atypical: the red clusters are much more numerous than the blue ones and the separation of the peaks is very pronounced. In fact, the fraction of blue clusters, with respect to the total GC population is low for the three systems, but even lower for NGC 1387 (43%, 45%, and 24% for NGC 1374, NGC 1379 and NGC 1387, respectively).

With regard to the radial distributions, the blue GCs in these systems show flatter distributions than the red ones, while the respective galaxy light profiles follow the density profiles for all GCs.

By means of the luminosity functions we estimate the total GC populations in NGC 1374, NGC 1379 and NGC 1387 between 200 and 400 clusters, and obtain specific frequencies $S_N = 2.4$, 1.4 and 1.8, respectively. These specific frequencies are rather small when compared to the typical value $S_N = 4$ found for elliptical galaxies in dense environments [9].

Figure 2 shows the specific frequencies and fractions of blue clusters versus distance from NGC 1399, including the data for NGC 1404 [6] and NGC 1427 [7] (see Fig. 1). The specific frequencies seem to decrease with decreasing distance from NGC 1399 and a similar trend is present in the fraction of blue clusters. These behaviours support the idea that galaxies closer to NGC 1399, in projected distance, might be losing their blue GCs as a result of some interaction process, like tidal stripping of clusters by the giant elliptical NGC 1399 (e.g. [5, 3, 11], etc.).

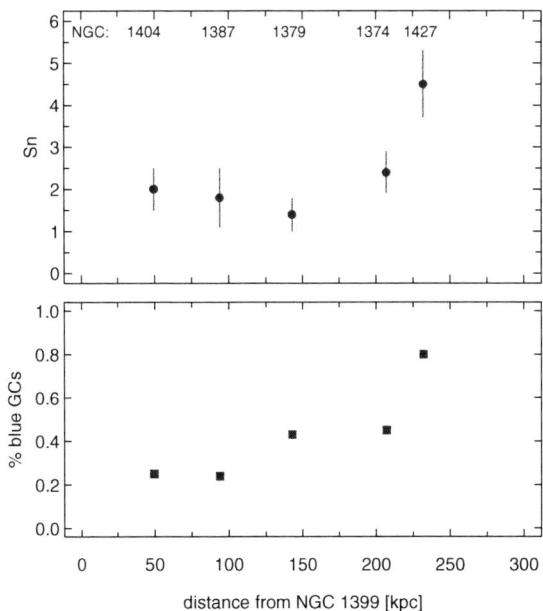

Fig. 2. Specific frequencies (*upper panel*) and fraction of blue GCs (*lower panel*) versus projected distance from NGC 1399.

4 GCS Around NGC 1399

The large-scale study of the projected radial distribution of GCs around NGC 1399 [2] shows that the blue clusters extend up to 250 kpc from the galaxy centre (45 ± 5 arcmin) while the red ones show a steeper radial profile that reaches a radius of 140 kpc (25 ± 5 arcmin). To our knowledge, this is one of the largest GCSs ever studied and, as our limiting magnitude is $T_1 = 23$, we cannot assure that there are no more GCs further out.

The colour distributions at different radial ranges confirm that the red GCs are more centrally concentrated than the blue ones, and show that the blue peak gets bluer with increasing galactocentric radius; a similar gradient was found in the NGC 1427 GCS by [7].

The colour distributions in different magnitude ranges show that the distribution is unimodal for the brightest bin (magnitudes similar to that of ω Cent), as already pointed out by [4]. However, we found no evidence for a "blue tilt" (e.g. [10]), i.e. the blue peak does not get redder with increasing luminosity.

The azimuthal distribution of the smoothed projected number density of blue clusters around NGC 1399 (Fig. 9 in [2]) shows two tails: one towards NGC 1404 and another towards NGC 1387. In the case of NGC 1404 one might wonder if this is just an overlapping of the two GCSs but it has been proposed, and tested by numerical simulations [3], that its GCs are probably being stripped by NGC 1399. The other tail, towards NGC 1387 cannot be just an overlapping of GCSs due to the small size of the GCS of NGC 1387 (r = 3 arcmin [1]) as compared to the projected distance to NGC 1399 (19 arcmin). Such overdensity of blue GCs may be understood as evidence that blue clusters, the less bound ones, are being lost by NGC 1387 due to some interaction process with the central cluster galaxy, like tidal stripping.

References

1. Bassino, L.P., Richtler, T., Dirsch, B.: MNRAS **367**, 156 (2006)
2. Bassino, L.P., Faifer, F.R., Forte, J.C., et al.: A&A **451**, 789 (2006)
3. Bekki K., Forbes D.A., Beasley M.A., Couch W.J.: MNRAS **344**, 1334 (2003)
4. Dirsch B., Richtler T., Geisler D., et al.: AJ **125**, 1908 (2003)
5. Forbes D.A., Brodie J.P., Grillmair C.J.: AJ **113**, 1652 (1997)
6. Forbes D.A., Grillmair C.J., Williger G.M., et al.: MNRAS **293**, 325 (1998)
7. Forte J.C., Geisler D., Ostrov P.G., et al.: AJ **121**, 1992 (2001)
8. Geisler D.: AJ **111**, 480 (1996)
9. Harris W.E.: in Kissler-Patig M., ed., Extragalactic Globular Cluster Systems, ESO Astrophysics Symposia, Springer-Verlag, Berlin, p. 317 (2003)
10. Harris, W.E., Whitmore, B.C., Karakla, D., et al.: ApJ **636**, 90 (2006)
11. Kissler–Patig M., Grillmair C.J., Meylan G., et al.: AJ **117**, 1206 (1999)

Globular Cluster Bimodality Revisited (and the Globulars-Galaxy Halo Connection)

Juan C. Forte[1], Favio Faifer[1], and Doug Geisler[2]

[1] FCAG Univ. de La Plata, and CONICET, Argentina
   forte@fcaglp.unlp.edu.ar
[2] Univ. de Concepcion, Chile dgeisler@astro-udec.cl

1 Introduction

The color of the "valley" in bimodal color distributions is frequently adopted as a separation between the so called "blue" and "red" globular cluster populations. In this way, the blue clusters usually exhibit shallower spatial distributions than those of the red ones. A further division of the so defined blue globulars in NGC 1399 [1] shows that the distribution becomes even flatter when only clusters bluer than the "blue peak" are considered. This work extends that analysis to both NGC 1399 and NGC 4486 (on the basis of a different data set described in [2]) and also discusses the connection between the globular clusters and the stellar population for the first of those galaxies.

2 GCs Areal Density Distribution

The areal density distributions for globular clusters in NGC 1399 and NGC 4486 are depicted in Figs. 1 and 2. These figures show the areal density run for globular clusters within four color windows: blue, green, yellow and red. These windows cover the color ranges defined between the bluest globulars and the blue peak, this last peak and the color valley, this valley and the red peak and this last peak and the reddest globulars, respectively. The galactocentric range (2 to 6 arcmin) was chosen seeking to include a large number of cluster candidates while keeping a low level of contaminating background objects. Common features in both galaxies are: (a) clusters within the "yellow" and "red" windows exhibit the same slope (suggesting that they are a single population within a given galaxy); (b) while clusters in the "blue" window show a very shallow distribution, those in the "green" domain exhibit intermediate slopes. As discussed in what follows, this distinctive behavior is consistent with a superposition of the two globular populations in the "green" window.

3 Monte Carlo Models

An attempt to match the shape of the color histograms of the globulars was made via a Monte Carlo model that, starting with a given (Z) abundance

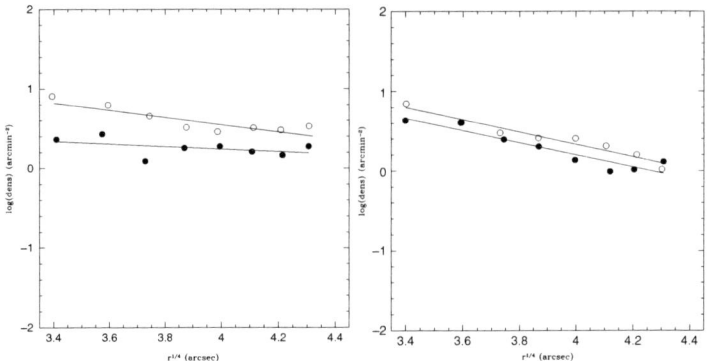

Fig. 1. *Left*: Logarithmic GC areal density vs $r^{1/4}$ (arcsecs) within the blue (*filled circles*) and green windows in NGC 1399. *Right*: idem for the yellow (*filled circles*) and red windows.

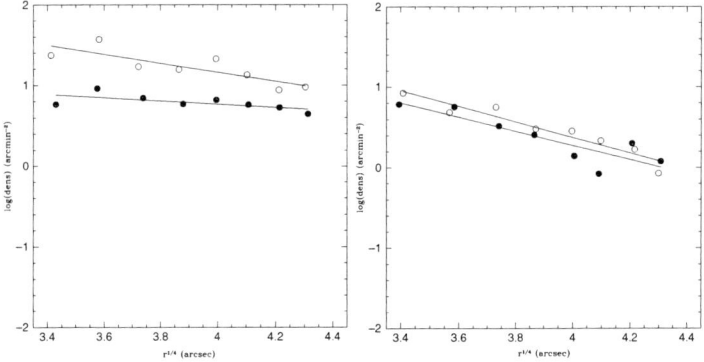

Fig. 2. Idem Fig. 1 but for NGC 4486. "Green" globulars here have been shifted upwards by 0.5.

distribution, yields an integrated globular cluster color (through an adopted abundance-metallicity-color relation). The adoption of color metallicity relations as those discussed in [3] always require two different cluster populations in order to obtain acceptable fits to the observed color histograms. Model clusters in each family are then spatially distributed (adopting the density slopes derived in the "yellow-red" window for the red clusters and in the "blue" window for the blue clusters). Once these populations are projected on the sky, the recovered density slope, within the "green" window, provides a good fit to the observed slopes in both galaxies. Figure 3 shows the GCs color histograms, that, once composed, give a good fit to the observed ones. The vertical lines (at $(C - T_1) \approx 1.5$) indicate the position of the color "valley". In both cases, the red globulars population shows a blue tail that "spills" into the formal color domain of the blue clusters.

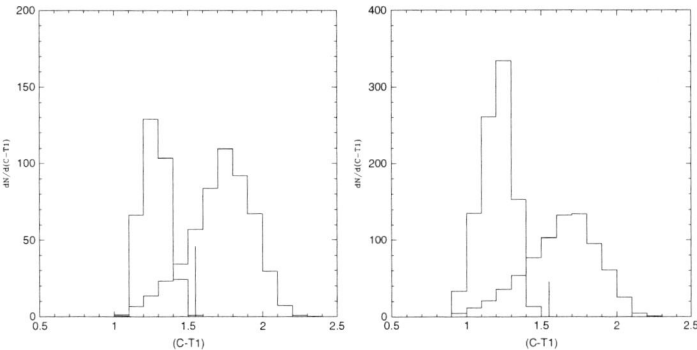

Fig. 3. Model Blue and Red globulars for NGC1399 (*left panel*) and NGC 4486.

4 The GCs-Galaxy Halo Connection

A tentative approach that aims at connecting the GCS features with those of the galaxy halo on a large scale has been presented in [1] for the case of NGC 1399. A further generalization of that approach assumes that every globular is formed with an associated "diffuse" stellar mass that shares the same metallicity (and, from an adopted mass-to-light ratio, yields a given integrated luminosity). Adding the diffuse stellar luminosity associated with each globular then leads to an integrated luminosity profile that should fit

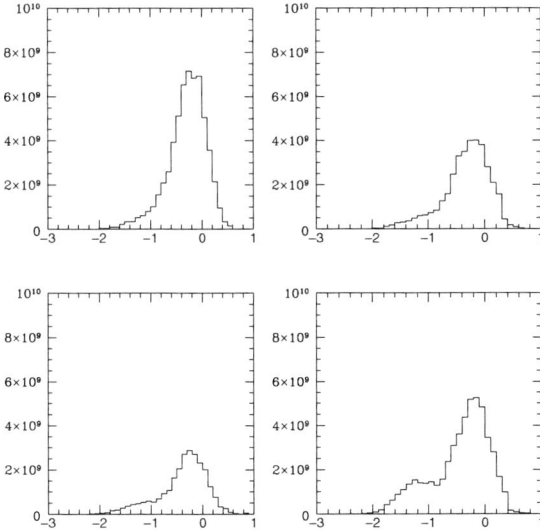

Fig. 4. Relative [Fe/H] distribution for stars in NGC 1399. *Upper left*: 0–5 kpc; *upper right*: 5–10 kpc; *lower left*: 10–15 kpc; *lower right*: 15–30 kpc

both the observed galaxy surface brightness and color gradient. The best fits are obtained by assuming an exponential dependence of the number of globular cluster per unit stellar mass (that increases with decreasing metallicity). This scenario, in turn, implies a globular cluster specific frequency that increases with decreasing metallicity (as suggested in [3] for NGC 5128). Figure 4 shows the inferred $[Fe/H]$ distribution for the stellar population in NGC 1399 within different ranges of galactocentric radius. These histograms strongly resemble the observed metallicity distribution in NGC 5128 [4], M31 [5] and in a number of edge-on spirals [6]. The model also reproduces another common feature in a number of galaxies: The color gradient of the galaxy halo and that of the globular clusters (as a whole) are roughly parallel and exhibit a blue off-set for the globulars (e.g. [7,8]).

As a summary: a simultaneous fit to the observed (bimodal) color histograms and spatial distributions of the globular clusters in NGC 1399 and NGC 4486 require two populations with remarkably different abundance and spatial distributions. In the case of NGC 1399, both the surface brightness of the galaxy and its color gradient are well matched with a two components stellar population (each one traced by the red or the blue globulars, respectively). The expected metallicity distributions for these field stars are compatible with results obtained from resolved stars in nearby systems.

References

1. J.C. Forte, F. Faifer, D. Geisler: Mon. Not. R.A.S. **357**, 56 (2005)
2. J.C. Forte, D. Geisler, E. Kim, M-G. Lee, P. Ostrov: IAUS **207**, 251 (2002)
3. W. Harris, G. Harris: Astron. J. **120**, 2423 (2000)
4. M. Rejkuba, L. Greggo, W. Harris, G. Harris, E. Peng: Astrophys. J. **631**, 262 (2005)
5. P. Durrell, W. Harris, Ch. Pritchet: Astron. J. **121**, 255 (2001)
6. M. Mohucine: APJ, **652**, 277 (2006)
7. J.C. Forte, S. Strom, K. Strom: Astrophys. J. **245**, L9 1981
8. A. Jordan, P. Cote, R. Marzke, D. Minniti, M. Rejkuba: Astron. J. **127**, 24 2004

Globular Cluster Systems, Diffuse Star Clusters, and Host Galaxies in the ACS Virgo Cluster Survey

Eric W. Peng

Herzberg Institute of Astrophysics, 5071 W. Saanich Road, Victoria, BC V9E 2E7, Canada

Abstract. We present the color distributions of globular cluster (GC) systems for 100 Virgo cluster early-type galaxies observed in the ACS Virgo Cluster Survey, the deepest and most homogeneous survey of this kind to date with over 11,000 observed GCs. On average, galaxies at all luminosities in our study ($-22 < M_B < -15$) appear to have bimodal or asymmetric GC color distributions. The colors of both subpopulations correlate with host galaxy luminosity and color, with the red GCs having a steeper slope. Using a preliminary nonlinear $(g–z)$-[Fe/H] relation for globular clusters to convert color to metallicity, we find for the first time that the metallicities of metal-poor and metal-rich GCs vary similarly with respect to galaxy luminosity and stellar mass. We also present an investigation of the diffuse star cluster systems in the Virgo galaxies. Compared to globular clusters (GCs), these star clusters have moderately low luminosities ($M_V > -8$) and a broad distribution of sizes ($3 < r_h < 30$ pc), but they are principally characterized by their low mean surface brightnesses which can be more than three magnitudes fainter than a typical GC ($\mu_g > 20$ mag arcsec^{-2}). These clusters are found in 12 galaxies, are typically red, and are largely found in lenticulars.

1 Introduction

The color distributions of globular cluster systems have played an important role in constraining the evolution of elliptical galaxies. Globular clusters trace each major epoch of star formation, and because they are single-age, single-metallicity systems with ages older than 10 Gyr, they provide a means to determine the distribution in metallicity of the early major star forming events that built their host galaxies. Bimodality in GC color distributions is now well-established in massive elliptical galaxies (e.g. [3]). This has led to the nomenclature of blue or "metal-poor", and red or "metal-rich" GC populations, which in the Milky Way would correspond to the halo GCs ($\langle$[Fe/H]$\rangle \sim -1.59$) and the bulge/thick disk GCs ($\langle$[Fe/H]$\rangle \sim -0.55$). Those dwarf elliptical galaxies that have been studied to date exhibit purely unimodal populations of metal-poor GCs [6]. The origin of these distinct GC subpopulations, and their precise variation as a function of galaxy properties is a key issue when trying to assemble a consistent picture of galactic spheroid formation.

In the second part of this proceeding, we present results on the nature of diffuse star cluster systems. Observations of nearby spiral, irregular, and lenticular galaxies have revealed a wide array of star cluster characteristics. *HST* imaging has uncovered a different population of faint and unusually extended old star clusters in two nearby lenticular galaxies [4,5]. Extended clusters that appear atypical for the Milky Way have also been identified in M31 and M33, and appear particularly numerous in M51 [1]. We extend a search for these kinds of star clusters to galaxies in the Virgo Cluster, and quantify their properties.

The ACS Virgo Cluster Survey (ACSVCS) is an HST/ACS imaging program of 100 early-type galaxies in the Virgo cluster, and is designed for studying GC systems in a deep and homogeneous manner. The breadth and depth of this survey makes it the most complete and homogeneous study of extragalactic globular cluster populations ever undertaken. Descriptions of this survey are given in ([2] and Jordán 2006, this volume). For details on GC photometry, selection, and cleaning of foreground and background see Peng et al. [7]. Ultimately, we produce a cleaned catalog of over 11,000 GCs across all 100 galaxies.

2 Color and Metallicity Distributions

GC color distributions are often described using a variety of two-Gaussian mixture models. When the data is of high enough quality, however, it is sometimes best to view the data itself in aggregate rather than struggle with parametrization. Figure 1(a) displays a nonparametric representation of the GC color distributions. Each column of this plot is the background cleaned color distribution of a single galaxy's GCs constructed with a Gaussian kernel, where the grayscale has been set such that zero density is black and the mode of the distribution is white. The galaxies are rank ordered by their luminosity with approximate M_B labeled on the x-axis. Immediately apparent from this image are the two GC subpopulations and their behavior with galaxy luminosity. Nearly all galaxies appear to possess a blue subpopulation of GCs, and the mean of this population varies slowly with galaxy luminosity. In addition, there is a population of red GCs whose color and number fraction increase in a continuous fashion across our entire sample—a red wing appears to exist even in some of our faintest galaxies. The mean color of the red GCs clearly increases with galaxy luminosity, as does the mean color of the blue GCs, although it is less obvious because the galaxies are rank ordered. The slope for the red GCs is 4.6 times steeper than for the blue GCs.

Because GCs are old, simple stellar populations, broadband color should be a good proxy for metallicity. We have completed a program to image the Milky Way GC system in these two bandpasses in order to derive an empirical $(g-z)$–[Fe/H] relation. We presented a preliminary relation in Peng et al. [7], while the final analysis will be presented in West et al. (2006, in prep). We find

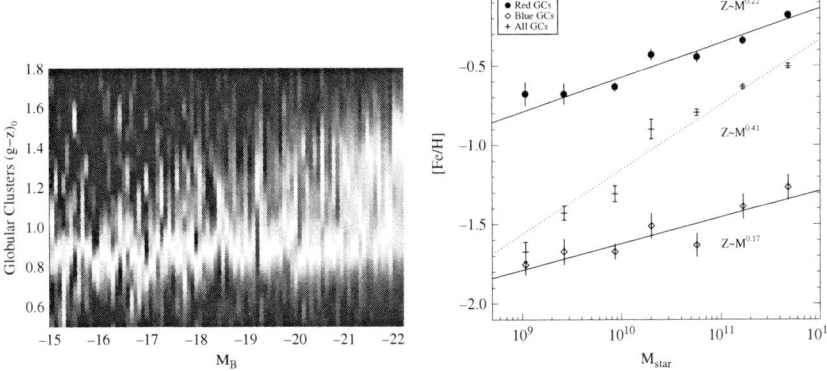

Fig. 1. (a) This image shows the GC color distributions of the ACSVCS galaxies ordered by host galaxy M_B. Each column is a kernel density estimation of a galaxy's GC color distribution, with white normalized to represent the peak density. (b) Mean metallicities of *blue, red*, and total GC populations as a function of galaxy stellar mass. Stellar mass has been translated from luminosity using an average, luminosity-weighted mass-to-light ratio. Metallicity is obtained using an empirical nonlinear color-metallicity relation

that the $(g-z)$–[Fe/H] relation is nonlinear with a steeper slope at the metal-poor end. We can then transform the colors of the binned GC populations to [Fe/H], and the compare them against one of the most fundamental galaxy properties—its stellar mass. Figure 1(b) exhibits the relationship between the metallicity of the GC subpopulations and galaxy stellar mass, showing the slopes for the two populations to be nearly identical. Across nearly four orders of magnitude in stellar mass, we find that $Z \propto M_\star^\beta$ where $\beta = 0.17 \pm 0.04$, 0.22 ± 0.03, and 0.41 ± 0.01 for the metal-poor, metal-rich, and total GC populations [7].

3 Diffuse Star Clusters

We used the same set of ACS imaging to investigate the nature and frequency of diffuse star clusters (DSCs) in early-type galaxies. Detailed results are presented in Peng et al. [8]. These star clusters have lower surface brightnesses than typical globular clusters. Figure 2(a) shows a comparison of the star cluster populations between the gE VCC 1226 (M49) and the luminous S0 galaxy VCC 798 (M85). While VCC 1226 has a GC population with a unimodal surface brightness distribution, VCC 798 has an extra population of diffuse star clusters. We find 12 galaxies in our sample that have a $> 3\sigma$ excess of diffuse sources, and nine of these galaxies are classified as lenticulars. The median colors of diffuse star cluster systems are red, $1.1 < g - z < 1.6$, which

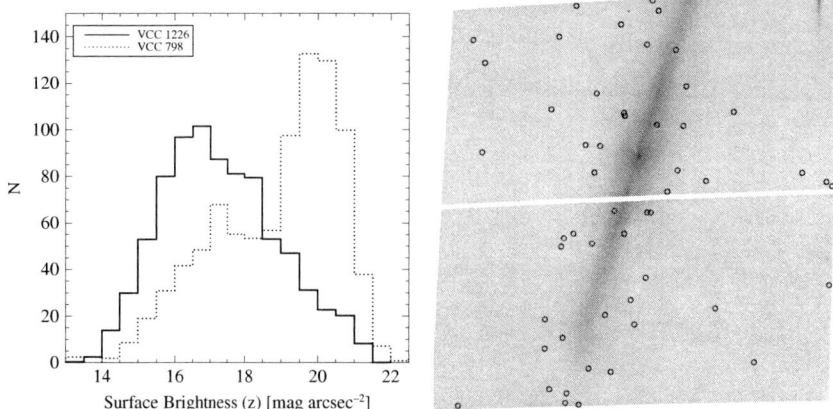

Fig. 2. (a) Histograms of z surface brightness for objects in VCC 1226 (*solid*) and VCC 798 (*dotted*). Expected background galaxy contamination as measured in the custom control fields has been subtracted. Compared to VCC 1226, VCC 798 has a large population of star clusters with lower surface brightnesses ($z > 19$ mag arcsec^{-2}). (b) Spatial distribution of red diffuse sources around VCC 2095, one of the most edge-on galaxies in our sample.

is redder than metal-rich GCs and often as red as the galaxy itself. Most DSC systems thus have mean ages older than 5 Gyr or else have super-solar metallicities implying that diffuse star clusters are likely to be long-lived. In cases where we see galactic disks edge-on, the DSCs appear to be aligned with the disk (Fig. 2(b)). The closest Galactic analogs to the DSCs are the old open clusters and we suggest that DSCs may be the remnants of disk formation episodes.

References

1. Chandar, R., Whitmore, B., & Lee, M. G. 2004, ApJ, 611, 220
2. Côté, P., et al. 2004, ApJS, 153, 223
3. Gebhardt, K., & Kissler-Patig, M. 1999, AJ, 118, 1526
4. Larsen, S. S., & Brodie, J. P. 2000, AJ, 120, 2938
5. Larsen, S. S., Brodie, J. P., Huchra, J. P., Forbes, D. A., & Grillmair, C. J. 2001, ApJ, 121, 2974
6. Lotz, J. M., Miller, B. W., & Ferguson, H. C. 2004, ApJ, 613, 262
7. Peng, E. W., et al. 2006a, ApJ, 639, 95
8. Peng, E. W., et al. 2006b, ApJ, 639, 838

Hot Populations in M87 Globular Clusters

S.T. Sohn[1,2], R.W. O'Connell[2], A. Kundu[3], W.B. Landsman[4],
D. Burstein[5], R.C. Bohlin[6], J.A. Frogel[7], and J.A. Rose[8]

[1] Korea Astronomy and Space Science Institute `tonysohn@gmail.com`
[2] Department of Astronomy, University of Virginia
[3] Department of Physics and Astronomy, Michigan State University
[4] NASA Goddard Space Flight Center
[5] Department of Physics and Astronomy, Arizona State University
[6] Space Telescope Science Institute
[7] Association of Universities for Research in Astronomy, Inc.
[8] Department of Physics and Astronomy, University of North Carolina

1 Introduction

The ultraviolet upturn (UVX), a sharp rise in flux at wavelengths shorter than 2000Å, is the most variable photometric feature of old stellar populations; UV-to-optical flux ratios vary by nearly two orders of magnitudes. This remarkable variation has led to the expectation that the UVX could be a sensitive probe of the star formation and chemical enrichment histories of early type galaxies.

Early UV observations by the *Astro* missions revealed that the stars responsible for the UVX are most likely old low-mass stars on the extreme horizontal branch (EHB) with $T_{\rm eff} > 15,000$ and tiny envelope masses ($\leq 0.05 M_\odot$), and their hot descendents [3]. Since the helium core mass of a low-mass HB star is insensitive to other parameters, the scatter in the initial surface temperature of HB stars is caused by the variance in pre-HB mass loss. Hence, the net UV-output of an EHB population is mainly governed by the envelope mass-loss on the red giant branch and helium abundance. Mass-loss efficiency, in turn, is strongly related to age and metal abundance, in the sense that it increases as a population gets older and more metal-rich.

Although we understand the basic physics of the stars responsible for the UVX as discussed above, we do not yet understand how the parent population's global characteristics (age, helium abundance, metal abundance, and dynamics) determine the temperature distribution of EHB stars and hence the strength and shape of the upturn. Undoubtedly, globular clusters (GCs) offer the best means of clarifying these global effects since they are (1) well understood from the standpoint of evolution; and (2) mostly simple populations with a small internal dispersion in age and abundance.

Unfortunately, the available UV data on GCs are limited and hard to interpret. Dorman et al. [1] showed that some Galactic GCs have FUV-V colors comparable to those of elliptical galaxies. However, there is no correlation between the FUV-V and the metallicity of GCs. In contrast, the FUV-V among the galaxies appear to get bluer with increasing metallicity.

The available sample for GCs does not overlap with the galaxies in metallicity. Hence, it is difficult to understand why the GCs differ from galaxies in the metallicity dependence of FUV-V. Consequently, the objects of primary interest will be GCs around giant elliptical galaxies like M87. In this study, we use *HST* STIS far-UV observations to determine how the M87 clusters compare to the Galactic GCs and the elliptical galaxies in their UV color properties.

2 *HST* STIS Far-Ultraviolet Observations

We have imaged three different fields in the inner ($r < 1.5$ arcmin) region of M87 using the *HST* STIS FUV-MAMA. This camera has a field of view of 25 arcsecond square with 0.024 arcsecond per pixel. The F25SRF2 longpass filter (1270–2000Å, pivot wavelength 1460Å) was used. Total exposure time was $\sim 10,000$ sec for each field.

An additional far-UV image, obtained with the same detector/filter combination as the other three images, was downloaded from the archive. This image was originally taken as part of the ongoing investigation of the M87 jet, and therefore include the jet and the diffuse background of the galaxy that increases rapidly toward the nucleus. The exposure time for this image was $\sim 3,500$ sec.

All of the STIS far-UV images were calibrated with the standard IRAF CALSTIS pipeline. Since the sources in far-UV images have poorly defined point spread function (PSF), we centroided on each object detected by eye and cross-identified them with objects listed in Kundu et al. ([2]; hereafter K99). When detected far-UV sources did not match any of the objects in K99 photometry, we classified them as "FUV-only sources." Aperture photometry on the far-UV sources was performed using a modified version of the IDL APER routine, an implementation of the standard DAOPHOT [4] photometry routine. Net flux of each source was measured through an aperture size of 8 pixels in radius and using a local sky background measured within an annulus of 20–30 pixels. All far-UV counts were converted into STMAG system.

With respect to the V and I photometry of K99, we detect 94%, 63%, and 48% of the clusters respectively for $V_0 \leq 22.0, 23.0$, and 24.0 in the far-UV.

3 Comparison of M87 to Galactic Globular Clusters

Figures 1a, b shows the extinction-corrected (FUV-V, M_V)$_0$ and (FUV-V, FUV)$_0$ color-absolute magnitude diagrams (CMDs) for M87 clusters and Galactic GCs. These CMDs show that we have sampled in M87 almost the full V-band luminosity range present in the Galactic UV-detected sample, although the average M87 detection is more luminous than in the Galaxy.

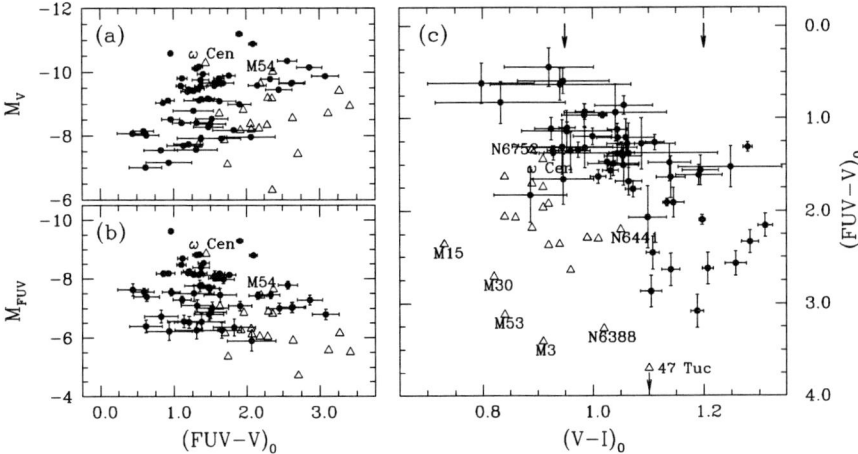

Fig. 1. (a) and (b): (FUV-V, M_V)$_0$ and (FUV-V, M_{FUV})$_0$ color-magnitude diagrams for M87 clusters (*filled circles*) and Galactic GCs (*open triangles*). (c): (FUV-V, M_V)$_0$ vs. ($V - I$)$_0$ color-color diagram for M87 clusters and Galactic GCs. Symbols are same as those in (a) and (b). The arrows at ($V - I$)$_0 = 0.95$ and 1.20 show the two peaks in the optical M87 color distribution found by K99.

There are three M87 clusters brighter than the most massive Galactic GC (ω Cen), in M_V, and two in M_{FUV}. We also find that the M87 clusters extend to a significantly bluer (FUV - V) color than the Galactic GCs.

In Fig. 1c, we show the ($V - I$, FUV - V)$_0$ color-color diagram for all data plotted in Fig. 1a, b. Seventeen of the UV-detected M87 clusters with ($V - I$)$_0 > 1.1$ are presumably more metal-rich than any Galactic sample cluster. On average, these objects have redder (FUV - V)$_0$ colors than the other M87 GCs, but they are considerably bluer than metal-rich Galactic GCs like 47 Tuc and NGC 6388. Thus, it is clear that M87 clusters with metallicities considerably in excess of those in the Galaxy are capable of producing large EHB star populations.

Despite strong overlap in V- and I- band properties, the M87 GCs have UV-optical properties that are distinct from clusters in the Galaxy. M87 clusters produce larger hot HB populations than do their Galactic analogs, and metal-poor GCs are better at this than are metal-rich GCs.

4 UV Colors and Metallicities in Clusters and Elliptical Galaxies

Figure 2 shows the far-UV versus Mg_2 correlation for M87 GCs, Galactic GCs and open clusters, and galaxies. For M87 GCs, we have estimated Mg_2 using color-[Fe/H]-Mg_2 relations. The galaxy measurement refer to the central $10'' \times 20''$ region covered by the *IUE* aperture.

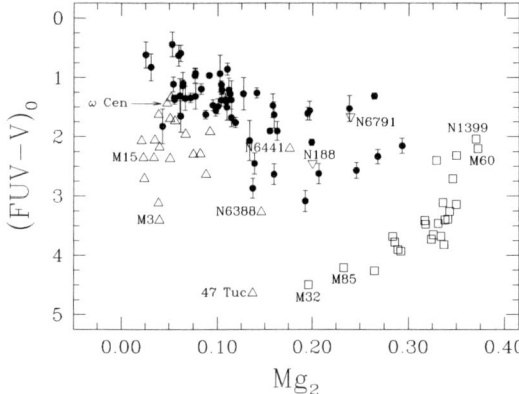

Fig. 2. Metallicity dependence of (FUV - $V)_0$ colors for M87 GCs (*circles*), Galactic GCs (*triangles*), two Galactic open clusters (*inverted triangles*), and galaxies (*squares*). The Mg_2 indices for M87 clusters were estimated from their $(V - I)_0$ colors.

About 10 metal-rich M87 clusters overlap with the lower metallicity end of the galaxy sequence, but the UV-detected M87 clusters produce much more far-UV light than do the galaxies at a given Mg_2. Also, the metal-rich clusters of M87 do not appear to represent a transition between Galactic-type clusters and elliptical galaxies. We note that this comparison is tentative because of the incompleteness in both samples and the uncertainty in the inferred Mg_2 values for the M87 clusters.

The two open clusters, NGC 188 and NGC 6791, fall together with the metal-rich M87 clusters and have much bluer colors than elliptical galaxies with similar Mg_2 values. Theoretical models for the UVX (e.g., [5]) predict that its strength is a step function of age, with the populations older than ~ 5 Gyr having a significant upturn due to rich EHB populations. The open-cluster observations can be taken as evidence that the transition age is younger than ~ 6 Gyr (the age of NGC 188). If so, Fig. 2 indicates that the light of most of the galaxies is dominated by populations younger than about 6 Gyr.

References

1. Dorman, B., O'Connell, R. W., & Rood, R. T. 1995, ApJ, 442, 105
2. Kundu, A., Whitmore, B. C., Sparks, W. B., Macchetto, F. D., Zepf, S. E., & Ashman, K. M. 1999, ApJ, 513, 733
3. O'Connell, R. W. 1999, ARA&A, 37, 603
4. Stetson, P. B. 1987, PASP, 99, 191
5. Yi, S., Lee, Y.-W., Woo, J.-H., Park, J.-H., Demarque, P., & Oemler, A., Jr. 1999, ApJ, 513, 128

A Subaru/Suprime-Cam Wide-Field Survey of Globular Cluster Populations around M87

Naoyuki Tamura[1,2], Ray M. Sharples[2], Nobuo Arimoto[3], Masato Onodera[3,4], Kouji Ohta[5], and Yoshihiko Yamada[3]

[1] Subaru Telescope, Hilo, HI 96720, USA naoyuki@naoj.org
[2] Department of Physics, University of Durham, Durham, DH1 3LE, UK
    r.m.sharples@durham.ac.uk
[3] National Astronomical Observatory of Japan, Tokyo 181-8588, Japan arimoto, monodera, yyamada@optik.mtk.nao.ac.jp
[4] Department of Astronomy, University of Tokyo, Tokyo 113-0033, Japan
[5] Department of Astronomy, Kyoto University, Kyoto 606-8502, Japan
    ohta@kusastro.kyoto-u.ac.jp

Abstract. We report a wide-field imaging survey of the globular cluster (GC) populations around M87 with Suprime-Cam on the 8.2 m Subaru telescope. By investigating the GC colour distributions and surface number densities as a function of distance from the host galaxy, we argue that the contribution of GCs associated with the Virgo cluster (i.e. intergalactic GCs) is insignificant around M87; most of the blue GCs around luminous ellipticals as well as the red GCs are presumably associated with the host galaxy.

1 Introduction

One of the most exciting recent developments in the study of extragalactic globular clusters (GCs) is the discovery that many luminous elliptical galaxies have bimodal or multimodal colour distributions of GCs. Since this is unlikely to be produced in a model of elliptical galaxy formation due to a simple monolithic collapse, more sophisticated models have been proposed [4, 1, 2]. However, most of the ellipticals whose GC populations have so far been well studied reside in high density environments. The GC properties upon which the proposed scenarios are based may therefore have been substantially modified by environmental effects. In fact, one scenario to explain the GC bimodality known as the accretion scenario [3] assumes that the blue (metal-poor) GCs have been captured from other (presumably less luminous) galaxies through tidal stripping and/or accretion. A population of intergalactic GCs ("i-GC" hereafter) could originate from such interactions and may naturally explain the fact that GCs tend to be extremely populous around central cluster galaxies, because they would be most likely to be surrounded by a significant population of i-GCs [13]. To directly survey for i-GCs, characterize this population, and test the validity of the accretion scenario, we performed a wide-field imaging survey in the BVI bands of the GC populations around M87 at the centre of the Virgo cluster.[6]

2 Observations and Data Analyses

Since the observations, data reductions and data analyses are explained in detail by [10], here we only highlight a couple of points. The imaging observations were carried out on 17 and 18 March 2004 with Suprime-Cam (SCam) on the Subaru telescope. The survey area is $\sim 2° \times 0°.5$ (560 kpc × 140 kpc). The limiting magnitudes are $B = 25.8$ mag, $V = 25.2$ mag, and $I = 24.5$ mag (5 σ, point source). We also analyzed BVI images of control fields, HDF-N and Lockman Hole taken with SCam/Subaru. These control field data are compatible with those in the M87 field in terms of the filter set, limiting magnitudes, and image quality, which minimizes the possibility of introducing any systematic errors into the subtractive corrections for fore- and background contamination in the GC sample.

To select GC candidates, we firstly detected objects using SExtractor (only objects detected in all the three bands were used for analysis). Second, we selected objects with `CLASS_STAR` indices larger than 0.6 as unresolved. The V-band image was used for this classification because of the best image quality. PSF-fitting photometry was performed to measure BVI magnitudes of these unresolved objects and a colour criterion was then imposed on them to isolate GC candidates, which was defined to include the Galactic GCs but minimize the contamination by foreground stars and background galaxies.

3 Colour and Spatial Distributions of GCs

The $V - I$ colour distributions of GCs as a function of distance from the host galaxy centre indicates that in both of M87 and NGC 4552, evidence of bimodality is more clear in the inner region (the peaks of the bimodality are at $V - I \sim 1.0$ and 1.2) and the fraction of red globular clusters tends to decrease at larger distances from the host galaxy centre. In other words, the red GCs tend to be more centrally concentrated than the blue GCs. This characteristic can also be described by the radial profiles of GC surface number densities (Fig. 1), where the GCs are divided at the middle of the peaks in the bimodal colour distribution into red GCs ($V - I \geq 1.1$) and blue ($V - I < 1.1$). In Fig. 1, the surface brightness distributions of the host galaxies are also plotted. This indicates that the red GC distribution is similar to the surface brightness distribution of the host galaxy, but the blue GC distribution is significantly more extended.

---

[6] An angular scale of 1' corresponds to 4.7 kpc and 4.5 kpc at the distance of M87 and NGC 4552, respectively, for our adopted distances [11].

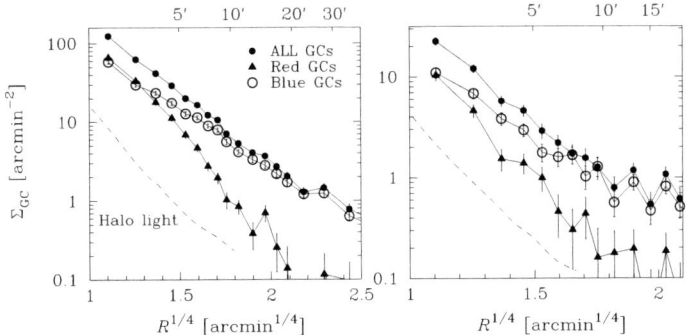

Fig. 1. Radial profile of GC surface number densities around M87 (*left panel*) and NGC 4552 (*right panel*). *Filled triangles* and *open circles* show the surface number densities of the *red* ($V-I > 1.1$) and *blue* ($V-I \leq 1.1$) GCs, respectively, and solid circles are those of the total GC population. Dashed lines show the host galaxy halo light distribution in the V band, which is arbitrarily shifted in the vertical direction for comparison with the GC profiles.

This extended nature of the blue GCs may imply that a large fraction of the blue GCs are not associated with any luminous galaxies and that there are a number of i-GCs. However, our data suggest that the blue GC distribution around M87 is significantly more concentrated toward the M87 centre compared to the radial profile of the dark matter surface mass density in the Virgo cluster [8] (Fig. 2(a)). It should also be pointed out that such

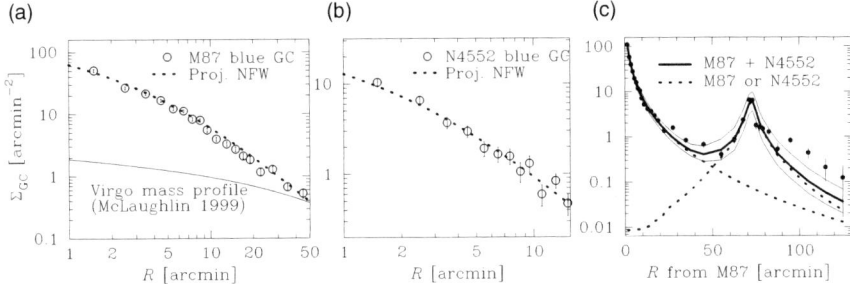

Fig. 2. (a) Surface densities of the M87 blue GCs are plotted with open circles. The dotted line indicates a projected NFW profile with a scale radius of 23 kpc. The solid line indicates the projected mass profile of the Virgo cluster [8]. (b) Same as (a), but for the NGC 4552 blue GCs. The *dotted line* shows a projected NFW profile with a scale radius of 16 kpc. (c) GC surface density is plotted over the full range of distance from M87 covered by our data. The *dotted lines* indicate the model profiles of GC densities associated with M87 and NGC 4552, and the sum of the two components is indicated by the *thick solid line*. The *thin solid lines* around this *thick line* show the 1 σ envelope due to the fitting errors of the $r^{1/4}$ laws to the measured GC densities around M87 and NGC 4552.

an extended distribution of blue GCs is seen not only around M87 but also NGC 4552 (Fig. 1), NGC 4472 [7] and NGC 4649 [5] which are located in the outer regions of the Virgo cluster. Both of these facts suggest that the contribution of i-GCs is insignificant around M87 and most of the blue GCs as well as the red GCs must be associated with the host galaxy.

Then, what does the blue GC distribution represent? One interesting possibility is it traces the dark matter halo of the host galaxy. In fact, gravitational lens analyses of early-type galaxies at intermediate redshifts (e.g., [12]) and X-ray hot gas analyses of nearby ellipticals [6] suggest that dark matter halos of luminous (massive) ellipticals are more spatially extended than their stars. Also, the blue GC distributions of M87 and NGC 4552 can be modeled by NFW profiles [9] (see Fig. 2(a), (b)). The similarity is remarkable and at least suggests there may be a physical link between the blue GC subpopulation and the dark matter content of the host galaxy.

4 Any Evidence for Intracluster GCs?

Provided that the majority of the blue GCs as well as the red GCs belong to the host galaxy, is there any evidence for an i-GC population in the GC spatial distribution? In Fig. 2(c), the GC surface densities for the full coverage of our data are plotted as a function of distance from M87. This radial profile is modeled by a superposition of two $r^{1/4}$ laws each of which is fitted to the radial profile of total GC surface densities within 20′ of M87 or NGC 4552 and is extrapolated to larger distances. This plot indicates that the GC surface densities around 100′ from M87 exceed those expected from just the M87 GCs and NGC 4552 GCs, suggesting that an additional i-GC population exists with a surface density of ~ 0.2 arcmin^{-2}. The estimated i-GC surface density however depends sensitively on the assumed radial distribution of GCs associated with M87 and NGC 4552. The possible statistical detection and estimated number density of i-GCs are therefore tentative and need to be confirmed with future surveys.

References

1. K. M. Ashman, & S. E. Zepf: ApJ, **384**, 50 (1992)
2. M. Beasley, C. M. Baugh, D. A. Forbes, et al.: MNRAS, **333**, 383 (2002)
3. P. Côté, R. O. Marzke, & M. J. West: ApJ, **501**, 554 (1998)
4. D. A. Forbes, J. P. Brodie, & C. J. Grillmair: AJ, **113**, 1652 (1997)
5. D. A. Forbes, F. R. Faifer, J. C. Forte, et al.: MNRAS, **355**, 608 (2004)
6. Y. Fukazawa, J. G. Betoya-Nonesa, J. Pu, et al.: ApJ, **636**, 698 (2006)
7. M. G. Lee, E. Kim, & D. Geisler: AJ, **115**, 947 (1998)
8. D. E. McLaughlin: AJ, **117**, 2398 (1999)
9. J. F. Navarro, C. S. Frenk, & S. D. M. While: ApJ, **490**, 493 (1997)

10. N. Tamura, R. M. Sharples, N. Arimoto, et al.: MNRAS, **373**, 588 (2006)
11. J. L. Tonry, A. Dressler, J. P. Blakeslee, et al.: ApJ, **546**, 681 (2001)
12. T. Treu, & L. V. E. Koopmans: ApJ, **611**, 739 (2004)
13. M. J. West, P. Côté, C. Jones, et al.: ApJ *Letters*, **453**, L77 (1995)

Stellar Populations of Globular Clusters in NGC 1407

A.J. Cenarro[1], J.P. Brodie[2], M.A. Beasley[3], and J. Strader[2]

[1] Universidad Complutense de Madrid, Avda. Complutense s/n, E-28040 Madrid, Spain cenarro@astrax.fis.ucm.es
[2] UCO/Lick Observatory, University of California, Santa Cruz, CA 95064, USA brodie@ucolick.org, strader@ucolick.org
[3] Instituto de Astrofísica de Canarias, Vía Láctea, E-38200, La Laguna, Tenerife, Spain beasley@iac.es

We present high-quality, Keck spectroscopic data for a sample of 20 globular clusters (GCs) of NGC 1407. On the basis of this data, ages, metallicities and $[\alpha/\mathrm{Fe}]$ ratios for the GC system and the integrated spectrum of the galaxy have been derived from their Lick/IDS indices, making use of the SSP model predictions by Maraston, Thomas and Bender [4] and Thomas, Maraston and Korn [3], the ones account for different $[\alpha/\mathrm{Fe}]$ ratios as well as for blue horizontal branch (BHB) effects at the low metallicity regime.

Figure 1 illustrates some interesting, index-index diagnostic diagrams that include Hγ_A, Mgb, <Fe>, [MgFe]' and some C and/or N dependent indices such as CN$_2$, C$_2$4668 and G4300. Lines correspond to the above model predictions at either constant $[\alpha/\mathrm{Fe}]$ (+0.3 dex; panel a), thus varying age (from 1 to 15 Gyr) and metallicity (from -2.25 to $+0.67$ dex) as given in the labels, or constant age (11 Gyr; panels $b-f$), thus varying metallicity and $[\alpha/\mathrm{Fe}]$ (0.0, +0.3 and +0.5). The filled star corresponds to the $R_{\mathrm{eff}}/8$ central spectrum of NGC 1407. Asterisks represent the sample of 41 Galactic GCs of Schiavon et al. [1], and the rest of symbols (circles and triangles) are GCs in NGC 1407: metal-poor (MP) GCs ($B - I < 1.87$) are illustrated as black-filled symbols, whilst open and grey-filled ones represent the metal-rich (MR) subpopulation ($B - I > 1.87$).

Concerning ages and mean metallicities, all GCs in NGC 1407 look old and follow a tight metallicity sequence that reaches values above solar (panel a). The old ages derived from all the Balmer Lick indices ($\sim 11 \pm 2$ Gyr) confirm the above statement for most GCs (circles) except for three GCs with mean, derived ages of $\sim 3-4$ Gyr (triangles). However, in the light of the spectral diagnostic proposed by Schiavon et al. [2] to detect BHB effects in the integrated spectrum of GCs, we find these three *young* candidates to be consistent with being old GCs hosting BHB stars.

As regards to element abundance ratios, in panel b we find GCs to pose mean [Mg/Fe] ratios of $\sim 0.33 \pm 0.17$ dex. There exists, however, a subset of MR GCs that systematically exhibit higher [Mg/Fe] values (~ 0.5 dex; open circles) as compared to the rest of MR GCs (grey-filled symbols). Interestingly, a similar systematic difference between the mean [C/Fe] values of both

MR GC subsets is reported in panel c using C_24668. This is the first evidence for a correlation between Mg and C abundances among extragalactic GCs. The possibility that such high [Mg/Fe] and [C/Fe] ratios could be just driven by a Fe depletion is ruled out since GCs exhibiting high [C/Fe] ratios indeed populate the high-metallicity regime of the C-sensitive diagram in panel d (G4300 – C_24668). Different star-formation time-scales could be driving the observed Mg and C overabundances.

We also find a striking CN enhancement all over the GC metallicity range (panel e). The behavior of C (C_24668) and N (which is thought to drive the CN variations) in MP GCs clearly deviates from the one exhibited by MR GCs (panel f). In particular, for MR GCs, N increases drastically at the time that C essentially saturates and becomes depleted as compared to the model predictions. This may be interpreted as a consequence of the increasing importance of the CNO cycle with the increasing metallicity.

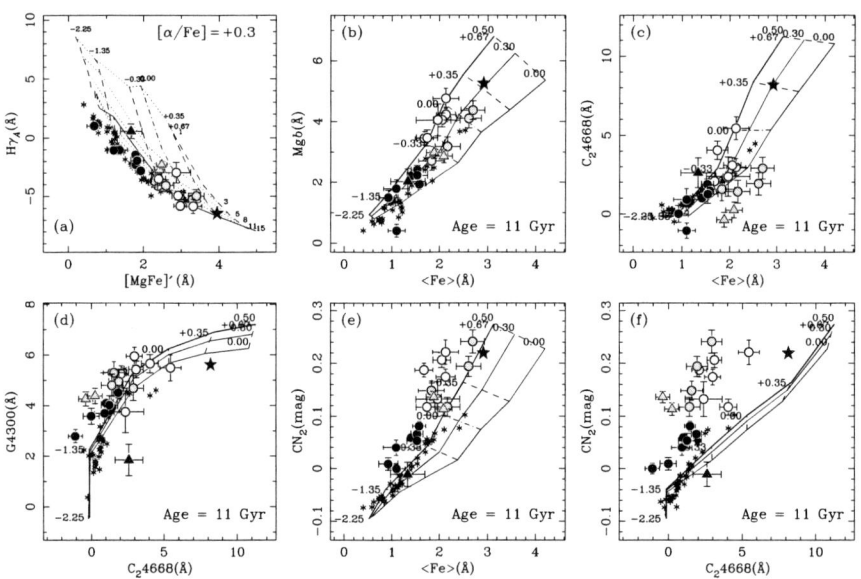

Fig. 1. Some age, metallicity and element abundance ratio diagnostic diagrams constructed on the basis of Lick indices. See details in the text.

References

1. R.P. Schiavon, J.A. Rose, S. Courteau et al.: ApJS **160**, 163 (2005)
2. R.P. Schiavon, J.A. Rose, S. Courteau et al.: ApJ **608**, L33 (2004)
3. D. Thomas, C. Maraston & A. Korn: MNRAS **351**, L19 (2004)
4. D. Thomas, C. Maraston & R. Bender: MNRAS **339**, 8987 (2003)

The Globular Cluster System of NGC 5846 Revisited: Colours, Sizes and X-Ray Counterparts *

Ana L. Chies-Santos[1], Basilio X. Santiago[1], Miriani G. Pastoriza[1], and Duncan A. Forbes[2]

[1] Departamento de Astronomia, Instituto de Física, UFRGS. Av. Bento Gonçalves 9500, Porto Alegre, RS, Brazil `ana.leonor@ufrgs.br`
[2] Centre for Astrophysics & Supercomputing, Swinburne University, Hawthorn VIC 3122, Australia

NGC 5846 is a giant elliptical galaxy with a previously well studied globular cluster system (GCS) (e.g. [1] and [2]), known to have a bimodal colour distribution with a remarkably high red fraction. Here we revisit the central galaxy regions searching for new globular cluster (GC) candidates using archival Hubble Space Telescope WFPC2 images, from which we modelled and subtracted the host light distribution and increased the available sample of GCs.

We measured V and I magnitudes following Forbes, Brodie and Huchra [1] and calibrated with their sample. The colour distribution shows a hint of becoming bluer in the very central galaxy regions. Figure 1 shows the colour distributions of GCs with V < 26.5 for two different rings from the center. The inner population is shifted slightly towards bluer colours relative to the one located further out to the center of the galaxy. The inner regions have a mean (V–I)= 1.17 ± 0.04 compared to (V–I) = 1.24 ± 0.07 in the next radial bin. Thus the difference (δ(V–I) = 0.07) is small enough to place both regions within the metal-rich GC subpopulation and it is of only marginal statistical significance. If real it could be due to a combination of age and/or metallicity effects.

We measured sizes following *ishape* [3]. This code convolves the WFPC2/HST PSF with model king profile in 2-D. The resulting model image is compared to real GCs and an estimate of the size is derived from the fit. Reliable sizes are obtained for about 60 GCs; their typical effective radii are in the range 3–5 pc. The largest clusters are located in the central regions. No clear evidence for a size-colour relation is found.

We also aim to find any possible link between GCs and X-ray point sources detected by Trinchieri and Goodfrooij [4]. We matched X-ray positions to our sample of GCs and found 7 GC/X-ray sources with offsets of 0.8″ in RA and 0.2″ in DEC, most of which are located in the central parts. Optically, they tend to be bright (V ≤ 23.5) and compact, since only one was clearly resolved. All of them have (V − I) > 1.1, which places them as members of the red

---
*A&A accepted, astro-ph/0604499

sub-population. Their X-ray luminosities are also among the highest in the Chandra sample, with 3 of them having $L_X > 10^{39} \mathrm{erg\,s^{-1}}$ and 2 others being also among the 25% most luminous. In Fig. 2 we show the on-sky distribution of both optical and X-Ray objects, with the positions already compensated for the offsets mentioned above.

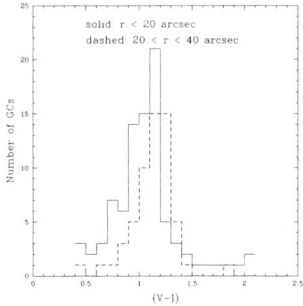

Fig. 1. Colour distributions at the central parts of NGC 5846. *Solid line*: $r < 20''$; *dashed line*: $20'' < r < 40''$.

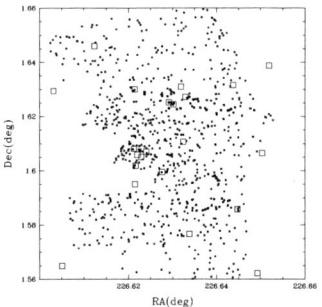

Fig. 2. On-sky map of X-ray sources and GCs in the direction of NGC 5846. The small points represent the GCs and the large squares correspond to the Chandra point sources.

References

1. Forbes, D. A., Brodie, J. P. & Huchra, J. 1996, AJ, 112, 2448
2. Forbes, D. A., Brodie, J. P. & Huchra, J. 1997, AJ, 113, 887
3. Larsen, S. S. 1999, A&AS, 139, 393
4. Trinchieri, G. & Goodfrooij, P., 2002, A&A, 386, 472

Globular Cluster Systems in Shell Ellipticals

Jaeil Cho[1] and Ray M. Sharples[2]

[1] Department of Physics, University of Durham, South Road, Durham, DH1 3LE, UK. `jaeil.cho@durham.ac.uk`
[2] Department of Physics, University of Durham, South Road, Durham, DH1 3LE, UK. `r.m.sharples@durham.ac.uk`

We investigate the properties of the globular cluster systems around six Os-hell ellipticals. The three most luminous galaxies show strong evidence of a bimodal colour distribution, whilst the three least luminous galaxies are all consistent with only having a single blue peak of clusters. These peak positions agree well with previous studies. We discuss the results from this pilot study in the context of current scenarios for the origin of globular cluster bimodal color distributions and shell ellipticals.

1 Introduction

Globular clusters (GCs) are widely used to trace the formation and evolution of their host galaxy, and are believed to form in concert with the majority of the stellar content of a galaxy. In addition, globular clusters are composed of millions of stars with the same age and metallicity, which allows us to compare with simple stellar population models. More than 20% of elliptical galaxies have shell features [1]. A shell structure is thought to originate in a galaxy merger (e.g. [2,3]) or in a weak interaction without mergers [4]. We have used archival HST/ACS data of six well known shell ellipticals in the F814W($\approx$I) and F606W($\approx$V) filters to study their globular cluster systems. Exposure times were typically 1000sec. SExtractor was used to detect potential GC candidates, and the IRAF PHOT task was used for aperture photometry.

2 Results

The colour distributions of the GC candidates are shown in Fig. 1. The thick black histograms are the colour distribution of GCs after applying a magnitude cut of V=25 mag (95% completeness limit) and a color cut of $0.5 < V - I < 1.5$. The histograms are sorted by host galaxy luminosity from top-left to bottom-right. Contamination by foreground stars and background galaxies has been estimated using a Milky Way star number density model and comparison with GOODS fields for background galaxies, and is negligible. A KMM test indicates that the three most luminous galaxies host a strong bimodal colour distribution (dashed lines), while the three least luminous

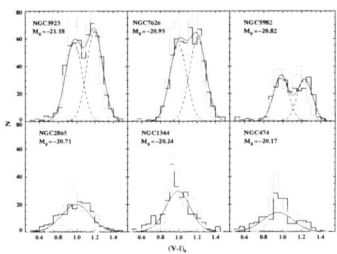 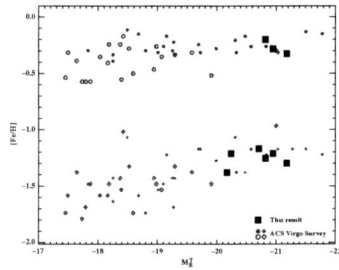

Fig. 1. Colour distributions of GC candidates.

Fig. 2. Positions of the blue and red peaks vs. host galaxy brightness.

galaxies are more likely to be unimodal. The thin histogram is a simulation following the prescription from [5] after applying our magnitude cut to their model. The overall trend of our colour distributions with galaxy luminosity are well reproduced by their model. Figure 2 shows positions of the blue and red peaks against host galaxy brightness, compared the ACS Virgo survey [6]. Our peak positions (filled/open squares) are consistent with both studies.

3 Conclusions

We have found a strong bimodal colour distribution in three luminous shell elliptical galaxies, with weak evidence in three less luminous examples. Application of models from [5] to our results reveals that the projection effect caused by a non-linear colour-metallicity relation could explain the general trend of the colour distributions. The peak positions of the colour distributions agree well with previous studies. No other peculiarities in terms of the properties of their GC populations has been found in these shell ellipticals. Global parameters (e.g. host galaxy brightness) appear to mainly govern their GC properties.

References

1. Malin, D. F., & Carter, D. 1980, Nature, 285, 643
2. Quinn, P. J. 1984, ApJ, 279, 596
3. Dupraz, C., & Combes, F. 1986, AAP, 166, 53
4. Thomson, R. C., & Wright, A. E. 1990, MNRAS, 247, 122
5. Yoon, S.-J., Yi, S. K., & Lee, Y.-W. 2006, Science, 311, 1129
6. Peng, E. W., et al. 2006, ApJ, 639, 838

GMOS Photometry of Five Globular Cluster Systems: NGC 4649, NGC 3923, NGC 524, NGC 3115 and NGC 3379

F.R. Faifer[1,2], J.C. Forte[1], M. Beasley[3], T. Bridges[4], D. Forbes[5], K. Gebhardt[6], D. Hanes[4], M. Norris[7], M. Pierce[5], R. Proctor[5], R. Sharples[7], and S. Zepf[8]

[1] Facultad de Ciencias Astronómicas y Geofísicas, Universidad Nacional de La Plata, Paseo del Bosque s/n, La Plata (B1900FWA), Argentina
favio@fcaglp.unlp.edu.ar
[2] IALP - Conicet, Argentina
[3] Instituto de Astrofísica de Canarias, Spain
[4] Department of Physics, Queen's University, Kingston, ON K7L 3N6, Canada
[5] Centre for Astrophysics & Supercomputing, Swinburne University, Hawthorn, VIC 3122, Australia
[6] Astronomy Department, University of Texas, Austin, TX 78712, USA
[7] Department of Physics, University of Durham, South Road, Durham DH1 3LE, UK
[8] Department of Physics and Astronomy, Michigan State University, East Lansing, MI 48824, USA

1 Introduction

Globular clusters (GCs) are useful tools to study the star formation and chemical evolution history as well as the dynamics and dark matter content of early-type galaxies. In this context, our group is involved in a major programme using the Gemini Multi-Object Spectrograph (GMOS) on the 8 m Gemini telescopes to obtain photometry and spectroscopy of GCs in a sample of early-type galaxies, covering a range of galaxy type, luminosity, and environment [3, 5, 6, 2]. Here we briefly describe the first photometric results of five galaxies: NGC 4649, NGC 3923, NGC 524, NGC 3115 and NGC 3379.

2 Observations

The observations were performed with the Gemini Multi-Object Spectrograph (GMOS) on Gemini North and Gemini South. The instrument consists of three 2048 × 4608 pixels CCDs, with a scale of 0.072 arcsec pixel^{-1}.

The images were taken using modified Sloan g', r' and i' filters [4], under excellent seeing conditions (~ 0.7 arcsec). Our observations, of 2–3 fields per galaxy, have total exposure times of 800 sec in g' and 400 sec in r' and i' filters.

The data were reduced using tasks within the Gemini IRAF package. The reduction included removal of CCD-to-CCD gain differences, subtraction of the bias pattern, division by a twilight sky flat, and mosaicing of the processed CCD output to form a single image. Interference fringes were removed from the GMOS-S i' data by subtracting a fringe frame that was created by combining observations of a blank field. The processed images were then aligned and co-added to reject cosmic rays and fill in the gaps between the CCDs.

The object search was carried out on the i' images, using the finding routines of the SExtractor program [1]. The selection of the GC candidates was implemented through a combination of the Stellarity Index from SExtractor and the i' mag vs. (aperture -psf magnitude) diagram. The psf photometry on the co-added images was made using Daphot task within IRAF.

3 Results

We obtained a sample of GC candidates consisting of 944, 834, 362, 306 and 182 objects in NGC 4649, NGC 3923, NGC 524, NGC 3115 and NGC 3379, respectively. The GC samples are very clean in terms of background contamination and they define the targets for the forthcoming spectroscopic study.

We found clear evidence for bimodality in the GC colour distributions in all galaxies of our sample, except for the S0 galaxy NGC 524. This last galaxy shows a broad unimodal color distribution.

The inner halos of the galaxies show colors remarkably similar to mean colors of the red GC sub-population. In the case of NGC 524, this color is comparable with the mean color of the total GC population.

In the bimodal systems, the red GCs show a more concentrated spatial distribution than the blue ones.

References

1. E. Bertin, S. Arnouts: A&AS **117**, 393 (1996)
2. T. Bridges, K. Gebhardt, R. Sharples, F. Faifer, J. C. Forte, M. Beasley, S. Zepf, D. Forbes, D. Hanes, M. Pierce: MNRAS, **373**, 157 (2006)
3. D. Forbes, F. Faifer, J. C. Forte, T. Bridges, M. Beasley, K. Gebhardt, D. Hanes, R. Sharples, S. Zepf: MNRAS, **355**, 608 (2004)
4. M. Fukugita, T. Ichikawa, J. Gunn, M. Shimasaku, D. Schneider: AJ, **111**, 1748, (1996)
5. M. Pierce, M. Beasley, D. Forbes, T. Bridges, K. Gebhardt, F. Faifer, J. C. Forte, S. Zepf, R. Sharples, D. Hanes, R. Proctor: MNRAS, **366**, 1253 (2006)
6. M. Pierce, T. Bridges, D. A. Forbes, R. Proctor, M. A. Beasley, K. Gebhardt, F. R. Faifer, J. C. Forte, S. E. Zepf, R. Sharples, D. A. Hanes: MNRAS, **368**, 325 (2006)

Structural Parameters from Ground-based Observations of Globular Clusters in NGC 5128

M. Gómez[1], D. Geisler[1], W.E. Harris[2], T. Richtler[1], G.L.H. Harris[3], and K.A. Woodley[2]

[1] Grupo de Astronomía, Depto. de Física, Univ. de Concepción, Casilla 160-C, Concepción, Chile `matias@astro-udec.cl`
[2] Department of Physics and Astronomy, McMaster University, Hamilton ON L8S 4M1, Canada
[3] Department of Physics, University of Waterloo, Waterloo, Ontario, N2L 3G1, Canada

We have investigated a number of globular cluster candidates from a recent wide-field study by Harris et al. [2] of the giant elliptical galaxy NGC 5128. We used the Magellan I telescope + MagIC camera under excellent seeing conditions ($0.3''$–$0.6''$) and obtained very high resolution images for a sample of 44 candidates. Of these 15 appear to be bonafide globular clusters in NGC 5128 while the rest are either foreground stars or background galaxies. We also serendipitously discovered 18 new cluster candidates in the same fields. Our images allow us to study the light profiles of the likely clusters, all of which are well resolved. This is the first ground-based study of structural parameters for globular clusters outside the Local Group. We compare the psf-deconvolved profiles with King models and derive structural parameters, ellipticities and surface brightnesses. We compare the derived structural properties with those of other well-studied globular cluster systems. In general, our clusters are similar in size, ellipticity, core radius and central surface brightness to their counterparts in other galaxies, in particular those in NGC 5128 observed with HST by Harris et al. [1]. However, our clusters extend to higher ellipticities and larger half-light radii than their Galactic counterparts, as do the Harris et al. sample. Combining our results with those of Harris et al. fills in the gaps previously existing in $r_h - M_V$ parameter space and indicates that any substantial difference between presumed distinct cluster types in this diagram, including for example the Faint Fuzzies of Larsen and Brodie [4] and the 'extended, luminous' M31 clusters of Huxor et al. [3] is now removed and that clusters form a continuum in this diagram. Indeed, this continuum now extends to the realm of the Ultra Compact Dwarfs. The metal-rich clusters in our sample have half-light radii that are almost twice as large on the average as their metal-poor counterparts, at odds with the generally observed trend. The possibility exists that this result could be due in part to contamination by background galaxies. We have carried out additional analysis to quantify this contamination. This shows that, although galaxies cannot be easily separated from clusters in some of the structural diagrams, the combination of excellent image quality and Washington photometry should limit the contamination to roughly 10% of the population of cluster candidates.

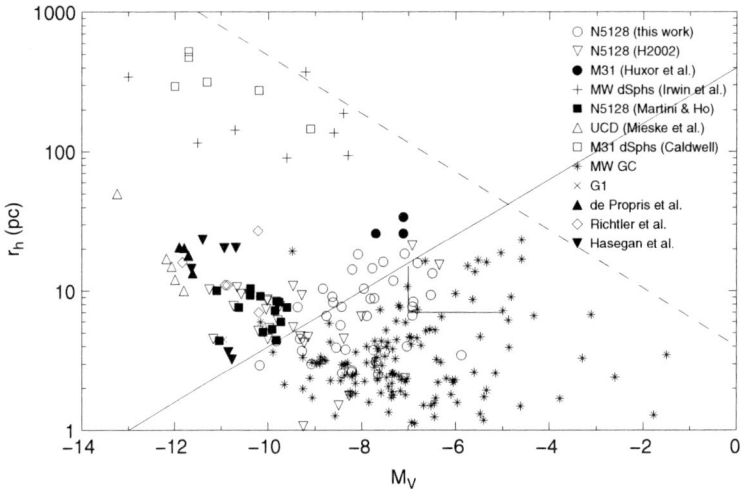

Fig. 1. Half-light radius r_h (in pc) versus M_V. A distance of 4 Mpc for NGC 5128 has been assumed. The equation $log r_h = 0.2 M_V + 2.6$ from van den Bergh and Mackey [5] is the *solid line* and the *dashed line* shows a value of constant average surface brightness with r_h. The solid L-shape indicates the region (above and to the right) where FFs are found. GCs form a continuum in this diagram and even approach the region occupied by UCDs.

References

1. Harris, W.E., Harris, G.L.H., Holland, S., & McLaughlin, D., 2002, AJ 124, 1435
2. Harris, G., Geisler, D., & Harris, W.E. 2004, AJ 128, 712
3. Huxor, A., Tanvir, N.R., & Irwin, M.J., et al., 2005, MNRAS 360, 1007
4. Larsen, S.S., & Brodie, J.P., 2000, AJ 120, 2938
5. van den Bergh, S., & Mackey, A.D., 2004, MNRAS 354, 713

Globular Cluster Populations in Early-Type Galaxies

M. Hempel[1], M. Kissler-Patig[2], T.H. Puzia[3], M. Hilker[4], S. Zepf[1], and A. Kundu[1]

[1] Michigan State University, Department of Physics and Astronomy, East Lansing, 48824 MI, USA hempel@pa.msu.edu
[2] European Southern Observatory, Karl-Schwarzschild Str. 2, 85748 Garching, Germany mkissler@eso.org
[3] Space Telescope Science Institute, Baltimore, USA tpuzia@stsci.edu
[4] Argelander Institut für Astronomie, Abteilung Sternwarte, Universität Bonn, Auf dem Hügel 71, 53121 Bonn, Germany mhilker@astro.uni-bonn.de

1 Introduction

We study the major star formation events in nearby early-type galaxies using globular clusters in order to probe the formation histories of the host galaxies. Globular clusters, bona fide single stellar populations are perfect tools for this research. In our study we use combined optical and near-infrared photometry to detect age sub-populations e.g. [6,3,5] and to set some constraints on their age and relative size. The increasing number of galaxies in our study allows us also to search for correlations between the age structure in early-type galaxies and various galaxy parameters, e.g. galaxy environment [1,2].

2 Data and Methodology

To derive the age structure in globular cluster systems we apply a semi-numerical method based on combined optical and near-infrared photometry of observed and simulated globular cluster systems [4]. The color-color distributions are used to derive the cumulative age distribution, which is later compared to the one in simulated systems with a given age structure (age and relative size of two populations) via a χ^2-test. Under the assumption of an old ($\approx$13 Gyr) and possibly one intermediate age population 1 and population 2, respectively, we quantify the age structure with the so-called *Methusalem* parameter MTH (see Eq. 1), using the age t and the relative size f of both populations. A GCS dominated by an old stellar population will have a MTH value close to 13 whereas for a mostly young population we expect a much lower MTH -value. This allows us to search for correlations between the age structure in a given galaxy and various selected galaxy parameters.

$$MTH = t_{\mathrm{pop1}} * f_{\mathrm{pop1}} + t_{\mathrm{pop2}} * f_{\mathrm{pop2}} \qquad (1)$$

3 Results

Previous studies [6, 3, 5] have shown that NGC 4365 and NGC 5846 host a significant fraction of intermediate age GCs in the central parts of the galaxy. As expected the MTH values for both galaxies are very small (4.6 and 5.8, respectively). In contrast we derive MTH values ≥7.5 for the remaining galaxies (see Fig. 1). Looking at the correlation between the age structure and galaxy parameters we note that NGC 4365 and NGC 5846 represent the group environment in our sample. If we divide the galaxy sample with respect to σ, as a mass dependent parameter, admittedly rather arbitrarily, into three intervals (e.g. 100–200 km/s, 200–300 km/s and >300 km/s) we can extract some correlation for the more numerous GC samples. Massive galaxies, such as NGC 1399 and M87 (also residing in the center of a galaxy cluster) have a considerably larger "Methuselah" value than smaller ones (NGC 4365, NGC 5846). Nevertheless, more data are required to investigate the correlations between the properties of GCSs (ages, chemical composition, spatial distributionradial properties, etc.) with those of their host galaxies.

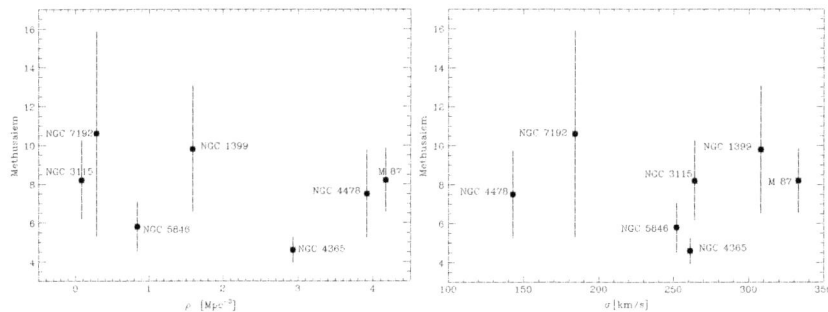

Fig. 1. The dependency between the age structure in globular cluster systems, represented by the *Methusalem* parameter (see Sect. 2) and: the galaxy density (*left*) and velocity dispersion as a proxy for the galaxy mass (*right*).

References

1. M.L. Balogh, S.M. Morris, H.K.C. Lee et al.: ApJ **527**, 54 (1999)
2. M.L. Balogh, J.F. Navarro, & S.M. Morris: ApJ **540**, 113 (2000)
3. M. Hempel, M. Hilker, M. Kissler-Patig, T.H. Puzia et al.: A&A **405**, 487 (2003)
4. M. Hempel & M. Kissler-Patig: A&A **419**, 863 (2004)
5. A. Kundu, S.E. Zepf, M. Hempel et al.: ApJ **634**, L41 (2005)
6. T. H. Puzia, S.E. Zepf, M. Kissler-Patig et al.: A&A **391**, 453 (2002)

The Low-Mass X-Ray Binary Globular Cluster Connection in the ACS Virgo Cluster Survey

A. Jordán[1], G.R. Sivakoff[2], C.L. Sarazin[3], J.P. Blakeslee[3], E.L. Blanton[4], P. Côté[5], L. Ferrarese[5], J.A. Irwin[6], A.M. Juett[2], S. Mei[7], E.W. Peng[5], and M.J. West[8,9]

[1] European Southern Observatory, Karl-Schwarzschild-Str. 2, Garching bei München, 85748 Germany ajordan@eso.org
[2] Department of Astronomy, Univ. of Virginia, Charlottesville, VA 22903, USA
[3] Department of Physics and Astronomy, Washington State University, Pullman, WA 99163, USA
[4] Astronomy Department, Boston University, Boston, MA 02215, USA
[5] Herzberg Institute of Astrophysics, Victoria, BC V9E 2E7, Canada
[6] Astronomy Department, University of Michigan, MI 48109, USA
[7] Department of Physics and Astronomy, Johns Hopkins University, Baltimore, MD 21218, USA
[8] Department of Physics and Astronomy, University of Hawaii, Hilo, HI 96720, USA
[9] Gemini Observatory, Casilla 603, La Serena, Chile

Chandra has revealed that nearby luminous early-type galaxies (E/S0s) harbor hundreds of bright ($> 10^{37}$ erg/s) Low-Mass X-Ray Binaries (LMXBs). The dense stellar environments of globular clusters (GCs) form LMXBs more efficiently than in the fields of galaxies by a factor of a few times 10^2, presumably due to increased efficiency of dynamical processes of LMXB formation in the dense cores of GCs.

With its wider field-of-view ($\approx 3.2' \times 3.2'$) than WFPC2 and much better resolution than ground-based observations, HST's Advanced Camera for Surveys (ACS) can identify large numbers of GCs and measure their half-light radii [1]. The ACS Virgo Cluster Survey (ACSVCS; [2]) observed the centers of 100 E/S0s, in part to measure GC properties. Using Chandra and HST observations of M87 [3] were the first to use structural parameters to find evidence that dynamical processes appear to affect LMXB formation efficiency in early-type galaxy GCs. By extending similar analysis to all early-type galaxies in the ACSVCS that have suitable Chandra data, the dependence of LMXB formation efficiency can be probed as a function of photometric parameters (luminosity as a proxy for mass M, and $(g-z)$ color) and structural parameters (r_h).

In agreement with previous studies (e.g., [4–6,3]) we find that luminous and metal-rich GCs are more likely to harbour an LMXB. The available r_h measurements let us additionally probe dynamical processes. The tidal capture and exchange interaction rates can be linearly traced by the observable quantity $\sigma \propto M^{1.5} r_h^{-2.5}$, assuming the virial theorem and a constant concentration for GCs. We have devised a new maximum-likelihood method to determine which of all the available GC variables (r_h, Z, M, σ) determine the presence of an LMXB if we assume simultaneous power-law dependencies in all variables. We find that the best description is afforded by an expression of the form $\Lambda \propto Z^{0.39\pm0.07}\sigma^{0.8\pm0.05}$. Note that σ is capable of capturing the combined effects of M and r_h, and thus provides further support for the dynamical origin of GC LMXBs. Note also that the dependence on encounter rate is shallower than predicted by simple theory. Below we contrast the observed distributions of various GC parameters (r_h, Z, M, σ) [histograms], versus the predicted distributions [points] according to our best solution (denoted Λ above). These results will be presented in Sivakoff et al. (2006, in preparation).

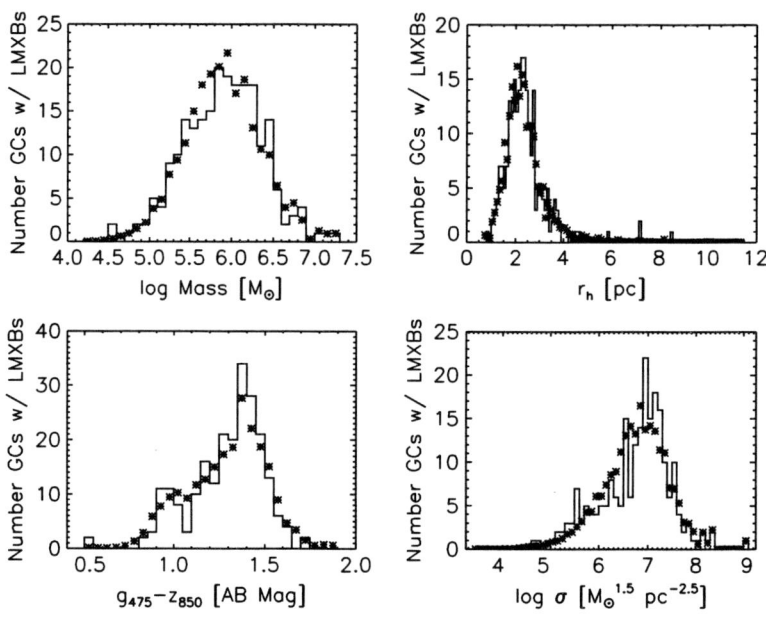

References

1. Jordán, A. et al. 2005, ApJ, 634, 1002
2. Côté, P. et al. 2004, ApJS, 153, 223
3. Jordán, A. et al. 2004, ApJ, 613, 279
4. Angelini et al. 2001, ApJ, 557, L35
5. Kundu et al. 2002, ApJ, 574, L5
6. Sarazin et al. 2003, ApJ, 595, 743

The Globular Cluster System of NGC 5128: Combining Broad-Band Color and Lick Index Analysis

Thomas Lilly[1], Uta Fritze-v. Alvensleben[2], and Richard de Grijs[3]

[1] Institut für Astrophysik Göttingen, Germany
   tlilly@astro.physik.uni-goettingen.de
[2] Institut für Astrophysik Göttingen, Germany
   ufritze@astro.physik.uni-goettingen.de
[3] Department of Physics & Astronomy, University of Sheffield, GB
   R.deGrijs@sheffield.ac.uk

Abstract. We present a mathematically advanced method for the determination of age and metallicity of individual members of globular cluster systems (CGSs) in galaxies by combining all the information inherent in broad-band color and Lick index measurements, and we present first results of our analysis of the GCS of the early type galaxy NGC 5128.

Using our well established spectral energy distribution analysis tool AnalySED ([2], MNRAS 347, 196), we can reveal the age, metallicity, mass (and possibly extinction) of individual GCs by comparing the observations with an extensive grid of SSP model colors (cf. [5], A&A 392, 1; [1], A&A 401, 1063). This is done in a statistically advanced and reasonable way, following a χ^2 approach. However, since for all colors the evolution slows down considerably at ages older than about 8 Gyr, even with several passbands and a long wavelength basis the results are severely uncertain for old clusters. Therefore, we incorporated empirical calibrations for Lick indices in our models and developed a Lick indices analysis tool that works in the same way as our SED analysis tool ([4], A&A, accepted). However, even when using spectral information, results still suffer from age-metallicity degeneracy: Since metallicities proved to be very reliable using Lick index analysis, 1 σ uncertainties in age can still be very high (up to more than 10 Gyrs).

In order to utilize *all* the information available, i.e. colors *and* Lick indices, we multiply the probabilities assigned to each model grid point during the analysis by both methods (AnalySED and Lick-Analysis), renormalize the probability space and determine the best model again (PRODUCT). Figure 1 shows the probability spaces, and gives the best models using all three methods for the example of pff_gc-006, a member of the GCS of the giant elliptical galaxy NGC 5128.

As a first application, we have analysed 215 GCs of NGC 5128 for which UBVRI photometry as well as Lick indices (Hβ, Mgb, Mg$_2$, Fe5270, Fe5335) are available (photometry taken from Peng et al. 2004, ApJS 150, 367; indices from Peng 2005, private communication).

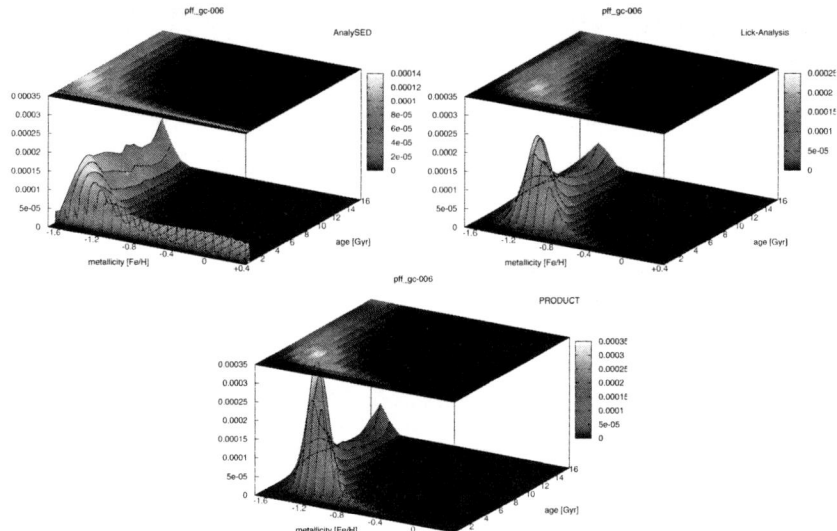

Fig. 1. Normalized probability space resulting from an analysis of broad-band magnitudes UBVRI (*upper left panel*; best model: age = $5.55^{+10.45}_{-4.32}$ Gyr, [Fe/H] = $-1.7^{+0.7}_{-0.0}$), the Lick indices Hβ, Mgb, Mg$_2$, Fe5270, Fe5335 (*upper right panel*; best model: age = $3.91^{+12.09}_{-1.84}$ Gyr, [Fe/H] = $-1.2^{+0.3}_{-0.3}$), and a combination of both methods (*bottom panel*; best model: age = $4.69^{+3.12}_{-2.75}$ Gyr, [Fe/H] = $-1.4^{+0.4}_{-0.2}$).

We find bimodal age *and* metallicity distributions with a surprisingly high amount of clusters younger than 8 Gyr (>50%); in addition to the formerly known cluster populations (old GCs of all metallicities, and young metal-rich GCs), we find, for the first time, a population of intermediate age and metal-poor clusters in NGC 5128. However, while age determinations are much improved using PRODUCT, clusters with very low and very high metallicities, respectively, still are highly degenerated in terms of age. Therefore, in order to further reduce the age-metallicity degeneracy, we plan to complement the dataset analysed so far with Washington CMT$_1$ photometry provided by Harris et al. ([3], AJ 128, 712).

More information can be found on the original poster (ask TL for a pdf-copy), and in Lilly et al. ([4], A&A, in preparation).

References

1. P. Anders, U. Fritze-v. Alvensleben: A&A, **401**, 1063 (2003)
2. P. Anders et al.: MNRAS, **347**, 196 (2004)
3. G. L. H. Harris et al.: AJ, **128**, 712 (2004)
4. T. Lilly, U. Fritze-v. Alvensleben: A&A, **457**, 467 (2006)
5. J. Schulz et al.: A&A, **392**, 1 (2002)

The Galaxy – Globular Cluster Connection in NGC 3115

Mark A. Norris[1], Ray M. Sharples[2], and Harald Kuntschner[3]

[1] Department of Physics, Science Laboratories, South Road, Durham, UK
 m.a.norris@durham.ac.uk
[2] Department of Physics, Science Laboratories, South Road, Durham, UK
[3] Space Telescope European Coordinating Facility, ESO, Garching, Germany

1 Introduction

Globular Clusters (GCs) provide important insights into the formation of galaxies, under the assumption that GCs form preferentially during major star formation events and hence act as good tracers of the properties of the overall stellar population of a galaxy. Using Gemini-GMOS longslit spectral observations we have examined the stellar populations of the isolated S0 NGC 3115 to explore this relationship. Here we present the main conclusions of this investigation and compare these to previously published results for the GC system of NGC 3115 [1,2].

2 Observations

The observations were carried out on the 18/19th December 2001 with the GMOS instrument on the Gemini-North telescope. A 1-arcsec wide longslit was used with the B600 grism giving a resolution of 4.4Å FWHM. Total integration times for each axis was 7200s. Data reduction was accomplished using a combination of the standard Gemini IRAF routines and custom IDL programs.

3 Results

We have obtained kinematic data and Lick/IDS absorption line strength indices for the major and minor axes of NGC 3115 out to around $2R_e$. Using simple stellar population models [3,4], we derive metallicities, abundance ratios and ages for the stellar population of NGC 3115. Our main conclusions are:

- NGC 3115 has a significant stellar disc component, which is both kinematically and chemically distinct from the surrounding spheroidal component.
- The major axis data displays clear evidence for contamination by the stellar disc, which is younger (5–8 Gyr old) and more chemically enriched than the spheroidal component.

- The spheroidal component of NGC 3115 is consistent with having a uniformly old ~10–12 Gyr age and [α/Fe] of 0.2–0.3.
- At large radii, the minor axis (which traces the spheroidal component more or less exclusively) is entirely consistent in age, [Z/H] and [α/Fe] with the red GC subpopulation (Fig. 1).
- We find that a simple two-component galaxy model with a combination of an old, relatively metal poor spheroidal component with [α/Fe]=0.3 and a young, more metal rich disc component with [α/Fe]=0.0 reproduces the observed trends of [α/Fe] with radius for both axes.

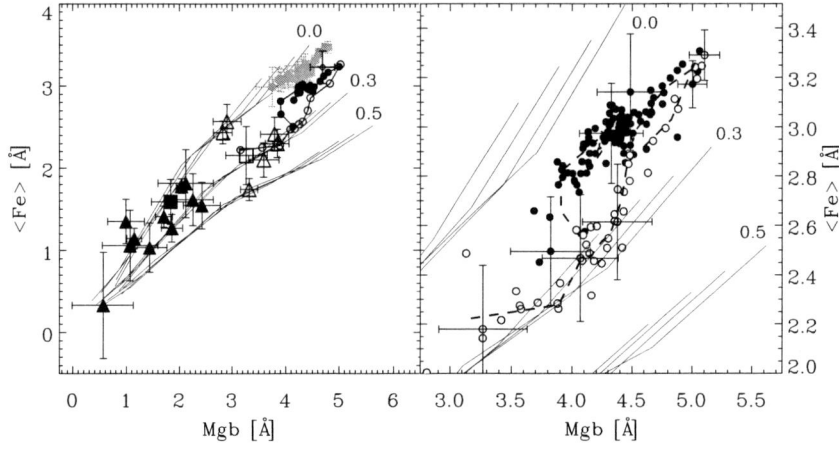

Fig. 1. Comparison of the [α/Fe] ratios of the galaxy data and GC population. In the left-hand panel, filled and open black circles are binned major and minor axis data from this study. Filled black triangles are the blue GC sample and unfilled triangles the red GC population [2]. Model grids ([3,4]) have [α/Fe] = 0.0, 0.3 and 0.5. Right-hand panel shows the unbinned galaxy data.

References

1. M. A. Norris, R. M. Sharples, H. Kuntschner : MNRAS, **367**, 815 (2006)
2. H. Kuntschner, B. L. Zeigler, R. M. Sharples et al.: A&A, **395**, 761 (2002)
3. D. Thomas, C. Maraston, R. Bender: MNRAS, **339**, 897 (2003)
4. D. Thomas, C. Maraston, A. Korn: MNRAS, **351**, L19 (2004)

Velocity Dispersions of Bright Globular Clusters in NGC 5128

Marina Rejkuba[1], Dante Minniti[2], and Georges Meylan[3]

[1] ESO, Karl-Schwarzschild-Str. 2, D-85748 Garching, Germany
mrejkuba@eso.org
[2] P. Univ. Católica, Vicuña Mackenna 4860, Santiago 22, Chile
dante@astro.puc.cl
[3] EPFL, Observatoire, CH-1290 Sauverny, Switzerland georges.meylan@epfl.ch

We have used UVES (UV-Visual Echelle Spectrograph) at the VLT to obtain the integrated spectra of 22 globular clusters belonging to the nearby giant elliptical galaxy NGC 5128 (=Centaurus A). Using cross-correlation technique on these high resolution spectra we have determined accurate radial velocities and projected core velocity dispersions of individual globular clusters. Our targets are among the brightest members of the globular cluster system of NGC 5128 and their projected velocity dispersion values range from 5 to 30 km/s. We compare them with globular clusters in the Local Group galaxies.

1 The Data and Results

The high resolution spectra of 22 globular clusters and globular cluster candidates in NGC 5128 were obtained using the 580 nm setup and red arm of UVES at the VLT in April 2002. The spectral range covered was 480–680 nm and the velocity resolution was 0.83 km/sec/pix.

In addition to 22 clusters, selected from the catalogues of [1], [2] and [3], we have also observed 17 different red giant stars with known radial velocities, spectral types from G0 to M0.5, and a wide range of metallicities ($-2.6 <$ [Fe/H] $< +0.3$ dex). Some of the stars were observed several times, thus making a total of 28 stellar templates that could be used for cross-correlation.

The data reduction was done using MIDAS based UVES pipeline and IRAF tasks. The cross-correlation of stellar spectra with globular clusters was done using *fxcor* task in IRAF package *rv*. The width of the cross-correlation function (CCF) is a function of the internal velocity dispersion of the cluster and of the instrumental width. The latter was calibrated by cross-correlating stars against stars. As a byproduct we also obtained precise radial velocities for all our targets. For 8 clusters these are the first radial velocity measurements, which have confirmed their nature and membership in the globular cluster system of NGC 5128.

In Fig. 1 we plot CCFs for all the clusters. On the x-axis is the radial velocity, which corresponds to the peak of the CCF, while the width of the CCF is related to the velocity dispersion. The derived velocity dispersions for each cluster are given in parenthesis in each panel.

The relation of the velocity dispersion and the absolute V magnitude for these bright globular clusters in NGC 5128 is remarkably similar to the relation derived for other, less luminous, globular clusters in the Milky Way, M31, M33 and Fornax dSph. The similarity of this relation indicates similarities in mass-to-light ratios for these clusters indicating perhaps similar formation mechanisms. The combination of the velocity dispersion measurements with the structural parameters for these clusters will be used to derive accurate mass-to-light ratios for all our targets (Rejkuba et al., in preparation).

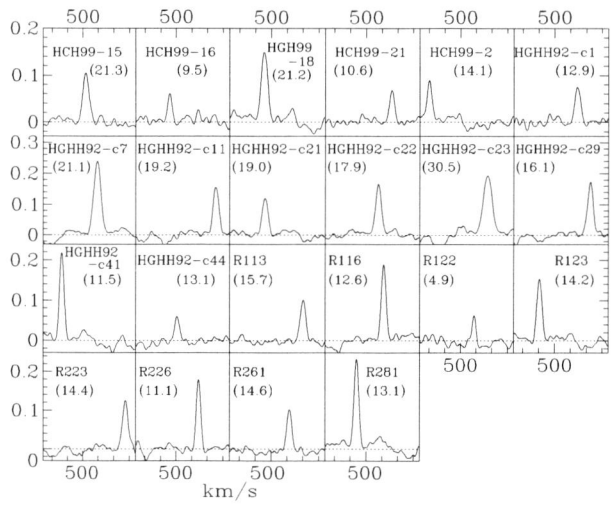

Fig. 1. Cross-correlation functions for all the clusters. The number in the parenthesis in each panel is the derived velocity dispersion in km/s.

References

1. G. L. H. Harris, D. Geisler, H. C. Harris & J. E. Hesser: AJ **104**, 613 (1992)
2. S. Holland, P. Côté & J. E. Hesser: A&A **348**, 418 (1999)
3. M. Rejkuba: A&A **369**, 812 (2001)

Part VII

Evolution of Cluster Systems
and their Host Galaxies

Imprint of Galaxy Formation and Evolution on Globular Cluster Properties

Kenji Bekki

School of Physics, University of New South Wales, Sydney 2052, Australia
bekki@bat.phys.unsw.edu.au

Abstract. We discuss the origin of physical properties of globular cluster systems (GCSs) in galaxies in terms of galaxy formation and evolution processes. Based on numerical simulations of dynamical evolution of GCSs in galaxies, we particularly discuss (1) the origin of radial density profiles of GCSs, (2) kinematics of GCSs in elliptical galaxies, (3) transformation from nucleated dwarf galaxies into GCs (e.g., omega Centauri), and (4) the origin of GCSs in the Large Magellanic Cloud (LMC).

1 Numerical Archeology

Based on penetrative analysis of metal-poor halo stars and globular clusters (GCs) in the Galaxy, two canonical Galaxy formation scenarios – the monolithic collapse scenario [1] and the accretion/merging one [2] – that have long been influential for later observational and theoretical studies of disk and elliptical galaxies. Although observational studies of stellar halos in galaxies beyond the Local Group of galaxies have just recently started revealing structural and chemical properties of the halos [3, 4], physical properties of globular cluster systems (GCSs) in these galaxies have long been investigated in much more details [5]. Wide-field imaging and spectroscopic studies with large ground-based telescopes (e.g., Keck 10m) have recently revealed GCS structures and kinematics in galaxies with different Hubble types [6]. Furthermore a growing number of theoretical/numerical studies have recently been accumulated which have investigated dynamical and chemical properties of GCs and GCSs based on admittedly realistic and self-consistent models of GC formation during galaxy formation and evolution [7, 8]. In this review paper, we therefore try to derive physical meanings from the selected four observed properties of GCs and GCSs by comparing our numerical simulations of GC and GCS formation with latest observations.

2 GCS Density Profiles

It has long been known that the radial density profiles of GCSs in elliptical galaxies (Es) vary with the total luminosities of their host galaxies [9]. If the projected GCS density profiles in Es with V-band absolute magnitudes of M_V

are fitted to the power-law ones like $\Sigma_{gc} \propto R^{\alpha_{gc}}$, where R is the distance from the center of the host galaxy of a GCS, the power-law index α_{gc} is smaller (i.e., the profiles are steeper) for larger M_V (i.e., fainter Es). Although two physical mechanisms – GC destruction by galactic tidal fields [10, 11] and dynamics of galaxy merging (BF06; [12]) – have been so far proposed for the origin of GCS density profiles, we focus on the latter case in this paper.

BF06 numerically investigated the structural properties of GCSs in Es formed from a sequence of major dissipationless galaxy merging and thereby found that the radial density profiles of GCSs in Es become progressively flatter as the galaxies experience more major merger events (See Fig. 1). The simulated profiles of GCSs are found to be well described as power-laws with α_{gc} ranging from -2.0 to -1.0 in Es. They are flatter than, and linearly proportional to, the slopes (α_s) of the stellar density profiles. By applying a

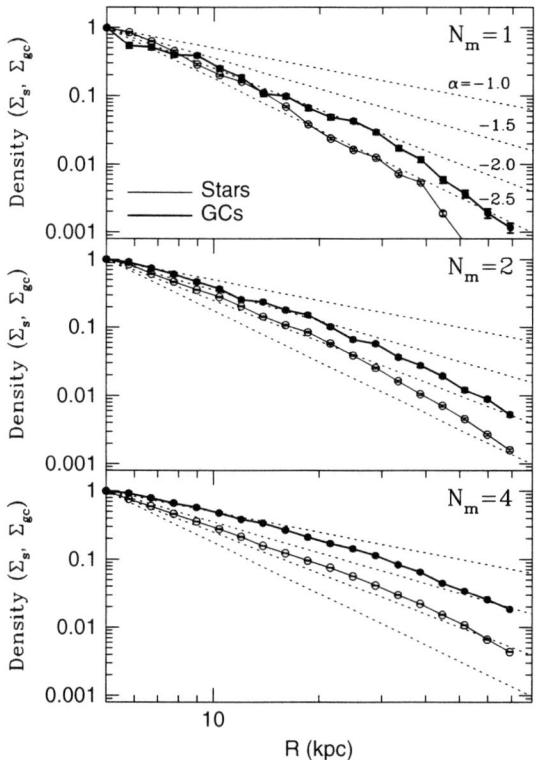

Fig. 1. Dependences of projected number distributions of stars (*thin*) and GCs (*thick*) in merger remnants (i.e., elliptical galaxies) on the total number of major merger events (N_m) which an elliptical experienced during its formation (BF06). For clarity, the density distributions are normalized to their central values. *Thin dotted lines* represent power-law slopes (α) of $\alpha = -2.5, -2.0, -1.5,$ and -1.0. Note that the density profiles of GCSs become flatter for larger N_m, i.e. more mergers.

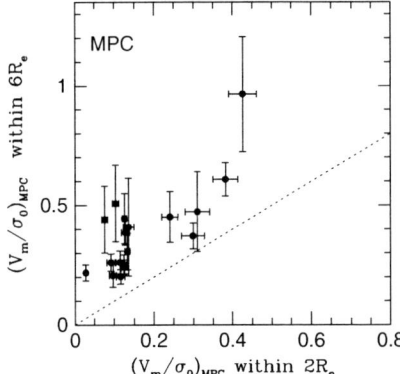

Fig. 2. Correlations between $(V_m/\sigma_0)_{\mathrm{MPC}}$ estimated for $R \leq 2R_e$ and those for $R \leq 6R_e$ for metal-poor clusters (MPCs). These correlations are derived from 18 results of 6 major merger models with three different projections [18].

reasonable scaling relation between luminosities and sizes of galaxies to the simulation results, BF06 showed that $\alpha_{\mathrm{gc}} \approx -0.36 M_V - 9.2$, $r_c \approx -1.85 M_V$, and $\alpha_{\mathrm{gc}} \approx 0.93 \alpha_s$. These correlations between GCS profiles and their host galaxy luminosities are consistent reasonably well observations [13,9] which suggests that the origin of structural non-homology of GCSs in Es can be understood in terms of the growth of Es via major dissipationless galaxy merging.

3 GCS Kinematics in E/S0s

Recent observations on GCS kinematics in E/S0s have revealed that GCS kinematics can be quite diverse: GCSs in some galaxies like M87 [14] NGC 4472 [15] and NGC 5128 [16] show rotation whereas those in some galaxies such as NGC 1399 [17] do not. Bekki et al. (B05; [18]) first tried to understand the observed diversity in GCS kinematics by numerically investigating GCS kinematics of E/S0s formed from major/minor galaxy merging. B05 demonstrated that both metal-poor cluster (MPCs) and metal-rich ones (MRCs) in Es formed from major mergers can exhibit significant rotation at large radii (∼20 kpc) due to the conversion of initial orbital angular momentum into intrinsic angular momentum of the remnant.

Based on a wide parameter study of galaxy mergers, B05 found that MPCs show higher central velocity dispersions than MRCs for most major merger models. V_m/σ_0 (where V_m and σ_0, are the GCS maximum rotational velocity and central velocity dispersion, respectively) ranges from 0.2–1.0 and 0.1–0.9 for the MPCs and MRCs respectively, within $6R_e$ for the remnant elliptical. Figure 2 shows an interesting result that does not depend on merger parameters: V_m/σ_0 of MPCs within $6R_e$ are greater than those of MPCs within $2R_e$.

B05 also revealed the alignment of the major axes in 2D distributions between stars, GCs, and dark matter halos in the simulated Es. The aligned major axis between stars, GCs, and dark matter appears to be one of the principal characteristics of Es formed by major merging, which implies that *observational studies on 2D distributions of GCSs in Es can tell us about the shapes of their host dark matter halos.* We also showed in this meeting that the total masses of E/S0s estimated from the GCS kinematics can be much closer to the real masses than those from the PNe systems owing to less anisotropic velocity dispersion in GCSs, in particular, for face-on S0s. This suggests that (1) GCSs are better mass-estimators in E/S0s and (2) kinematical data sets of PNe systems in E/S0s [19] should be more carefully interpreted for the total masses of E/S0s. Although these simulations are not based on cosmological N-body ones, these results may well provide new insight on the origin of the observed diversity in GCS kinematics of E/S0s.

4 The Age Gap Problem in the LMC

Possible candidates of young and metal-rich GCs were discovered in interacting and merging galaxies [20], and physical properties of these GCs have been discussed in different contexts of galaxy and GC formation, such as the observed color bimodality in GCSs in Es [9], the birth rate of young GCs as a function of time in M82 [21] and the age-metallicity relation of GCs in nearby interacting galaxies like the LMC and the SMC (e.g., B04; [22]). In this paper, we focus on the LMC's GCS in which nearly all GCs are either very old (~ 13 Gyr) or younger than 3–4 Gyr – the "age-gap" problem [23, 24].

B04 and Bekki & Chiba [25] challenged this age gap problem by investigating chemodynamical evolution of the LMC interacting both with the Galaxy and the SMC for a long time scale (~ 9 Gyr). They found that the first close encounter between the LMC and the SMC about 4 Gyr ago was the beginning of a period of strong tidal interaction witch likely induced dramatic gas cloud collisions, leading to an enhancement of the GC formation which has been sustained by strong tidal interactions to the present day. Figure 3, showing the simulated age distributions of field stars and GCs in a model, reveals that GC formation can be reactivated about 3–4 Gyr ago, when the LMC can start its strong tidal interaction with the SMC. These results imply that the origin of the age gap can be closely associated with interaction histories of the LMC, the SMC, and the Galaxy.

5 Very Massive Star Clusters

Very massive star clusters (VMSCs) such as ω Centauri, ultra-compact dwarfs (UCDs), and massive nuclear star clusters have unique characteristics that are quite different from those of "normal" GCs. For example, UCDs discovered

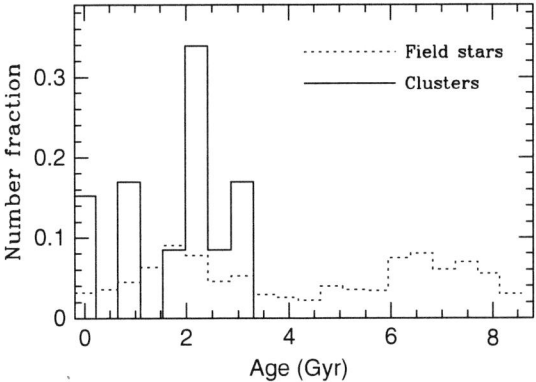

Fig. 3. The simulated age distributions of field stars (*dotted*) and clusters (*solid*) in the LMC at the present epoch [25]. For convenience, the normalized fraction of stars in each age bin is shown.

in the Fornax and the Virgo clusters of galaxies [26–28] have intrinsic sizes of less than 100 pc, and have absolute B−band magnitudes ranging from -13 to -11 mag, which is more than 2 magnitudes brighter than the most massive GC in the Galaxy (i.e., ω Cen). Two physical mechanisms for the VMSC formation have been so far proposed: The "Galaxy threshing" scenario [29, 30] in which VMSCs originate from nuclei of nucleated dwarf galaxies and the "cluster merging" one in which VMSCs are formed from merging of smaller star clusters in tidal tails of merging galaxies [31]. Although more details on physical properties of VMSCs have been recently revealed [32, 33], it remains unclear which of the two scenarios is more convincing for the origin of VMSCs.

Here we discuss what we can learn from VMSC properties *if they were previously nuclei of nucleated galaxies*. Both dissipationless [34, 35] and dissipative [36] formation scenario of stellar galactic nuclei have provided some interesting predictions on nuclear properties of galaxies [37]. Figure 4 shows the scaling relations between different physical parameters of merger remnants of star clusters initially with the observed GC scaling relations [38]. The fact that the simulated relations deviate from the GC's ones implies that *the scaling relations of VMSCs can be used for understanding whether the stellar nuclei can be formed from merging of many smaller clusters in the central regions of galaxies* [39]. The dissipative nucleus formation model [37] has predicted the spread of ages and metallicities in stellar populations of stellar galactic nuclei and accordingly can be discussed in the context of the observed spread of ages and metallicities in ω Cen [32]. Thus dynamical and chemical properties of VMSCs can tell us about nucleus formation histories in galaxies.

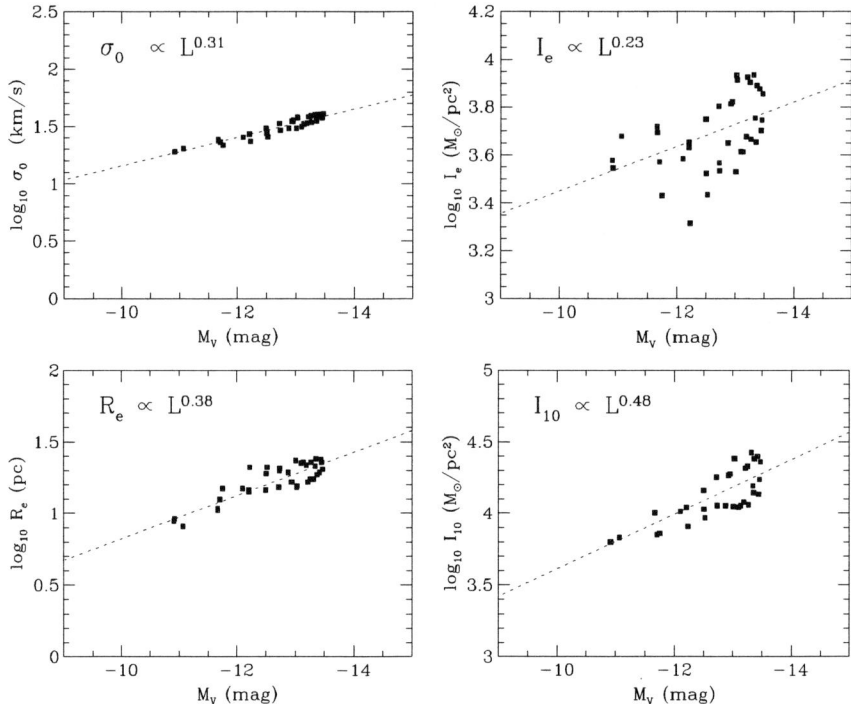

Fig. 4. Correlations of structural and kinematical parameters with M_V (V−band absolute magnitude) for the VMSCs in 40 models [22]. Projected central velocity dispersion (σ_0; *upper left*), half-light-averaged surface brightness (I_e; *upper right*), effective radius (R_e; *lower left*), and central surface brightness (I_{10}; *lower right*) are plotted against M_V. Here the central surface brightness I_{10} is expressed as $0.1L/\pi/R_{10}^2$, where L, R_{10} are the total luminosity of a VMSC and the radius within which 10% of L is included, respectively. The best fit scaling relation for the VMSCs is derived for each panel using the least square fitting method and described as a *dotted* line with the derived relation (e.g., $\sigma_0 \propto L^{0.31}$).

6 Future Works: Hierarchical Galaxy Formation and GCSs

Thus, structural, kinematical, and chemical properties of GCSs in galaxies have fossil information on dynamics of major/minor galaxy merging (e.g., angular momentum redistribution processes in merging), interaction histories of galaxies, and the formation histories of stellar galactic nuclei in dwarfs. Previous theoretical/numerical studies however did not discuss so extensively the observed correlations between GCS properties and their host ones [6] in the context of a hierarchical clustering scenario of galaxy formation. Several authors just recently have started their investigation on the GCS-host relations based on semi-analytic models [40] and high-resolution numerical simulations in ΛCDM models [41, 42]. A number of observed GCS-host relations, such as the positive correlation between GCS metallicities and their host galaxy luminosities [6] have not been clearly explained by any galaxy formation scenarios. Since these GCS-host relations may have profound physical meanings on galaxy formation and evolution, it is doubtlessly worthwhile for future numerical simulations of GCS formation to explore the origin of these relations.

References

1. O.J. Eggen, D. Lynden-Bell, A.R. Sandage: ApJ, **136**, 748 (1962)
2. L. Searle, R. Zinn: ApJ, **225**, 357 (1978)
3. J. Dalcanton, R.A. Bernstein: AJ, **123**, 1328 (2002)
4. S. Zibetti, S.D.M. White, J. Brinkmann: MNRAS, **347**, 556 (2004)
5. W.E. Harris: ARA&A, **29**, 543 (1991)
6. J.P. Brodie, J. Strader: ARA&A, textbf44, 193 (2006)
7. K. Bekki, D.A. Forbes, M.A. Beasley, W.J. Couch: MNRAS, **335**, 1176 (2002)
8. O.Y. Gnedin: these proceedings (2008)
9. K.M. Ashman, S.E. Zepf: in Globular cluster systems, Cambridge, U. K.; New York: Cambridge University Press (1998)
10. H. Baumgardt: A&A, **330**, 480 (1998)
11. E. Vesperini, S.E. Zepf, A. Kundu, K.M. Ashman: ApJ, **593**, 760 (2003)
12. K. Bekki, D.A. Forbes: A&A, **445**, 485 (2006)
13. W.E. Harris: AJ, **91**, 822 (1986)
14. M. Kissler-Patig, K. Gebhardt: AJ, **116**, 2237 (1998)
15. S.E. Zepf, M.A. Beasley et al: AJ, **120**, 2928 (2000)
16. E.W. Peng, H.C. Ford, K.C. Freeman: ApJ, **602**, 705 (2004)
17. T. Richtler et al: AJ, **127**, 2094 (2004)
18. K. Bekki, M.A. Beasley, J.P. Brodie, D.A. Forbes, MNRAS, **363**, 1211 (2005)
19. A.J. Romanowsky, N.G. Douglas et al.: Science, **301**, 1696 (2003)
20. B.C. Whitmore, F. Schweizer: AJ, **109**, 960 (1995)
21. R. de Grijs, N. Bastian, H.J.G.L. Lamers: MNRAS, **340**, 197 (2003)
22. K. Bekki, W.J. Couch, W. J. et al.: **610**, L93 (2004)
23. G.S. Da Costa: in Haynes R., Milne D., eds, Proc. IAU Symp. 148, The Magellanic Clouds, Kluwer, Dordrecht, p. 183 (1991)
24. D. Geisler: these proceedings (2008)
25. K. Bekki, M. Chiba: MNRAS, **356**, 680 (2005)
26. M.J. Drinkwater, M.D. Gregg et al.: Nature, **423**, 519 (2003)
27. J.B. Jones, M.J. Drinkwater et al.: AJ, **131**, 312 (2006)
28. S. Mieske, M. Hilker, L. Infante: A&A, **418**, 445 (2004)
29. K. Bekki, W.J. Couch, M.J. Drinkwater: ApJ, **552**, L105 (2001)
30. K. Bekki, W.J. Couch, M.J. Drinkwater, Y. Shioya: MNRAS, **344**, 399 (2003)
31. M. Fellhauer, P. Kroupa: MNRAS, **330**, 642 (2002)
32. M. Hilker, A. Kayser, T. Richtler, P. Willemsen: A&A, **422**, L9 (2004)
33. M. Haşegan et al.: ApJ, **627**, 203 (2005)
34. S.D. Tremaine, J.P. Ostriker, L. Spitzer, Jr: ApJ, **196**, 407 (1975)
35. R. Capuzzo-Dolcetta, A. Tesseri: MNRAS, **308**, 961 (1999)

36. K. Bekki, W.J. Couch, Y. Shioya: ApJ **642**, L133 (2006)
37. K. Bekki, W.J. Couch, M.J. Drinkwater, Y. Shioya: ApJ, **610**, L13 (2004)
38. S.G. Djorgovski, R.R. Gal et al.: ApJ, **474**, L19 (1997)
39. P. Côte et al.: 2006, ApJS **165**, 57 (2006)
40. M.A. Beasley, C.M. Baugh et al.: MNRAS, **333**, 382 (2002)
41. K.L. Rhode, S.E. Zepf, M.R. Santos: ApJ, **630**, L21 (2005)
42. K. Bekki, H. Yahagi, D.A. Forbes: ApJ, **645**, L29 (2006)

Formation of Globular Clusters in Hierarchical Cosmology: ART and Science

Oleg Y. Gnedin and José L. Prieto

The Ohio State University, Department of Astronomy, Columbus, OH 43210, USA
ognedin@astronomy.ohio-state.edu

Abstract. We test the hypothesis that globular clusters form in supergiant molecular clouds within high-redshift galaxies. Numerical simulations demonstrate that such large, dense, and cold gas clouds assemble naturally in current hierarchical models of galaxy formation. These clouds are enriched with heavy elements from earlier stars and could produce star clusters in a similar way to nearby molecular clouds. The masses and sizes of the model clusters are in excellent agreement with the observations of young massive clusters. Do these model clusters evolve into globular clusters that we see in our and external galaxies? In order to study their dynamical evolution, we calculate the orbits of model clusters using the outputs of the cosmological simulation of a Milky Way-sized galaxy. We find that at present the orbits are isotropic in the inner 50 kpc of the Galaxy and preferentially radial at larger distances. All clusters located outside 10 kpc from the center formed in the now-disrupted satellite galaxies. The spatial distribution of model clusters is spheroidal, with a power-law density profile consistent with observations. The combination of two-body scattering, tidal shocks, and stellar evolution results in the evolution of the cluster mass function from an initial power law to the observed log-normal distribution. However, not all initial conditions and not all evolution scenarios are consistent with the observed mass function.

1 Giant Molecular Clouds at High Redshift

The outcomes of many proposed models of globular cluster formation depend largely on the assumed initial conditions. The collapse of the first cosmological $10^6\ M_\odot$ gas clouds, or the fragmentation of cold clouds in hot galactic corona gas, or the agglomeration of pressurized clouds in mergers of spiral galaxies could all, in principle, produce globular clusters, but only if those conditions realized in nature. Similarly, while observational evidence strongly suggests that all stars and star clusters form in molecular clouds, the initial conditions for cloud fragmentation are a major uncertainty of star formation models.

The only information that we actually have about the initial conditions comes from the early universe, when primordial density fluctuations set the seeds for structure formation. These fluctuations are probed directly by the anisotropies of the cosmic microwave background radiation. Cosmological numerical simulations study the growth of these fluctuations via gravitational instability, in order to understand the formation of galaxies and all other

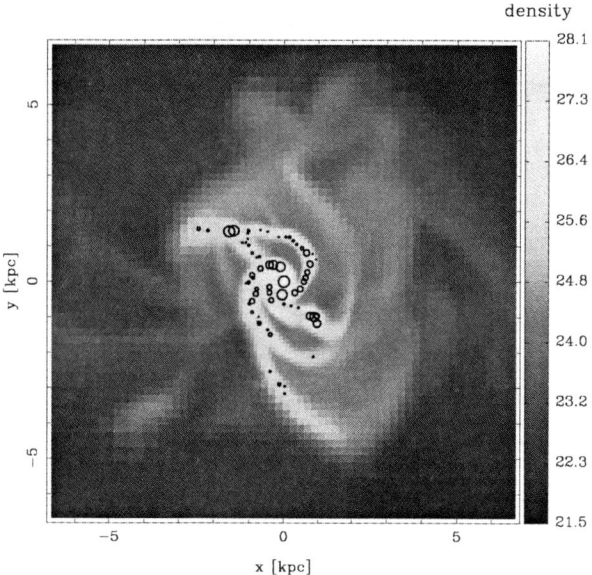

Fig. 1. A massive gaseous disk with prominent spiral arms, seen face-on at redshift $z = 4$ in the process of active merging. The gas density is projected over a 3.5 kpc slice. In our model star clusters form in giant gas clouds, shown by circles with the sizes corresponding to the cluster masses. From [6].

structures in the Universe. The simulations begin with tiny deviations from the Hubble flow, whose amplitudes are set by the measured power spectrum of the primordial fluctuations while the phases are assigned randomly. Therefore, each particular simulation provides only a statistical description of a representative part of the Universe, although current models successfully reproduce major features of the observed galaxies.

Reference [6] attempted to construct a first self-consistent model of star cluster formation, using an ultrahigh-resolution gasdynamics cosmological simulation with the Adaptive Refinement Tree (ART) code. They identified supergiant molecular clouds in high-redshift galaxies as the likely formation sites of globular clusters. These clouds assemble during gas-rich mergers of progenitor galaxies, when the available gas forms a thin, cold, self-gravitating disk. The disk develops strong spiral arms, which further fragment into separate molecular clouds located along the arms as beads on a string (see Fig. 1).

In this model, clusters form in relatively massive galaxies, with the total mass $M_{\mathrm{host}} > 10^9\ M_\odot$, beginning at redshift $z \approx 10$. The mass and density of the molecular clouds increase with cosmic time, but the rate of galaxy mergers declines steadily. Therefore, the cluster formation efficiency peaks at a certain extended epoch, around $z \approx 4$, when the Universe is only 1.5 Gyr old. The host galaxies are massive enough for their molecular clouds to be shielded from the extragalactic UV radiation, so that globular cluster formation is unaffected by the reionization of cosmic hydrogen. As a result of the mass-metallicity correlation of progenitor galaxies, clusters forming at the same epoch but in different-mass progenitors have different metallicities, ranging between 10^{-3} and 10^{-1} solar. The mass function of model clusters is consistent with a power law $dN/dM \propto M^{-\alpha}$, where $\alpha = 2.0 \pm 0.1$, similar to the observations of nearby young star clusters.

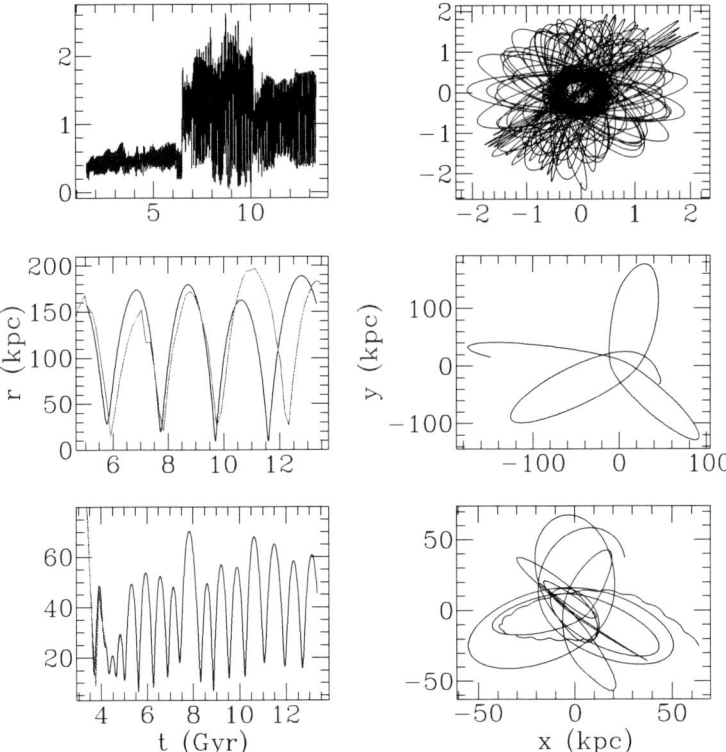

Fig. 2. Three types of globular clusters orbits. *Left panels* show the distance to the center of the main halo, right panels show orbits in the plane of the main disk. *Top:* cluster formed in the main halo, on an initially circular orbit but was later scattered by accreted satellites. *Middle:* cluster formed in a satellite halo, which survived as a distinct galaxy (*thick red line*). *Bottom:* cluster formed in a satellite that was tidally disrupted at $t \simeq 4$ Gyr.

2 Orbits of Globular Clusters

We adopt this model to set up the initial positions, velocities, and masses for our globular clusters. We then calculate cluster orbits using a separate collisionless N-body simulation described in [7]. This is necessary because

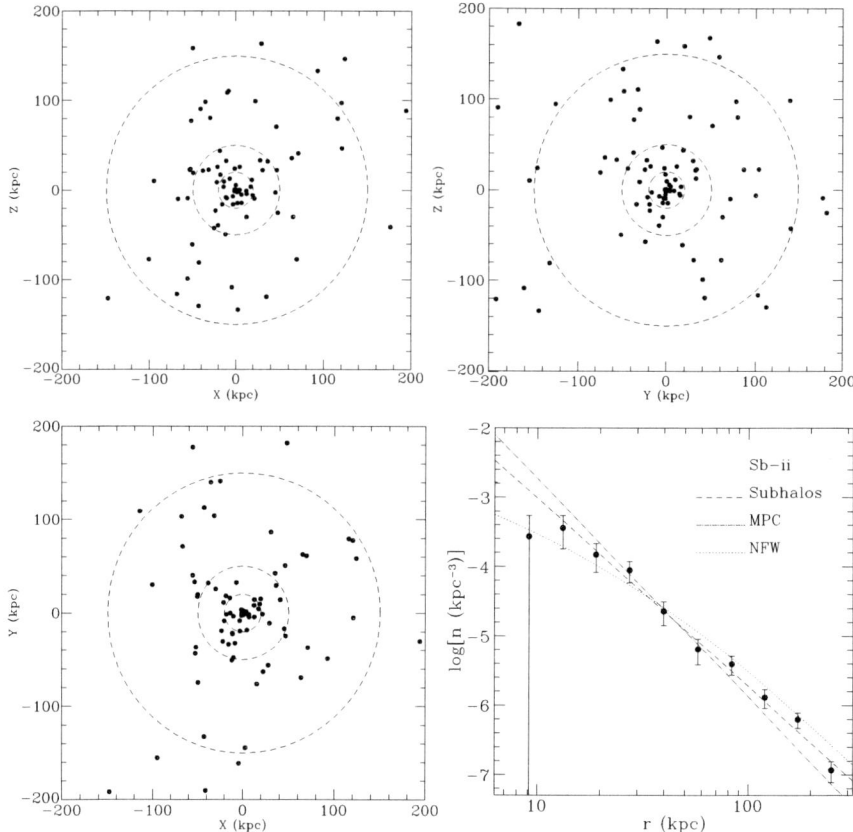

Fig. 3. Spatial distribution of surviving model clusters in the Galactic frame. Dashed circles are at projected distances of 20, 50, and 150 kpc. The number density profile (*bottom right*) can be fit by a power law, $n(r) \propto r^{-2.7}$. The distribution of model clusters is similar to that of surviving satellite halos (*dashed line*) and smooth dark matter (*dotted line*). It is also consistent with the observed distribution of metal-poor globular clusters in the Galaxy (*solid line*), plotted using the data from the catalog of [5].

the original gasdynamics simulation was stopped at $z \approx 3.3$, due to limited computational resources. By using the N-body simulation of a similar galactic system, but complete to $z = 0$, we are able to follow the full dynamical evolution of globular clusters until the present epoch. We use the evolving properties of all progenitor halos, from the outputs with a time resolution of $\sim 10^8$ yr, to derive the gravitational potential in the whole computational volume at all epochs. We convert a fraction of the dark matter mass into the analytical flattened disks, in order to model the effect of baryon cooling and star formation on the galactic potential. We calculate the orbits of globular clusters in this potential from the time when their host galaxies accrete

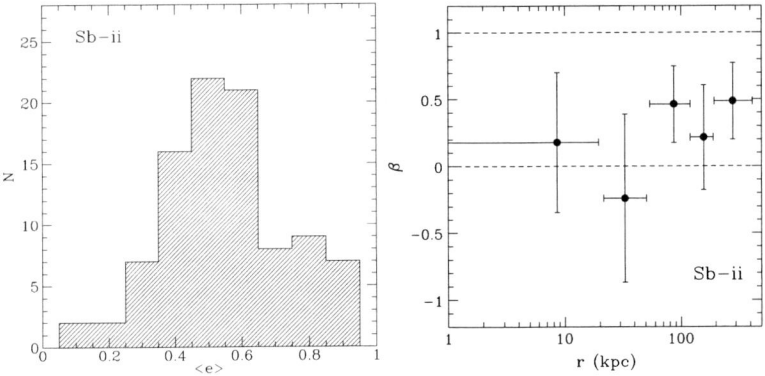

Fig. 4. *Left panel:* average eccentricity distribution of the surviving model clusters. *Right panel:* Anisotropy parameter β as a function of radius. Vertical errorbars represent the error of the mean for each radial bin, while horizontal errorbars show the range of the bin. Horizontal dashed lines illustrate an isotropic ($\beta = 0$) and a purely radial ($\beta = 1$) orbital distributions.

onto the main (most massive) galaxy. Using these orbits, we calculate the dynamical evolution of model clusters, including the effects of stellar mass loss, two-body relaxation, tidal truncation, and tidal shocks.

We consider several possible scenarios, one with all clusters forming in a short interval of time around redshift $z = 4$, and the others with a continuous formation of clusters between $z = 9$ and $z = 3$. Below we discuss the spatial and kinematic distributions of globular clusters for the best-fit model with the synchronous formation at $z = 4$.

In our model, all clusters form on nearly circular orbits within the disks of progenitor galaxies. Present globular clusters in the Galaxy could either have formed in the main disk, have come from the now-disrupted progenitor galaxies, or have remained attached to a satellite galaxy. Figure 2 shows the three corresponding types of cluster orbits. Even the clusters formed within the inner 10 kpc of the main Galactic disk do not stay on circular orbits. They are scattered to eccentric orbits by accreted satellites, while the growth of the disk increases the average orbital radius. Triaxiality of the dark halo (not included in present calculations) would also scatter the cluster orbits. The clusters left over from the disrupted progenitor galaxies typically lie at larger distances, between 20 and 60 kpc, and belong to the inner halo class. Their orbits are inclined with respect to the Galactic disk and are fairly isotropic. The clusters still associated with the surviving satellite galaxies are located in the outer halo, beyond 100 kpc from the Galactic center. Note that these clusters may still be scattered away from their hosts during close encounters with other satellites and consequently appear isolated.

Mergers of progenitor galaxies ensure the present spheroidal distribution of the globular cluster system (Fig. 3). Most clusters are now within 50 kpc from the center, but some are located as far as 200 kpc. The azimuthally-averaged space density of globular clusters is consistent with a power law, $n(r) \propto r^{-\gamma}$, with the slope $\gamma \approx 2.7$. Since all of the distant clusters originate in progenitor galaxies and share similar orbits with their hosts, the distribution of the clusters is almost identical to that of the surviving satellite halos. This power law is similar to the observed distribution of the metal-poor ([Fe/H] < −0.8) globular clusters in the Galaxy. Such comparison is appropriate, for our model of cluster formation at high redshift currently includes only low metallicity clusters ([Fe/H] $\leq$ −1). Thus the formation of globular clusters in progenitor galaxies with subsequent merging is fully consistent with the observed spatial distribution of the Galactic metal-poor globulars.

Figure 4 shows the kinematics of model clusters. Most orbits have moderate average eccentricity, $0.4 < \langle e \rangle < 0.7$, expected for an isotropic distribution. The anisotropy parameter, $\beta = 1 - v_t^2/2v_r^2$, is indeed close to zero in the inner 50 kpc from the Galactic center. At larger distances, cluster orbits tend to be more radial. There, in the outer halo, host galaxies have had only a few passages through the Galaxy or even fall in for the first time.

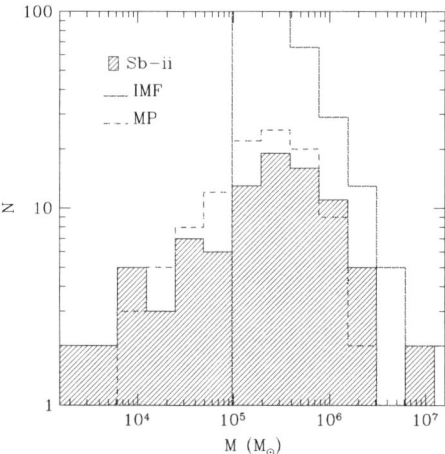

Fig. 5. Evolution of the mass function of model clusters from an initial power law (*solid line*) to a peaked distribution at present (*histogram*), including mass loss due to stellar evolution, two-body relaxation, and tidal shocks. For comparison, dashed histogram shows the mass function of metal-poor globular clusters in the Galaxy.

3 Evolution of the Globular Cluster Mass Function

Using these realistic orbits, we can now calculate the cluster disruption rates. Sophisticated models of the dynamical evolution of globular clusters have been developed using direct N-body simulations as well as the orbit-averaged

Fokker-Planck and Monte Carlo models. They are described and referenced in many good reviews, including [8, 4, 3, 2, 1]. Several processes combine and reinforce each other in removing stars from globular clusters: stellar mass loss, two-body scattering, external tidal shocks, and dynamical friction of cluster orbits. The last three are sensitive to the external tidal field and therefore, to cluster orbits. While a general framework for all these processes has been worked out already, the knowledge of realistic cluster orbits is essential for accurate calculations of the disruption.

Figure 5 shows the transformation of the cluster mass function from an initial power law, $dN/dM \propto M^{-2}$, into a final bell-shape distribution. In this model all globular clusters form at the same redshift, $z = 4$, or about 12 Gyr ago. The half-mass radii, R_h, are set by the condition that the median density, M/R_h^3, is initially the same for all clusters and remains constant as a function of time. Over the course of their evolution, numerous low-mass clusters are disrupted by two-body relaxation while the high-mass clusters are truncated by tidal shocks. The present mass function is in excellent agreement with the observed mass function of the Galactic metal-poor clusters.

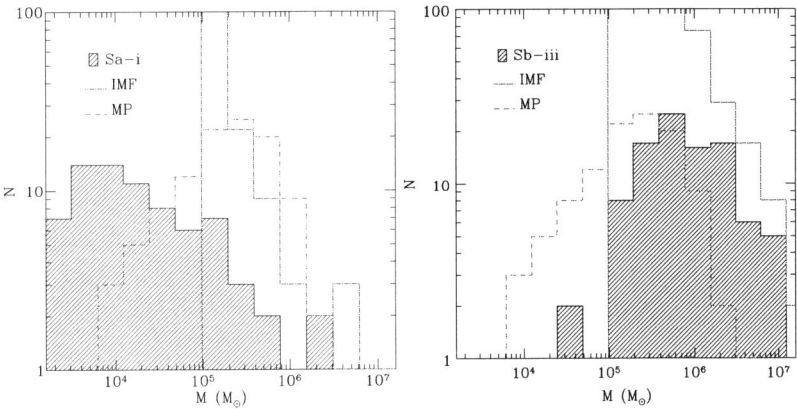

Fig. 6. Models that fail to reproduce the observed mass function of metal-poor globular clusters: with $R_h(t) = $ const (*left*) and with $R_h(t) \propto M(t)$ (*right*).

This result by itself is not new. Previous studies of the evolution of the cluster mass function have found that almost any initial function can be turned into a peaked distribution by the combination of two-body relaxation and tidal shocks. However, the efficiency of these processes depends on the cluster mass and size, $M(t)$ and $R_h(t)$. *The new result is that we find that not all initial conditions and not all evolutionary scenarios are consistent with the observed mass function.*

Figure 6 provides two examples. In the first, the half-mass radius is kept fixed at $R_h = 2.4$ pc (median value for Galactic globulars) for clusters of all masses and at all times. The median density $M(t)/R_h^3$ thus decreases as the clusters lose mass. Two-body scattering becomes less efficient and spares many low-mass clusters, while tidal shocks become more efficient and disrupt most high-mass clusters. The final distribution is severely skewed towards small clusters.

In the second example, the median density is initially fixed, as in our main model, but the size is assumed to evolve in proportion to the mass, $R_h(t) \propto M(t)$. In this case the cluster density increases with time. As a result, all of the low-mass clusters are disrupted by the enhanced two-body relaxation, while the high-mass clusters are unaffected by the weakened tidal shocks. The final distribution is skewed towards massive clusters.

Only our best-fit model (Figs. 2–5) successfully reproduces the observed mass function and spatial distribution of metal-poor globular clusters in Galaxy. In future work we will investigate the predicted properties of metal-rich globular clusters and their dependence on galaxy formation history.

Acknowledgements. OYG acknowledges the support of the American Astronomical Society and the National Science Foundation in the form of an International Travel Grant.

References

1. H. Baumgardt & J. Makino: MNRAS **340**, 227 (2003)
2. S.M. Fall & Q. Zhang: ApJ **561**, 751 (2001)
3. O.Y. Gnedin, H.M. Lee & J.P. Ostriker: ApJ **522**, 935 (1999)
4. O.Y. Gnedin & J.P. Ostriker: ApJ **474**, 223 (1997)
5. W.E. Harris: AJ **112**, 1487 (1996)
6. A.V. Kravtsov & O.Y. Gnedin: ApJ **623**, 650 (2005)
7. A.V. Kravtsov, O.Y. Gnedin & A.A. Klypin: ApJ **609**, 482 (2004)
8. L. Spitzer: Dynamical Evolution of Globular Clusters, Princeton University Press (1987)

Globular Cluster Formation in Mergers

François Schweizer

Carnegie Observatories, 813 Santa Barbara Street, Pasadena, CA 91101, USA
schweizer@ociw.edu

Abstract. Mergers of gas-rich galaxies lead to gravitationally driven increases in gas pressure that can trigger intense bursts of star and cluster formation. Although star formation itself is clustered, most newborn stellar aggregates are unbound associations and disperse. Gravitationally bound star clusters that survive for at least 10–20 internal crossing times ($\sim$20–40 Myr) are relatively rare and seem to contain <10% of all stars formed in the starbursts. The most massive young globular clusters formed in present-day mergers exceed ω Cen by an order of magnitude in mass, yet appear to have normal stellar initial mass functions.

In the local universe, recent remnants of major gas-rich disk mergers appear as proto-elliptical galaxies with subpopulations of typically 10^2–10^3 young metal-rich globular clusters in their halos. The evidence is now strong that these "second-generation" globular clusters formed from giant molecular clouds (GMC) in the merging disks, squeezed into collapse by large-scale shocks and high gas pressure rather than by high-velocity cloud–cloud collisions. Similarly, first-generation metal-poor globular clusters may have formed during cosmological reionization from low-metallicity GMCs squeezed by the universal reionization pressure.

1 On the Nature of Young Globular Clusters

When studying the myriads of point-like luminous sources brighter than any individual star on *HST* images of *ongoing* mergers (e.g., NGC 4038/39, NGC 3256), one would like to know which ones—or at least what fraction—will survive as globular clusters (GC). Yet, it is very difficult to distinguish gravitationally bound young star clusters from unbound OB associations or even spurious asterisms. As it turns out, the adopted operational definition for "cluster" may determine the answers to the scientific questions we ask about these objects.

Modern astronomical dictionaries universally include in their definition of "star cluster" (open or globular) the requirement that it be *gravitationally bound*, thus distinguishing it from any looser, expanding "stellar association" (e.g., [17, 27]). As I explain in Sect. 2 below, I believe that our present inability to make this distinction for many stellar aggregates younger than 10–20 t_{cr} (internal crossing times) in ongoing mergers leads to a notion of "infant mortality" that is seriously exaggerated.

In recent merger *remnants*, where the merger-induced starburst has subsided (e.g., NGC 3921, NGC 7252), the definition of a *young globular cluster (YGC)* is more easy and secure. Any young compact stellar aggregate older than 10–20 t_{cr} (~20–40 Myr), more massive than a few $10^4 M_\odot$, and with a half-light radius R_{eff} comparable to that of a typical Milky-Way globular (say, $R_{eff} \lesssim 10$ pc) is most likely gravitationally bound and, hence, a YGC. It is the size requirement that places stringent upper limits on any possible expansion velocity ($\lesssim 0.2$–0.5 km s^{-1}) and thus guarantees that the cluster is gravitationally bound.

An important result to emerge from recent *HST* and follow-up studies of YGCs concerns their masses. These masses do not only cover the full range observed in old Milky-Way GCs (~10^4 – $5 \times 10^6 M_\odot$), but also extend to nearly $10^8 M_\odot$ or ~20× the mass of ω Cen at the high-mass end. The most massive YGCs are invariably found in remnants of gas-rich major mergers such as NGC 7252 [31, 21], NGC 1316 [5], and NGC 5128 [22]. Interestingly, dynamical masses determined from velocity dispersions agree well with photometric masses based on cluster-evolution models with normal (e.g., Salpeter, Kroupa, or Chabrier) initial mass functions (IMFs). Therefore, some earlier worries that YGCs formed in mergers may have highly unusual stellar IMFs (e.g., [6]) seem now unfounded.

Relatively little work has been done so far on the brightness profiles and detailed structural parameters (core and tidal radii) of YGCs in mergers. Yet, the subject looks promising. Radial profiles of selected YGCs in NGC 4038 suggest that the initial power-law envelopes of YGCs may be tidally stripped within the first few 100 Myr, while the core radii may grow [39]. Correlations between core radius and cluster age are known to exist for the young cluster populations of the Magellanic Clouds (e.g., [20]) and deserve further study via the rich cluster populations of ongoing mergers and merger remnants.

2 Formation and Early Evolution

Star clusters form in giant molecular clouds (GMC), where optical extinction can be very significant. Hence the question arises what fraction of all young clusters "optical" surveys made with *HST* ($0.3 \lesssim \lambda \lesssim 1.0\,\mu$) may miss.

This question has been addressed by Whitmore and Zhang [38] for the "Overlap Region" of NGC 4038/39, which is known to harbor some of the most IR-luminous young clusters, yet appears heavily extincted at optical wavelengths and brightly emitting at 8μ [34]. A comparison between optical clusters and strong thermal radio sources shows that 85% of the latter have optical counterparts, whence even in this extreme region only ~15% of all clusters have been missed by *HST* surveys [38]. Measured cluster extinctions lie in the range $0.5 \lesssim A_V \lesssim 7.6$ mag and diminish to $A_V \lesssim 1.0$ mag for clusters 6 Myr and older. This suggests that cluster winds disperse most of the natal gas rapidly, and that optically-derived luminosity functions for clusters older than ~6 Myr should not be too incomplete.

2.1 Cluster Luminosity Functions

To first order, the luminosity functions (LF) of young-cluster systems in merger galaxies are well approximated by a power law of the form $\Phi(L)dL \propto L^{-\alpha}dL$ with $1.7 \lesssim \alpha \lesssim 2.1$ [37, 23, 35]. The similarities between this power law and the power-law mass function of GMCs, including the similar observed mass ranges, strongly suggest that young clusters form from GMCs suddenly squeezed by a rapid increase in the pressure of the surrounding gas [18, 16, 11] (see also Sect. 2.3).

Deep *HST* observations of mergers with rich cluster systems suggest that the cluster LFs may have a break ("knee") whose position varies from merger to merger (NGC 4038/39 [39]; NGC 3256 [40]; M51 [14]). Figure 1 displays for NGC 4038/39 both the original cluster LF [37] and two versions of the deeper LF [39] showing a break around $M_V = -10.0$ to -10.3. The interpretation of these breaks is presently controversial. Either the breaks reflect brightness-limited-selection effects (Whitmore et al., in prep.) or they may indicate a maximum cluster mass [14]. In the latter case, the measured LF breaks in the above three mergers would seem to suggest that the maximum mass increases with the vehemence of the merger, presumably indicating that under increased gas pressure GMCs coagulate into more massive aggregates.

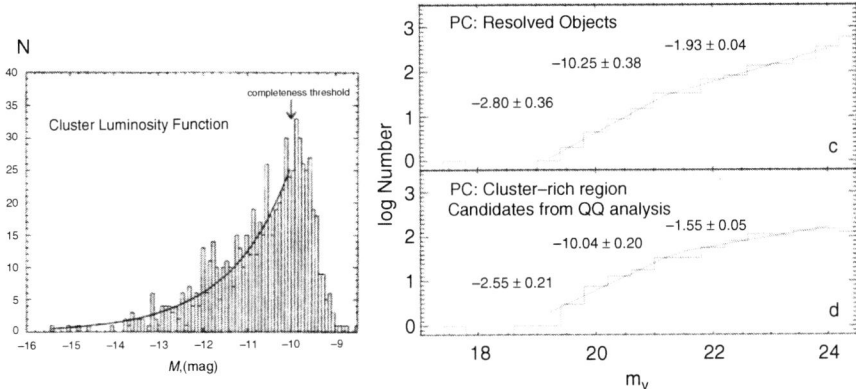

Fig. 1. Luminosity functions for candidate young star clusters in NGC 4038/39 from *HST* observations with (*left*) WFC1 [37] and (*right*) WFPC2 [39].

2.2 Star-Cluster Formation vs Clustered Star Formation

The *age distribution* of young clusters in NGC 4038/39 has recently been derived for two mass-limited subsamples defined by $M > 3 \times 10^4 M_\odot$ and

$M > 2 \times 10^5 M_\odot$ [13]. The masses themselves are estimates based on *HST* photometry in *UBVI* and Hα plus Bruzual-Charlot [7] cluster evolution models. The number distributions for both subsamples decline steeply with age τ, approximately as $dN/d\tau \propto \tau^{-1}$. Thus, it would seem that ~90% of all clusters disrupt during each age decade. The median age of the clusters is a mere ~10^7 yr, which Fall et al. interpret as evidence for rapid disruption, dubbed "infant mortality." These authors guess that "very likely ... most of the young clusters are not gravitationally bound and were disrupted near the times they formed by the energy and momentum input from young stars to the ISM of the protoclusters."

In my opinion, it is unfortunate that this loose, non-astronomical use of the word "cluster" may reinforce an increasingly popular view that most stars form in clusters. By the traditional astronomical definition of star clusters as gravitationally bound aggregates, most of the objects tallied by Fall et al. in The Antennae are not clusters, but likely young stellar associations. It seems to me in much better accord with a rich body of astronomical evidence gathered during the past 50 years to state that—although *star formation is clearly clustered*—even in mergers gravitationally bound clusters (open and globular) form relatively rarely and *contain <10% of all newly-formed stars.*

I believe that only with such careful distinction can we hope to study the true disruptive effects that affect any gravitationally bound star cluster over time, including mass loss due to stellar evolution and evaporation by two-body relaxation and gravitational shocks.

Further reason for caution is provided by the recent discovery that even in nearby M31, four of six claimed YGCs have turned out to be spurious asterisms when studied with adaptive optics [10]. Clearly, there is considerable danger in calling all luminous point-like (at *HST* resolution) sources in the distant NGC 4038/39 young "clusters"!

2.3 Shocks and High Pressure

Shocks and high pressure have long been suggested to be the main drivers of GC formation in gas-rich mergers and responsible for the increased specific frequency S_N of GCs observed in descendent elliptical galaxies [28, 18, 1].

Much new evidence supports this hypothesis. *Chandra* X-ray observations of the hot ISM in merger-induced starbursts, and especially in NGC 4038/39 [12], show that the pressure in the hot, 10^6–10^7K ISM of a merger can exceed 10^{-10} dyn cm^{-2} and is typically 10–100 times higher than it is in the hot ISM of our local Galactic neighborhood (e.g., [2, 33]). Thus GMCs in mergers do indeed experience strongly increased pressure from the surrounding gas.

The principal source of general pressure increase are gravitational torques between the gas and stellar bars, which tend to brake the gas and lead to rapid inflows and density increases (e.g., [4, 24]).

What has become clearer only recently is how much accompanying *shocks* may affect the spatial distribution of star and cluster formation. As Barnes

[3] shows via numerical simulations, star-formation recipes that include not only the gas density (i.e., Schmidt–Kennicut laws), but also the local rate of energy dissipation in shocks, lead to spatially more extended star and cluster formation that tends to occur earlier during the merger. A model with mainly shock-induced star formation for The Mice (NGC 4676) leads to significantly better agreement with the observations of H II regions and young clusters than one with only density-dependent star formation. Shock-induced star formation may also explain why cluster formation is already so vehement and wide-spread in The Antennae, where the two disks—currently on their second approach—are still relatively intact.

Are the shocks in mergers generated by high-velocity, 50–100 km s^{-1} cloud–cloud collisions [19] or more by large-scale gas motions? A high-resolution study with *HST*/STIS of the radial velocities of many dozens of young clusters in 7 regions of The Antennae shows that the average cluster-to-cluster radial-velocity dispersion is $\sigma_{v,cl} < 10$–12 km s^{-1} [36], as illustrated in Fig. 2. This relatively low velocity dispersion argues strongly against high-velocity cloud–cloud collisions and in favor of the general pressure increase being what triggers GMCs into forming clusters [18,11].

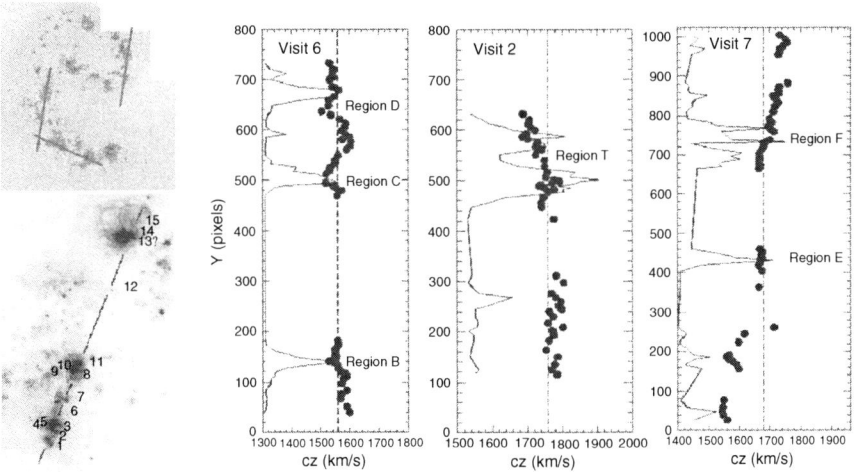

Fig. 2. Radial velocities of young clusters in NGC 4038/39, measured with *HST*/STIS (at Hα) along three lines crossing 7 major regions, each with many clusters. The three slit positions are shown at *upper left*, while *lower left panel* shows slit position across regions D, C, and B in more detail. After gradient subtraction, the cluster-to-cluster velocity dispersion is <10–12 km s^{-1} [36].

3 Young Metal-Rich Halo Globulars

There are several advantages to studying YGCs in relatively recent, about 0.3–3 Gyr old merger remnants: (1) Dust obscuration is much less of a problem than in ongoing mergers. (2) Most point-like luminous sources in such remnants are true GCs, since time has acted to separate the wheat from the chaff (= expanding associations), and clusters are now typically >100 Myr or >25–50 $t_{\rm cr}$ old. And (3), the remnants themselves appear to be evolving into bona fide early-type galaxies. Therefore, YGCs formed during the mergers can provide key evidence on processes that must have shaped GC populations in older E and S0 galaxies as well.

HST studies of recent merger remnants such as NGC 3921 [30], NGC 7252 [25], and NGC 3597 [8] show that these galaxies typically host about 10^2–10^3 point-like sources that appear to be mostly *young* GCs ($\lesssim 1$ Gyr old). (It is not that there are no old GCs in these relatively distant remnants, only that the YGCs are much brighter and more easily studied.) Age-dating based both on broad-band photometry and spectroscopy shows that the majority of these YGCs formed in relatively short, 100–200 Myr time spans during the mergers. The YGCs appear strongly concentrated toward their hosts' centers, half of them lying typically within $\lesssim 5$ kpc from the nucleus.

The few spectroscopic studies that have so far been made of such YGCs invariably show them to be of approximately solar metallicity: $[Z] = 0.0 \pm 0.1$ in NGC 7252 [31], 0.0 ± 0.5 in NGC 3921 [32], and—for the intermediate-age, ∼3–5 Gyr old GCs in more advanced remnants—$[Z] = 0.0 \pm 0.15$ in NGC 1316 [15] and -0.1 ± 0.2 in NGC 5128 [26].

Such near-solar metallicities in recently formed GCs are, of course, not unexpected and might not seem worth emphasizing, were it not for the fact that the YGCs with these metallicities all show *halo* kinematics (see refs. above). Therefore, the inevitable conclusion is that major mergers of gas-rich disk galaxies produce *young metal-rich halo GCs*. The existence of significant populations of such clusters in merger remnants ranging from ∼0.5 Gyr to 4–5 Gyr in age, together with observational and theoretical evidence that the remnants themselves are young to intermediate-age ellipticals, provides a strong link to the old metal-rich GC populations observed in virtually all E and many S0 galaxies (see [29] and Goudfrooij's contribution in this volume for further details).

4 Implications for Old Metal-Poor Globular Clusters

Perhaps the main result from studies of GC formation in mergers is that the process is driven by strong pressure increases that squeeze GMCs into rapid cluster formation. Observations show that the pressures in the ISM can exceed 10^{-10} dyn cm^{-2} already early on in a merger (Sect. 2.3), while simulations of gas-rich mergers demonstrate that most of the pressure increase is driven gravitationally [4, 24, 3].

These facts beg the question whether some nearly universal pressure increase may have caused the formation of the old metal-poor GCs that are so omnipresent in all types of galaxies and environments.

Cen [9] points out that the cosmological reionization at $z \approx 15$–7 may have provided just such a universal pressure increase. Ionization fronts driven by the external radiation field may have generated inward convergent shocks in gas-rich sub-galactic halos, which in turn triggered GMCs into forming clusters. If so, the formation of metal-poor GCs from early GMCs in many of these halos may have been nearly synchronous.

If Cen's hypothesis is correct, most GCs in the universe may have formed from shocked GMCs. The first-generation GCs formed near-simultaneously from low-metallicity GMCs shocked by the pressure increase accompanying cosmological reionization. Later-generation ("second-generation") GCs formed during subsequent galaxy mergers from metal-enriched GMCs present in the merging components and shocked by the rapid, gravitationally-driven pressure increases of the mergers. Major disk mergers, some of which occur to the present time, led to elliptical remnants with a mixture of first- and second-generation GCs that can still be traced by their bimodal color distributions. Finally, a minority of second-generation GCs seem to form sporadically from occasional pressure increases in calmer environments, such as in interacting irregulars and barred spirals.

5 Conclusions

During mergers, increased gas pressure leads to much *apparent* cluster formation, but most of the stellar aggregates are unbound and disperse. Gravitationally bound globular and open clusters are relatively *rare* and seem to contain <10% of all stars formed in the starbursts.

Major gas-rich mergers form not only E and S0 galaxies, but also their metal-rich "second-generation" GCs. Specifically, in the local universe young remnants of major such mergers appear as proto-elliptical galaxies with subpopulations of young metal-rich halo GCs (NGC 3921, NGC 7252; later NGC 1316, NGC 5128). The evidence is now strong that these second-generation GCs form from giant molecular clouds in the merging disks, squeezed into collapse by large-scale shocks and high gas pressure rather than by high-velocity cloud–cloud collisions.

Similarly, first-generation metal-poor GCs may have formed during cosmological reionization from low-metallicity giant molecular clouds squeezed by the reionization pressure.

Acknowledgements. I thank Brad Whitmore for his permission to reproduce some figures.

References

1. Ashman, K.M., & Zepf, S.E.: ApJ **384**, 50 (1992)
2. Baldi, A., Raymond, J.C., Fabbiano, G., et al.: ApJ **636**, 158 (2006)

3. Barnes, J.E.: MNRAS **350**, 798 (2004)
4. Barnes, J.E., & Hernquist, L.: ApJ **471**, 115 (1996)
5. Bastian, N., Saglia, R.P., Goudfrooij, P., et al.: A&A **448**, 881 (2006)
6. Brodie, J.P., Schroder, L.L., Huchra, J.P., et al.: AJ **116**, 691 (1998)
7. Bruzual, A.G., & Charlot, S.: MNRAS **344**, 1000 (2003)
8. Carlson, M.N., Holtzman, J.A., Grillmair, C.J., et al.: AJ **117**, 1700 (1999)
9. Cen, R.: ApJ **560**, 592 (2001)
10. Cohen, J.G., Matthews, K., & Cameron, P.B.: ApJ **634**, L45 (2005)
11. Elmegreen, B.G., & Efremov, Y.N.: ApJ **480**, 235 (1997)
12. Fabbiano, G., Baldi, A., King, A.R., et al.: ApJ **605**, L21 (2004)
13. Fall, S.M., Chandar, R., & Whitmore, B.C.: ApJ **631**, L133 (2005)
14. Gieles, M., Larsen, S.S., Scheepmaker, R.A., et al.: A&A **446**, L9 (2006)
15. Goudfrooij, P., Mack, J., Kissler-Patig, M., et al.: MNRAS **322**, 643 (2001)
16. Harris, W.E., & Pudritz, R.E.: ApJ **429**, 177 (1994)
17. Hopkins, J.: *Glossary of Astronomy and Astrophysics* (University of Chicago, Chicago 1976) p. 145
18. Jog, C.J., & Solomon, P.M.: ApJ **387**, 152 (1992)
19. Kumai, Y., Basu, B., & Fujimoto, M.: ApJ **404**, 144 (1993)
20. Mackey, A.D., & Gilmore, G.F.: MNRAS **338**, 85 & 120 (2003)
21. Maraston, C., Bastian, N., Saglia, R.P., et al.: A&A **416**, 467 (2004)
22. Martini, P., & Ho, L.C.: ApJ **610**, 233 (2004)
23. Meurer, G.R., Heckman, T.M., Leitherer, C., et al.: AJ **110**, 2665 (1995)
24. Mihos, J.C., & Hernquist, L.: ApJ **464**, 641 (1996)
25. Miller, B.W., Whitmore, B.C., Schweizer, F., & Fall, S.M.: AJ **114**, 2381 (1997)
26. Peng, E.W., Ford, H.C., & Freeman, K.C.: ApJ **602**, 705 (2004)
27. Ridpath, I.: *A Dictionary of Astronomy* (Oxford University Press, Oxford 1997) p. 450
28. Schweizer, F.: Star Formation in Colliding and Merging Galaxies. In: *Nearly Normal Galaxies*, ed. by S.M. Faber (Springer, New York 1987) p. 18
29. Schweizer, F.: Formation of Globular Clusters in Merging Galaxies. In: *New Horizons in Globular Cluster Astronomy*, ed. by G. Piotto et al. (ASP, San Francisco 2003) p. 467
30. Schweizer, F., Miller, B.W., Whitmore, B.C., & Fall, S.M.: AJ **112**, 1839 (1996)
31. Schweizer, F., & Seitzer, P.: AJ **116**, 2206 (1998)
32. Schweizer, F., Seitzer, P., & Brodie, J.P.: AJ **128**, 202 (2004)
33. Veilleux, S., Cecil, G., & Bland-Hawthorn, J.: ARA&A **43**, 769 (2005)
34. Wang, Z., Fazio, G.G., Ashby, M.L.N., et al.: ApJS **154**, 193 (2004)
35. Whitmore, B.C.: The Formation of Star Clusters. In: *A Decade of HST Science*, ed. by M. Livio et al. (Cambridge University Press, Cambridge 2003) p. 153
36. Whitmore, B.C., Gilmore, D., Leitherer, C., et al.: AJ **130**, 2104 (2005)
37. Whitmore, B.C., & Schweizer, F.: AJ **109**, 960 (1995)
38. Whitmore, B.C., & Zhang, Q.: AJ **124**, 1418 (2002)
39. Whitmore, B.C., Zhang, Q., Leitherer, C., et al.: AJ **118**, 1551 (1999)
40. Zepf, S.E., Ashman, K.M., English, J., et al.: AJ **118**, 752 (1999)

The Formation Histories of Metal-Rich and Metal-Poor Globular Clusters

Stephen E. Zepf

Department of Physics and Astronomy, Michigan State University
zepf@pa.msu.edu

Abstract. This review presents the results of ongoing studies of the formation histories of metal-poor and metal-rich globular clusters and their host galaxies. I first discuss the strong observational evidence that the globular cluster systems of most elliptical galaxies have bimodal metallicity distributions. I then focus on new results for metal-poor and metal-rich globular cluster systems. Metal-poor globular clusters are often associated with early structure formation, and I review new constraints on their formation epoch based on the "bias" of the number of metal-poor clusters with host galaxy mass. For metal-rich globular clusters, I discuss new results from ongoing optical to near-infrared photometric studies which both confirm an intermediate-age population in NGC 4365 and generally reveal a variety of formation histories for now quiescent ellipticals.

1 Introduction

As noted by several speakers at this meeting, many of us begin our talks and proposals giving reasons why we study globular clusters (GCs) to learn about the formation of their host galaxies. However, even if standard, listing these reasons is important because the characteristics that make globular clusters useful for studying the formation of their host galaxies are often revisited when comparing the results of the many different programs aimed at constraining galaxy formation and evolution. In this spirit, the following are some of the key reasons globular clusters are valuable tools for understanding the formation history of their host galaxies.

(1) Because the age and metallicity can be determined for each globular cluster individually, the distribution of ages and metallicities within a galaxy population can be constrained.
(2) Globular clusters are the best example we have of a simple stellar population, so determining the age and metallicity of individual GCs is much simpler than studies of integrated light in which the stars of many different metallicities and ages are seen as one luminosity weighted average.
(3) Globular clusters are observed to form in all major star formation events in galaxies, making them tracers of the major formation episodes of their host galaxies.
(4) Some globular clusters are among the oldest known objects, so they may provide constraints on early structure formation.

(5) As dense concentrations of $\sim 10^6$ stars, globular clusters can be observed in galaxies across the local universe. This enables the formation history of a representative sample of galaxies to be studied, including those of normal elliptical galaxies which are not present at very nearby distances.

In this review, I will focus on two new results regarding the formation history of globular cluster systems (GCSs) and their host galaxies. One of these new results is a constraint on the formation epoch of metal-poor globular clusters from the determination of their cosmological "bias" with host galaxy mass (see also Rhode, Zepf, and Santos [14], and Katherine Rhode's talk at this meeting). The second set of results are constraints on the formation history of elliptical galaxies from the age distributions of their globular cluster systems. I specifically review new work on the optical to near-infrared colors of globular cluster systems, which includes deep NICMOS data confirming previously identified intermediate age populations in the giant elliptical galaxy NGC 4365, and new PANIC data showing an effectively completely old population in another giant elliptical, NGC 4472. I then discuss the overall status of work on ages, and look to future samples and comparisons with elliptical galaxy formation histories estimated in other ways.

2 A Note About Metallicity and Color Distributions

The primary results discussed below involve using metal-poor globular clusters to probe early structure formation and metal-rich ones to probe the formation epochs of their host galaxies. While clearly some globular clusters have higher or lower metallicities than others, it is still natural to consider the basis for dividing globular cluster systems into metal-poor and metal-rich populations.

The obvious observational starting point is that most globular cluster systems of elliptical galaxies are observed to have bimodal color distributions. This was first noted many years ago now [17] and much subsequent work has made it clear that the globular cluster systems of most, although not all, elliptical galaxies have bimodal color distributions (e.g. [6,9]). It was also realized early on that for most elliptical galaxies GCSs, the optical color primarily traces metallicity. This is because metallicity is the primary driver for optical colors at ages greater than a few Gyr, and elliptical galaxies generally do not have huge amounts of recent star formation. It turns out the minority of ellipticals which have unimodal GCS color distribution may be an interesting exception to this rule as several of these have been found to have younger GC populations (see Sect. 4), but for typical bimodal systems, the GC color primarily traces metallicity.

Given that GC color primarily traces metallicity in typical bimodal systems, the next question is what is the detailed relationship between these two, and specifically whether the bimodal color distribution observed for most elliptical galaxy GCSs reflects a truly bimodal metallicity distribution. One common way to determine the relationship between color and metallicity is to fit the observed relation for the Milky Way GCS. Data for Galactic globular clusters is fundamental for this question, because only for these GCs are there true abundances from high resolution spectra for individual stars, as well as detailed color-magnitude diagrams for age and metallicity determination. This approach to the relationship between color and metallicity is then limited by the Galactic globular cluster sample. Its primary limitation for this purpose is the absence of high metallicity clusters. Additional concerns are that some of the few with higher metallicities also have large and uncertain reddenings, that the Galactic GCs do not extend to young ages, and simply the modest total number of Galactic GCs.

Another common approach is to use stellar populations models. However, these are not a panacea, because they are tested on Galactic data, and therefore have many of the same concerns at old ages and high metallicities as empirical fits of the Galactic GCS color-metallicity relation. Another possibility is using absorption-line indices for extragalactic systems to investigate the behavior of colors at higher metallicities. However, these indices are only related to actual abundances through stellar populations models or through Galactic GCs, and are therefore also strongly dependent on uncertainties in the models and any gaps in the Galactic sample on which these are tested.

There is a large body of work on determining color-metallicity relations using the above approaches and applying these to extragalactic globular cluster systems. These cover many color indices, including B−I and C−T1 which have large wavelength baselines and for which there are extensive datasets, and V−I for which there is significant HST archival data. The conclusion of these many studies is that while the exact shape of the metallicity distribution varies somewhat depending on the color-metallicity relation, the general bimodal nature of the metallicity distribution remains. It is very difficult to create a strong dip right in middle of the color distribution without a bimodal-like metallicity distribution. A variety of kinematical studies also find differences between the metal-poor and metal-rich globular cluster systems (e.g. [18, 4]), providing independent evidence for the presence of these two subpopulations, similar to the way in which kinematics provides further evidence for two populations of GCs in the Milky Way. One contrary claim has recently been put forward and was presented at this meeting by Yoon et al. (see these proceedings). They use a stellar populations model with a specific treatment of the horizontal branch, and show their model gives a very sharp feature in the color-metallicity relation at the precise place to mimic a metallicity bimodality. They also claim that this sharp feature is consistent with the preliminary g−z Galactic GC data in Peng et al. [11], although no comparison is made to other well-established Galactic GC colors with a similar wavelength baseline.

A powerful argument that the observed color bimodality is due to a truly bimodal metallicity distribution is that the Galactic GCS is known without question to be bimodal in metallicity. While one might say we happen to live in a galaxy with a complicated history, it would seem quite strange for us to have ended up in a complicated galaxy, while other galaxies have the virtue of uncomplicated pasts. This "Copernican" argument for metallicity bimodality is strongly supported by a number of other observational results.

One observation that provides a clear indication of a bimodal metallicity distribution is bimodality in the I−H color distribution for the M87 GC system [8]. The near-infrared I−H color is dependent almost completely on metallicity, and it is very hard to image how a horizontal branch effect as proposed by Yoon et al. can account for bimodality at these near-infrared wavelengths. This evidence for bimodality in metallicity is supported by a wide variety of other data. Perhaps the simplest is the analysis of many different colors vs metallicities for Galactic GCs, including the B−I and C−T1 shown in talks at this meeting, both of which have broad wavelength ranges, does not show the very sharp and precisely located jump in color vs. metallicity required to produce the observed bimodal color distribution from a unimodal metallicity distribution. As noted in Smits et al. [16], even linear fits of these Galactic GC color-metallicity relations do not have much more scatter than the observational uncertainties. Cohen et al. (2003) also noted in their fits of color to Galactic GC metallicities and absorption-line indices for extragalactic systems that a second-order fit is only slightly better than a linear fit, and their Figs. 13 and 15 do not provide evidence for sharp jumps in the color-metallicity relation. In this context is it worth noting that most absorption line studies of extragalactic GCSs are not fair samples of the cluster metallicity distribution, as many intentionally select their GC targets to have colors in the middle of the distribution in an attempt to decrease contamination of the sample by non-clusters. Therefore, while the comparison of colors and absorption-line indices for individual objects is valuable (e.g. [3]), the distribution of absorption-line indices in extragalactic samples is not generally useful because of the preferential selection of GCs with intermediate colors.

Thus, there is both direct evidence for bimodal metallicity distributions as described above and the simple argument that it seems unlikely we are privileged to live in one of the few galaxies that has GC system with a unimodal metallicity distribution. These strongly indicate that the bimodal color distributions observed for GC systems arise from bimodal metallicity distributions. Therefore, for the remainder of this review, we will consider the metal-poor and metal-rich GC populations separately.

3 The Formation Epoch of Metal-Poor Globular Clusters from Their Bias with Galaxy Mass

It is natural to associate metal-poor globular clusters with early epochs of structure formation when the overall enrichment in galaxies was not high. Moreover, for many Galactic globular clusters, age determinations also indicate an early epoch of formation, although the uncertainties in GC ages even for Galactic clusters are large when converted to redshift in the early universe. Therefore, an independent method to assess the formation epoch of metal-poor globular clusters would be very interesting.

One idea for constraining the general formation epoch of the metal-poor globular cluster population is to take advantage of the feature of hierarchical structure formation models that more massive objects have a greater fraction of their mass collapsed at higher redshift, and that this "bias" increases fairly steeply at very high z. This has recently been pursued by Rhode, Zepf, and Santos [14], based in part on the discussion in Santos [15], and is also discussed in this meeting in Katherine Rhode's talk. Specifically, we used new wide-field multi-color imaging for both ellipticals and spirals in a range of environments to determine the total number of metal-poor globular clusters around these galaxies. We then determined the galaxy-mass normalized number of metal-poor globular clusters for each galaxy (their "T_{blue}" value, and plotted this versus galaxy mass. If the metal-poor globular clusters formed at moderate redshift when most of the mass has collapsed for typical galaxies, there would be little or no bias and all galaxies would have similar T_{blue} values regardless of galaxy mass. With an extremely high redshift of formation for the metal-poor globular clusters, the mass-normalized number of metal-poor GCs would show a steep rise to higher galaxy masses, as only the most massive galaxies had even a small fraction of their mass collapse at very early epochs.

Our results are that the mass-normalized number of metal-poor GCs (T_{blue}) increases with increasing galaxy mass [14]. However, the increase is not particularly steep, suggesting extremely high redshifts for the typical formation epoch of metal-poor globular clusters are unlikely (see Katherine Rhode's talk for relevant plots). Moreover, most systematics tend to steepen the relation, hinting that the current result may be an upper limit to the cosmological "bias" of the metal-poor GC population and its formation redshift. Further work in this area will include more detailed modeling of the theoretical expectations (see [10] for such effort), use of the radial profile of the GCSs as a constraint, and of course needed increases in the size of the galaxy sample with data that provides sufficient areal coverage and depth to obtain a reliable determination of the total number of metal-poor GCs.

4 The Formation Epoch(s) of Elliptical Galaxies from the Ages of Their Metal-Rich Globular Clusters

Determining the formation history of massive early-type galaxies is one of the primary challenges of current extragalactic astronomy. These elliptical and S0 galaxies make up a significant fraction of the stellar mass in the local universe, but there is not yet a consensus on when they formed. Because the age of individual globular clusters can be determined, and globular cluster formation is observed to be a ubiquitous feature of starbursts, determining the age distribution of their globular clusters is a valuable way to constrain the formation history of elliptical galaxies.

There are several different observational approaches for determining the age and metallicity of extragalactic GCs. Although I have been involved in some way with nearly all of them, in this review I will focus on the optical to near-infrared color approach, as it is both very promising and not the subject of other talks at this meeting. The optical to near-infrared color technique is very promising because it relies on straightforward stellar physics to determine the age and metallicity, and because it is a photometric technique and therefore observationally efficient. The technique solves the age-metallicity degeneracy for unresolved simple stellar populations because the infrared color (e.g. I−H) is primarily sensitive to metallicity, while the optical color (e.g. g−I) has a greater sensitivity to age. Thus in plots of optical to near-infrared colors age and metallicity are separated (see Fig. 1). This separation is about 0.3 mag between intermediate ($\sim$ 3 Gyr) and old ($\sim$ 15) Gyr. A comparable example from another area is that 0.3 mag is similar to the difference in SN Ia brightnesses in an accelerating universe compared to an open one.

One of the first results to come from the application of the optical to near-infrared technique was the discovery of a substantial population of intermediate age globular clusters in the elliptical galaxy NGC 4365 ([13], hereafter P02). This result was originally thought to be surprising, in part because

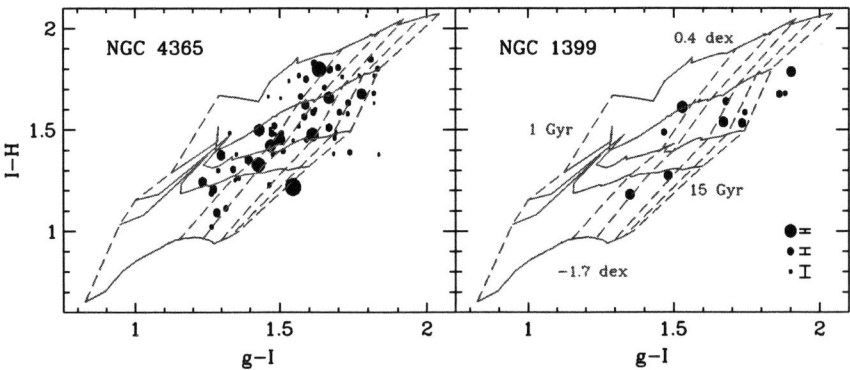

Fig. 1. The g–I vs. I–H plots of globular clusters in NGC 4365 and NGC 1399 from NICMOS and ACS data from Kundu et al. [7]. These show a substantial population of GCs with optical to near-IR colors indicative of intermediate age in NGC 4365, in agreement with earlier work (P02), while NGC 1399 has few such GCs. Only GCs with uncertainties less than 0.1 mag in each axis are shown. The lines trace age (1, 3, 5, 8, 11, and 15 Gyr from the left) and metallicity ([Fe/H] = −1.7, −0.7, −0.4, 0.0, and 0.4 dex from the bottom) contours from BC03 models [2]. The size of the points are inversely proportional to the I–H uncertainty as shown at the bottom right. The g–I uncertainties are comparable to the I–H values in NGC 4365 and are much smaller than the I–H uncertainties in NGC 1399.

spectroscopy of the integrated light of NGC 4365 was originally thought to indicate an old age and thus presented a puzzle. However, recent calculations using alpha-enhanced isochrones necessary for NGC 4365 and its large observed [Mg/Fe] give younger ages ([1], hereafter B05). Extant spectroscopic studies of the globular clusters themselves are inconclusive as one finds intermediate ages and the other does not, even for objects in common between the two studies (B05).

Therefore, to pin down the ages and metallicities of the GCs around NGC 4365, we obtained new, very deep NICMOS near-infrared photometry in three fields, and combine this with new ACS data. As shown in Fig. 1, these independent and much higher signal-to-noise data confirm the previous result that NGC 4365 has a substantial population of GCs with optical to near-infrared colors that can only be accounted for by intermediate ages of 2–7 Gyr (Kundu et al. 2005). We also show in this paper that the result is independent of which stellar population models are used, although the exact age of the intermediate age population does vary with model. In addition, Kundu et al. [7] used archival data for NGC 1399 to show that its GCS is predominantly old with a small younger component, in agreement with previous spectroscopic work.

A comparison of optical to near-infrared photometry and the absorption-line spectroscopy for different galaxy GCSs reveals generally good agreement (see [7]). Another example shown in Fig. 2 is new PANIC data we have obtained for NGC 4472, which shows an almost exclusively old GC system [5]. This is in agreement with extant spectroscopic studies ([3], Beasley et al.

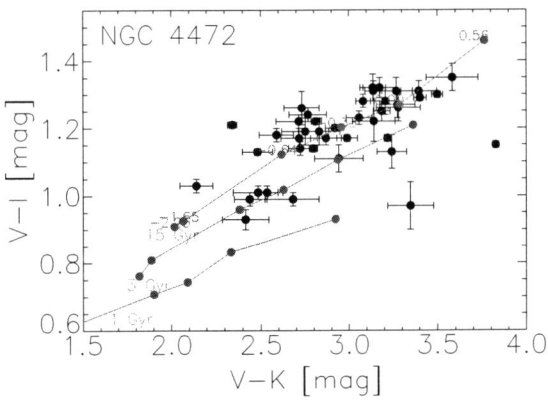

Fig. 2. The V−K, V−I plots for globular clusters around NGC 4472, indicating that the GCS of NGC 4472 appears to be nearly completely old. In this plot, the model ages decrease left to right, going from 15 Gyr on the upper left to 3 Gyr on the lower right, and metallicity increases from [Fe/H] = −2.2 in the lower left part to 0.5 on the upper right part of the plot. All of these are BC03 models, and the data are from our ongoing work [5].

2000). The key advantages of the optical to near-infrared approach are that as a photometric technique it is more efficient to obtain age and metallicity estimates of a given uncertainty in the same amount of observing time, and that it relies on straightforward stellar astrophysics. The advantages of absorption-line indices are that they enable estimates of abundance ratios, which are interesting probes of timescales of formation (e.g. [12]).

Acknowledgements. The work described here has been made possible by my many excellent collaborators. I acknowledge support for this research from NSF award AST-0406891, NASA LTSA grant NAG5-11319, STScI grant HST-GO-09878.01-A, and a NSF travel grant.

References

1. Brodie, J.P., et al. 2005, AJ, 129, 2643 (B05)
2. Bruzual, G., & Charlot, S. 2003, MNRAS, 344, 1000 (BC03)
3. Cohen, J.G., Blakeslee, J.P., & Côté, P. 2003, ApJ, 592, 866
4. Côté, P., McLaughlin, D.E., Cohen, J.G., & Blakeslee, J.P. 2003, ApJ, 591, 850
5. Hempel, M., Zepf, S.E., Kundu, A., & Geisler, D. 2006, in preparation
6. Kundu, A., & Whitmore, B.C. 2001, AJ, 121, 2950
7. Kundu, A., et al. 2005, ApJ, 634, L41
8. Kundu, A., & Zepf, S.E., 2007, ApJ, 660, L109
9. Larsen, S.S. Brodie, J.P., Huchra, J.P., Forbes, D.A., & Grillmair, C. 2001, AJ, 121, 2974
10. Moore, B., Diemand, J., Madau, P. Zemp, M., & Stadel, J. 2006, MNRAS, 368, 563
11. Peng, E.W., et al. 2006, ApJ, 639, 95
12. Pierce, M., et al. 2006, MNRAS, 368, 325
13. Puzia, T. H., Zepf, S. E., Kissler-Patig, M., Hilker, M., Minniti, D., & Goudfrooij, P. 2002, A&A, 391, 453
14. Rhode, K.L., Zepf, S.E., & Santos, M.R. 2005, ApJL, 630, 21
15. Santos, M.R., in Extragalactic Globular Cluster Systems, p. 348
16. Smits, M., Maccarone, T.J., Kundu, A., & Zepf, S.E. 2006, A&A, 458, 477 MNRAS
17. Zepf, S.E., & Ashman, K.M. 1993, MNRAS, 264, 611
18. Zepf, S.E., et al. 2000, AJ, 120, 2928

Globular Cluster System Evolution in Early Type Galaxies

R. Capuzzo-Dolcetta

Dipartimento di Fisica, Universitá di Roma "La Sapienza", P.le Aldo Moro, 2, I-00185 – Rome, Italy `roberto.capuzzodolcetta@uniroma1.it`

Abstract. Globular clusters (GCs) constitute a system which is evolving because of various interactions with the galactic environment. Evolution may be the explanation of many observed features of Globular Cluster Systems (GCSs); the different radial distribution of the GCS and the stellar component of early type galaxies are explained by dynamical friction and tidal effects, the latter acting both on the large scale (that of the bulge-halo stars) and on the small scale (that of the nucleus, often containing a central massive black hole). Merging of quickly orbitally decayed massive GCs leads to formation of a Super Star Cluster (SSC) which enriches the galactic nucleus and is a reservoire of mass-energy for a centrally located black hole.

1 Introduction

The Hubble Space Telescope and large ground based telescopes are providing a continuously increasing amount of data concerning GCSs in galaxies, mainly of the early types, since the pioneering work [16].
    Two are the most debated points: (i) the difference in the GCS and galaxy light spatial distribution, and, (ii) the existence of a bimodal color distribution for GCSs, and the possible differences between the *blue* and the *red* population,
The two points are, likely, related; in any case, here I will not discuss about point (ii) (see the recent [21] paper) but just about point (i) which is better observationally stated and deserves a correct interpretation.

2 The GCS and Stellar Radial Distributions in Galaxies

It is nowadays clear that the majority of galaxies shows a radial profile of their GCS shallower than that of the stars toward the galactic centre. Ellipticals show a more or less peaked stellar profile toward the galactic center (actually, many have a 'cuspy' profile), while the GCS radial distribution has, usually, a core. The related literature is so vast that we limit to recall [13, 12]. The explanation of this difference in terms of formation and evolution of elliptical galaxies (see [14, 12, 1, 3], or in terms of evolution of the GCS itself see [4, 10, 11]) is still debated. The interpretation on the basis of GCS evolution is more appealing, because much simpler and not based on qualitative and arbitrary modelizations of GC formation in galaxies (remember the Occam's razor...). Moreover, it has other important astrophysical implications.

Why is it simple? Because it is based just on the conservative assumption that the GCS and the halo-bulge galactic stars are coeval and had initially the same spatial distribution; the presently observed difference can be caused by evolution of the GCS. That GCSs in galaxies undergo evolution is beyond doubt, because they are evolving aggregates of stars moving in an external potential which influences the system also by tidal distortion and by dynamical friction. A detailed analysis of the GCS radial profile evolution in early type galaxies has been presented in [10] where a convincing explanation of the observed comparative features of GCS and stellar light profiles is given.

Some researchers have invoked one observational feature, the GCS radial distribution being shallower for brighter galaxies than for faint [12], as evidence against the 'evolutionary' explanation.

Apart from that the claimed correlation is not universal (for instance, [2] found a quite shallow GCS radial distribution in the Virgo dE VCC 1087), the evolution of a GCS due to the combined role of dynamical friction, acting on the large scale, and nuclear tidal distortion, on a smaller scale, leads to a correlation between the slope of the GCS radial profile and the galaxy integrated luminosity exactly as observed (see Fig. 1, left panel). This is because of the existing correlation between the two scales through the positive

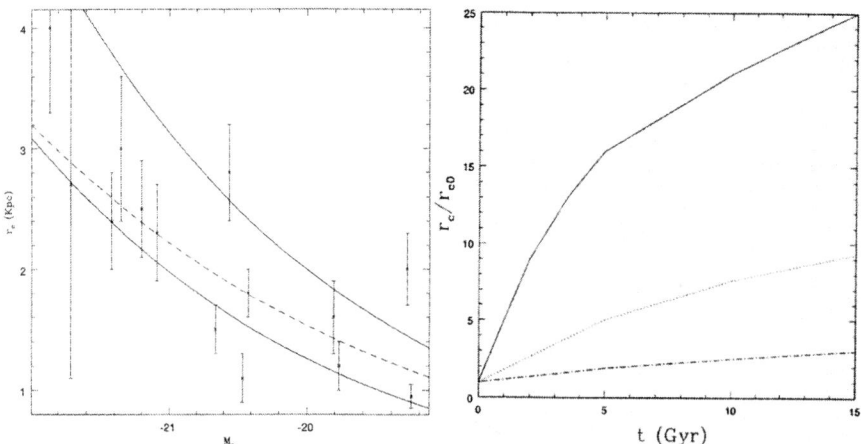

Fig. 1. *Left panel*: GCS core radius as a function of the absolute integrated V mag. of the parent galaxy. *Dots* refer to data in [12], with their best fit as *dashed line*. Solid lines are two evolutionary models, with two different initial value of the GCS core radius (see Fig. 3 in [10]). Right panel: time evolution of the core radius of the GCS in a triaxial galaxy containing a central BH of mass (from bottom up) 10^7, 10^8, 10^9 $M_\odot$ (Fig. 6 in [9]).

galaxy mass-central black hole (BH) mass relation in galaxies. As example of observational output of the large (GCS core radius) and small (BH tidal destruction radius) scale-correlation see right panel of Fig. 1. In conclusion, the GCS slope vs. galaxy luminosity correlation is not, unfortunately, a way to distinguish between the two above mentioned hypotheses (compare left panel of Fig. 1 with Fig. 4 in [3]).

3 Super Star Cluster Formation and Nucleus Accretion

There is growing evidence of the presence of very massive young clusters, as extensively discussed in this Conference, up to the extremely large mass of W3 in NGC 7252 ($M = 8 \pm 2 \times 10^7$ M$_\odot$, [18]). Massive clusters are not an insignificant fraction of the GCs in galaxies; on the contrary, [15] indicates how up to a 40% of the total mass in the GCS of brightest cluster galaxies is contributed by massive GCs (p.d. mass $> 1.5 \times 10^6$ M$_\odot$), in good agreement with recent theoretical results by [17].

The initial presence of massive clusters in a galaxy makes particularly intriguing the GCS evolutionary frame sketched in Sect. 2, for the presence of some massive primordial clusters may have had very important consequences on the initial evolution of the parent galaxy. Actually, the GCS evolution in an elliptical galaxy naturally suggests the following *scenario*:

(i) massive GCs on box orbits (in triaxial galaxies) or on low angular momentum orbits (in axisymmetric galaxies) lose their orbital energy rather quickly;
(ii) after ~ 500 Myr many GCs, sufficiently robust to tidal deformation, are limited to move in the inner galactic region where they merge and form an SSC;
(iii) stars of the SSC buzz around the nucleus where some of them are captured by a BH sitting there, partly increasing the BH mass;
(iv) part of the energy extracted from the SSC gravitational field goes into e.m. radiation inducing a high nuclear luminosity up to AGN levels.

Point (i) has been carefully studied in [22] and [11] in self consistent models of triaxial core-galaxies, and presently under study in triaxial cuspy-galaxies with dark matter halo [8]; the validity of point (ii) has been demonstrated by first results of [4], while the resistance to galactic tidal forces of sufficiently compact GCs confirmed by [19] and the actual formation of an SSC via orbitally decayed cluster merger has been proved by detailed N-body simulations [7, 20]. Points (iii) and (iv) deserve a deeper investigation by mean of accurate modeling, even if they seem reasonably well supported by previous studies [4, 6].

4 Conclusions

Various papers by our research group have shown that many of the observed GCS features find a natural explanation in terms of evolution of a GCS in the galactic field, assuming the (very conservative) hypothesis that it was initially radially distributed as the galactic stellar component and coeval to it. In other words, no *ad hoc* assumptions are needed to explain, for instance, the difference, observed in many galaxies, among the GCS-halo star profiles. The initial presence of some massive GCs ($M \geq 5 \times 10^6$ M$_\odot$) leads to the formation of a central SSC via merger of these orbitally decayed massive clusters. The SSC mixed up with the galactic nucleus in which it was embedded and constituted a mass reservoire to fuel and accrete a massive object therein. Observationally, this latter picture is supported by the observed positive correlation between the estimated quantity of mass lost by a GCS in galaxies and the mass of their central BHs (see Fig. 1 in [5]). On the theoretical side, the modes of mass accretion onto the BH via star capture from the merged SSC still remain to be carefully investigated.

Acknowledgements. Thanks are due to P. Di Matteo, L. Leccese, P. Miocchi, A. Tesseri, and A. Vicari, whose collaboration has been precious in a significant part of the work on the topics discussed here.

References

1. Ashman, K. M., & Zepf, S. E.: *Globular Cluster Systems*, Cambridge astrophysics series, 30 (Cambridge University Press, New York 1998)
2. Beasley, M. A., Strader, J., Brodie, J. P., Cenarro, A. J., & Geha, M.: AJ **131**, 814 (2006)
3. Bekki, K., & Forbes, D. A.: A& A **445**, 485 (2006)
4. Capuzzo Dolcetta, R.: ApJ **415**, 616 (1993)
5. Capuzzo Dolcetta, R.: in *The Central Kiloparsec of Starbursts and AGN: The La Palma Connection*, ASP Conference Proceedings Vol. 249, p. 237, J.H. Knapen et al. eds. (ASP, San Francisco, 2001)
6. Capuzzo-Dolcetta, R.: Ap&SS **294**, 95 (2004)
7. Capuzzo-Dolcetta, R., Di Matteo, P., & Miocchi, P.: in preparation (2006)
8. Capuzzo-Dolcetta, R., Leccese, L., Merritt, D., & Vicari, A.: ApJ, **666**, 165 (2007)
9. Capuzzo-Dolcetta, R., & Tesseri, A.: MNRAS **292**, 808 (1997)
10. Capuzzo-Dolcetta, R., & Tesseri, A.: MNRAS **308**, 961 (1999)
11. Capuzzo-Dolcetta, R., & Vicari, A.: MNRAS **356**, 899 (2005)
12. Forbes, D. A., Franx, M., Illingworth, G. D., & Carollo, C. M.: ApJ **467**, 126 (1996)
13. Grillmair, C. J., Faber, S. M., Lauer, Tod R., et al.: AJ **108**, 102 (1994)
14. Harris, W. E.: AJ **91**, 822 (1986)
15. Harris, W. E., Whitmore, B. C., Karakla, F., et al.: ApJ **636**, 90 (2006)

16. Harris, W. E., & Racine, R.: ARA& A **17**, 241 (1979)
17. Kravtsov, A. V., & Gnedin, O. Y.: ApJ **623**, 650 (2005)
18. Maraston, C., Bastian, N., Saglia, R. P., Kissler-Patig, M., Schweizer, F., & Goudfrooij, P.: A&A **416**, 467 (2004)
19. Miocchi, P., Capuzzo-Dolcetta, R., Di Matteo, P., & Vicari, A.: ApJ, in press, astro-ph/0501618 (2006)
20. Miocchi, P., Capuzzo-Dolcetta, R., Di Matteo, P., & Vicari, A.: ApJ **644**, 940 (2006)
21. Peng, E. W., Jordn, A., Ct, P., et al.: ApJ **639**, 95 (2006)
22. Pesce, E., Capuzzo-Dolcetta, R., & Vietri, M.: MNRAS **254**, 466 (1992)

Star Cluster Evolution: From Young Massive Star Clusters to Old Globulars

Richard de Grijs

Department of Physics & Astronomy, The University of Sheffield, Hicks Building, Hounsfield Road, Sheffield S3 7RH, UK R.deGrijs@sheffield.ac.uk

Abstract. Young, massive star clusters are the most notable and significant end products of violent star-forming episodes triggered by galaxy collisions, mergers, and close encounters. The question remains, however, whether or not at least a fraction of the compact YMCs seen in abundance in extragalactic starbursts, are potentially the progenitors of globular cluster (GC)-type objects. However, because of the lack of a statistically significant sample of similar nearby objects we need to resort to either statistical arguments or to the painstaking approach of case by case studies of individual objects in more distant galaxies. Despite the difficulties inherent to addressing this issue conclusively, an ever increasing body of observational evidence lends support to the scenario that GCs, which were once thought to be the oldest building blocks of galaxies, are still forming today.

1 Young Massive Star Clusters as Proto-Globular Clusters

The production of luminous, massive yet compact star clusters (YMCs; often with masses $m_{\rm cl} \geq 10^5 {\rm M}_\odot$) seems to be a hallmark of the most intense starbursts. YMCs are therefore important as benchmarks of cluster formation and evolution. They are also important as tracers of the (stellar) initial mass function (IMF) and other physical characteristics in starbursts. The key properties of YMCs have been explored in starburst regions in several dozen galaxies, both in normal spirals and in gravitationally interacting galaxies.

The question remains, however, whether or not at least some of the YMCs observed in extragalactic starbursts might survive for a Hubble time. If we could settle this issue convincingly, one way or the other, the implications would be far-reaching for a wide range of astrophysical questions, including (but not limited to) our understanding of the process of galaxy formation, assembly and evolution, and the process and conditions required for star and star cluster formation.

2 Resolution of the Evolutionary Question

The evolution of young clusters depends crucially on their stellar IMF: if the IMF is too shallow, i.e., if the clusters are significantly depleted in low-mass stars compared to, e.g., the solar neighbourhood, they will likely disperse within about a Gyr of their formation (e.g., [10,11,16,15]). At present, there are two principal approaches in which one can attempt to address the underlying IMFs of extragalactic YMCs [but see also de Grijs et al. [6] for an alternative approach].

2.1 The Cluster Luminosity Function: The Case of M82

In de Grijs et al. [3, 4] we reported the discovery of an approximately lognormal cluster luminosity and mass function (CLF, CMF) for the roughly coeval star clusters at the intermediate age of ~ 1 Gyr in M82's fossil starburst region "B". This provided the first deep CLF (CMF) for a star cluster population at intermediate age, which thus serves as an important benchmark for theories of the evolution of star cluster systems [see also Goudfrooij et al. [12] for a related important result for NGC 1316, at ~ 3 Gyr].

A substantial series of papers on the young Large Magellanic Cloud cluster system (with ages $\leq 2 \times 10^9$ yr), starting with the seminal work by Elson and Fall [9], seem to imply that the CLF of YMCs is well described by a power law [but see de Grijs and Anders [8] for caveats]. On the other hand, for old globular cluster (GC) systems with ages $\geq 10^{10}$ yr, the CLF shape is well established to be roughly log-normal, and almost universal among local galaxies. This type of observational evidence has led to the popular, but thus far mostly speculative theoretical prediction that not only a powerlaw, but *any* initial CLF (CMF) will be rapidly transformed into a log-normal distribution because of (i) stellar evolutionary fading of the lowest-luminosity (mass) objects to below the detection limit; and (ii) disruption of the low-mass clusters due both to interactions with the gravitational field of the host galaxy, and to internal two-body relaxation effects leading to enhanced cluster evaporation.

From our detailed analysis of the expected evolution of CMFs starting from initial log-normal and initial power-law distributions [7], we conclude that our observations of the M82 B CMF are inconsistent with a scenario in which the 1 Gyr-old cluster population originated from an initial power-law mass distribution. This applies to a large range of "characteristic" cluster disruption time-scales. Our conclusion is supported by arguments related to the initial density in M82 B, which would be unphysically high if the present cluster population were the remains of an initial power-law distribution (particularly in view of the effects of cluster "infant mortality", which require large excesses of low-mass unbound clusters to be present at the earliest times).

In de Grijs et al. [5] we showed that the CMFs of YMCs in many different environments are well approximated by power laws with slopes $\alpha \simeq -2$. However, except for the intermediate-age cluster systems in M82 B and NGC 1316 [12], the *expected* turn-over (or peak) mass (based on comparisons with present-day GC systems and taking evolutionary fading into account) in most YMC systems observed to date occurs close to or below the observational detection limit, simply because of their greater distances and shallower observations. As such, these results are not necessarily at odds with each other, but merely hindered by observational selection effects.

2.2 High-Resolution Spectroscopy: Individual Cluster Analysis

With the ever increasing number of large-aperture ground-based telescopes equipped with state-of-the-art high-resolution spectrographs and the wealth of observational data provided by the *Hubble Space Telescope*, we may now finally be getting close to resolving the potentially far-reaching issue of YMC-to-GC evolution conclusively. To do so, one needs to obtain (i) high-resolution spectroscopy, in order to obtain dynamical mass estimates, and (ii) high-resolution imaging to measure their sizes (and luminosities). As a simple first approach, one could then construct diagnostic diagrams of YMC mass-to-light ratio vs. age, and compare the YMC locations in this diagram with models of "simple stellar populations" (SSPs) using a variety of IMF descriptions (cf. [16, 15, 1]). However, such an approach, while instructive, has serious shortcomings:

(i) In this simple approach, the data can be described by *both* variations in the IMF slope *and* variations in a possible low-mass cut-off; the models are fundamentally degenerate for these parameters.
(ii) While the assumption that these objects are approximately in virial equilibrium is probably justified at ages greater than a few $\times 10^7$ yr (at least for the stars dominating the light), the *central* velocity dispersion (as derived from luminosity-weighted high-resolution spectroscopy) does not necessarily represent a YMC's total mass. It is now well-established that almost every YMC exhibits significant mass segregation from very early on, so that the effects of mass segregation must be taken into account when converting central velocity dispersions into dynamical mass estimates (see also [13] J.J. Fleck et al., in prep.).
(iii) With the exception of a few studies (e.g., M82-F; [16]), the majority of YMCs thus far analysed in this way have ages around 10 Myr. Around this age, however, red supergiants (RSGs) appear in realistic stellar populations. Unfortunately, the model descriptions of the RSG phase differ significantly among the various leading groups producing theoretical stellar population synthesis codes (Padova vs. Geneva vs. Yale), and therefore the uncertainties in the evolutionary tracks are substantial.

3 The Current Verdict?

It may appear that a fair fraction of the ~ 10 Myr-old YMCs that have been analysed thus far may be characterised by unusual IMFs, since their loci in the diagnostic diagram are far removed from any of the "standard" SSP models (see, e.g., [1]). However, Bastian and Goodwin [2] recently showed that this is most likely an effect of the fact that the velocity dispersions of these young objects do not adequately trace their masses. They are instead strongly affected by the effects of gas expulsion due to supernova activity and massive stellar winds. In this respect, it is encouraging to see that the older clusters (i.e., older than M82-F, a few $\times 10^7$ yr) seem to conform to "normal" IMFs; by those ages, the clusters' velocity dispersions seem to represent the underlying gravitational potential much more closely.

We recently reported the discovery of a extremely massive, but old (12.4 ± 3.2 Gyr) GC in M31, 037-B327, that has all the characteristics of having been an exemplary YMC at earlier times [14]. In order to have survived for a Hubble time, we conclude that its stellar IMF cannot have been top-heavy, i.e., characterized by a low-mass cut-off at $m_\star \geq 1$ M$_\odot$, as sometimes advocated for current YMCs (e.g., [16]). Using this constraint, and a variety of SSP models, we determine a photometric mass for 037-B327 of $M_{\rm GC} = (3.0 \pm 0.5) \times 10^7$ M$_\odot$, somewhat depending on the SSP models used, the metallicity and age adopted and the IMF representation. In view of the large number of free parameters, the uncertainty in our photometric mass estimate is surprisingly small. This mass, and its relatively small uncertainties, make this object the most massive star cluster of any age in the Local Group. As a surviving "super" star cluster, this object is of prime importance for theories aimed at describing massive star cluster evolution.

References

1. Bastian N., Saglia R.P., Goudfrooij P., Kissler-Patig M., Maraston C., Schweizer F., Zoccali M., 2006, A&A, 448, 881 (astro-ph/0511033)
2. Bastian N., Goodwin S.P., 2006, MNRAS, 369, L9 (astro-ph/0602465)
3. de Grijs R., Bastian N., Lamers H.J.G.L.M., 2003a, ApJ, 583, L17
4. de Grijs R., Bastian N., Lamers H.J.G.L.M., 2003b, MNRAS, 340, 197
5. de Grijs R., Anders P., Lynds R., Bastian N., Lamers H.J.G.L.M., O'Neill E.J., Jr., 2003c, MNRAS, 343, 1285
6. de Grijs R., Wilkinson M.I., Tadhunter C.N., 2005a, MNRAS, 361, 311
7. de Grijs R., Parmentier G., Lamers H.J.G.L.M., 2005b, MNRAS, 364, 1054
8. de Grijs R., Anders P., 2006, MNRAS, 366, 295
9. Elson R.A.W., Fall S.M., 1985, PASP, 97, 692
10. Gnedin O.Y., Ostriker J.P., 1997, ApJ, 474, 223
11. Goodwin S.P., 1997, MNRAS, 286, 669
12. Goudfrooij P., Gilmore D., Whitmore B.C., Schweizer F., 2004, ApJ, 613, L121
13. Lamers H.J.G.L.M., Anders P., de Grijs R., 2006, A&A, 452, 131
14. Ma J., de Grijs R., Yang Y., Zhou X., Chen J., Jiang Z., Wu Z., Wu J., 2006, MNRAS, 368, 1443
15. Mengel S., Lehnert M.D., Thatte N., Genzel R., 2002, A&A, 383, 137
16. Smith L.J., Gallagher J.S., 2001, MNRAS, 326, 1027

A Wide-Field Survey of the Globular Cluster Systems of Giant Galaxies

Katherine L. Rhode

Astronomy Department, Wesleyan University, Middletown, CT 06459, USA
kathy@astro.wesleyan.edu

Abstract. I present selected results from a wide-field CCD survey of the GC systems of giant galaxies, including showing how measurements of the specific frequency of metal-poor GCs can constrain the redshift of their formation.

1 Introduction to the Survey

1.1 Motivation and Design

Measurements of the ensemble properties of the globular cluster (GC) systems of massive galaxies can provide a direct test of theories for how such galaxies form. Making these global measurements requires observations of all or most of a galaxy's GC system, which in turn requires **wide-field imaging**. In recent years Steve Zepf (Michigan State) and I have been collaborating on a wide-field CCD survey of the GC systems of giant galaxies. Some motivations for the survey are that: the total numbers of GCs in giant galaxies are uncertain; few spiral galaxies have been studied; and the outer regions of GC systems are largely unexplored, at least by modern CCD studies.

To date we have BVR or BVI imaging of four E/S0 galaxies from the Kitt Peak 4m telescope and Mosaic Imager, and nine spiral galaxies from large-format CCDs on the WIYN 3.5 m telescope. At the distances of the targets (10–20 Mpc), the GCs are unresolved so we use three-color photometry and good image resolution to select them and to exclude stars and background galaxies. Our aim is to derive reliable global properties of the GC systems and to use them to study galaxy formation and test models for the origin of ellipticals, such as spiral mergers [1], multi-phase collapse [6], and collapse with accretion [5].

1.2 Data and Analysis Methods

The spiral galaxy images cover $7' \times 7'$ or $10' \times 10'$, depending on the WIYN detector used. This translates to radial coverage of 30–40 kpc. Early-type galaxies often have extended GC systems, so the images of the E/S0s are $38' \times 38'$ and provide radial coverage of 60–120 kpc. To find GCs, we create a deep, stacked image in each filter; remove the diffuse galaxy light; find sources above a chosen S/N level; discard extended objects; and select point sources with BVR or BVI magnitudes and colors like GCs. We also run completeness tests and fit the GCLF to determine what fraction of GCs are missing given our detection limits. We use Galactic star count models and archival HST data to quantify the contamination that remains from stars and galaxies.

2 Selected Results

The survey has so far yielded positions and BVR photometry of hundreds to thousands of GC candidates in eight galaxies. In all cases we have imaged the full radial extent of the systems. The derived properties—e.g., number of GCs, spatial and color distributions, color gradient—are thus global ones and provide an important comparison to model predictions. We find that all the models mentioned in Sect. 1.1 have inconsistencies with the data (see [9]). Below I highlight two general results from the survey.

2.1 Total Numbers and Specific Frequencies

We calculate the total number of GCs (N_{GC}) in each galaxy by integrating the derived radial distribution of the GC system from the galaxy center to the radius at which the GC surface density goes to zero within the errors. Table 1 gives N_{GC} and specific frequency (S_N from [8]) for the eight galaxies analyzed. Columns (1)–(3) are galaxy name, type, and magnitude, columns (4)–(5) are N_{GC} and S_N from the survey, and columns (6)–(7) list, when applicable, S_N from past work and a reference. (S_N from past work combines N_{GC} from the study with the M_V we assumed.) Our S_N values are 20–75% lower for four of the six galaxies in Table 1 that were studied previously, and our errors on S_N are 2–4 times smaller for all six galaxies. The smaller values are due at least in part to reduced contamination levels. Also, observing the full extent of the GC systems yields smaller, better-constrained total numbers than observing the inner GC system and extrapolating to an arbitrary outer radius, as is typical in past CCD work.

Table 1. Total number and S_N for eight survey galaxies

Galaxy	Type	M_V	N_{GC}	S_N	Previous S_N	References
NGC 4472	E2	−23.1	5900	3.6±0.6	4.5±1.3	[7]
NGC 4406	E3	−22.3	2900	3.5±0.5	4.6±1.1	[7]
NGC 4594	S0	−22.4	1900	2.1±0.3	2±1	[4]
NGC 3379	E1	−20.9	270	1.2±0.3	1.1±0.6	[7]
NGC 7814	Sab	−20.4	170	1.3±0.4	5.2±1.7	[3]
NGC 3556	Sc	−21.2	290	0.9±0.4	...	...
NGC 2683	Sb	−20.5	120	0.8±0.4	2.0±0.7	[12]
NGC 4157	Sb	−20.4	80	0.6±0.3	...	...

2.2 Mass-Normalized Number of Blue (Metal-Poor) GCs

Ashman & Zepf [1] proposed that elliptical galaxies form in spiral galaxy mergers and that the metal-poor GCs in ellipticals come directly from the progenitor spirals. If this is true, the galaxy-mass-normalized number of metal-poor GCs (called $T_{\rm blue}$) should be similar for spiral and elliptical galaxies. The well-determined global values from the survey allow us to test this prediction. Figure 1 shows $T_{\rm blue}$ vs. log of the stellar mass for eight survey galaxies, plus the Milky Way, M31, and three galaxies from the literature (see [10] for a list). $T_{\rm blue}$ is defined as $N_{GC}({\rm blue})/(M_G/10^9~{\rm M}_\odot)$. $T_{\rm blue}$ for the spiral galaxies is too small to account for the blue GC populations of the giant cluster ellipticals, implying that the merger model cannot explain their formation. The data also show a trend of increasing $T_{\rm blue}$ with increasing galaxy mass. This is generally consistent with a galaxy and GC system formation scenario described by Santos [11] that combines biased structure formation with hierarchical merging. In this model, metal-poor GCs form over a finite period in the early Universe, when gas-rich protogalactic fragments merge into larger structures. Structure formation is temporarily suppressed at high-z, perhaps due to cosmic reionization. Meanwhile, stellar evolution enriches the intergalactic medium, so any GCs formed later are metal-rich compared to the first generation of GCs. Today's massive galaxies in high-density regions of the Universe began assembling first, so formed relatively more metal-poor GCs during the initial formation epoch.

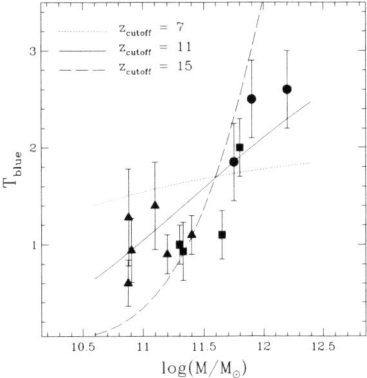

Fig. 1. $T_{\rm blue}$ vs. log of the galaxy mass for 8 survey galaxies and 5 from the literature. Circles are cluster elliptical galaxies, squares are field E/S0s, and triangles are field spirals. The curves show the expected trend if metal-poor GCs form prior to z of 7, 11, or 15 (details in Sect. 2.2).

Figure 1 shows the expected trend if the formation cutoff for the first generation of GCs occurred at $z = 7$, 11, or 15. The curves come from an extended Press-Schechter calculation by G. Bryan (Columbia) that assumes that T_{blue} is proportional to the fraction of a galaxy's mass that has collapsed into halos of at least 10^8 $M_\odot$ by the truncation redshift. Although a biased, hierarchical scenario appears generally consistent with our T_{blue} data, more modeling and data are needed to determine whether it can reproduce all of the properties of the galaxy GC systems we observe.

3 Ongoing and Future Work

Imaging. We are analyzing WIYN data of more spiral galaxies, have begun using an 8K×8K mosaic CCD on the MDM-2.4m telescope to image more targets, and also plan to use the WIYN One Degree Imager, which will be commissioned in 2009 and provides $0.5''$ resolution over a 1° field.

Follow-Up Spectroscopy. With our collaborators, we are using telescopes like the AAT, VLT, Keck, and WIYN to obtain spectra and derive radial velocities of the GC candidates, in order to study the kinematics of the GC systems and measure the mass profiles of the host galaxies to $10-15$ R_{eff}. Our analysis of the mass distribution of NGC 3379 is published in Bergond et al. [2].

New Model Predictions. The ideas about blue GCs and biasing in Sect. 2.2 were introduced in Santos [11] and based on models of the formation of the Milky Way and its GC system. M. Santos (STScI) is modeling the formation of galaxies of varying masses and environments and we are working with him to develop meaningful comparisons between our data and the simulations.

Acknowledgements. The author is supported by an NSF Astronomy & Astrophysics Postdoctoral Fellowship under award AST-0302095.

References

1. Ashman, K.M., & Zepf, S.E. 1992, ApJ, 384, 50
2. Bergond, G., et al. 2006, A&A, 448, 155
3. Bothun, G.D., Harris, H.C., & Hesser, J.E. 1992, PASP, 104, 1220
4. Bridges, T.J., & Hanes, D.A. 1992, AJ, 103, 800
5. Côté, P., Marzke, R.O., & West, M.J. 1998, ApJ, 501, 554
6. Forbes, D.A., Brodie, J.P., & Grillmair, C.J. 1997, AJ, 113, 1652
7. Harris, W.E. 1991, ARAA, 29, 543
8. Harris, W.E., & van den Bergh, S. 1981, AJ, 86, 1627
9. Rhode, K.L., & Zepf, S.E. 2004, AJ, 127, 302
10. Rhode, K.L., Zepf, S.E., & Santos, M.R. 2005, ApJ, 630, L21
11. Santos, M.R. 2003, in Extragalactic Globular Cluster Systems, ed. M. Kissler-Patig (New York: Springer-Verlag), p. 348
12. Harris, H.C., et al. 1985, AJ, 90, 2495

IGCs in the Virgo Cluster

Marianne Takamiya[1], Michael West[1,2], Patrick Côté[3], Andrés Jordán[4], Eric Peng[3], Laura Ferrarese[3], and the ACS VCS Team

[1] University of Hawaii Hilo 640 N. A'ohoku Place, Hilo, HI 96720
   takamiya@hawaii.edu
[2] Gemini Observatory, La Serena, Chile mwest@gemini.edu
[3] HIA, Victoria, Canada Patrick.Cote@nrc-cnrc.gc.ca,
   Eric.Peng@nrc-cnrc.gc.ca, Laura.Ferrarese@nrc-cnrc.gc.ca
[4] ESO, Garching, Germany ajordan@eso.org

Abstract. Preliminary results of the ACS Virgo Cluster Survey WFPC2 parallel observations are presented. We searched for intergalactic globular clusters in 100 WFPC2 fields within the Virgo Cluster. We find globular clusters preferentially near the most luminous cluster galaxies (M87, M49, and M60). For M87, the number density of globular clusters out to 70 kpc is consistent with that determined by ground based studies of the globular cluster systems. The intergalactic nature of the population of globular clusters we are detecting is still a matter of controversy.

1 Introduction

The existence of material residing between galaxies was first pointed out by Zwicky [21] in his seminal study of the Coma Cluster of galaxies. He drew attention to the fact that the boundaries of galaxies are ill-defined and that light spills into the space between Coma cluster galaxies. Numerous studies have since confirmed the existence of diffuse intergalactic or intracluster light in different clusters: Virgo [14], Coma [19, 16, 7, 4, 1], Fornax [15] Centaurus [4], Abell 2029 [17], Abell 3888 [12], Abell 2390 and CL 1613+31 [18] clusters, among others. Recently, deep optical imaging down to $\mu_V = 28.5$ mag/arcsec^2 (1σ) of the center of the Virgo Cluster has revealed a myriad of tidal tails and streams [14], forming a complex array which suggests that the cluster is still evolving and that the creation of intracluster components is still ongoing. The nature of the intracluster light has not been well established but there is ample evidence suggesting that the main origin of the intergalactic component is stellar (see contribution of West in these proceedings).

2 Parallel Observations of the ACSVCS

The ACS Virgo Cluster Survey program (GO-9401, hereafter ACSVCS) described in detail by Côté et al. [5] contained a parallel program using the WFPC2 instrument to survey the *field* in the Virgo Cluster in search for intergalactic globular clusters (hereafter IGC). In total 100 WFPC2 fields in F814W and F606W were observed; each WFPC2 field was positioned ∼5–6 arcmin (24–29 kpc) from the main ACS target galaxies. A total area of $23\rlap{.}''8 \times 23\rlap{.}''8$, approximately $116 \times 116\,\mathrm{kpc}^2$ is covered by the WFPC2 observations. The 10σ PSF limiting magnitudes are V=25.8 and I=24.5.

3 Selection of Candidates

We select candidate globular clusters based on magnitude, color, and, taking advantage of the superb image quality of the WFPC2 data, image morphology. The GC luminosity function (GCLF) peaks at V=23.7 at the distance of Virgo which is well within the 10σ point-source sensitivity of the WFPC2 data. We choose GC candidates having magnitudes in the range 19.6 to 25.2 in V, 19.6 to 24.2 in I, and colors between $0.7 < $ V-I $ < 1.4$. We select all sources with light profiles well fit by a King model [11, 10] that have half-light radii between 2 and 10 pc. We find a few sources that have similar magnitudes and colors but have larger half-light radii in the range 10 to 150 pc. These brighter GCs are likely to be the dwarf-globular cluster transition objects (DGTO) of Haşegan et al. [8] or ultra-compact dwarf galaxies (UCD) of Drinkwater et al. [6].

Properties of DGTO/UCDs: We find 7 candidate DGTOs, one of which is a confirmed member of the cluster. The absolute magnitudes of these sources range from –10.6 to –13.6 in M_V if at the distance of Virgo. Their V-I colors are typically 1.3. The location of the DGTOs within the cluster are presented in Fig. 1 (left panel). The DGTOs lie close to the most luminous galaxies (M87 and M49) and also along the so-called principal axis of the Virgo Cluster [20].

Properties of GCs: There are 947 GC candidates in the *field* of Virgo. The GCs are predominantly close to the brightest galaxies as seen in Fig. 1 (right panel). In the case of M87, the number density profile of GCs detected between 30 and 70 kpc ($\sim$ 400–850 arcsec) from the center of M87 in the WFPC2 fields are consistent with the GC profile of McLaughlin [13] (see Fig. 2).

4 Conclusions

The detection of globular clusters in the field of Virgo out to 2–3 Mpc from the center of the cluster are presented. The GCs are predominantly found near the most luminous galaxies of the cluster. Whether these are bona-fide intergalactic globular clusters or members of M87 is still open to debate. An excess of GCs of 0.2 arcmin^{-2} as detected in the wide-field survey by Naoyuki Tamura (see contribution in this proceedings) would be hard to distinguish in the log-log plot of the number density profiles (Fig. 2). However the fact that the density and distribution of the GCs appears to be tied more to the

IGCs in the Virgo Cluster 363

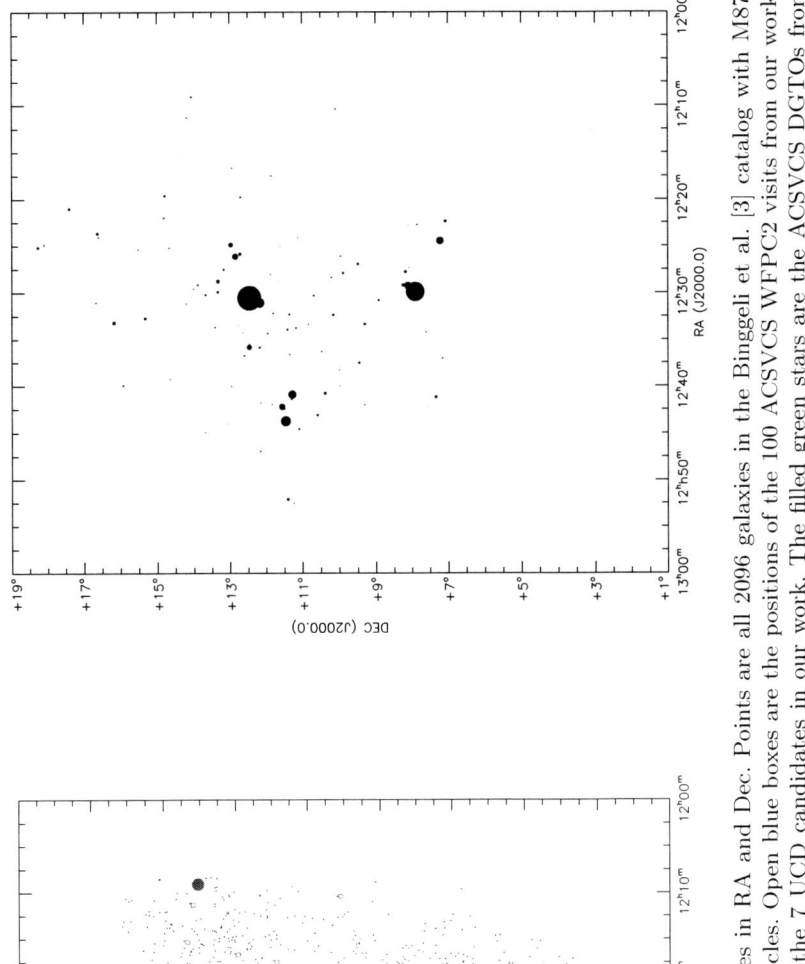

Fig. 1. (*Left*) Location of UCD candidates in RA and Dec. Points are all 2096 galaxies in the Binggeli et al. [3] catalog with M87, M49, and M60 identified as large open circles. Open blue boxes are the positions of the 100 ACSVCS WFPC2 visits from our work. The filled red circles are the positions of the 7 UCD candidates in our work. The filled green stars are the ACSVCS DGTOs from Hasegan et al. citehasegan05. The magenta stars near M87 are the UCDs from Jones et al. [9]. (*Right*) Location of GC candidates in RA and Dec. Each point is proportional to the number of GCs candidates detected in the WFPC2 data.

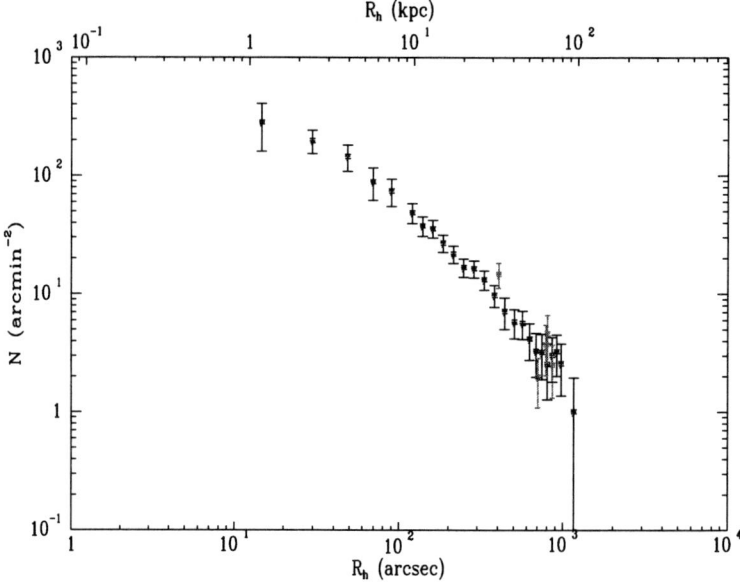

Fig. 2. Number density of globular clusters of M87. Black dots are from McLaughlin [13] and references therein. The red dots are the WFPC2 counts from our study.

galaxies than to the cluster itself (Fig. 1), probably suggests that a large percentage of the IGCs have already been captured by the cluster galaxies and the small number of IGCs detected could reflect either a lack of recent significant merging events or a recent merger event the debris of which was rapidly assimilated by M87. The number of UCDs we find is also small compared to all-field searches of UCDs in Fornax (e.g. [2]). While the difference in the number of sources between Fornax and Virgo may reflect different selection criteria, more interestingly, it could indicate a difference in the evolutionary stages between these two clusters. An added complication is the very existence of such a massive galaxy as M87 in the center of Virgo; there is no M87 counterpart in Fornax. M87 dominates the cluster's potential and its role may be such that most of the intergalactic material is doomed to fall onto M87 in a short time scale. To substantiate either one of these possibilities, these types of studies would have to be extended to a large number of clusters (e.g. beginning with Coma) and obtain high quality imaging and spectra of GCs to be able to precisely characterise the different GC populations (e.g. age, dynamics).

References

1. C. Adami, A. Biviano, F. Durret, A. Mazure: A&A **443**, 17 (2005)

2. L.P. Bassino, F.R. Faifer, J.C. Forte, B. Dirsch, T. Richtler et al.: A&A **451**, 789 (2006)
3. B. Binggeli, A. Sandage, G.A. Tammann: AJ **90**, 1759 (1985)
4. C. Calcáneo-Roldán, B. Moore, J. Bland-Hawthorn, D. Malin, E.M. Sadler: MNRAS **314**, 324 (2000)
5. P. Côté, J.P. Blakeslee, L. Ferrarese, A. Jordán,S. Mei, et al.: ApJS **153**, 223 (2004)
6. M.J. Drinkwater, M.D. Gregg, M. Hilker, K. Bekki, W.J. Couch, et al. : Nature **423**, 519 (2003)
7. M.D. Gregg, M.J. West: Nature **396**, 549 (1998)
8. M. Haşegan, A. Jordán, P. Côté, S.G. Djorgovski, D.E. McLaughlin et al.: ApJ **627**, 203 (2005)
9. J.B. Jones, M.J. Drinkwater, R. Jurek, S. Phillipps, M.D. Gregg, et al.: AJ **131**, 312 (2005)
10. A. Jordán, P. Côté, J.P. Blakeslee, L. Ferrarese, D.E. McLaughlin et al.: ApJ **634**, 1002 (2005)
11. I.R. King: AJ **71**, 64 (1966)
12. J.E. Krick, R.A. Bernstein, K.A. Pimbblet: AJ **131**, 168 (2006)
13. D.E. McLaughlin: AJ **109**, 2034 (1995)
14. J.C. Mihos, P. Harding, J. Feldmeier, H. Morrison: ApJ **631**, L41 (2005)
15. T.Theuns, S.J. Warren: MNRAS **284**, L11 (1997)
16. T.X. Thuan, J. Kormendy: PASP **89**, 466 (1977)
17. J.M. Uson, S.P. Boughn, J.R. Kuhn: ApJ **369**, 46 (1991)
18. R. Vílchez-Gómez, R. Pelló, N. Sanahuja: ApJ **283**, 37 (1994)
19. G.A. Welch, G.N. Sastry: ApJ **169**, L3 (1971)
20. M.J. West, J.P. Blakeslee: ApJ **543**, L27 (2000)
21. F. Zwicky: PASP **63**, 61 (1951)

A New Explanation of Globular Cluster Color Distributions

Suk-Jin Yoon[1,2,3], Sukyoung Ken Yi[1,2], and Young-Wook Lee[1]

[1] Department of Astronomy & Center for Space Astrophysics, Yonsei U., Seoul, Korea
[2] Astrophysics, University of Oxford, Keble Road, Oxford OX1 3RH, UK
[3] sjyoon@galaxy.yonsei.ac.kr

Abstract. The colors of globular clusters (GCs) in most large early-type galaxies are bimodal. This is generally taken as evidence for the presence of two GC subpopulations that have different geneses. However we show that, because of an inflection along the metallicity-color relation, a coeval group of old (>10 Gyr) GCs can exhibit bimodal color distributions *even when the intrinsic metallicity distribution is unimodal*. The inflection is a result of two complementary effects: (i) the integrated color of main-sequence and giant-branch is a mild non-linear function of metallicity, and (ii) the rapid change in color due to the onset of the hot horizontal-branch further strengthens the non-linearity. We demonstrate that the hypothesis gives remarkably simple and cohesive explanations for all the key observations, including the close link of the GC color distributions to the host galaxy properties.

1 Introduction

One of the most outstanding discoveries from observations of early-type galaxies over the last decade is the bimodal color distribution of globular clusters (GCs) [1, 2]. This phenomenon has received profuse attention because of its broad implication to galaxy formation and widely interpreted as evidence of two GC sub-systems with distinct geneses within individual galaxies. We however propose a new solution that does not necessarily invoke distinct GC sub-systems and has a sound basis both on the empirical and theoretical relations between metallicity and colors of GCs [3].

2 The Nonlinear Metallicity-to-Color Transformations

The advent of the high-resolution observations by the HST/ACS has enabled progress in testing GC stellar population models. A recent observation [4] reveals that the [Fe/H]–$(g-z)$ relation of the GCs in our Galaxy and two giant ellipticals (M49 and M87) shows a significant departure from linearity (Fig. 1A). A closer inspection suggests that the GCs follow an inverted S-shaped "wavy" curve with a possible quasi-inflection point at [Fe/H] ≈ -0.8. Our population model predictions are highly consistent with the observation, suggesting that the wavy feature is real. The observed wavy feature in the

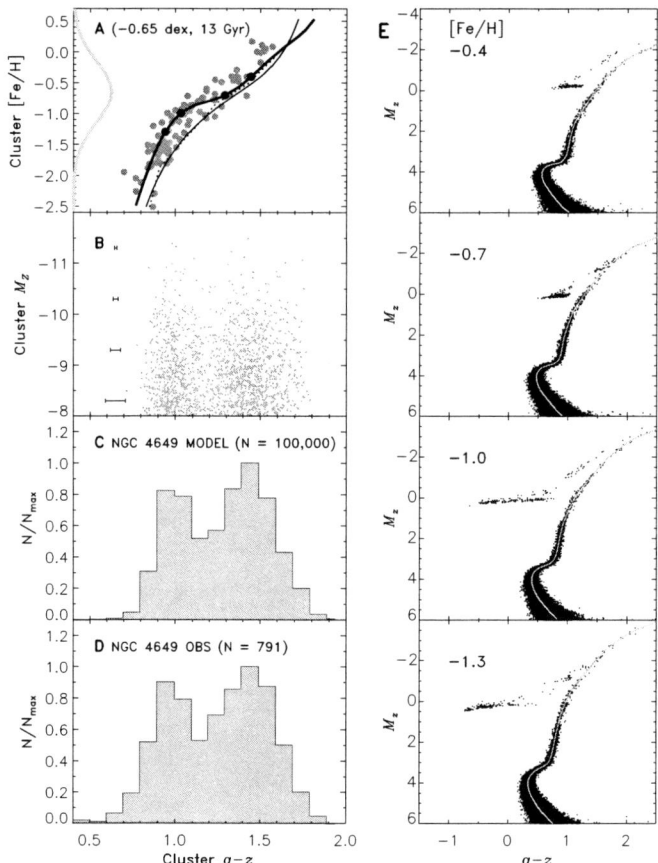

Fig. 1. (**A**) Correlation between [Fe/H] and $(g-z)$ for GCs in our Galaxy, M49, and M87 [4]. Our 13-Gyr simple stellar population model (*thick solid line*) is overlaid. Circles along the line denote the points for which the color-magnitude diagrams are shown in (E). *Thin solid line* is for our model without inclusion of the HB variation, whereas *thin dotted line* represents the prediction from [5]. The metallicity distribution of 10^5 model GCs is shown along the y-axis (thick gray line), and their mean metallicity and age are denoted in parenthesis. (**B**) Color-magnitude diagram of 2000 randomly selected model GCs of 13 Gyr. Cluster $(g-z)$ is transformed from [Fe/H] by using the theoretical relation shown in (**A**). (**C**) The color histogram of 10^5 model GCs of 13 Gyr. (**D**) The observed color histogram for 791 GCs in NGC 4649 [4]. (**E**) Synthetic color-magnitude diagrams for 13-Gyr GCs with various [Fe/H] values. Gray loci are the model isochrones.

metallicity-color relation is a result of two complementary effects: (i) the integrated color of the stars prior to the HB stage (i.e., the main-sequence and red-giant-branch) is a non-linear function of metallicity at given ages, showing a mild departure from linearity at lower metallicity, and (ii) the color of the

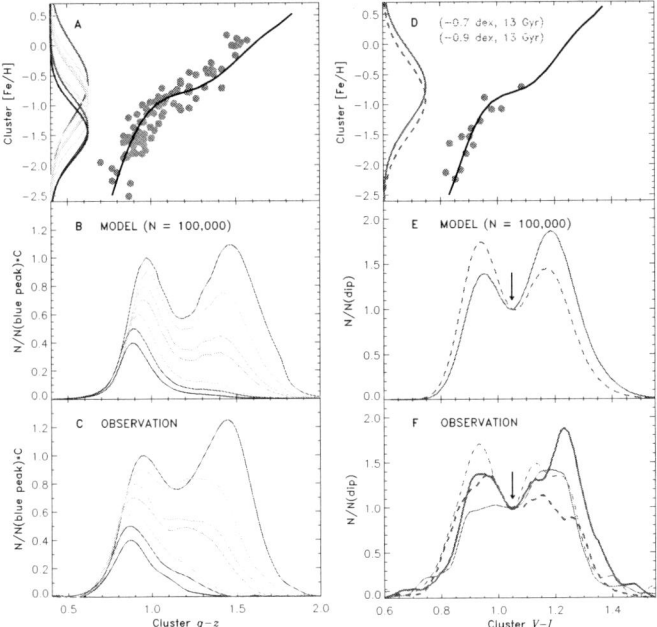

Fig. 2. (*Left*) Monte Carlo simulations of GC color distributions for various host galaxy luminosities. (**A**) Various metallicity distributions of 10^5 model GCs for seven bins of host galaxy magnitude are shown along the y-axis. The values of $\langle[\text{Fe/H}]\rangle$ are adopted from [4]. (**B**) Model GC color histograms. The histograms are multiplied by constants, C, for clarity. The magnitude bins are 1 magnitude wide and extend from $M_B = -21$ (top, $C = 1.0$) to -15 (*bottom*, $C = 0.4$). (**C**) Observed GC color histograms [4]. (*Right*) Monte Carlo simulations for GC color distributions for different radial bins. (**D**) Circles are for the Galactic GCs [9]. Thick solid line represents our theoretical [Fe/H]–$(V - I)$ relation for 13-Gyr GCs. Two metallicity distributions used in the simulation are shown along the y-axis. (**E**) Simulated color distributions, scaled to 1.0 at the dips (*arrows*). Solid and dashed lines correspond to the solid and dashed metallicity distribution in (A). (**F**) The observed color distributions for NGC 4649 (*thick lines*) [10] and for M87 (*thin lines*) [11]. For NGC 4649, the original $(g - i)$ color has been transformed to $(V - I)$ and solid (*dashed*) line is for the radial bin of $\sim$90 arcsec (90$\sim$200 arcsec). For M87, solid (*dashed*) line represents the radial bin inside (outside) of $\sim$60 arcsec.

horizontal-branch changes at a brisk pace between [Fe/H] ≈ -0.6 and -0.9 (Fig. 1E), further strengthening the departure from linearity [6–8]. A polynomial fit to our theoretical [Fe/H]–$(g - z)$ relation is given by

$$[\text{Fe/H}] = 104.02 - 624.33(g-z) + 1414.65(g-z)^2 - 1613.08(g-z)^3$$
$$+ 991.20(g-z)^4 - 313.74(g-z)^5 + 40.20(g-z)^6. \quad (1)$$

3 A New Explanation of GC Color Distributions

This non-linear nature of the relation between intrinsic metallicity and its proxy, colors, may hold the key to understanding the color bimodality phenomenon: the wavy feature brings about the bimodality by projecting equidistant metallicity intervals near the quasi-inflection point onto larger color intervals. We have performed a Monte Carlo simulation using 100,000 coeval GCs (Fig. 1B–D). In the color-magnitude diagram of 2,000 randomly selected GCs (Fig. 1B), two vertical bands of GCs are immediately visible at $(g-z) \approx$ 0.9 and 1.4. The resultant color histogram of 100,000 GCs (Fig. 1C) clearly show a prominent dip near its center, successfully reproducing the observed histogram of NGC 4649 (Fig. 1D).

There is growing evidence that the color distributions of GC systems are closely linked to the host galaxy luminosity [1,2]. The number fraction of red GCs and the mean colors of both blue and red GCs increase progressively for more luminous galaxies. To simulate the color distributions as a function of host galaxy B-band luminosity (M_B), we adopted the mean GC metallicity ($\langle[Fe/H]\rangle$) of different M_B from the observation [4] (Fig. 2A). The resulting color histograms (Fig. 2B) show that, in proceeding to lower luminosities, the distributions are more consistent with being predominantly blue-peaked. This is in good accordance with the observation [4] (Fig. 2C). It is likely that that the projection effect and the $\langle[Fe/H]\rangle$ variation are the main drivers behind the link of the GC color distributions to the host galaxy luminosity.

It is now well established that red GCs are more centrally concentrated and of lower velocity dispersion within individual galaxies [1,2]. Given the fact that observations indicate higher $\langle[Fe/H]\rangle$ toward the galaxy center [11,12], the projection effect naturally causes the red GCs to be more popular toward the galaxy center (Fig. 2D–F). Besides, kinematic studies indicate that the systematically lower velocity dispersion of red GCs is likely a consequence of the red GC number density that is higher in the galaxy center [13]. Thus, it appears that the differences between blue and red GCs in spatial distribution and kinematics are also consistent with our projection effect explanation.

4 Conclusion

It has been a popular view that the presence of two discrete GC subsystems is responsible for the color bimodality. With true metallicity being bimodal, color bimodality would be strengthened further. But the essence of our explanation is that one does not need to invoke two distinct metallicity groups to explain the observed level of color bimodality. Further spectroscopic observations are needed to obtain true metallicity distributions of GCs.

References

1. M.J. West, P. Côté, R.O. Marzke, A. Jordán: Nature **427**, 31 (2004)
2. J.P. Brodie, J. Strader: astro-ph/0602601 and references therein.
3. S.-J. Yoon, S.K. Yi, Y.-W. Lee: Science **311**, 1129 (2006)
4. E. Peng et al.: ApJ **639**, 95 (2006)
5. G. Bruzual, S. Charlot: MNRAS **344**, 1000 (2003)
6. Y.-W. Lee, P. Demarque, R. Zinn: ApJ **423**, 248 (1994)
7. S.-J. Yoon, Y.-W. Lee: Science **297**, 578 (2002)
8. H.-C. Lee, Y.-W. Lee, B. Gibson: AJ **124**, 2664 (2002)
9. W.E. Harris: AJ **112**, 1487 (1996)
10. D.A. Forbes et al.: MNRAS **355**, 608 (2004)
11. A. Kundu et al.: ApJ **513**, 733 (1999)
12. J.G. Cohen, J.P. Blakeslee, A. Ryzhov: ApJ **496**, 808 (1998)
13. T. Richtler et al.: AJ **127**, 2094 (2004)

Formation of Intracluster and Intercluster Globular Clusters

Kenji Bekki[1] and Hideki Yahagi[2]

[1] School of Physics, University of New South Wales, Sydney 2052, Australia
  bekki@bat.phys.unsw.edu.au
[2] Department of Astronomy, University of Tokyo, 7-3-1 Hongo, Bunkyo ward,
  Tokyo 113-0033, Japan hyahagi@astron.s.u-tokyo.ac.jp

Abstract. Based on high-resolution cosmological simulations with models of GC formation at high redshifts ($z > 6$), we investigate structural and kinematical properties of intracluster globular clusters (ICGCs) that are located outside virialized galaxy-scale halos yet within clusters of galaxies with the total masses of $M_{\rm CL}$. We find that these ICGCs are formed as a result of tidal stripping of GCs initially within galaxy-scale halos during hierarchical growth of clusters via halo merging. We also find that ICGCs comprise 20–40% of all GCs in clusters with $1.0 \times 10^{14} {\rm M}_\odot \leq M_{\rm CL} \leq 6.5 \times 10^{14} {\rm M}_\odot$. Furthermore, the projected radial density profiles ($\Sigma_{\rm GC}$) of all GCs in clusters with different $M_{\rm CL}$ are found to be quite diverse. If $\Sigma_{\rm GC}(R) \propto R^\alpha$, α ranges from ≈ -1.5 to ≈ -2.5 for GCs in clusters with the above mass range. $\Sigma_{\rm GC}$ is demonstrated to be dependent on the truncation epoch of GC formation at high redshifts by cosmological processes such as reionization. About $\sim 1\%$ of all GCs formed before $z > 6$ can not be within any galaxy-scale halos and thus identified as intercluster or intergroup (or intergalactic) GCs.

1 ICGC Formation in a Hierarchical Clustering Scenario

Previous observational studies of globular clusters (GCs) in clusters of galaxies have suggested that there can be a population of GCs that are bounded by cluster gravitational potentials rather than those of cluster member galaxies [1–3]. Since the presence and the absence of intracluster GCs (ICGCs) are now being investigated by a number of observational projects ([4–6]) it would be important for numerical studies of GCs to provide some useful predictions on physical properties of ICGCs. Previous simulations on GC stripping processes in clusters of galaxies [7] did not provide predictions on *entire distributions of GCs across the clusters*. Yahagi & Bekki (2005; YB05; [8]) first investigated ICGC properties based on high-resolution cosmological simulations with models of GC formation at high redshifts ($z > 6$) and the truncation of GC formation at $z_{\rm trun}$ (> 6). Therefore, the simulated GC properties can be compared with old, metal-poor GCs in a more self-consistent way.

YB05 revealed that (1) the distributions of ICGCs are quite inhomogeneous, asymmetric, and elongated (See Fig. 1 for a simulated massive cluster model), (2) the radial surface density profiles of the simulated intracluster

GCs are highly likely to be flatter than those of GCs within cluster member galaxies, and (3) ICGCs comprise 20–40% of all GCs in clusters. YB05 also revealed that the spatial distributions of GCs in clusters can depend on $z_{\rm trun}$ in such a way that GC distributions are more compact and steeper in the models with higher $z_{\rm trun}$. There appears to be no remarkable kinematical differences between ICGCs and GCs within galaxy-scale halos in clusters. These predictions are going to be compared with ongoing observational studies based on wide-field imaging of GCs in clusters of galaxies ([9]). Metallicity distributions and luminosity function of ICGCs and physical relations of ICGCs to ultra-compact dwarfs are given in [9].

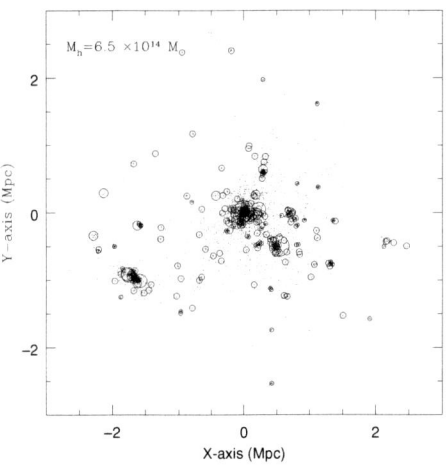

Fig. 1. Distributions of GCs projected onto the x-y plane at $z = 0$ for the model with the total mass of $6.5 \times 10^{14} M_\odot$. GCs within circles represent those within tidal radii of galaxy-scale halos and the radii of the circles represent the tidal radii. GCs that are not within any circles are regarded as intracluster GCs (ICGCs).

References

1. M.J. West, P. Côte et al: ApJ, **453**, L77 (1995)
2. L.P. Bassino, S.A. Cellone, J.C. Forte, B. Dirsch: A&A, **399**, 489 (2003)
3. A. Jordán, W.J. West et al: AJ, **125**, 1642 (2003)
4. B. Dirsch, T. Richtler et al: AJ, **125**, 1908 (2003)
5. Côte, P. et al: ApJS in press (astro-ph/0603252) (2006)
6. N. Tamura et al: in this volume (2006)
7. K. Bekki, D.A. Forbes, M.A. Beasley, W.J. Couch: MNRAS, **344**, 1334 (2003)
8. H. Yahagi, K. Bekki: MNRAS, **364**, L86 (2005)
9. K. Bekki, H. Yahagi: MNRAS, 372, 1019 (2006)

The Effect of Giant Molecular Clouds on Star Clusters

M. Gieles[1,2], S.F. Portegies Zwart[2], and E. Athanassoula[3]

[1] Utrecht University, Utrecht The Netherlands gieles@astro.uu.nl
[2] University of Amsterdam, Amsterdam, The Netherlands spz@science.uva.nl
[3] Observatoire de Marseille, Marseille, France lia@oamp.fr

1 When a Star Cluster Meets a Cloud

We study the encounters between stars clusters and giant molecular clouds (GMCs) [3]. The effect of these encounters has previously been studied analytically for two cases: (1) head-on encounters, for which the cluster moves through the centre of the GMC [1] and (2) distant encounters, where the encounter distance $p > 3R_n$, with p the encounter parameter and R_n the radius of the GMC [6]. We introduce an expression for the energy gain of the cluster due to GMC encounters valid for all values of p and R_n of the form

$$\Delta E \simeq \frac{4.4\, r_h^2}{\left(p^2 + \sqrt{r_h\, R_n^3}\right)^2} \left(\frac{GM_n}{V_{\max}}\right)^2 M_c. \qquad (1)$$

Here $V_{\max}$ is the maximum relative velocity, M_n is the mass of the GMC, G is the gravitational constant and r_h and M_c are the half-mass radius and mass of the cluster, respectively. We perform N-body simulations of encounters with different p and compare the resulting ΔE of the cluster to Eq. (1). Figure 1 shows the very good agreement between simulations and predictions of Eq. (1). Snapshots of one simulation are shown in Fig. 2.

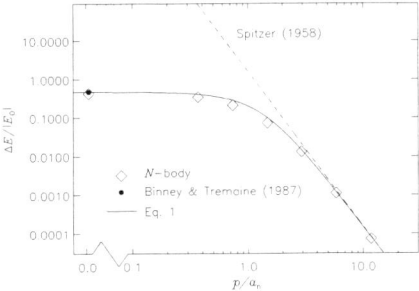

Fig. 1. $\Delta E/|E_0|$ of a cluster as a function of p. The N-body results are shown with diamonds. The result of [1] and [6] for head-on and distant encounters are shown as a filled circle and as a dashed line, respectively. Equation (1) is shown as a full line.

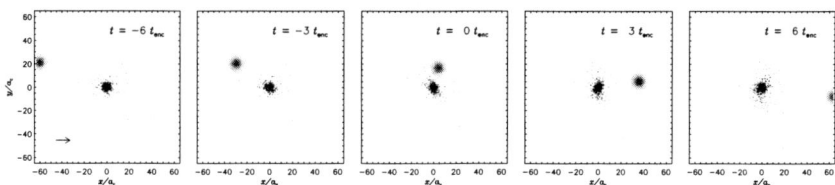

Fig. 2. Simulation of a close encounter between a GMC (grey shades) and a star cluster. The snapshots are viewed in the centre-of-mass frame of the cluster. a_c is the Plummer radius of the cluster.

2 The Cluster Disruption Time

From the simulations we find that the fractional mass loss ($\Delta M/|M_0|$) is only 25% of $\Delta E/|E_0|$. This is because stars escape with velocities much higher than the escape velocity. Defining the cluster disruption time as $t_{\rm dis} = M_{\rm c}/\dot{M}_{\rm c}$, we find a cluster disruption time of the form

$$t_{\rm dis} = 2.0\, S\, \left(M_{\rm c}/10^4\, M_\odot\right)^\gamma\, {\rm Gyr}, \qquad (2)$$

with $S \equiv 1$ for the solar neighbourhood and scales with the global GMC density ($\rho_{\rm n}$) as $S \propto \rho_{\rm n}^{-1}$. The index γ is defined as $\gamma = 1 - 3\lambda$, with λ the index that relates the cluster half-mass radius to its mass ($r_{\rm h} \propto M_{\rm c}^\lambda$). The observed shallow relation between cluster radius and mass (e.g. $\lambda \simeq 0.1$), makes the index ($\gamma = 0.7$) similar to the index found both from observations [4] and from simulations of clusters dissolving in tidal fields ($\gamma \simeq 0.62$). The constant of 2.0 Gyr, which is the disruption time of a $10^4\, M_\odot$ cluster in the solar neighbourhood, is about a factor of 3.5 shorter than found from earlier simulations of clusters dissolving under the combined effect of the galactic tidal field and stellar evolution. It is only slightly higher than the observationally determined value of 1.3 Gyr [4], suggesting that the combined effect of tidal field and encounters with GMCs can explain the lack of old open clusters in the solar neighbourhood [5]. GMC encounters can also explain the (very) short disruption time that was observed for star clusters in the central region of M51 [2], since there $\rho_{\rm n}$ is an order of magnitude higher than in the solar neighbourhood.

References

1. J. Binney, S. Tremaine, Galactic dynamics (1987)
2. M. Gieles, N. Bastian, H.J.G.L.M. Lamers, J.N. Mout, A&A **441**, 949 (2005)
3. M. Gieles, S.F. Portegies Zwart, H. Baumgardt, E. Athanassoula, H.J.G.L.M. Lamers, M. Sipior, J. Leenaarts, MNRAS, **371**, 793 (2006)
4. H.J.G.L.M. Lamers, M. Gieles, N. Bastian, et al., A&A **441**, 117 (2005)
5. J.H. Oort, Ricerche Astronomiche, **5**, 507 (1958)
6. L.J. Spitzer, ApJ **127**, 17 (1958)

Metal-rich Globular Clusters: An Unaccounted Factor Responsible for Their Formation?

V.V. Kravtsov

Instituto de Astronomía, Universidad Católica del Norte, Avenida Angamos 0610, Casilla 1280, Antofagasta, Chile; Sternberg Astronomical Institute, University Avenue 13, 119899 Moscow, Russia `vkravtsov@ucn.cl`

Abstract. A presently unaccounted but quite probable "chemical factor" may be responsible for the formation of old metal-rich globular clusters (MRGCs) in spheroids, as well as of their conterparts, young (intermediate–age) massive star clusters (MSCs) in irregulars. Their formation presumably occurs ∼ at the same stage of the host galaxies' chemical evolution and is related to the essentially increased SF activity in the hosts around the same metallicity, $\sim Z_\odot/3$ ([Fe/H]∼ −0.5). It is achieved very soon in massive spheroids, later in lower-mass spheroids, and (much) more later in irregulars.

1 MSCs as Young Counterparts of Old MRGCs

Are merger of gas–rich spirals and multiphase collapse the only contributors to the formation of old MRGC populations? I argue that MSCs (compact populous and super–star clusters with $M \geq 10^4 M_\odot$) in the LMC and other irregulars are counterparts of the old MRGCs and that another reason (quantitative and qualitative changes of the dust?) leads to (favors) their formation.

Peak metallicities of MRGCs in early–type galaxies with stellar masses differing by nearly 2.5 order of magnitude are estimated by [1] to fall between $-0.7 \leq$[Fe/H]≤ -0.2. The MRGC populations are assumed to be coeval, and their color trend is fully attributed to their metallicity trend. However, this is not supported by data on timing of spheroids' formation: the more massive spheroid, the shorter timescale of its formation [2–4]. Real scatter of the MRGC peak metallicities around mean, [Fe/H]∼ −0.5, may be at least twice as lower, by accepting conservative estimate of possible systematic age difference of ∼5 Gyr between MRGCs in spheroids of the range of mass.

According to [5], a mean metallicity of the populous star clusters formed in the LMC 1-3 Gyr ago is close to [Fe/H]= -0.5, irrespective of their age and location in the galaxy. However, metallicity of the field stars near these clusters exhibits obvious dependence on age (see Fig. 1, where squares and asterisks are data from [6] and [7], respectively). Moreover, the MDF for the disk stars of the LMC is virtually identical with that for the old red giant stars in the halo of NGC 5128 and M31 ([8]), reaching its maximum somewhere near [Fe/H]∼ −0.5. Published data on generic (mean) metallicities of populations of MSCs and/or on their hosts in irregulars with increased (bursting) SF activity (NGC1140, NGC1156, NGC1313, NGC1569,

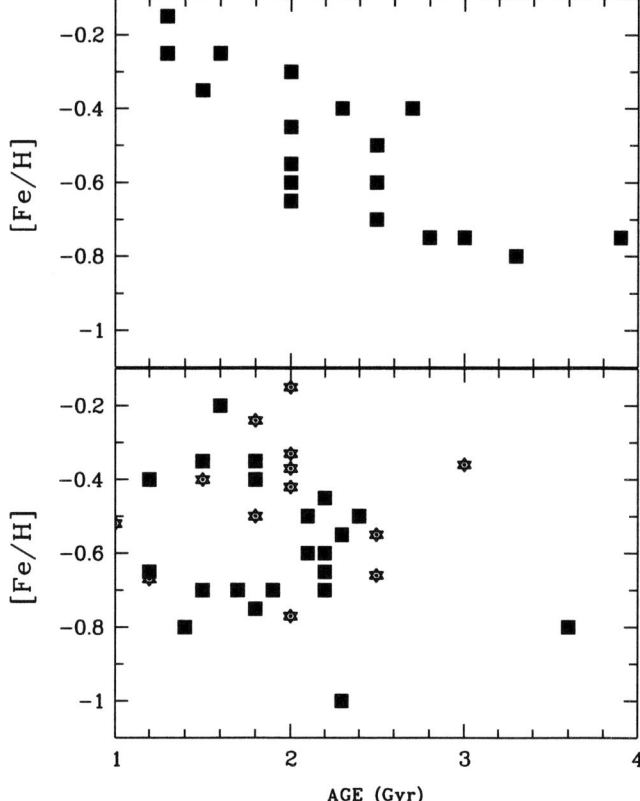

Fig. 1. *Upper panel*: the age–metallicity relation for the LMC intermediate–age field stars; *lower panel*: the same for the LMC intermediate–age populous star clusters.

NGC1705, NGC4214, NGC4449, NGC5253, NGC6745, and IC 10) show that the metallicities fall, as those of MRGCs do, around $\sim Z_\odot/3$, between $0.004 \leq Z \leq 0.008$ ($-0.7 \leq$ [Fe/H] ≤ -0.3). For details and references, see [9].

2 Implications

Both the most probable formation of GCs (MSCs) and internally regulated SF activity increasing in the hosts near $\sim Z_\odot/3$ may shed more light on: the same metallicity value of the intracluster gas in galaxy clusters; starburst phenomenon in isolated galaxies; formation of MSCs in the disks of isolated spirals, etc. The difference between the age-metallicity relations for MSCs and stars in the LMC implies (provided it is the same in spheroids) different concentrations of MRGCs and stars to the centers of spheroids even under negligible GC disruption in the central parts and no merger of gas-rich spirals.

References

1. E.W. Peng, A. Jordán, P. Côté et al.: ApJ **639**, 95 (2006)
2. L.G. Granato, L. Silva, P. Monaco et al.: MNRAS, **324**, 757 (2001)
3. D. Thomas, C. Maraston, R. Bender: ApSS, **281**, 371 (2002)
4. N. Caldwell, J.A. Rose, K.D. Concannon: AJ, **125**, 2891 (2003)
5. D. Geisler, A.E. Piatti, E. Bica et al.: MNRAS, **341**, 771 (2003)
6. A.E. Piatti, E. Bica, D. Geisler et al.: MNRAS, **344**, 965 (2003)
7. E.W. Olszewski, R.A. Schommer, N.B. Suntzeff et al.: AJ, **101**, 515 (1991)
8. W.E. Harris, G.L.H. Harris: AJ, **122**, 3065 (2001)
9. V.V. Kravtsov: AJ, **132**, 1248 (2006)

On the Globular Cluster Color Distributions

Suk-Jin Yoon[1,2] and Chul Chung[1]

[1] Department of Astronomy & Center for Space Astrophysics, Yonsei U., Seoul, Korea
[2] sjyoon@galaxy.yonsei.ac.kr

Abstract. The bimodal color distribution of globular clusters (GCs) is an outstanding discovery from observations of early-type galaxies [1,2]. Yoon, Yi, & Lee have recently presented a new theoretical metallicity–color relationship that has a significant inflection and showed that such a relation can produce bimodal color distribution of an old GC system *even when the intrinsic metallicity distribution is unimodal* [3]. We have applied this hypothesis to (1) the considerable variation in color histogram morphologies from color to color, and (2) the broad *unimodal* color distributions of GCs found in some galaxies. Our population model suggests the different shapes of the [Fe/H]–color relations depending on the color used. As a result, the [Fe/H] distribution of a GC system can be transformed into color distributions that significantly vary from color to color, consistent with the currently available data (Fig. 1). On the other hand, the simulation shows that the presence of a metal-rich, young GC population in addition to the old population may be responsible for the broad unimodal color distributions (Fig. 2). The results may be taken as a further support for the Yoon, Yi, & Lee hypothesis.

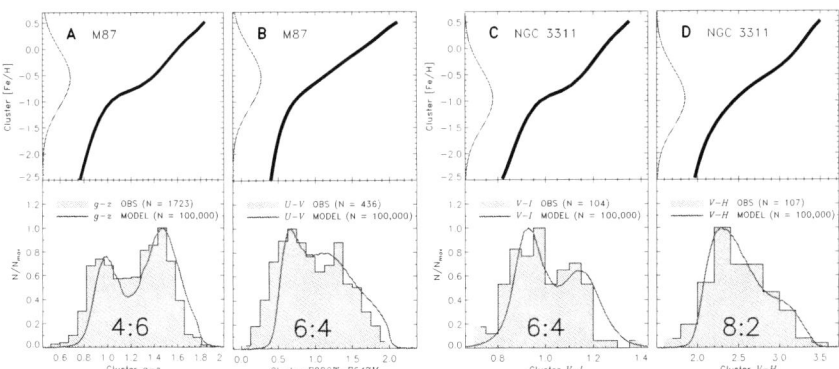

Fig. 1. Comparison between the observed and simulated GC color distributions for various colors. The morphology of observed color histograms, most conspicuously the relative fraction of *blue* and *red* GCs, varies depending on the color in use. The simulations indicate that this is simply due to the [Fe/H]–color relations that differ from color to color. (**A**, *top*) Thick line represents our theoretical [Fe/H]–$(g-z)$ relation for the 13-Gyr GCs. The metallicity distribution of 10^5 model GCs is shown along the y-axis (thin line). (**A**, *bottom*) Comparison between the observed [4] and simulated color distributions for M87 GC system. The ralative fraction of blue and red GCs is denoted. (**B**) Same as (A), but for $(F336W - F547M)$ for the same galaxy [5]. (**C–D**) Same as (A), but for $(V-I)$ and $(V-H)$ for NGC 3311 [6].

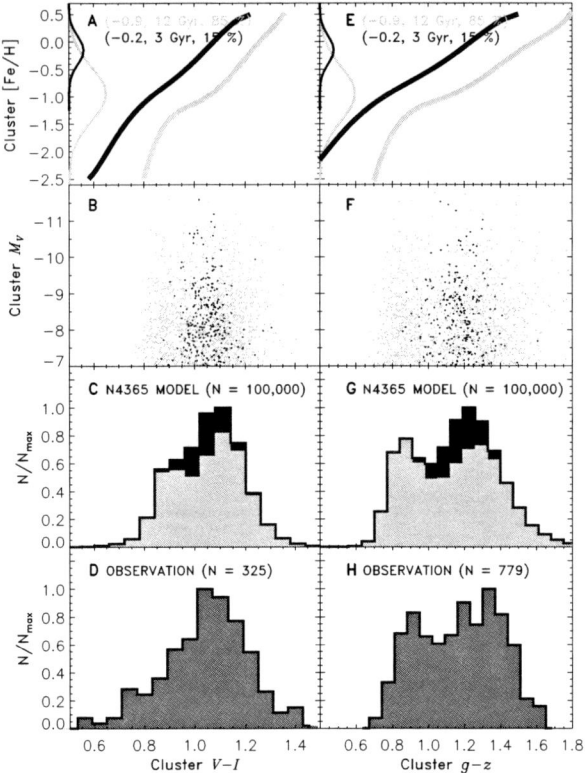

Fig. 2. Monte Carlo simulations for the GC color distribution of NGC 4365. (**A**) The theoretical [Fe/H]–$(V-I)$ relations for old and young GCs and their [Fe/H] distributions. Their mean metallicity, age, and number fraction are denoted with the same color code in parenthesis. The mean metallicity and age of the young GCs are adopted from [7]. (**B**) Color-magnitude diagram of 2000 randomly selected model GCs. Light (dark) gray dots are for the old (young) GCs. Cluster $(V-I)$ is transformed from [Fe/H] by using the theoretical relations shown in (A). (**C**) The simulated color histogram (N = 100,000). Young GCs (dark gray) on top of old GCs (light gray) make a broad unimodal distribution. (**D**) The observed color histogram (N = 325) of the NGC 4365 GCs [8]. (**E–H**) Same as (A–D), but for $(g-z)$ distribution [9]. Unlike the single broad $(V-I)$ distribution, the $(g-z)$ distribution appears bimodal (or trimodal), which can be explained by the projection effect.

References

1. M.J. West, P. Côté, R.O. Marzke, A. Jordán: Nature **427**, 31 (2004)
2. J.P. Brodie, J. Strader: ARA&A **44**, 193 (2006)
3. S.-J. Yoon, S.K. Yi, Y.-W. Lee: Science **311**, 1129 (2006)
4. E. Peng et al.: ApJ **639**, 95 (2006)

5. A. Jordán, P. Côté, M.J. West, R.O. Marzke: ApJ **576**, L113 (2002)
6. M. Hempel, D. Geisler, D.W. Hoard, W.E. Harris: A&A **439**, 59 (2005)
7. S.S. Larsen et al.: ApJ **585**, 767 (2003)
8. A. Kundu, B.C. Whitmore: AJ **121**, 2950 (2001)
9. J.P. Brodie et al.: AJ **129**, 2643 (2005)

Part VIII

Dynamical Evolution of Star Clusters

Dissolution of Globular Clusters

Holger Baumgardt

Argelander Institute for Astronomy, Auf dem Hügel 71, 53121 Bonn, Germany
holger@astro.uni-bonn.de

1 Introduction

Globular clusters are among the oldest objects in galaxies, and understanding the details of their formation and evolution can bring valuable insight into the early history of galaxies. Until the late 1970s, globular clusters were thought of to be relatively static stellar systems, a view which was supported by the fact that most observed density profiles of globular clusters can be fitted with equilibrium models like e.g. King [17] profiles. This view has changed significantly over the last twenty years. On the observational side, the evidence for differences in the stellar mass-functions of globular clusters [28, 24], which are believed to be at least partly the result of their dynamical evolution, and the discovery of extratidal stars surrounding globular clusters [15, 26] are strong indications for the ongoing dynamical evolution and dissolution of globular clusters.

On the theoretical side, N-body simulations of star cluster evolution have become increasingly sophisticated, due to both progresses in simulation techniques (e.g. [25, 1]) and the development of the GRAPE series of special purpose computers [31, 23], which allows to simulate the evolution of star clusters with increasingly larger particle numbers.

This review summarises the current knowledge about the dissolution of star clusters and discusses the implications of star cluster dissolution for the evolution of the mass function of star cluster systems in galaxies.

2 Dissolution Mechanisms

Star clusters evolve due to a number of dissolution mechanisms, the most important of which are:

(1) Primordial gas loss
(2) Stellar evolution
(3) Relaxation
(4) External tidal perturbations

The importance of the different processes changes as a star cluster ages and depends also on the position of the cluster in its parent galaxy. The combined effect of all dissolution mechanisms can dramatically alter the properties of star cluster systems, so it is important to understand them.

2.1 Primordial Gas Loss

Star formation is typically less than 40% efficient and the gas not turned into stars is lost within a few 10^5 to 10^6 Myrs due to stellar winds from massive stars or supernova explosions. N-body simulations have shown that the loss of the primordial gas can easily cause star clusters to lose a large fraction of their stars [14, 18, 20] or unbind them completely. Together with the loss of a large mass fraction within a short timescale, star clusters also undergo significant expansion. Connected to the problem of primordial gas loss is the question whether star clusters form in virial equilibrium, since, in addition to gravity, molecular clouds are also held together by magnetic fields and the pressure from the ambient gaseous medium. Since these forces do not act on stars, some clusters might already be unbound at birth and disperse within a few crossing times. As a result, a significant fraction of clusters might not survive the first 10 Myrs ("infant mortality problem", [21]).

These considerations are supported by observations which show that a large fraction of clusters dissolve at an early stage. Fall et al. [9] for example found that in the Antennae galaxies the number of clusters decreases strongly with cluster age and that the median age of clusters is only 10^7 yrs, which they interpreted as evidence for rapid cluster disruption. Similarly, only a small fraction of star-forming embedded clusters in the Milky Way evolve to become open clusters [21] implying that the majority of clusters must dissolve within a few Myrs.

2.2 Stellar Evolution

For clusters which survive the early evolutionary stages, the next dissolution mechanism is stellar evolution. For a standard stellar IMF like for example the Kroupa [19] IMF, about 30% of the mass of a star cluster is lost due to the stellar evolution of the member stars within a few Gyrs [4] and the fraction can be significantly higher for star clusters starting with a top-heavy IMF. Mass-loss from stellar evolution causes the clusters to expand while at the same time decreasing the tidal radius of star clusters. Although the effect is less strong than the loss of the primordial gas since the mass loss happens on a longer timescale, N-body simulations have shown that the effect is strong enough to unbind low-concentration clusters surrounded by an external tidal field.

Fukushige and Heggie [10] for example found that if the stellar IMF follows a power-law with index $\alpha = -1.5$ between $0.4 < m < 15$ $M_\odot$, clusters less concentrated than King $W_0 = 7$ models were easily disrupted by stellar evolution mass-loss. For Salpeter-like IMFs, clusters had to have an initial concentration $W_0 \geq 5$ to be stable against disruption. Similar results were also found by Giersz [12] and Joshii et al. [16] in Monte-Carlo simulations.

2.3 Relaxation

Relaxation arises due to mutual encounters between stars in a globular cluster, and causes a slow drift of stars in energy space. Dynamical processes like the segregation of heavy mass stars into the cluster center or the core collapse of star clusters are driven by relaxation. For a star cluster containing N stars with mean mass $<m>$ and half-mass radius r_h, the relaxation time at r_h is given by (Spitzer [30], eq. 2-63):

$$t_{rh} = 0.138 \frac{\sqrt{N} \; r_h^{3/2}}{\sqrt{<m>} \sqrt{G} \; \ln(\gamma N)} \; , \qquad (1)$$

where G is the constant of gravity and γ the Coulomb logarithm.

Due to mutual encounters between cluster stars, stars can also gain enough energy to leave the cluster completely, which causes a slow evaporation of

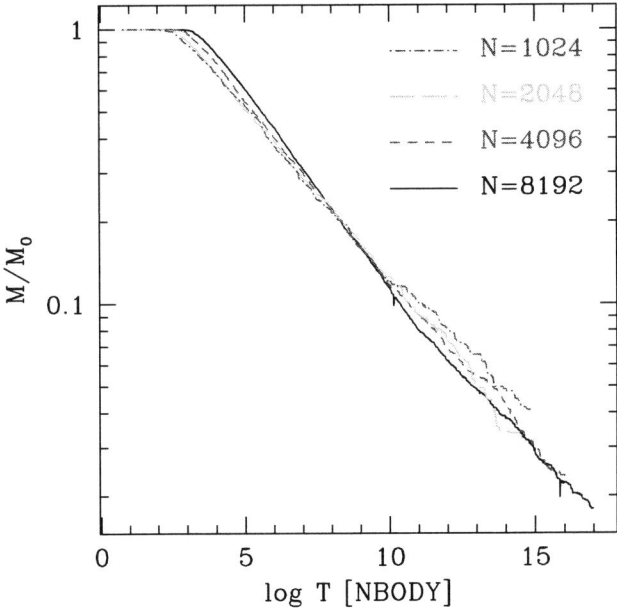

Fig. 1. Dissolution of isolated clusters starting with different initial particle numbers N. Irrespective of N, complete dissolution takes about 10^{15} N-body times since relaxation causes a strong cluster expansion, which slows down the overall evolution (from Baumgardt et al. [3]).

the whole cluster. N-body simulations (Baumgardt et al. [3] and Fig. 1) have shown that for *isolated* clusters this process is inefficient in dissolving star clusters since even low-mass clusters would need of order 10^{15} initial crossing times to completely evaporate, i.e. they would not dissolve within the lifetime of the universe. The reason is that relaxation is dominated by the cumulative effect of many distant encounters, so stars change their energies smoothly and do not jump around in energy space. As the energy of a star approaches $E \to 0$, it inevitably moves through the outskirts of the cluster for most of the time where the stellar density is low and the star has only few encounters with other stars. As a result, relaxation alone causes clusters to expand but does not dissolve them.

2.4 External Tidal Fields

Star clusters are usually not isolated but move in the gravitational field of their parent galaxy. The external galaxy can influence the evolution of a star cluster in two ways: A constant tidal field, which arises for example if clusters move on circular orbits through axisymmetric potentials, confines the cluster stars into a certain volume around the cluster centre, outside of which the stars are unbound to the cluster. This prevents clusters from expanding indefinitely and accelerates the escape of stars since the energy necessary to escape from the cluster is lowered.

Variable external fields arise if clusters move on elliptic orbits or through a galactic disc. In the first case they experience disruptive tidal shocks since stars are accelerated away from the cluster centre, in the latter case the shocks are compressive. In both cases the internal energy of the clusters is increased.

For star clusters moving in circular orbits, stars with an energy E_* only slightly higher than the critical energy E_C needed for escape can escape only through small apertures around the lagrangian points L1 and L2 connecting the centre of the galaxy with the centre of the star cluster. Fukushige and Heggie [11] have shown that this leads to a scaling of the average time needed to escape according to $T_{Esc} \propto 1/(E_* - E_C)^2$, i.e. for small energy differences

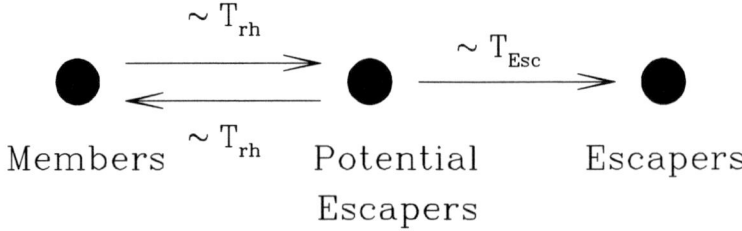

Fig. 2. Escape of stars from clusters in circular orbits. Bound members are scattered on relaxation time scales to become potential escapers with $E_* > E_C$. Potential escapers either escape or lose energy and become bound members again.

the escape time can become of the order of the relaxation time or even larger. This leads to a complication of the whole escape process, as illustrated in Fig. 2: Bound members are scattered on a relaxation time to become potential escapers with $E_* > E_C$. Potential escapers can either escape or are scattered to lower energies and become bound members again.

Baumgardt [2] has shown that this influences the scaling of the lifetimes of star clusters with the number of cluster stars or the total mass. While the lifetimes of single-mass clusters surrounded with a tidal boundary scale with the relaxation time, clusters moving in circular orbits through their parent galaxy show a scaling of their lifetimes according to $T_{rh}^{0.75}$. If lifetimes of star clusters are estimated by scaling the results of low-N models to higher particle numbers, this causes a reduction of the lifetimes of globular clusters by a factor of a few. Interestingly, a similar slow increase of the lifetime of star clusters with the particle number was also found in observational studies of open cluster systems [6, 22].

3 Evolution of Realistic Clusters

Since the pioneering study by Chernoff and Weinberg [7], a number of papers have studied the evolution of multi-mass star clusters evolving under the combined effects of stellar evolution, relaxation and an external tidal field (e.g. [13, 32, 18]). In the following I will concentrate on the results of Baumgardt and Makino [4], who have so far performed the largest set of N-body calculations of the evolution of multi-mass star clusters in external tidal fields. Their clusters all started with Kroupa [19] IMFs, but varying initial particle numbers, orbital types and density profiles.

Figure 3 summarises their results for the dissolution times. Independent of orbital type and initial cluster concentration, the lifetimes always scaled with the relaxation time as T_{rh}^x where $x \approx 0.7$. Clusters starting from King models with a larger central concentration (open circles) show a slightly steeper scaling of the lifetimes. Interestingly, the exponent does not change if one goes from circular to elliptic orbits (triangles), indicating that although tidal shocks help in removing stars, the general picture of stellar escape is not changed significantly. Star clusters moving at smaller galactocentric distances R_G have smaller lifetimes since the tidal field is stronger and confines the clusters to smaller volumes. Baumgardt and Makino [4] summarised their results for the lifetimes with the following formula:

$$\frac{T_{Diss}}{[Myr]} = \beta \left(\frac{N}{ln(\gamma N)}\right)^x \frac{R_G}{[kpc]} \left(\frac{V_G}{220\ km/s}\right)^{-1} (1-\epsilon) \qquad (2)$$

where $\beta = 1.91$, $\gamma = 0.02$ and $x = 0.75$ are constants, V_G is the circular velocity of the external galaxy and ϵ the orbital eccentricity of the cluster. This formula did fit the N-body results for star clusters surviving for a Hubble time to within 10%. It slightly overpredicted the lifetimes of star clusters dissolving in less than 1 Gyr since such clusters contain massive stars for a larger fraction of their lifetime, which reduces their relaxation times.

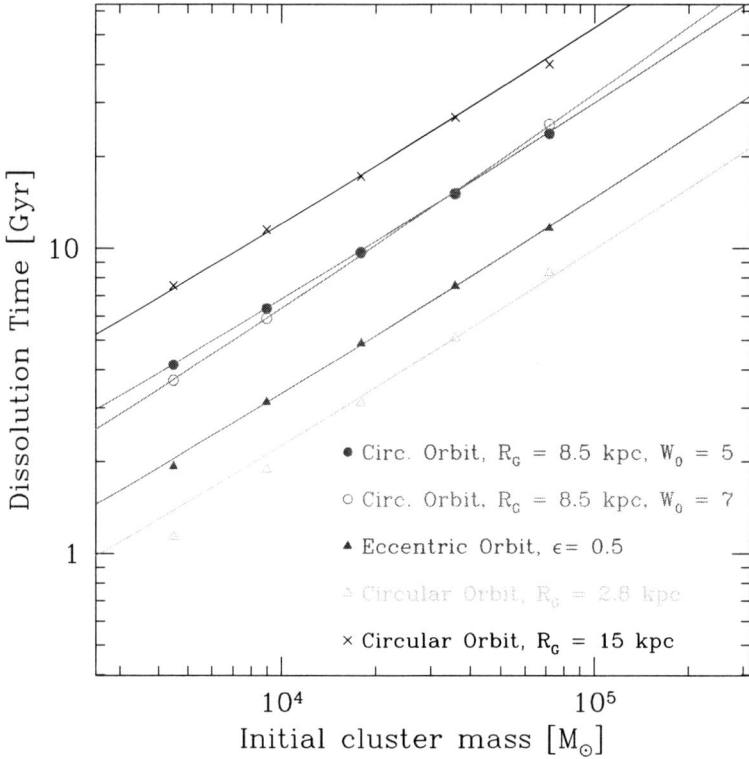

Fig. 3. Lifetimes of star clusters in dependence of their mass for clusters moving in different orbits and starting from different initial King models. In all cases, the lifetimes show a scaling with the mass close to $T_{Diss} \sim M^{0.75}$.

4 Evolution of Globular Cluster Systems

The globular cluster system of the Milky Way as we observe it today is characterised by a Gaussian distribution in absolute magnitudes, with mean $M_V = -7.4$ and scatter $\sigma_M = 1.15$, similar to what is observed for globular clusters in other galaxies. In contrast, young massive clusters in interacting and starburst galaxies follow power-law distributions over mass $N(M) \sim M^{-\beta}$, with slopes close to $\beta = 2$, so the question arises whether the globular cluster system of the Milky Way started with a similar mass-function and has lost the low-mass clusters due to dissolution. Stars lost from these clusters could nowadays form the halo field stars.

In order to study this question, I have performed a number of Monte-Carlo simulations in which clusters were assumed to start with a power-law IMF $N \sim M_C^{-\beta}$ with power-law index $\beta = 2.0$ and followed a radial distribution in the galaxy according to $\rho \sim R_G^{-\alpha}$ with $\alpha = 4.5$ and core radius $R_C = 1$ kpc. The galaxy was modelled as an isothermal sphere with a circular velocity of $V_G = 200$ km/s and the cluster system was set-up such that the clusters had an isotropic velocity dispersion at all radii. The lifetimes derived by Baumgardt and Makino [4] and given in Eq. (2) were used to dissolve clusters.

In addition to the dissolution mechanisms already considered by Baumgardt and Makino [4], the simulations also included dynamical friction and disc shocks. Dynamical friction was modeled as a steady shrinking of the cluster orbit according to eq. 7–25 of Binney and Tremaine [5], while disc shocks were included according to Spitzer and Chevalier [29], assuming that a fractional increase of the cluster energy by $\Delta E/E_C$ corresponds to a fractional mass loss of the same amount.

Figure 4 shows the resulting evolution of the mass function of the galactic globular cluster system. Inside $R_G = 10$ kpc, the dissolution mechanisms are strong enough to evolve an initial power-law MF into a bell-shaped MF. The mean and dispersion of the surviving clusters agree rather well with the observed MF of galactic globular clusters. The situation changes if one considers clusters at larger galactocentric radii. For outer clusters the tidal field is much weaker, meaning less destruction of the cluster system. As a result, the MF of surviving clusters is still increasing towards the lowest masses

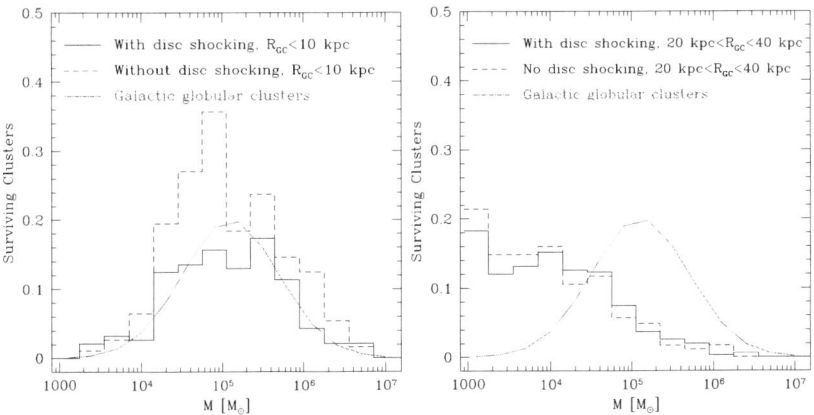

Fig. 4. Evolution of the mass function of galactic globular clusters for inner ($R_G <$ 10 kpc, *left panel*) and and outer (20 kpc$< R_G <$ 40 kpc, *right panel*) clusters. While inside 10 kpc, power-law IMFs can be turned into gaussians, the mass-function of outer clusters is increasing towards the lowest masses and is in contradiction with the observed MF.

considered, in contrast with what is observed for the galactic globulars. This agrees qualitatively with results derived by Vesperini [33] and Parmentier and Gilmore [27] but is in contrast to what Fall and Zhang [8] found for the outer clusters. Further research is necessary to constrain the starting condition of the galactic GC system.

References

1. S. Aarseth: PASP **111**, 1333 (1999)
2. H. Baumgardt: MNRAS **325**, 1323 (2001)
3. H. Baumgardt, P. Hut, D.C. Heggie: MNRAS **336**, 1069 (2002)
4. H. Baumgardt, J. Makino: MNRAS **340**, 227 (2003)
5. J. Binney, S. Tremaine: *Galactic Dynamics*, (Princeton University Press, Princeton 1987)
6. S.G. Boutloukos, H.J.G.L.M. Lamers: MNRAS **338**, 717 (2003)
7. D.F. Chernoff, M.D. Weinberg: ApJ **351**, L21 (1990)
8. S.M. Fall, Q. Zhang: ApJ **561**, 751 (2001)
9. S.M. Fall, R. Chandar, B.C. Whitmore: ApJ **631**, L133 (2005)
10. T. Fukushige, D. Heggie: MNRAS **276**, 206 (1995)
11. T. Fukushige, D. Heggie: MNRAS **318**, 753 (2000)
12. M. Giersz: MNRAS **324**, 218 (2001)
13. O.Y. Gnedin, J.P. Ostriker: ApJ **474**, 223 (1997)
14. S. Goodwin: MNRAS **284**, 785 (1997)
15. C.J. Grillmair et al.: AJ **109**, 2553 (1995)
16. K.J. Joshii, C.P. Nave, F.A. Rasio: ApJ **550**, 691 (2001)
17. I. King: AJ **71**, 64 (1966)
18. P. Kroupa, S. Aarseth, J. Hurley: MNRAS **321**, 699 (2001)
19. P. Kroupa: MNRAS **322**, 231 (2001)
20. P. Kroupa, C.M. Boily: MNRAS **338**, 673 (2003)
21. C.J. Lada, E.A. Lada: ARAA **41**, 57 (2003)
22. H.J.G.L.M. Lamers et al.: A&A **441**, 117 (2005)
23. J. Makino, T. Fukushige, M. Koga, & K. Namura: PASJ **55**, 1163 (2003)
24. G. de Marchi, et al.: A&A **343**, 9 (1999)
25. S. Mikkola, S. Aarseth: CeMDA **57**, 439 (1993)
26. M. Odenkrichen et al.: AJ **126**, 2385 (2003)
27. G. Parmentier, G. Gilmore: MNRAS **363**, 326 (2005)
28. G. Piotto, A. Cool, I. King: AJ **113**, 1345 (1997)
29. L. Spitzer Jr., R.A. Chevalier: ApJ **183**, 565 (1973)
30. L. Spitzer Jr.: *Dynamical Evolution of Globular Clusters*, (Princeton University Press, Princeton 1987)
31. D. Sugimoto et al.: Nature **345**, 33 (1990)
32. E. Vesperini, D.C. Heggie: MNRAS **289**, 898 (1997)
33. E. Vesperini: MNRAS **299**, 1019 (1998)

Dynamical Masses of Young Star Clusters: Constraints on the Stellar IMF and Star-Formation Efficiency

Nate Bastian[1] and Simon P. Goodwin[2]

[1] University College London, London, UK bastian@star.ucl.ac.uk
[2] University of Sheffield, Sheffield, UK S.Goodwin@sheffield.ac.uk

1 The Stellar Initial Mass Function in Clusters

Many recent works have attempted to constrain the stellar initial mass function (IMF) inside massive clusters by comparing their dynamical mass estimates (found through measuring the velocity dispersion and effective radius) to the measured light. These studies have come to different conclusions, with some claiming standard Kroupa-type [11] IMFs (e.g. [14, 13]) while others have claimed extreme non-standard IMFs (e.g. the top or bottom of the IMF is over-populated with respect to a Kroupa IMF [17]). However, the results appear to be correlated with the age of the clusters, as older clusters (>80 Myr) all appear to be well fit by a Kroupa-type IMF whereas younger clusters display significant scatter in their best fitting IMF [3]. This has led to the suggestion that the younger clusters are out of Virial equilibrium, thus undercutting the fundamental assumption which is necessary to derive dynamical masses. We will return to this point in Sects. 2 and 3. Focusing on the older clusters, we see that they all have standard IMFs (see Fig. 2), arguing that at least in massive clusters the IMF does not vary significantly.

2 Dynamical Equilibrium of Young Clusters

One explanation of why the youngest clusters are not in dynamical equilibrium is that young clusters are expected to expel their remaining gas (left over from the star-formation process) on extremely rapid timescales, which will leave the cluster severely out of equilibrium (e.g. [7]). In order to search for such an effect we compared the luminosity profiles of three young clusters with that of N-body simulations of clusters which are undergoing violent relaxation due to rapid gas loss [2]. The simulations (Fig. 1, right panel) make the generic prediction of excess light at large radii (with respect to the best fitting EFF profile [5]), due to an unbound expanding halo of stars which stays associated with the cluster for $\sim 20-50$ Myr. These stars are unbound due to the rapid decrease of potential energy as the gas is removed on timescales shorter than a crossing time (e.g. [7]). Observations of the three young clusters also show excess light at large radii (Fig. 1, left panel),

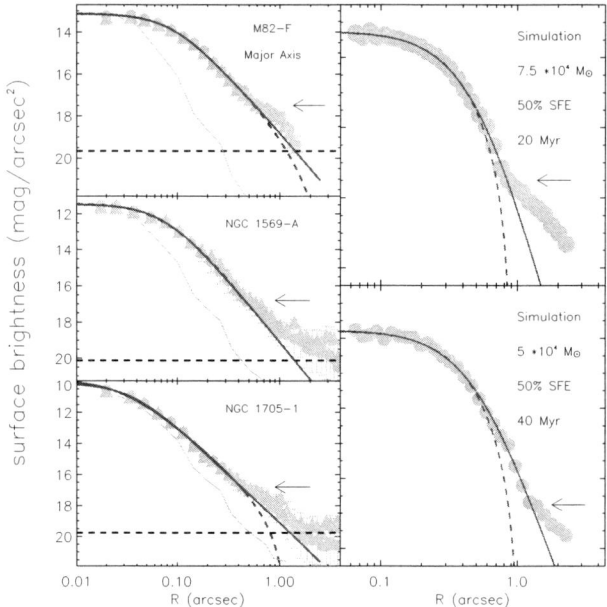

Fig. 1. Taken from [2]: Surface brightness profiles for three young clusters (left - M82-F, NGC 1569-A, and NGC 1705-1) and two N-body simulations which include the rapid removal of gas which was left over from a non-100% star-formation efficiency (*right*). The solid (*red*) and dashed (*blue*) lines are the best fitting EFF [5] and King [10] profiles respectively. Note the excess of light at large radii with respect to the best fitting EFF profile in both the observations and models. This excess light is due to an unbound expanding halo of stars caused by the rapid ejection of the remaining gas after the cluster forms. *Hence, excess light at large radii strongly implies that these clusters are not in dynamical equilibrium.* For details of the modelling and observations see [2,8].

strongly suggesting that they are experiencing violent relaxation [2]. Hence these clusters are not in dynamical equilibrium.

3 The Star Formation Efficiency and Infant Mortality

Assuming that young clusters are out of equilibrium due to rapid gas loss (the extent of which is determined by the star-formation efficiency – SFE one can fold these effects (see Fig. 3 in [2]) into SSP models [8]. The results are shown as solid and dashed red lines in Fig. 1 for various SFEs, where we have assumed all gas is lost instantaneously at 2 Myr. The dashed lines show the results for SFEs below 30% for which the cluster will become completely unbound. Solid lines represent SFEs above 30% where a bound core may remain. Note that the observed SFEs of the clusters range from 10% to 60% [8].

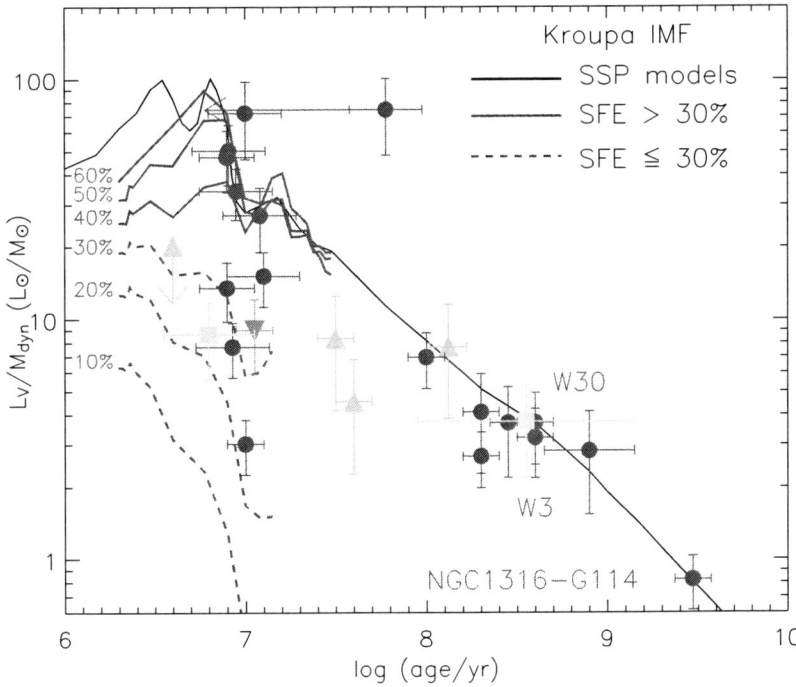

Fig. 2. Taken from [8]: The light-to-mass ratio of young clusters. The circles (*blue and red*) are taken from [3] and [14] and references therein, the triangles with errors (*green*) are LMC clusters [15], the upside down triangle (*brown*) is for NGC 6946-1447 corrected for internal extinction [13], and the squares (*cyan*) are from [16]. The arrow extending from M82F [17] is a possible correction to its age (see [3]). The triangle without errors is the tentative upper limit for cluster R136 in 30 Dor [4, 9]. The solid (*black*) line is the prediction of simple stellar population models (SSPs) with a Kroupa [11] stellar IMF. The red lines are the SSP model tracks folded with the effects of rapid gas removal following non-100% star-formation efficiencies (SFE) [2]. Dashed lines represent the SFEs where the clusters will become completely unbound. The SFE in the simulations measures the degree to which the cluster is out-of-virial equilibrium after gas loss, and so is an *effective* SFE (see [2, 8]).

We also note that 7 out of the 12 clusters with ages below 20 Myr appear unbound (i.e. SFE < 30%), suggesting that ∼ 60% of clusters will become unbound in the first 20–50 Myr of their lives [8], i.e. what has been termed "infant mortality". This is in close agreement with cluster population studies of M51 which found an infant mortality rate of 68% [1] and comparable to the open cluster dispersal rate of ∼ 87% [12] (see also [18]).

4 Conclusions

Through detailed comparisons of the luminosity profiles of young clusters with N-body simulations of clusters including the effects of rapid gas loss, we argue that young clusters are not in Virial equilibrium. This undercuts the fundamental assumption needed to determine dynamical masses. This suggests that the claimed IMF variations are probably due to the internal dynamics of the clusters and not related to the IMF. By limiting the sample to the oldest clusters (which appear to be in equilibrium) we see that they are all well fit by a Kroupa-type IMF arguing that, at least in massive star clusters, the IMF does not vary significantly.

By combining the above N-body simulations with SSP models we can derive the (effective) SFE of clusters. From this we find that $\sim 60\%$ of young clusters appear to be unbound, in good agreement with other estimates of the infant mortality rate. Note however that even if a cluster survives this phase it may not survive indefinitely due to internal and external effects (e.g. [6]).

Acknowledgements. NB gratefully thanks his collaborators Roberto Saglia, Paul Goudfrooij, Markus Kissler-Patig, Claudia Maraston, Francois Schweizer, and Manuela Zoccali on dynamical mass studies.

References

1. Bastian, N., Gieles, M., Lamers, H.J.G.L.M., Scheepmaker, R. A., and de Grijs, R. 2005, A&A 431, 905
2. Bastian, N. and Goodwin, S.P. 2006, MNRAS, 369, 9
3. Bastian, N., Saglia, R.P., Goudfrooij, P., Kissler-Patig, M., Maraston, C., Scwheizer, F., Zoccali, M. 2006, A&A, 448, 881
4. Bosch, G., Selman, F., Melnick, J., and Terlevich, R. 2001, A&A, 380, 137
5. Elson, R.A.W., Fall, M.S., and Freeman, K.C. 1987, ApJ 323, 54 (EFF)
6. Gieles, M., Bastian, N., Lamers, H.J.G.L.M. and Mout, J.N. 2005, A&A, 441, 949
7. Goodwin, S.P. 1997, MNRAS, 284, 785
8. Goodwin, S.P. and Bastian, N. 2006, in prep.
9. Hunter, D.A., Shaya, E.J., Holtzman, J.A., et al. 1995, ApJ, 448, 179
10. King, I. 1962, AJ 67, 471
11. Kroupa, P. 2002, Science, 295, 82
12. Lada, C.J., and Lada, E.A. 1991, in The formation and evolution of star clusters, ed. K. Joes, ASP Conf. Ser., 13, 3
13. Larsen, S.S., Brodie, J.P., Hunter, D.A. 2006, AJ, 131, 2362L (erratum)
14. Maraston, C., Bastian N., Saglia R.P., Kissler-Patig, M., Schweizer, F., Goudfrooij, P. 2004, A&A, 416, 467
15. McLaughlin, D.E. and van der Marel, R.P. 2005, ApJSS, 161, 304 Kissler-Patig M., Schweizer F., and Goudfrooij P. 2004, A&A, 416, 467
16. Östlin, G., Cumming, R.J., Bergvall, N. 2007, A&A, 461, 471
17. Smith, L.J. and Gallagher, J.S. 2001, MNRAS, 326, 1027
18. Whitmore, B. C. 2003, in A Decade of HST Science, ed. M. Livio, K. Noll, and M. Stiavelli (Cambridge: Cambridge University Press), 153

Dynamical Evolution of Rotating Globular Clusters with Embedded Black Holes

José Fiestas[1] and Rainer Spurzem[2]

[1] Astronomisches Rechen-Institut Heidelberg, Heidelberg, Germany
  fiestas@ari.uni-heidelberg.de
[2] Astronomisches Rechen-Institut Heidelberg, Heidelberg, Germany
  spurzem@ari.uni-heidelberg.de

Evolution of self-gravitating dense stellar systems (e.g. globular clusters, galactic nuclei) with embedded black holes is investigated. The interaction between the stellar and black hole component is followed in a way, different from most other investigations in this field, as flattening of the system due to differential rotation is allowed. The interplay between velocity diffusion due to relaxation and black hole star accretion is followed together with cluster rotation. The results show how angular momentum transport and star accretion support the development of central rotation in relaxation time scales. Gravogyro and gravothermal instabilities conduce the system to a faster evolution leading to shorter collapse times with respect to models without black hole, and a faster cluster dissolution in the field of a parent galaxy.

1 Method

The model uses the solution of the Fokker Planck equation: Boltzmann-equation + collisional term. The latter is given through the -local- Fokker Planck approximation

$$\left(\frac{\partial f_a}{\partial t}\right)_{\text{coll}} = -\frac{\partial}{\partial v^\mu}(f_a \langle \Delta v^\mu \rangle_a) + \frac{1}{2}\frac{\partial^2}{\partial v^\mu \partial v^\nu}(f_a \langle \Delta v^\mu \Delta v^\nu \rangle_a) \qquad (1)$$

where $\mu = 1,2,3$ and $\nu = 1,2,3$ (tensor notation). v^μ gives the velocity in Cartesian coordinates. The first order diffusion coefficients $\langle \Delta v^\mu \rangle_a$ describe the dynamical friction and the second order ones $\langle \Delta v^\mu \Delta v^\nu \rangle_a$ give the real velocity diffusion.

The distribution of stars is represented by a single-particle system that is initially axisymmetric in space and anisotropic in velocity space [1]. No stellar spectrum is included in this model (but see [6]). The initial BH mass, M_{BH}, is much smaller than the cluster mass M_{cl} and no binaries and stellar evolution is considered.

A star is disrupted if its z-component of angular momentum is less than the loss-cone limit (minimum angular momentum for an orbit of energy E):

$$J_z^{min}(E) = r_d\sqrt{2(E - GM_{\text{bh}}/r_d)} \qquad (2)$$

where r_d is the disruption radius of the BH, calculated as $r_d \propto r_*(M_{bh}/m_*)^{1/3}$ ([2]). r_* and m_* are the stellar radius and mass, respectively.

The central BH grows due to star accretion, which is driven by angular momentum diffusion. Using the time scales of replenishment $t_{in} \propto (Y^{min})^2$, and loss-cone depletion $t_{out} \propto (Y^{diff})^2$,[3] the probability that a loss cone orbit ($Y < Y_{min}$) diffuse an amount $\Delta Y < (Y - Y_{min})$ is calculated assuming a Gaussian distribution of orbits in Y with the dispersion Y^{diff}, centered at each (X, Y) grid cell.

2 Results

As initial cluster configurations, truncated King models with added bulk motion are used: $f(E, J_z) \propto \exp(-\beta\Omega_0 J_z) \cdot [\exp(-\beta(E - E_{tid})) - 1]$, ($E < E_{tid}$), where $\beta = 1/\sigma_c^2$ and Ω_0 is an angular velocity. The initial conditions of each model are fixed by the familiar King parameter W_0 and the initial rotational parameter $\omega_0 = \sqrt{9/(4\pi G n_c)}\Omega_0$, where G is the gravitational constant and n_c the central cluster density.

The results show that gravo-gyro effects, coupled to gravo-thermal instabilities, drive core collapse, with an acceleration of collapse occuring due to the presence of the BH. Furthermore, the BH acts as an energy source and reverses collapse, leading to post-collapse phase. Stars in the core are heated via the consumption of stars in bound, high energetic orbits. The outward flux of energy is achieved by small angle, two-body encounters in the cusp, by which some stars lose energy and move closer to the BH being eventually consumed, while the interacting stars gain energy and move outward from the cusp. At the same time, gravogyro effects carry out angular momentum from the core. As in models without rotation, they achieve final steady state solutions of the density cusp ($n \propto r^\lambda$, $\lambda = -1.75$) and the vel.disp.-cusp ($\sigma \propto r^\gamma$, $\gamma = -0.5$) inside the BH influence radius r_a, corresponding to Keplerian bounded orbits, independent of initial concentration and rotation.

The final M_{BH} stalls at $\sim 0.01 M_{cl}$ at collapse time, and remains nearly constant in the post-collapse phase, while BH mass accretion rate (dM_{BH}/dt) reaches a maximum at collapse time, due to the higher density of orbits in the core, and falls afterwards (see Fig. 1). Cluster mass (M_{cl}) loss in the tidal field of the parent galaxy is very strong during the re-expansion of the core. The M_{BH}-limit for a typical cluster of $5 \cdot 10^6 M_\odot$, varies between $1.9 \cdot 10^3 M_\odot$ and $9.5 \cdot 10^4 M_\odot$ for models of different parameters. This is a good approximation of the mass of IMBH estimated by theoretical studies and observations (see below).

Random vs. ordered motion during the evolution of BH-models, develops a growing V_{rot}/σ already at early times and a further increasing up to

---

[3] the square root dependence of J_z^{min} on time reflects the fact, that entry into the loss cone is a diffusive process

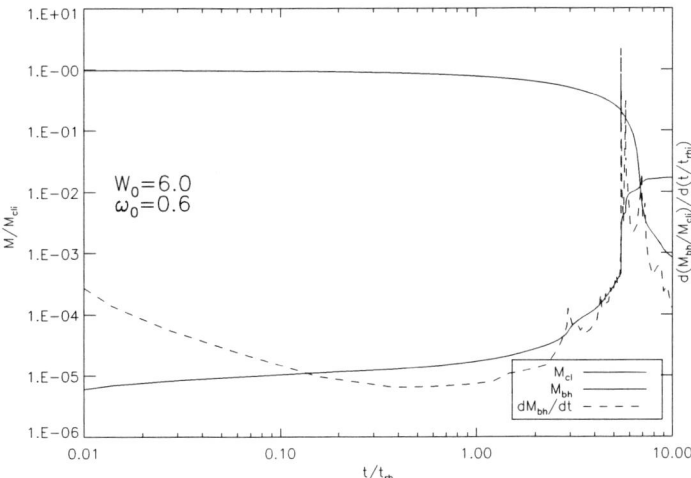

Fig. 1. Evolution of BH and cluster mass for a model $(6.0, 0.6, 5 \cdot 10^{-6})$. The *dot-dashed line* shows the mass of the cluster and the *solid line* the BH mass. Mass accretion rate is shown as a *dashed line* in units of $d(M_{\rm BH}/M_{\rm cl_i})/d(t/t_{\rm rh_i})$. The time is given in units of initial half-mass relaxation time ($t_{\rm rhi}$). Initial $M_{\rm cl} = 1$.

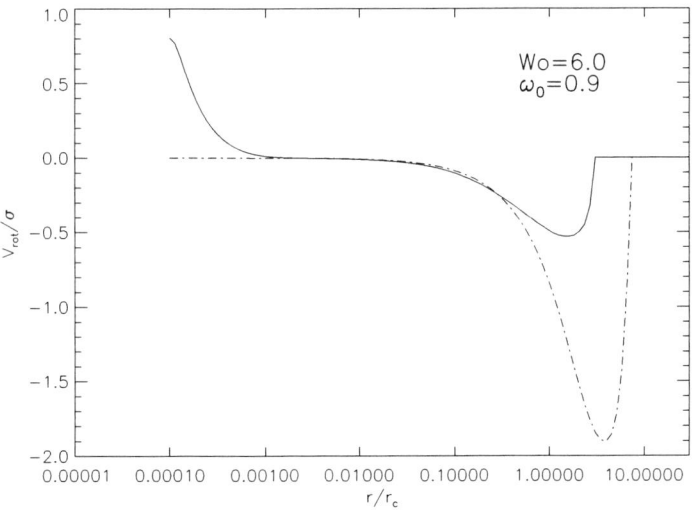

Fig. 2. $V_{\rm rot}/\sigma$ as a function of radius in units of initial core radius for the model B3 $(6.0, 0.6, 5 \cdot 10^{-6})$ at collapse time. The *dotted-dashed line* represents the initial profile. The location of the initial maximum $V_{\rm rot}/\sigma$ moves inward and drops in time. A second maximum grows in the core and in opposite direction of initial rotation.

collapse time. This behavior depends on an increasing central rotation due to the interplay of an efficient angular momentum diffusion and BH star accretion. ($W_0 = 6.0$, $\omega = 0.9$) models develop $V_{\rm rot}/\sigma \lesssim 1$ at the time of maximum rotation ($\sim t_{\rm cc}$, see Fig. 2). This result is comparable with the central $V_{\rm rot}/\sigma \sim 1$ profile of the galactic GC M15 [3–5].

Although the simplified assumptions of the present simulations need to include constraints in the evolution of rotating clusters like mass spectrum, stellar evolution, galactic tidal mass loss or shocks, which are targets for further development of the code, the results presented agree with the estimation of BH masses and rotational profiles given by observations and with current theoretical studies. Thus, rotation needs to be taken into account for the understanding of the formation and evolution of GCs, specially at early evolutionary times (e.g. in the young clusters of the LMC).

A detailed set of model data, covering a wide range of rotation rates and initial concentrations of GCs, with and without BH, is ready to use by observers for comparisons with their data.[4]

References

1. C. Einsel and R. Spurzem, Mon. Not. Royal Astron. Soc., **302**, 81 (1999)
2. J. Frank and M. Rees, Mon. Not. Royal Astron. Soc., **176**, 633 (1976)
3. K. Gebhardt et al.: Astrophys. J., **578**, 41 (2002)
4. J. Gerssen et al.: Astron. J., **124**, 3270 (2002)
5. J. Gerssen et al.: Astron. J., **125**, 376 (2003)
6. E. Kim, H.M. Lee and R. Spurzem, Mon. Not. Royal Astron. Soc., **351**, 220 (2004)

---

[4]The cluster database can be found on the web, at http://www.ari.uni-heidelberg.de/clusterdata/

The Dynamical Evolution of Young Clusters and Galactic Implications

Pavel Kroupa

Argelander Institute for Astronomy, Auf dem Hügel 71, D 53121 Bonn, Germany
pavel@astro.uni-bonn.de

Star clusters are observed to form in a highly compact state and with low star-formation efficiencies. If the residual gas is expelled on a dynamical time the clusters disrupt thereby (i) feeding a hot kinematical stellar component into their host-galaxy's field population, and (ii) if the gas-evacuation time-scale depends on cluster mass, then a power-law embedded-cluster mass function transforms within ten to a few dozen Myr to a mass function with a turnover near $10^5\,M_\odot$, thereby possibly explaining this universal empirical feature.

1 Early Cluster Evolution

The star-formation efficiency (sfe), $\epsilon \equiv M_{\rm ecl}/(M_{\rm ecl}+M_{\rm gas})$, where $M_{\rm ecl}, M_{\rm gas}$ are the mass in freshly formed stars and residual gas, respectively, is $0.2 \lesssim \epsilon \lesssim 0.4$ [8] implying that the physics dominating the star-formation process on scales < 10 pc is stellar feedback. Within this volume, the pre-cluster cloud core contracts under self gravity thereby forming stars ever more vigorously, until feedback energy suffices to halt the process (*feedback-termination*), [13]. This occurs on one to a few crossing times ($\approx 10^6$ yr), and since each protostar needs about 10^5 yr to accumulate about 95% of its mass [15], the assumption may be made that the very young cluster is mostly virialised at feedback-termination. Its stellar velocity dispersion, $\sigma \approx \sqrt{G\,M_{\rm ecl}/(\epsilon\,R)}$, may then reach $\sigma = 40$ pc/Myr if $M_{\rm ecl} = 10^{5.5}\,M_\odot$ which is the case for $\epsilon\,R < 1$ pc. This is easily achieved since the radius of one-Myr old clusters is $R \approx 1$ pc with a weak, if any dependence on mass.

The above exercise demonstrates that the possibility may be given that a *hot kinematical component* could add to a galactic disk as a result of clustered star formation for reasonable physical parameters. But this depends on (i) ϵ, (ii) R (cluster concentration) and (iii) the ratio of the gas-expulsion time-scale to the dynamical time of the embedded cluster, $\tau_{\rm gas}/t_{\rm cross}$.

1.1 Empirical Constraints

The first (i) of these is clearly fulfilled: $\epsilon < 40$ % [8]. The second (ii) also appears to be fulfilled such that clusters with ages $\lesssim 1$ Myr have $R \lesssim 1$ pc independently of their mass. Some well-studied cases are tabulated and discussed in [5]. Finally, the ratio $\tau_{\rm gas}/t_{\rm cross}$ (iii) remains uncertain but critical.

The well-observed cases discussed in [5] do indicate that the removal of most of the residual gas does occur within a cluster-dynamical time, $\tau_{\rm gas}/t_{\rm cross} \lesssim 1$. Examples noted are the Orion Nebula Cluster (ONC) and R136 in the LMC both having significant super-virial velocity dispersions. Other examples are the Treasure-Chest cluster and the very young starbursting clusters in the massively-interacting Antennae galaxy which appear to have HII regions expanding at velocities such that the cluster volume may be evacuated within a cluster dynamical time.

A simple calculation of the amount of energy deposited by an O star into its surrounding cluster-nebula also suggests it to be larger than the nebula binding energy [5]. Furthermore, [2] note that many young clusters have a radial-density profile signature expected if they are expanding rapidly.

Thus, the data suggest the ratio $\tau_{\rm gas}/t_{\rm cross}$ to be near one, but much more observational work needs to be done to constrain this number. Measuring the kinematics in very young clusters would be an extremely important undertaking, because the implications of $\tau_{\rm gas}/t_{\rm cross} \lesssim 1$ are dramatic.

To demonstrate these implications it is now assumed that a cluster is born in a very compact state ($R \approx 1$ pc), with a low sfe ($\epsilon < 0.4$) and $\tau_{\rm gas}/t_{\rm cross} \lesssim 1$. As noted in [5], "in the presence of O stars, explosive gas expulsion may drive early cluster evolution independently of cluster mass".

2 Implications

2.1 Heating Galactic-Field Populations

As one of the important implications, a cluster in the age range of $\approx 1-50$ Myr will have an unphysical M/L ratio because it is out of dynamical equilibrium rather than having an abnormal stellar IMF [2]. Another implication would be that a Pleiades-like open cluster would have been born in a very dense ONC-type configuration and that, as it evolves, a "moving-group-I" is established during the first few dozen Myr which comprises roughly 2/3rds of the initial stellar population and is expanding outwards with a velocity dispersion which is a function of the pre-gas-expulsion configuration [7]. These computations were in fact the first to demonstrate, using high-precision N-body modelling, that the re-distribution of energy within the cluster during the embedded phase and the expansion phase leads to the formation of a substantial remnant cluster despite the inclusion of all physical effects that are disadvantageous for this to happen (explosive gas expulsion, Galactic tidal field and mass loss from stellar evolution). A "moving-group-II" establishes later as the "classical" moving group made-up of stars which slowly diffuse/evaporate out of the re-virialised cluster remnant with relative kinetic energy close to zero.

Thus, the moving-group-I would be populated by stars that carry the initial kinematical state of the birth configuration into the field of a galaxy. Each generation of star clusters would, according to this picture, produce overlapping moving-groups-I (and II), and the overall velocity dispersion of the new field population can be estimated by adding in quadrature all expanding populations. This involves an integral over the embedded-cluster mass function, $\xi_{ecl}(M_{ecl})$, which describes the distribution of the stellar mass content of clusters when they are born [4,5]. Because the embedded cluster mass function is known to be a power-law this integral can be calculated for a first estimate. The result is that for reasonable upper cluster mass limits in the integral, $M_{ecl} \lesssim 10^5 \, M_\odot$, the observed age–velocity dispersion relation of Galactic field stars can be re-produced.

This theory can thus explain the much debated "energy deficit": that the observed kinematical heating of field stars with age cannot, until now, be explained by the diffusion of orbits in the Galactic disk as a result of scattering on molecular clouds, spiral arms and the bar [3]. Because the age–velocity-dispersion relation for Galactic field stars increases with stellar age, this notion can also be used to map the star-formation history of the Milky-Way disk by resorting to the observed correlation between the star-formation rate in a galaxy and the maximum star-cluster mass born in the population of young clusters [14].

2.2 Structuring the Initial Cluster Mass Function

Another potentially important implication from this theory of the evolution of young clusters is that *if* the gas-expulsion time-scale and/or the sfe varies with initial (embedded) cluster mass, then an initially featureless power-law mass function of embedded clusters will rapidly evolve to one with peaks, dips and turnovers at "final" cluster masses that characterize changes in the broad physics involved, such as the gas-evacuation time-scale.

As an example, [6] assumed that the function $M_{icl} = f_{st} M_{ecl}$ exists, where M_{ecl} is as above, M_{icl} is the "classical initial cluster mass" and $f_{st} = f_{st}(M_{ecl})$. The "classical initial cluster mass" is that mass which is inferred by classical N-body computations without gas expulsion (i.e. in effect assuming $\epsilon = 1$, which is however, unphysical). Thus, for example, for the Pleiades, $M_{cl} \approx 1000 \, M_\odot$ at the present time (age: about 100 Myr), and a classical initial model would place the initial cluster mass to be $M_{icl} \approx 1500 \, M_\odot$ by using standard N-body calculations to quantify the secular evaporation of stars from an initially bound and virialised "classical" cluster [10]. If, however, the sfe was 33% and the gas-expulsion time-scale was comparable or shorter than the cluster dynamical time, then the Pleiades would have been born in a compact configuration resembling the ONC and with a mass of embedded stars of $M_{ecl} \approx 4000 \, M_\odot$ [7]. Thus, $f_{st}(4000 \, M_\odot) = 0.38$.

By postulating that there exist three basic types of embedded clusters, namely clusters without O stars (type I: $M_{ecl} \lesssim 10^{2.5} \, M_\odot$, e.g. Taurus-Auriga pre-main sequence stellar groups, ρ Oph), clusters with a few O stars (type II: $10^{2.5} \lesssim M_{\rm ecl}/M_\odot \lesssim 10^{5.5}$, e.g. the ONC) and clusters with many O stars and with a velocity dispersion comparable to the sound velocity of ionized gas (type III: $M_{\rm ecl} \gtrsim 10^{5.5}\,M_\odot$) it can be argued that $f_{\rm st} \approx 0.5$ for type I, $f_{\rm st} < 0.5$ for type II and $f_{\rm st} \approx 0.5$ for type III. The reason for the high $f_{\rm st}$ values for types I and III is that gas expulsion from these clusters may be longer than the cluster dynamical time because there is no sufficient ionizing radiation for type I clusters, or the potential well is too deep for the ionized gas to leave (type III clusters). Type II clusters undergo a disruptive evolution and witness a high "infant mortality rate" [8], therewith being the pre-cursors of OB associations and open Galactic clusters.

Under these conditions and an assumed functional form for $f_{\rm st} = f_{\rm st}(M_{\rm ecl})$, the power-law embedded cluster mass function transforms into a cluster mass function with a turnover near $10^5\,M_\odot$ and a sharp peak near $10^3\,M_\odot$ [6]. This form is strongly reminiscent of the initial globular cluster mass function which is inferred by e.g. [11, 12, 9, 1] to be required for a match with the evolved cluster mass function that is seen to have a universal turnover near $10^5\,M_\odot$.

This ansatz may thus bear the solution to the long-standing problem that the deduced initial cluster mass function needs to have this turnover, while the observed mass functions of young clusters are feature-less power-law distributions.

References

1. H. Baumgardt, this volume (2006)
2. N. Bastian, S.P. Goodwin, MNRAS, 369, 9 (2006)
3. A. Jenkins, MNRAS, **257**, 620 (1992)
4. P. Kroupa, MNRAS, **330**, 707 (2002)
5. P. Kroupa, ESA SP-576: The Three-Dimensional Universe with Gaia, 629 (astro-ph/0412069) (2005)
6. P. Kroupa, C.M. Boily 2002, MNRAS, **336**, 1188 (2002)
7. P. Kroupa, S.J. Aarseth, J. Hurley, 2001, MNRAS, **321**, 699 (2001)
8. C.J. Lada, E.A. Lada, ARA&A, **41**, 57 (2003)
9. G. Parmentier, G. Gilmore, MNRAS, **363**, 326 (2005)
10. S.F. Portegies Zwart, S.L.W. McMillan, P. Hut, J. Makino, 2001, MNRAS, **321**, 199 (2001)
11. E. Vesperini, MNRAS, **299**, 1019 (1998)
12. E. Vesperini, MNRAS, **322**, 247 (2001)
13. C. Weidner, P. Kroupa, MNRAS, **365**, 1333 (2006)
14. C. Weidner, P. Kroupa, S.S. Larsen, 2004, MNRAS, **350**, 1503 (2004)
15. G. Wuchterl, W.M. Tscharnuter, A&A, **398**, 1081 (2003)

Simulations of Globular Clusters Merging in Galactic Nuclear Regions

P. Miocchi[1,2], R. Capuzzo Dolcetta[2], and P. Di Matteo[3,2]

[1] INAF - Osserv. di Teramo, Via M. Maggini, 64100 – Teramo, Italy
   miocchi@uniroma1.it
[2] Dipartimento di Fisica, Universitá di Roma "La Sapienza", P.le Aldo Moro, 2, I00185 – Rome, Italy
   roberto.capuzzodolcetta@uniroma1.it
[3] LERMA - Observ. de Paris, 61, Av. de L'Observatoire, 75014 – Paris, France
   paola.dimatteo@obspm.fr

Abstract. We present the results of detailed N-body simulations regarding the interaction of four massive globular clusters in the central region of a triaxial galaxy. The systems undergo a full merging event, producing a sort of 'Super Star Cluster' (SSC) whose features are close to those of a superposition of the individual initial mergers. In contrast with other similar simulations, the resulting SSC structural parameters are located along the observed scaling relations of globular clusters. These findings seem to support the idea that a massive SSC may have formed in early phases of the mother galaxy evolution and contributed to the growth of a massive nucleus.

1 Introduction

In the paper [11], we analyzed the head-on collision between two globular clusters (GCs) moving on quasi-radial orbits in the galactic central region, with the aim, also, to understand how effective is the cluster tidal destruction. One of the main findings was that sufficiently compact clusters (initial King concentration parameter $c \geq 1.6$) keeps bound a substantial amount of their mass up to the complete orbital decaying. Another important result is that the orbital energy dissipation due to the tidal interaction is of the same order of that caused by dynamical friction (df). In light of these results, and given that df was shown to be important indeed in segregating massive GCs in triaxial potentials [13,4], a natural further step in our investigation program is to study the possible merging of a set of GCs decayed in the central galactic region and the relevant characteristics of the super star cluster (SSC) they eventually form.

Furthermore, this topic is connected with the problem of the origin and formation history of the various kinds of SSCs recently observed in form of ultracompact dwarf galaxies [5], young massive clusters in starburst regions of interacting galaxies [9], and nuclear clusters in the central regions of early-type spirals [3,14]. One of the most debated questions is whether they formed from the condensation of primordial gas clouds or through the 'dissipationless' merging of smaller subunits. Recent indications coming from some N-body simulations [7,1,2] are in favour of this latter hypothesis. In this work we studied the merging process among massive GCs already decayed in the central region of a triaxial galaxy, including tidal distortion and df.

2 The Model

The galactic environment in which GCs are embedded is represented by an analytical potential derived from the self-consistent *triaxial* model described in [6] (see also [11]). The galactic core radius is 200 pc, the core mass is 3×10^9 $M_\odot$ with a central density $\rho_{g0} = 375$ $M_\odot \text{pc}^{-3}$. The deceleration due to df on the stellar motion is considered through a generalization of the Chandrasekhar formula to the triaxial case [12]. The four simulated GCs have an initial internal distribution sampled from King isotropic models with total mass, central velocity dispersion and half-mass radius ranging, rispectively, in $M \simeq 42\text{--}54 \times 10^6$ $M_\odot$, $\sigma \simeq 25\text{--}36$ km s^{-1}, $r_c \simeq 11\text{--}19$ pc. The reference frame has the origin at the galactic center and the x and z axes are, respectively, along the maximum and minimum axis of the triaxial ellipsoid. The clusters were initially located within 100 pc from the galactic center. We represented each GC with $N = 2.5 \times 10^5$ 'particles', simulating their dynamics by means of the parallel 'ATD' treecode [10, 11].

3 Results

From Fig. 1 it can be clearly seen that the merging occurs rather quickly, i.e. in less than 20 galactic core crossing time (~ 14 Myr), after which the resulting system attains soon a dynamical equilibrium configuration, as confirmed by the evolution of the lagrangian radii that are nearly constant throughout the duration of the simulation (~ 35 Myr)

The final SSC morphology is that of an axisymmetric ellipsoid (axial ratios 1.4 : 1.4 : 1, ellipticity $\simeq 0.3$) without figure rotation. Nearly all the progenitors mass ends up into the SSC, which shows a central velocity dispersion $\sigma_0 \simeq 150$ km s^{-1} and a half-mass radius of $r_h \simeq 40$ pc. This means that the SSC is located much closer to the σ_0–M scaling relation followed by GCs than to that of elliptical galaxies (Fig. 2 lower panel), contrarily to what found by [2,7]. This is probably due to that our system is located deep inside the galaxy potential well. Indeed, if one uses the virial relation to estimate the system total mass, $M = 7.5\sigma_0^2 r_h/G$ ([8]), then the location approaches the elliptical galaxies scaling law (Fig. 2).

The radial density profile of the SSC (Fig. 3) is similar to that given by the 'sum' of the initial profiles of the progenitors, with a central density a factor only 1.7 smaller and a slightly shallower profile outwards. This is in the direction of supporting the validity of the nuclear accretion model, suggested

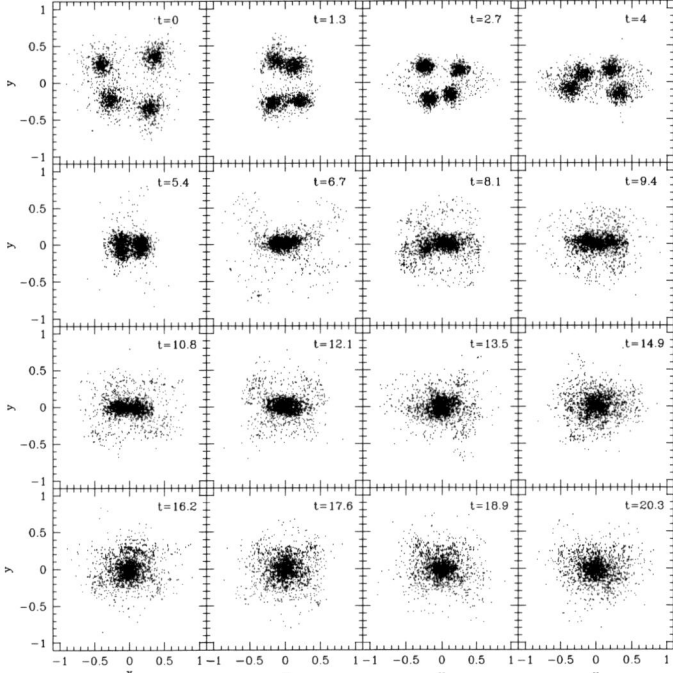

Fig. 1. Snapshots of the merging event (projection on the x–y plane). The labelled time is in unit of 0.8 Myr. Each snapshot size is ~ 400 pc.

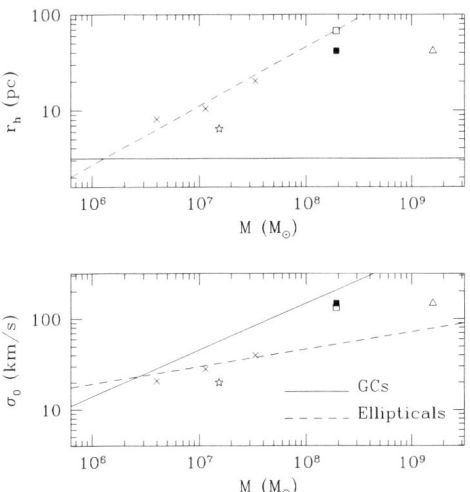

Fig. 2. Scaling relations for the last configuration of our SSC. The best fit relations for GCs and elliptical galaxies are also reported. The squares indicate the location of our SSC in the case where df is present (*filled*) or not (*open*), and when M is estimated by the virial relation (open triangles). The merging remnants locations for some of the cases simulated by [2] (*crosses*) and by [7] (*stars*) are also plotted.

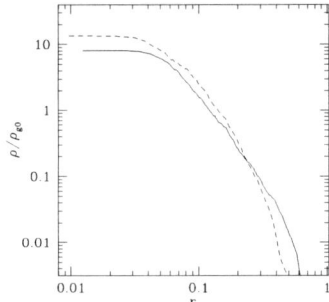

Fig. 3. Radial density profile of the SSC (*solid line*), compared with the profile given by the sum of the initial density distributions of the progenitors clusters (*dashed line*). The distance r (in units of 200 pc) is to the galactic center.

by [4]. Finally, we found that the SSC formed in the simulation without df, reaches an equilibrium state with similar morphological features, though in a nearly doubled time-scale and with a ~ 3 times lower central density.

References

1. H. Baumgardt, J. Makino, P. Hut et al.: ApJ **589**, L25 (2003)
2. K. Bekki, W.J. Couch, M.J. Drinkwater, Y. Shioya: ApJ **610**, L13 (2004)
3. T. Böker, M. Sarzi, D.E. McLaughlin et al.: AJ **127**, 105 (2004)
4. R. Capuzzo Dolcetta: ApJ **415**, 616 (1993)
5. M.J. Drinkwater et al.: A&A **355**, 900 (2000)
6. T. de Zeeuw, D. Merritt: ApJ **267**, 571 (1983)
7. M. Fellhauer, P. Kroupa: MNRAS **330**, 642 (2005)
8. M. Kissler-Patig, A. Jordán, N. Bastian: astro-ph/0512360 (2005)
9. C. Maraston, N. Bastian, R.P. Saglia et al.: A&A **416**, 467 (2004)
10. P. Miocchi, R. Capuzzo Dolcetta: A&A **382**, 758 (2002)
11. P. Miocchi, R. Capuzzo Dolcetta, P. Di Matteo, A. Vicari: ApJ, **644**, 940 (2006)
12. E. Pesce, R. Capuzzo Dolcetta, M. Vietri: MNRAS **254**, 466 (1992)
13. S. Tremaine, J.P. Ostriker, L. Spitzer Jr.: ApJ **196**, 407 (1975)
14. C.J. Walcher, R.P. van der Marel, D. McLaughlin et al.: ApJ **635**, 741 (2005)

The Origin of the Gaussian Initial Mass Function of Globular Cluster Systems

Geneviève Parmentier and Gerard Gilmore

Institute of Astronomy, University of Cambridge, Madingley Road, Cambridge CB3 0HA, UK gparm@ast.cam.ac.uk

Abstract. Evidence favouring a Gaussian initial mass function (IMF) for systems of old globular clusters (GC) has accumulated over recent years. We show that a bell-shaped mass function may be the imprint of expulsion from the protocluster of the leftover star forming gas due to supernova activity. Owing to the corresponding weakening of its gravitational potential, the protocluster retains a fraction only of its newly formed stars. The mass fraction of bound stars extends from zero to unity depending on the star formation efficiency (SFE) achieved by the protoglobular cloud. We investigate how such wide variations affect the mapping of the protoglobular cloud mass function to the GC IMF. An initial power-law cloud mass function generates cluster IMFs which are bell-shaped, with a turnover whose location is mostly driven by the lower limit of the cloud mass range.

1 The GC Initial Mass Function: Gas Removal Impact

The cluster mass function (i.e. the number of clusters per logarithmic mass interval $dN/d\log m$) is one of the primary characteristics of any GC system hosted by a massive galaxy. Intriguingly, it proves to be almost independent of the size, the morphological type or the environment of the host galaxy. This universal GC mass function is well fitted by a Gaussian with a mean of $\log m/M_\odot \simeq 5.2$. GCs having evolved over a Hubble-time in their galactic environment, their IMF has however remained controversial, with two competing hypotheses. It may have been a featureless power-law with a slope of ~ -1, the Gaussian function characteristic of old GC populations then resulting from a purely evolutionary effect, namely, the preferential removal of the more vulnerable low-mass clusters [2]. Yet, the present-day mass function represents an equilibrium state so that the initial one may also have been a Gaussian similar to that today [10]. *If so, the Gaussian shape is the fossil imprint of the GC formation process, this holding the clue to the universality of the observed GC mass function.*

Parmentier and Gilmore [8] and de Grijs, Parmentier and Lamers [1] provide evidence for a Gaussian cluster IMF in the Galactic halo and in the M82 B 1 Gyr-old starburst site, respectively (see also Vesperini et al. [11] for the case of the giant elliptical M87). Theoretical support for a bell-shaped cluster IMF has been missing so far however (although see [6]). As a result of the high SFE required to form a bound cluster, the GC IMF has often been postulated to mirror the mass function of the GC gaseous progenitors. Actually, numerous studies (e.g. [5,7,3]) point out that star forming clouds must be better than 30–50% efficient in converting gas into stars to produce bound stellar clusters. The limited variations in the SFE (i.e., less than a factor of 3) may then guarantee that the IMF of the clusters is that of their parent clouds.

However, following the dispersal of the residual star forming gas by the combined actions of stellar winds and supernova explosions, the newly formed stars suddenly find themselves in a shallower gravitational potential, resulting into either the escape of some of them or even the complete disruption of the protocluster. Therefore, the initial mass of a stellar cluster is not determined by the SFE only. It depends on the mass fraction of the cluster parent cloud which is turned into stars *remaining bound after the dispersal of the gaseous component*. Specifically, the cluster initial mass m_{init} obeys $m_{init} = F_{bound} \times SFE \times m_{cloud}$, where m_{cloud} is the mass of the gaseous progenitor and F_{bound} is the mass fraction of stars remaining bound after gas removal. The bound fraction F_{bound} ranges from 0 (when the SFE is smaller than a threshold value $SFE_{th} \simeq 0.35$) up to 1 (when the $SFE \lesssim 1$, so that gas removal is just a small perturbation of the stellar system). As a result, the assumed mirroring effect between the mass function of the cluster forming clouds on the one hand and the IMF of the clusters on the other hand can no longer be taken for granted, the latter depending on the former *and* on gas removal through the variations in the quantity $F_{bound} \times SFE$.

We now investigate whether one factor dominates the other.

2 From a Power-Law to a Bell-Shaped Mass Function

We assume that the protoglobular cloud mass spectrum obeys a power-law $dN \propto m^\alpha dm$, with α varying between -2.5 up to -1.5, as is observed for giant molecular clouds in the Local Group (e.g. [9]). As for the cloud mass range, we adopt the Jeans mass range, say, $[4 \times 10^5 M_\odot, 10^7 M_\odot]$. Star forming clouds being characterized by a range in their respective SFE, the cloud mass spectrum is convolved with an SFE probability distribution, which we describe as a decreasing power-law of slope δ and core r_c. Since star formation proceeds inefficiently on the average, these two parameters are bounded so that the mean SFE (i.e. averaged over the whole SFE probability distribution) is one per cent. We then account for the impact of gas removal by combining each SFE value to the corresponding fraction F_{bound} of bound stars (see [3], their Fig. 2). Finally, the initial mass m_{init} of a GC is derived following $m_{init} = F_{bound} \times SFE \times m_{cloud}$. The corresponding cluster IMFs, along with their parent cloud mass functions, are shown in Fig. 1. It is worth noting that, although the simulations were started with power-law cloud mass functions, the newly formed gas-free bound star clusters show

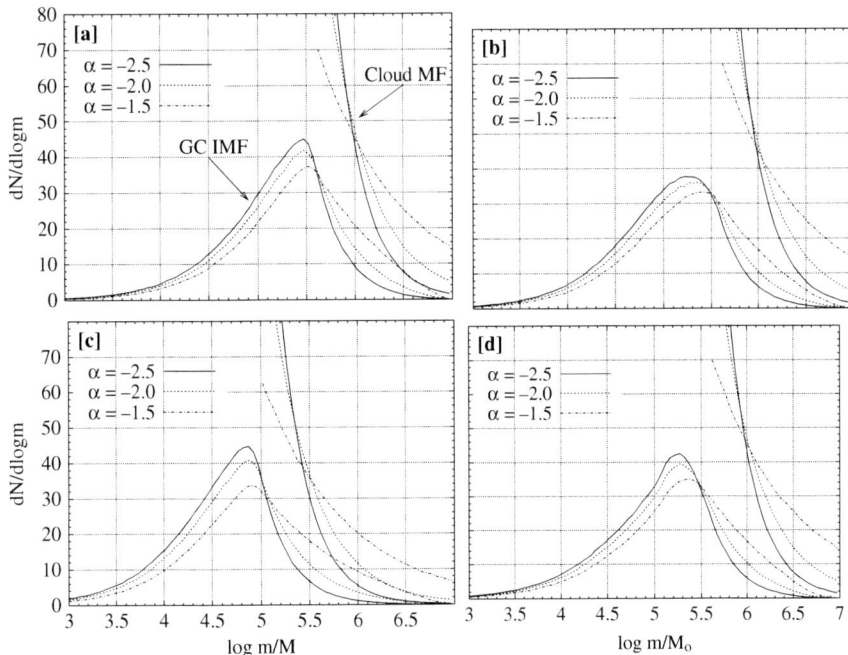

Fig. 1. Power-law protoglobular cloud mass functions and resulting bell-shaped GC IMFs. **(a)** $\delta = -2$, $m_{low} = 4 \times 10^5 M_\odot$, $m_{up} = 10^7 M_\odot$, $SFE_{th} = 0.35$, **(b)** same as (a) but $\delta = -4$, **(c)** same as (a) but $m_{low} = 10^5 M_\odot$, **(d)** same as (a) but $SFE_{th} = 0.15$ (see text for the meaning of the various symbols).

bell-shaped mass functions, that is, little memory of the cloud mass function is retained. These results highlight gas removal as a prime candidate mechanism responsible for generating bell-shaped (although not necessarily Gaussian) GC IMFs.

If the GC IMF is actually a bell-shape/Gaussian similar to that today, then the origin of the universal turnover location is locked into the cluster formation process. In the frame of this model, we have therefore explored what the cluster mass at the turnover depends on. Figure 1 illustrates the effect of varying [1] the spectral index α of the cloud mass spectrum (panels a-d), [2] the slope δ of the SFE probability distribution (panel b vs. panel a), [3] the lower limit m_{low} of the cloud mass range (panel c vs. panel a) and [4] the SFE threshold SFE_{th} (i.e. the minimum SFE requested for a protocluster to retain a bound core of stars) (panel d vs. panel a). We found no dependence of the turnover location on the upper limit m_{up} of the cloud mass range. [1] If $\delta = -2$, the location of the peak of the GC IMF is almost insensitive to α. Memory of the slope of the cloud mass spectrum is mostly retained in the skewness of the cluster IMF. If $\delta = -4$ or if $SFE_{th} = 0.15$, then the peak location depends moderately on α ($\Delta \log m \lesssim 0.2$). [2] The SFE probability distribution being very poorly constrained, we now consider a significantly steeper distribution and we adopt $\delta = -4$. This downward-shifts the turnover location moderately, that is, less than 0.2 in $\log m$. [3] Unlike α and δ, the lower limit of the cloud mass range is a prime controlling parameter of the turnover location, as highlighted by Fig. 1c. Specifically, a four times smaller lower limit (i.e. 10^5 M$_\odot$ instead of 4×10^5 M$_\odot$) results in a turnover shifted by -0.6 in $\log m$. [4] In all the simulations presented so far, we have assumed that gas removal proceeds instantaneously (i.e. on a time-scale τ_{gas} shorter than a protocluster crossing time τ_{cross}), which leads to $SFE_{th} \simeq 0.35$. A gas removal time-scale longer than a protocluster crossing time decreases this value however. Figure 1d shows the case where the ratio τ_{gas}/τ_{cross} is on the order of 10, corresponding to $SFE_{th} = 0.15$ ([4], their Fig. 4). This shifts the turnover downward by 0.2 at most.

As a summary, the turnover universality would originate mostly from a common value among galaxies for the lower mass limit of the protoglobular clouds (specifically those achieving an efficiency larger than the threshold), possibly with second-order variations driven by differences in the slope of the cloud mass spectrum, that of the SFE distribution, as well as by differences in the gas removal time-scale. We point however that our model still lacks a crucial ingredient, namely, the tidal field of the host galaxy since the F_{bound} vs. SFE relation we use has been derived in the case of isolated GCs. The tidal radius of a cluster depending on its mass, the tidal field may also contribute to the shape of the GC IMF. Whether the turnover location is affected as well remains to be investigated.

Acknowledgements. This work has been supported by a *Marie Curie* Intra-European Fellowship within the 6*th* European Community Framework Programme.

References

1. de Grijs R., Parmentier G., Lamers H.J.G.L.M.: MNRAS **364**, 1054 (2005)
2. Fall S.M., Zhang Q.: ApJ **561**, 751 (2001)
3. Fellhauer M., Kroupa P.: MNRAS **630**, 879 (2005)
4. Geyer M.P., Burkert A.: MNRAS **323**, 988 (2001)
5. Hills J.G.: ApJ **225**, 986 (1980)
6. Kroupa P., Boily, C.M.: MNRAS **336**, 1188 (2002)
7. Lada C.J., Margulis M., Dearborn D.: ApJ **285**, 141 (1984)
8. Parmentier G., Gilmore G.: MNRAS **363**, 326 (2005)
9. Rosolowski E.: PASP **117**, 1403 (2005)
10. Vesperini E.: MNRAS, **299**, 1019 (1998)
11. Vesperini E., Zepf S.E., Kundu A., Ashman K.M.: ApJ, **593**, 760 (2003)

Evolution of Globular Cluster Systems

E. Vesperini

Drexel University, Philadelphia, PA, USA vesperin@physics.drexel.edu

1 Introduction

Many of the theoretical predictions concerning the effects of dynamical evolution of globular clusters have been confirmed by observational studies (see e.g. [1, 2, 8]); however some of the observed global properties of GCS are apparently in conflict with what one would expect if dynamical evolution played an important role in shaping the current GCS properties. In particular observations show that the overall shape of GCMF and the mean mass of clusters in galaxies with different structure are very similar to each other and that, within individual galaxies, the mean mass of clusters does not significantly depend on the galactocentric distance.

Here I summarize the main results of a number of simulations aimed at studying the evolution of GCSs in elliptical galaxies and at exploring whether it is possible to reconcile the observed GCMF properties with the effects of evolutionary processes predicted by theoretical studies.

2 Evolution of GCS in Elliptical Galaxies

The left panel of Fig. 1 shows the final GCS mean mass $\overline{\log M}_f$, versus the effective mass of the host galaxy from our simulations ([9, 10]) following the evolution of GCS in a number elliptical galaxies with structural parameters equal to those determined by observational studies. We have investigated the evolution of GCS starting with a power-law GCMF as observed in young cluster systems in merging galaxies (see e.g. [13]) and with a log-normal mass function with initial mean mass and dispersion similar to those observed in old GCSs. For simulations with a power-law initial GCMF, the galaxy-to-galaxy dispersion of $\overline{\log M}_f$ is large and not consistent with the approximate universality of the GCSs mean mass reported by observations; moreover, the values of $\overline{\log M}_f$ are smaller than those observed. For models with a log-normal initial GCMF, on the other hand, both the values of $\overline{\log M}_f$ and the galaxy-to-galaxy dispersion are perfectly consistent with observations. The results obtained with a log-normal initial GCMF do not significantly depend on the exact shape of the GCMF and they are similar to those obtained for different bell-shaped (t-Student, two-slope power-law) initial GCMFs with a similar initial mean mass.

The right panel of Fig. 1 shows the fraction of clusters surviving after 15 Gyr versus the effective mass of the host galaxy for the simulations starting with a log-normal GCMF.

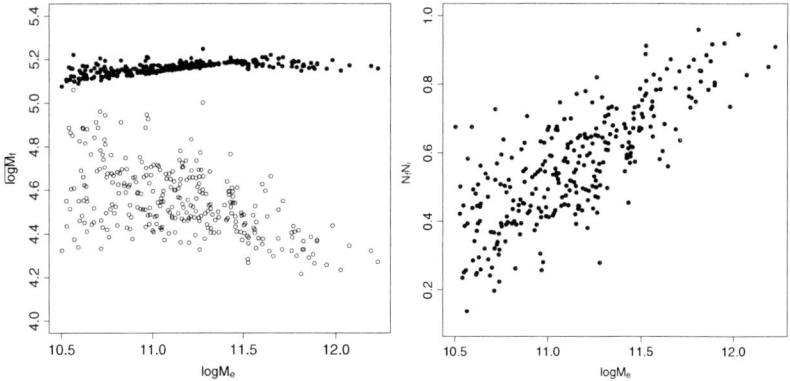

Fig. 1. (*Left panel*) Final GCS mean mass, $\overline{\log M}_f$, versus the logarithm of the effective mass of the host galaxy; open dots are from simulations with a power-law initial GCMF, filled dots are from simulations with a log-normal initial GCMF; (*right panel*) fraction of the initial population of clusters surviving after 15 Gyr versus the logarithm of the effective mass of the host galaxy from simulations with a log-normal initial GCMF.

As for the radial variation of the mean mass within individual galaxies, models starting with a log-normal initial GCMF have negligible radial gradients consistent with observations whereas in models starting with a power-law initial GCMF, the final mean mass strongly depends on the galactocentric distance and the radial gradients found are not consistent with observations.

3 Dynamical Evolution of the M87 GCS

In order to explore possible solutions to the problems found in models with a power-law initial GCMF, we have carried out a detailed study of the dynamical evolution of the M87 GCS ([11]) for which good observational constraints on the kinematics, on the spatial distribution and on the GCMF properties are available.

For the initial GCMF, we have considered a power-law function and a two-slope power-law with $f(M) \sim M^{-1.8}$ for $10^{5.25} < M/M_\odot < 10^7$ and $f(M) \sim M^{-0.2}$ for $10^4 < M/M_\odot < 10^{5.25}$ which is bell-shaped if binned in $\log M$ and it is similar to the current GCMF of old GCSs.

The initial GCS velocity distribution adopted is the same used in [3] in an investigation of the evolution of the Galactic GCS and it is characterized by a radial anisotropy increasing with the galactocentric distance as in the Osipkov-Merritt models ([6]); a number of simulations with different initial anisotropy have been carried out.

For the initial spatial distribution of the GCS, we have adopted a Navarro, Frenk & White profile [7] with scale radius, r_s, equal to 9.1 kpc which is the value required to fit the current spatial distribution of the M87 GCS (see [5]). In order to explore whether the observed inner flattening in the M87 GCS spatial distribution could result from the disruption of inner clusters we have also considered an initial NFW profile with $r_s = 0.1$ kpc.

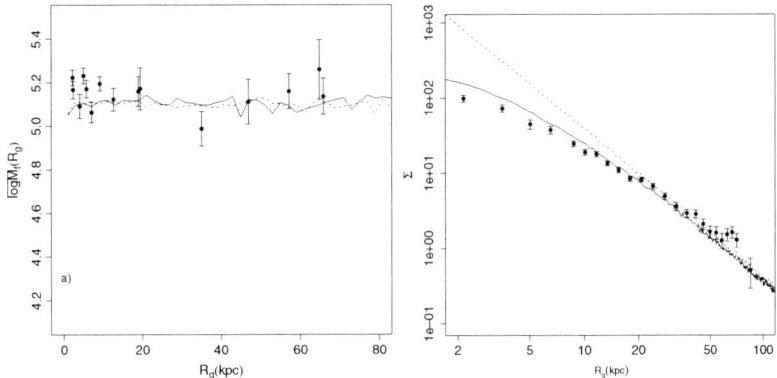

Fig. 2. (*Left panel*) Final M87 GCS mean mass vs projected galactocentric distance for models ([11]) with a two-slope power-law initial GCMF and initial anisotropy radius equal to 3 (solid line), 50 (dashed line) and 100 kpc (*dotted line*). Filled dots show observational data. (*Right panel*) Initial (*dashed line*) and final (*solid line*) surface number density profile from a simulation with a two-slope power-law initial GCMF. Filled dots show the M87 observed surface number density profile (from [5]).

The main results are:

(1) For a power-law initial GCMF, only in models with a strong initial radial anisotropy the final mean mass radial profile fits the flat mean mass profile observed. However, the final properties of these models have a very strong radial anisotropy inconsistent with observational kinematical data.
(2) For a two-slope power-law initial GCMF, the final mean mass radial profile is approximately flat and consistent with observations for all the initial spatial and velocity distributions considered (see left panel of Fig. 2). A number of these models have also final kinematical properties and final spatial distributions consistent with observations and they, therefore, satisfy all the observational constraints available.

(3) A centrally peaked initial spatial distribution can be significantly flattened in the inner regions by the preferential disruption of inner clusters and evolve into a final surface number density consistent with the observed spatial distribution of the M87 GCS (see right panel of Fig. 2).

4 Effect of the Early Dissolution of Low-Concentration Clusters on the GCMF Evolution

In [12] we have explored the role of early dissolution of clusters due to mass loss associated to stellar evolution on the evolution of the GCMF.

As shown in a number of theoretical investigations (see e.g. [4]), if the initial concentration of a cluster is not high enough, a cluster will expand and dissolve as a result of the mass loss due to stellar evolution. We have studied the evolution of GCSs in elliptical galaxies with effective masses and effective radii equal to those already considered for the simulations discussed in Sect. 2 and we have assumed an initial trend between cluster concentration and mass similar to that observed for Galactic clusters.

The results of our simulations show that the dissolution of low-concentration clusters can transform a power-law GCMF into a bell-shaped GCMF with a mean mass similar to that observed in old GCSs before the effects of disruption by two-body relaxation and dynamical friction become dominant. Two-body relaxation and dynamical friction then lead to the additional disruption of a significant fraction of the population of clusters unscathed by this process and alter the properties of the surviving clusters; nevertheless, as discussed in Sect. 2, when these other evolutionary processes act on a bell-shaped GCMF with a mean mass similar to that of old GCSs, the final values of the mean mass of GCSs, the mean mass galaxy-to-galaxy dispersion and the dependence of the mean mass on the galactocentric distance within individual galaxies are perfectly consistent with observations.

References

1. D.F.Chernoff, S.G. Djorgovski; ApJ, **339**, 339, (1989)
2. S.G. Djorgovski G. Meylan; AJ, **108**, 1292 (1994)
3. S.M. Fall, Q. Zhang: ApJ **561**, 751 (2001)
4. T. Fukushige, D.C. Heggie: MNRAS **276**, 206 (1995)
5. D.E. McLaughlin: AJ **117**, 2398 (1999)
6. D. Merritt: AJ **90**, 1027 (1985)
7. J. Navarro, C. Frenk, S. White: ApJ **462**, 563 (1996)
8. M. Odenkirchen, et al.; ApJ, **548**, L165 (2001)
9. E. Vesperini: MNRAS **318**, 841 (2000)
10. E. Vesperini: MNRAS **322**, 247 (2001)
11. E. Vesperini, S.E. Zepf, A. Kundu, K.M. Ashman: ApJ,**593**, 760 (2003)
12. E. Vesperini, S.E. Zepf: ApJ, ApJ,**587**, L97 (2003)
13. Q. Zhang, S.M. Fall, ApJ, **527**, L81, (1999)

Tidal Disruption and the Tale of Three Clusters

Guido De Marchi[1], Luigi Pulone[2], and Francesco Paresce[3]

[1] ESA, Space Science Department, Noordwijk, Netherlands `gdemarchi@esa.int`
[2] INAF, Observatory of Rome, Monte Porzio, Italy `pulone@mporzio.astro.it`
[3] INAF, Rome, Italy `fparesce@inaf.it`

Abstract. How well can we tell whether a cluster will survive the Galaxy's tidal forces? This is conceptually easy to do if we know the cluster's total mass, mass structure and space motion parameters. This information is used in models that predict the probability of disruption due to tidal stripping, disc and bulge shocking [7,6,2]. But just how accurate is the information that goes into these models and, therefore, how reliable are their predictions? To understand the virtues and weaknesses of these models, we have studied in detail three clusters (NGC 6397, NGC 6712, NGC 6218) whose predicted interaction with the galaxy is very different. We have used deep HST and VLT data to measure the luminosity function (LF) of stars throughout the clusters in order to derive a solid global mass function (GMF). The latter is the best tell-tale of the strength and extent of tidal stripping operated by the Galaxy. This is because the evaporation of stars from the cluster causes a progressive flattening of the IMF [9] and this effect is enhanced by tidal stripping and disc/bulge shocking. Therefore, at any time, the shape of the GMF must reflect the past interaction of the cluster with the Galaxy. Since the three clusters that we have studied have widely different probabilities of disruption (see the predicted times to disruption $T_{\rm d}$ in Table 1), we expected to find widely different GMFs. We indeed found that the GMF of the three clusters is different, but not in the way predicted by the models, as we explain below.

To derive a reliable GMF, it is necessary to measure the stellar LF at various locations in the cluster, in order to build a solid model of the cluster's dynamics. Our accurate measurements in several bands [4,1,5] have allowed us to determine the GMF of the three clusters for stars down to $\sim 0.2\,{\rm M}_\odot$, where it is most sensitive to tidal stripping (Fig. 1). While the GMF of NGC 6397 is that typical of globular clusters, with a peak at $\sim 0.3\,{\rm M}_\odot$, NGC 6218 and NGC 6712 have very flat GMF. Near the half-mass radius, the LF that we observed for these two clusters is severely depleted at the low-mass end. But while the orbit of NGC 6712 is compatible with "ferocious" stripping, that of NGC 6218 is not [3]. However, our analysis shows that the orbit of NGC 6218 used in these models is not realistic. In fact, more recent work based on the Hipparcos reference system [8] indicates for NGC 6218 an irregular orbit with shorter perigalactic distance, more prone to extensive stripping. We conclude that existing models of cluster disruption are limited by the lack of information on the precise cluster space motions. However, they benefit from an accurate knowledge of the GMF. Before the advent of GAIA, which will make

---

Space limitations do not allow us to present our work in full detail. We, therefore, include here a brief summary of the main results. For the complete version of this paper, the reader should refer to De Marchi et al. [5].

it possible to measure reliably the orbits of many clusters, the only way forward to understand the evolution of the globular cluster system rests on the accurate measurement of the GMF of a large number of these objects. A telescope such as the VLT is perfectly suited for this task.

Table 1. Predicted times to disruption T_d in Gyr

Reference	NGC 6397	NGC 6218	NGC 6712
[7]	2.1	14.5	0.3
[6]	3.9	29.4	3.7
[2]	11.3	16.3	9.0

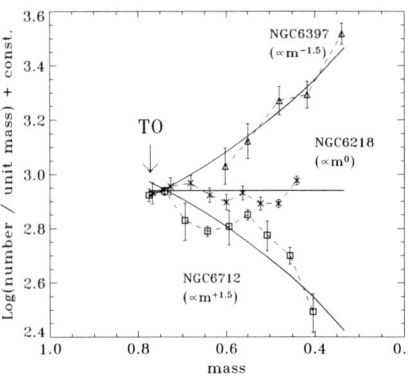

Fig. 1. The *solid lines* show the best fitting power-law GMF as derived using our dynamical models applied to a set of radial LFs. The symbols joined by the dashed lines show the local MF as measured near the half-mass radius of each cluster, proving that the latter is a good approximation to the GMF. While the time to disruption (T_d) of NGC 6712 justifies its falling GMF, NGC 6218 should have a GMF as steep as that of NGC 6397, given its very long T_d.

References

1. G. Andreuzzi, G. De Marchi, F. Ferraro, F. Paresce, L. Pulone, R. Buonanno: A&A, **372**, 851 (2001)
2. H. Baumgardt, J. Makino: MNRAS, **340**, 227 (2003)
3. B. Dauphole, M. Geffret, J. Colin, C. Ducourant, M. Odenkirchen, H. Tucholke: A&A, **313**, 119 (1996)
4. G. De Marchi, F. Paresce, L. Pulone: ApJ, **530**, 342 (2000)
5. G. De Marchi, L. Pulone, F. Paresce: A&A, **449**, 161 (2006)
6. D. Dinescu, T. Girard, W. van Altena: AJ, **117**, 1792 (1999)
7. O. Gnedin, J. Ostriker: ApJ, **486**, 581 (1997)
8. M. Odenkirchen, P. Brosche, M. Geffert, H. Tucholke: NewA, **2**, 477 (1997)
9. E. Vesperini, D. Heggie: MNRAS, 289, **898** (1997)

Tidal Tails Around Globular Clusters: Are they Good Tracers of Cluster Orbits?

P. Di Matteo[1,2], R. Capuzzo Dolcetta[2], P. Miocchi[3,2], and M. Montuori[4,2]

[1] Obs. de Paris, 61, Av. de L'Observatoire 75014 Paris, France
   paola.dimatteo@obspm.fr
[2] Dip. di Fisica, Universitá di Roma La Sapienza, P.le Aldo Moro 2, 00185 Roma, Italia roberto.capuzzodolcetta@uniroma1.it
[3] INAF - Oss. di Teramo, via M. Maggini, 64100 Teramo, Italia
   miocchi@uniroma1.it
[4] CNR - Ist. Sistemi Complessi, Roma, Italia montuorm@roma1.infn.it

1 Introduction

In the last decade, observational studies have shown the existence of tidal streams in the outer part of many galactic globular clusters [6,7]. The most striking examples of clusters with well defined tidal tails are represented by Palomar 5 [9] and NGC 5466 [2] (both observed in the framework of the Sloan Digital Sky Survey), which show structures elongated for 4 kpc and 1 kpc in length, respectively. Unfortunately, most of the observational studies about globular clusters (GCs) do not cover such a large field of the sky as the SDSS does.

In this framework, by mean of a parallel, adaptive tree-code [8], we performed detailed N-body simulations of GCs moving in a realistic three-components (bulge, disk and halo) Milky Way potential [1], in order to clarify whether and to what extent tails in the clusters outer regions (few tidal radii) are tracers of the local orbits and, also, if some kind of correlation exists among the cluster orbital phase and the orientation of such streams.

2 Results

While the outer part of the tails are good tracers of GC trajectory, the inner part is never aligned with the GC path, except for more eccentric orbits, when the GC moves towards the pericenter [3,4]. For a given orbit, a strong correlation exists between tails alignment and GC orbital angular velocity (see Fig. 1): when $|\omega|$ increases (moving to pericenter), tails are stretched along the GC path (the angle between tails and GC orbital velocity decreases and reaches its minimum value just before the pericenter passage); when $|\omega|$ decreases (moving to apocenter) tails deviate from the GC path and, in turn, the angle between tails and galactic center (black curve in Fig. 1) diminishes, indicating that, in this phase, tails are more radially pointing. For a discussion, see also [3–5].

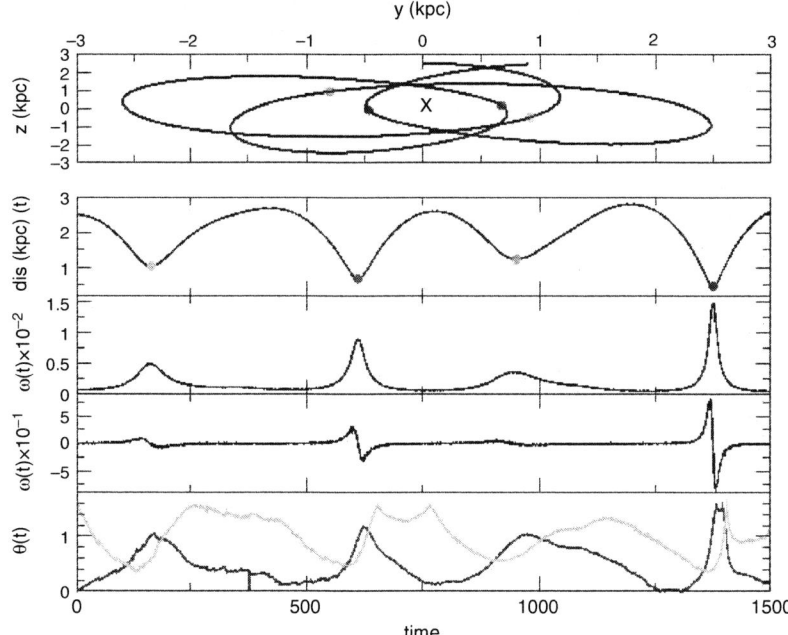

Fig. 1. From *top* to the *bottom*: 1. Plot of the GC orbits; 2. Distance of the GC from the Galaxy center as a function of time; 3. $|\omega|$, as a function of time; 4. Derivative with time of $|\omega|$; 5. Angles formed by the inner part of the tails with the galactic center direction (*black curve*) and with the GC local orbit (*grey curve*). A strong correlation exists between tails alignment and GC orbital angular velocity.

References

1. C. Allen, A. Santillan: RevMexAA, **22**, 255 (1991)
2. V. Belokurov, N.W. Evans, M.J. Irwin et al.: ApJL, **637**, 29 (2006)
3. R. Capuzzo Dolcetta, P. Di Matteo, P. Miocchi: AJ, **129**, 1906 (2005)
4. P. Di Matteo, R. Capuzzo Dolcetta, P. Miocchi: CelMecDynAstr, **91**, 59 (2005)
5. P. Di Matteo, R. Capuzzo Dolcetta, A. Lepinette et al.: in preparation
6. C. Grillmair, K. C. Freeman, M. Irwin, et al.: AJ, **109**, 2553 (1995)
7. S. Leon, G. Meylan, F. Combes: A&A, **359**, 907 (2000)
8. P. Miocchi, R. Capuzzo Dolcetta: A&A, **382**, 758 (2002)
9. M. Odenkirchen, E. K. Grebel, W. Dehnen, et al.: AJ, **126**, 2385 (2003)

Modelling the Tidal Tails of NGC 5466

M. Fellhauer[1], V. Belokurov[1], M.I. Wilkinson[1], N.W. Evans[1], and G. Gilmore[1]

Institute of Astronomy, University of Cambridge, Madingley Road, Cambridge CB3 0HA, UK madf, vasily, markw, nwe, gilmore @ast.cam.ac.uk

1 Observational Evidence

In a recent paper, Belokurov et al. ([1], B06) report the discovery of tidal tails around the globular cluster (GC) NGC 5466 using wide-field data from Sloan Digital Sky Survey (SDSS). The tails extend ≈ 4 degree on the sky, corresponding to about 1 kpc in projected length. Later, Grillmair and Johnson [7] reported the discovery of a $45°$ tail associated with NGC 5466. For a detailed description of the properties and data of NGC 5466 we refer to our journal paper ([5]F07) and references therein.

2 Results

In our theoretical study, we use the observed position [8] and velocity (Dinescu et al. 1999) as well as the internal properties [8–10] of NGC 5466 and adopt two standard potentials for the Milky Way, namely a Miamoto-Nagai disc combined with a logarithmic halo and a Hernquist bulge as described in Dinescu et al. [3] and a Dehnen and Binney [2] model of the Milky Way. With this information, we calculate the orbit 10 Gyr backwards in time. At this point, we insert a Plummer sphere with 1000000 particles representing our initial model of NGC 5466 and simulate the evolution of the GC until today using the particle-mesh code Superbox [4]. For a detailed description of the setup, we refer to our journal paper (F07).

The results from our theoretical investigation of NGC 5466 are two-fold. First, we showed that the reported proper motion yields tidal tails misaligned with the observations. The leading tail always emerges from the side pointing towards the Galactic Centre (GalC) and turns backs to the orbital path from 'within', while the trailing tail emerges from the side opposite to the GalC and turns back to the orbital path from 'outside'. This is not the case in Fig. 1 of B06. With our models, we showed that this is not a projection effect, but rather that the proper motion of the globular cluster has to be smaller in declination or larger in right ascension to account for the position of the tidal tails. The surface density of the tidal tails falls off along the innermost tails very steeply and stays at a very low density of 20–50 $M_\odot \, \mathrm{deg}^{-2}$ throughout the tails.

Second, we verified the theoretical prediction by Gnedin et al. [6] that a globular cluster of mass, size and orbit of NGC 5466 should mainly be affected by tidal shocks at each disc crossing. Internal effects play a less important role in the mass-loss, even though they are not negligible. If the observationally determined mass-to-light ratio of 1 is correct, the initial mass of this globular cluster was of the order $7 \cdot 10^4$ $M_\odot$. This initial mass refers to a point in evolution after the GC has lost its gaseous content and the violent stellar evolution of the most massive stars is over, i.e. the GC has reached a quasi-equilibrium state.

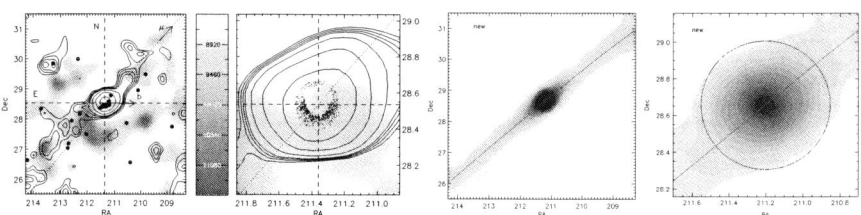

Fig. 1. Contour plots of NGC 5466. *Left*: Observational data from B06. *Arrow* shows the proper motion vector of Dinescu et al. [3]. *Right*: Our best fit model with slightly altered proper motion to match the tails of B06. *Solid line* shows the orbit, circle marks the tidal radius.

References

1. Belokurov V., Evans N.W., Irwin M.J., Hewett P.C., Wilkinson M.I., 2006, ApJL, 637, 29L
2. Dehnen W., Binney J., 1998, MNRAS, 294, 429
3. Dinescu D.I., Girard T.M., van Altena W.F., 1999, AJ, 117, 1792
4. Fellhauer M., Kroupa P., Baumgardt H., Bien R., Boily C.M., Spurzem R., Wassmer N., 2000, NewA, 5, 305
5. Fellhauer, M., Evans, N.W., Belokurov, V., Wilkinson, M.I., Gilmore, G., 2007, MNRAS, 380, 749
6. Gnedin O.Y., Lee H.M., Ostriker J.P., 1999, ApJ, 522, 935
7. Grillmair C.J., Johnson R., 2006, ApJL, 639, 17L
8. Harris W.E., 1996, AJ, 112 1487
9. Lehmann I., Scholz R.-D., 1997, A&A, 320, 776
10. Pryor C., McClure R.D., Fletcher J.M., Hesser J.E., 1991, AJ, 102, 1026

The Search for Tidal Tails of Globular Clusters: NGC4147

Katrin Jordi and Eva K. Grebel for the SDSS Collaboration

Astronomical Institute, Department of Physics & Astronomy, University of Basel, 4102 Binningen, Switzerland `jordi@astro.unibas.ch`, `grebel@astro.unibas.ch`

Globular clusters are stellar systems undergoing internal and external evolution. As a consequence member stars gain or lose energy and leave the gravitational potential of the cluster, so clusters are slowly dissolving. The major cause for this mass loss is the Galactic tidal field. Especially globular clusters on highly inclined orbits experience tidal shocks during their passage through the Galactic disk or close by the Galactic bulge.

In this project we are searching in the Sloan Digital Sky Survey (SDSS) for the tidally disrupted cluster members which are spread along the orbital path of the cluster and thus forming tidal structures of various dimensions.

The SDSS is the largest photometric and spectroscopic survey in the optical range. It is mapping out more than one quarter of the entire sky, and is measuring positions and magnitudes for over 100 million celestial objects. For technical details on the SDSS see [6]. The SDSS covers 14 globular clusters which we can study, [2].

We applied the same method as [4] to map cluster stars. We used color-magnitude-selected star counts in a $3°$ by $3°$ field on the sky. We defined a cluster sample, stars within $4'$ from the center, and a field sample, stars at least $30'$ away from the center of NGC4147. From the cluster sample we defined new orthogonal color indices, c_1 and c_2, as functions of the SDSS colors g-r and g-i.

The cluster sample contains stars of a common origin, and thus they occupy a very distinct region in the color-magnitude diagram (CMD). Here the CMD is the plane spanned by c_1 and i or c_2 and i . In our counting algorithm the single stars are weighted according to their position in the CMD and these weights are summed for all stars in 4 arcmin^2 bins on the sky. This gives us the estimated number of cluster stars in each bin. For the inner part of the cluster we used stars from the photometric standards from [5]; their BVRI photometry was transformed into SDSS photometry with the transformation equations from [3].

With a distance of 21 kpc, NGC4147 is a member of the Galactic halo. It lies near the north Galactic pole, at b=$77°$, on an orbit that is highly inclined with respect to the Galactic plane. The cluster will feel tidal shocks every time it passes through the Galactic disk. [1] concluded from 2MASS data that NGC4147 is immersed in the Sgr Stream and physically associated with the

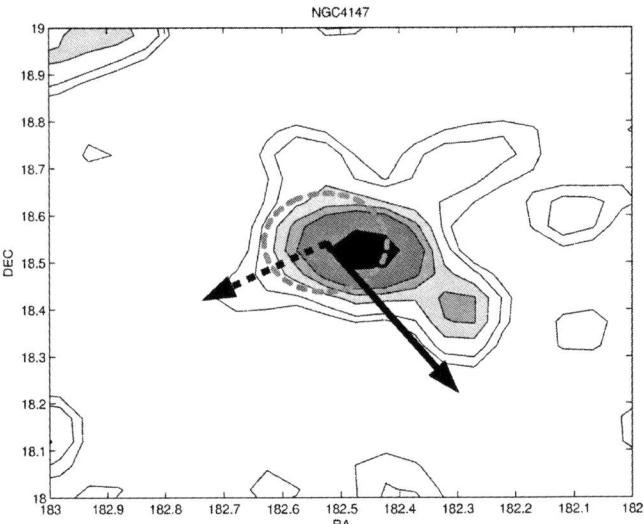

Fig. 1. Map of the surface density of the 1° by 1° field around NGC4147. The *shaded areas* represent regions of different cluster star densities, 1.5, 2, 3, 4, 5 and 10 times the average field star density of 0.37 arcmin^{-2}. The *dashed ellipse* marks the nominal tidal radius of NGC4147. The *arrow* points in the direction of the proper motion derived by [7], the *dashed arrow* points towards the Galactic Center.

structure. Figure 1 maps the surface density of stars that are photometrically concordant with the stellar population of NGC4147 in a 1° by 1° field centered on the cluster. One can clearly see an elongated density enhancement which stretches out about three tidal radii.

Funding for the SDSS and SDSS-II has been provided by the Alfred P. Sloan Foundation, the Participating Institutions, the National Science Foundation, the U.S. Department of Energy, the National Aeronautics and Space Administration, the Japanese Monbukagakusho, the Max Planck Society, and the Higher Education Funding Council for England. The SDSS Web Site is http://www.sdss.org/. KJ and EKG gratefully acknowledge support by the Swiss National Science Foundation.

References

1. Bellazzini, M., Ibata, R., Ferraro,F.R., and Testa, V., 2003, A&A, 405, 577
2. Harris, W.E., 1996, AJ, 112, 1487
3. Jordi, K. and Grebel, E.K., 2005, AN., 326, 657
4. Odenkirchen, M. et al., 2003, AJ, 126, 2385
5. Stetson, P.B., 2000, PASP, 112, 925
6. York, D.G., Adelman, J., Anderson, J.E., et al., 2000, AJ, 120, 1579
7. Wang, J.-J. and Wu, Z.-y., 2000, ChA&A, 24, 61

Internal Rotation of Young Globular Clusters

E. Vesperini[1] and S.E. Zepf[2]

[1] Drexel University, Philadelphia, PA, USA vesperin@physics.drexel.edu
[2] Michigan State University, East Lansting, MI, USA zepf@pa.msu.edu

1 Introduction

We are currently carrying out a survey of N-body simulations to study the early and long-term evolution of clusters with initial conditions similar to those suggested by observations of young clusters.

Here we summarize some of the results of N-body simulations of the very early evolution of clusters and show how the kinematical properties of equilibrium clusters emerging at the end of violent relaxation are affected by the motion of the cluster in its host galaxy and by the host galaxy tidal field.

2 Results

The results discussed here are from N-body simulations following the violent relaxation of a cluster evolving in an external tidal field equal to that of the Milky Way in the solar neighborhood. The cluster is initially cold ($Q = T/V < 0.5$), spherical, homogeneous and with a limiting radius equal to 0.5 the tidal radius. All the results of these simulations and their dependence on the initial conditions adopted will be discussed in detail in [5].

Figure 1 (left panel) shows the radial profile of the ratio of the rotation to the dispersion velocity, V_{rot}/σ, at the end of violent relaxation. The Coriolis term in the equations of motion couples the internal dynamics of stars in the cluster with the external dynamics of the cluster motion around its host galaxy. As a consequence of this coupling, in a cold stellar system which collapses while it rotates around the center of the host galaxy, the radial motion of stars is deflected tangentially and a cluster, as the left panel of Fig. 1 shows, can acquire a significant internal rotation. Figure 1 (right panel) shows the orientation of the rotation axis: as expected, the rotation axis is perpendicular to the orbital plane of the cluster (which for these simulations lies on the Galactic disk plane). Such a rotation is likely to have a significant impact on the subsequent evolution of clusters (see e.g. [1]) and we are currently exploring their long-term evolution as driven by two-body relaxation and the differences in the long-term evolution between rotating and non-rotating clusters.

The rise of V_{rot}/σ in the very inner regions of the simulated cluster ($r/r_h < 0.2$) resembles a similar feature found in the observational studies of the kinematics of M15 (see e.g. [2,3]) whose origin is still unexplained and has also been interpreted as a possible signature of a central massive black hole (see e.g. [4]). The internal rotation we find is due to statistical fluctuations in the measured value of V_{rot}/σ in the very inner regions of the cluster where the number of stars per bin is increasingly small. The dashed lines show the radial profile of the average V_{rot}/σ ($\pm 1\sigma$) expected for a non-rotating cluster with the same total number of stars in each bin as used to calculate the profile of the simulated cluster. We believe this effect might explain also the observed rise of V_{rot}/σ found in the innermost regions of M15 and we expect that a similar feature is likely to be found in all the observational studies of the kinematics of the innermost regions of globular clusters as the number of stars sampled to probe the kinematics decreases.

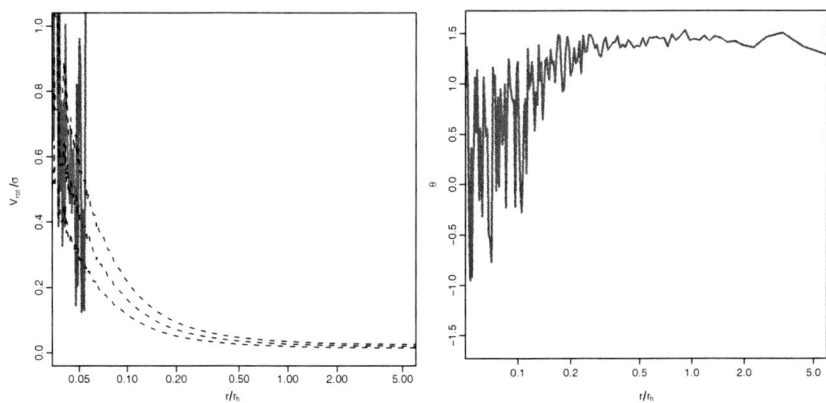

Fig. 1. *Left panel*: radial profile of V_{rot}/σ for the simulated cluster; the dashed lines show the expected V_{rot}/σ profile for a non-rotating cluster (see text for further discussion). *Right panel*: radial profile of the orientation of the rotation axis: with the exception of the innermost regions which are dominated by small number statistical noise, the rotation axis is perpendicular ($\theta \sim \pi/2$) to the orbital plane of the cluster.

References

1. C. Einsel, R. Spurzem, MNRAS **302**, 81, (1999)
2. K. Gebhardt et al., ApJ **119**, 1268 (2000)
3. J. Gerssen et al., AJ **124**, 3270 (2002)
4. M.C. Miller, E.J.M. Colbert, Int. J. Mod. Phys. **D13**, 1 (2004)
5. E. Vesperini, S. Zepf, in preparation. (2006)

Mass Segregation in Young Star Clusters

E. Vesperini[1], S.L.W. McMillan[2], and S.F. Portegies Zwart[3]

[1] Drexel University, Philadelphia, PA, USA vesperin@physics.drexel.edu
[2] Drexel University Philadelphia, PA, USA steve@physics.drexel.edu
[3] University of Amsterdam, Amsterdam, The Netherlands spz@science.uva.nl

1 Introduction

A number of observational studies have shown the presence of mass segregation in young clusters with ages less than the timescale needed to produce the observed segregation by two-body relaxation (e.g. [3]). In addition, some studies have suggested that the observed mass segregation could be primordial and related to the higher gas accretion rate for stars in the centers of young clusters ([1,2]). Here we report some results from an extensive numerical study aimed at studying the evolution of mass segregation during the very early evolution of cold ($Q = T/V < 0.5$) star clusters as they undergo violent relaxation and evolve toward virial equilibrium.

2 Results

The left panel of Fig. 1 shows the evolution of the ratio of the half-mass radius of the light stars ($0.1 < m < 0.12 m_\odot$) to that of the massive stars ($m > 4 m_\odot$) in two of our simulations. These simulations contain $N = 64k$ particles drawn from a Kroupa ([4]) stellar mass function between $m_{min} = 0.1 m_\odot$ and $m_{max} = 100 m_\odot$. The system is initially homogeneous and far from virial equilibrium ($Q = 0.001$): it collapses, reaching maximum contraction after about one dynamical time, then re-expands and eventually settles into a final quasi-equilibrium state. The solid curve in the left panel of Fig. 1 shows how mass segregation is produced during the maximum contraction stage ($t \sim 8$ N-body units): when the system re-expands after this stage, the massive stars are significantly more concentrated than the light stars (see also the right panel of Fig. 1). After the jump around the time of maximum contraction, $R_h(\text{light})/R_h(\text{massive})$ stays approximately constant for the rest of the simulation and subsequent mass segregation occurs on a much longer timescale, proportional to the half-mass relaxation time of the final equilibrium system. Also included in the left panel of Fig. 1 is the evolution of $R_h(\text{light})/R_h(\text{massive})$ for a simulation with the same initial conditions but with softening chosen to suppress the effects of two-body relaxation during the maximum contraction stage: the fact that the mass segregation produced in this case is substantially smaller suggests that mass segregation in the unsoftened simulation is due mostly to the effects of two-body relaxation around the time of maximum contraction.

We are currently studying the dependence of the mass segregation produced during the early stages of a cluster evolution on the cluster initial conditions (e.g. virial ratio, density profile, number of particles, number of clumps and filling factor for clumpy initial conditions). We are also studying the implications of early mass segregation for the long-term evolution of clusters. The results of this investigation will be discussed in a series of papers currently in preparation.

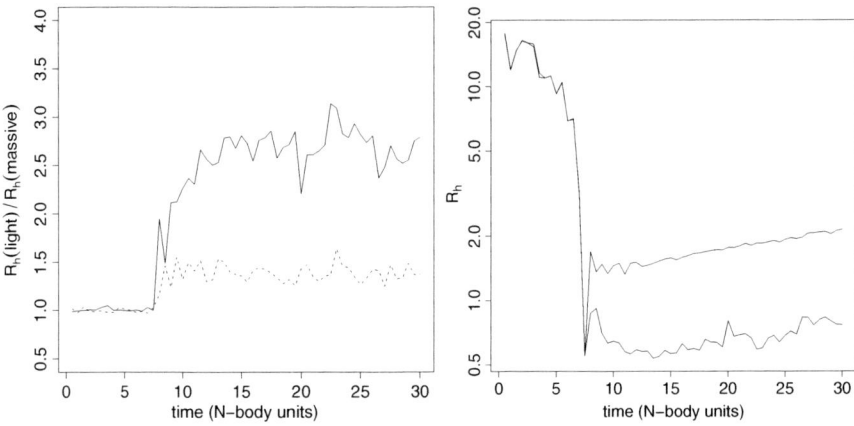

Fig. 1. (*Left panel*) Time evolution of the ratio of the half-mass radius of light stars ($0.1 < m/m_\odot < 0.12$) to the half-mass radius of massive stars ($m > 4m_\odot$) during violent relaxation of an initially cold homogeneous star cluster. The *solid line* shows the results of a simulation without softening while *dashed line* is from a simulation with softening. (*Right panel*) Time evolution of the half-mass radius of light stars ($0.1 < m/m_\odot < 0.12$) (*upper curve*) and of massive stars ($m > 4m_\odot$) for the simulation without softening.

References

1. I. A. Bonnell, M.R. Bate, C.J. Clarke, J.E. Pringle MNRAS, **323**, 785, (2001)
2. I. A. Bonnell, M.R. Bate, MNRAS, 370, 488 (2006)
3. R. de Grijs, G.F. Gilmore, R.A. Johnson, A.D. Mackey, MNRAS **331**, 245, (2002)
4. P. Kroupa, MNRAS **322**, 231, (2001)

Part IX

Dynamics of Globular Cluster Systems

Kinematics of Globular Cluster Systems

Aaron J. Romanowsky

Departamento de Física, Universidad de Concepción, Casilla 160-C, Concepción, Chile `romanow@astro-udec.cl`

I review the field of globular cluster system (GCS) kinematics, including a brief primer on observational methods. The kinematical structures of spiral galaxy GCSs so far appear to be broadly similar. The inferred rotation and mass profiles of elliptical galaxy halos exhibit a diversity of behaviors, requiring more systematic observational and theoretical studies.

1 Introduction

Globular clusters (GCs) are well known as useful probes of the formation histories of galaxies, especially at early times. *Kinematics* of globular cluster systems (GCSs) brings an extra dimension to these probes, helping in the first place to *distinguish GC subsystems* by their distinctive kinematical signatures. The need for such information is highlighted by recent doubts about bimodalities in GC metallicity distributions. Kinematical parameters can also help identify *correlations between GCs and field stars* within galaxies, and so determine which objects had a common origin. The present-day motions of GCs can provide *signatures of their formation and evolution,* as different physical processes will lead to different orbital properties. Last but not least, the residence of GCs at very large radii around galaxies makes them excellent *halo mass tracers.*

The informative capacity of GCS kinematics can be illustrated by the well-known case of the Milky Way, where there are at least two subpopulations (the bulge/disk and halo GCs). The cornerstone characteristics of these subpopulations are their spatial distributions, metallicities, and kinematics (e.g. [1–3]): without all three clues, the distinction would be much less clear. One may make further decompositions into "thin/thick" disk GCs, and "young/old" halo GCs—which is a level of detail difficult to achieve in external galaxies. Note that these nomenclatures are not arbitrary, but connect to the picture that GC and field star formation are associated. GCs have also proved valuable for measuring the mass of the Milky Way halo (e.g. [4–6]).

I will hereafter review the study of GC kinematics, with an emphasis on GCs in elliptical galaxies. Observations are discussed in Sect. 2, spiral GCS kinematics in Sect. 3, and GCS rotation in ellipticals in Sect. 4. Section 5 covers GCS dynamics (mass and anisotropy profiles), and Sect. 6 summarizes.

2 GCS Kinematics Observations

The study of GCS kinematics beyond the Local Group began in the 1980s with the use of 4-m-class telescopes, providing dozens of velocities in the bright Virgo ellipticals M87 and M49 [7–9], and expanded in the 1990s with the use of 8-m-class telescopes to obtain hundreds of velocities out to distances of ~ 20 Mpc [10–13]. Today, the sample of well-studied galaxies is still small, so it is worth reviewing the observational challenges.

2.1 Observational Requirements and Issues

The ideal starting point for GCS kinematics observations is *wide-field imaging*, which allows candidate GCs to be identified far out into a galaxy's halo. Imaging that reaches beyond the extent of the GCS can further be used to measure the "contamination" from stars and galaxies. Although relatively shallow images are enough to find the bright GCs which are accessible to spectroscopy, it is important to obtain much deeper images, since GCS dynamical models need a well-known spatial distribution. There have so far not been many GCS imaging studies published with the $\sim 0.5°$ field of view needed for nearby galaxies (e.g. [14–19]).

Contamination levels increase rapidly as one moves to studying outer halo GCs, so any available countermeasure should be employed. *High spatial resolution* imaging is profitable but not usually obtainable. Multiple ACS pointings can provide a reasonably wide field-of-view around a galaxy, in the case of NGC 4697 resulting in only 1% contamination [20]. From the ground, good seeing can permit GCs to be resolved out to 5 Mpc [21, 22].

The standard tool for GC identification is *color selection*. The Washington C filter is especially powerful in identifying unresolved faint blue galaxies by their UV excess [14]. Three-band photometry helps even more with object discrimination [16], and broader color baselines are better at resolving GC bimodality: $(C - R)$ has three times the resolution of $(V - I)$.

For acquiring the GC velocities, efficient wide-field multi-plexed spectrographs are needed. For astrometric reasons it is ideal to have *pre-images* of the GCs taken on the same instrument, which entails an imaging spectrograph. However, many observatories cannot provide pre-imaging without introducing a year's delay in the spectroscopy. Another observational hurdle is *competition*: acquiring hundreds of GC velocities involves ~ 10 h of 8-m dark time, often in an especially oversubscribed period (R.A. ~ 12 h).

A key issue in GC spectroscopy is *sky subtraction*, which is normally more accurate with slits than with fibers. It is not yet clear that the benefits of nod-and-shuffle outweigh the complications. A stable high-resolution fiber spectrograph such as UT+FLAMES/GIRAFFE can provide velocities as accurate as 5 km s^{-1}, which is useful for galaxies with very low mass or low rotation. The wavelength range used is typically ~ 4000–6000 Å, but an option for brighter skies is the Ca II triplet at ~ 8600Å [23, 24]. Well-matched

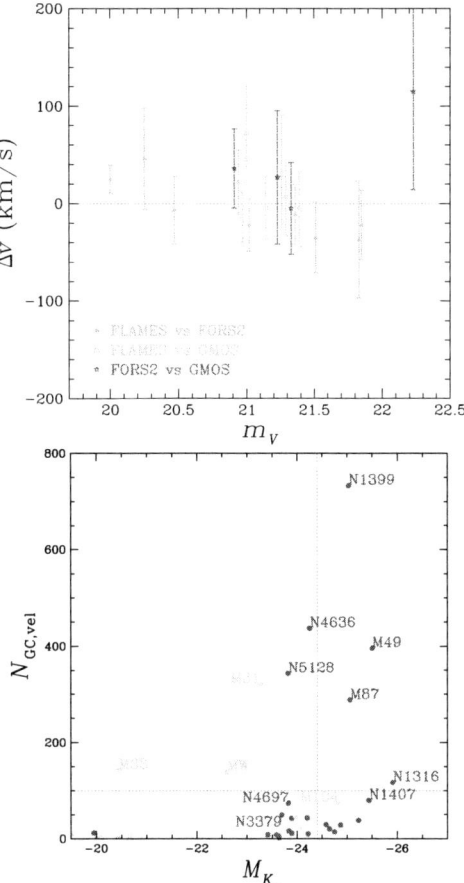

Fig. 1. *Left:* GC velocity comparisons around the galaxy NGC 3379, as a function of magnitude, from different instruments [26–28]. *Right:* Number of GC velocities acquired as a function of galaxy magnitude. Different colors and symbols show early-type and late-type galaxies. A "critical" value of 100 GCs is shown by a horizontal line, and a "typical" early-type galaxy luminosity L^* is shown by a vertical line.

template spectra are essential to finding accurate velocities. Synthetic spectra or observations of Galactic stars can be used, but better matches come from Local Group GCs, or early-type dwarf galaxies. A final impediment to progress is the rarity of fully-operational *data-reduction pipelines* for multi-object spectrographs.

With improvements in telescopes, instruments, and techniques, the observational productivity and accuracy of GC kinematics have increased dramatically in recent years. Example at GC magnitudes of $R \sim 20$, the velocity uncertainties have improved from ~ 100 km s^{-1} to ~ 20 km s^{-1} [25]. The reliability of such measurements is shown by the consistency within the stated errors of the *absolute* velocities (no systematic shift applied) when comparing 8-m telescope observations of GCs around NGC 3379 down to $V = 22$ (see Fig. 1, left). The most distant galaxy yet reached with GCS spectroscopy is NGC 3311 at 50 Mpc, using UT3+VIMOS [29].

It should be remembered that if one would like to compare GC and stellar kinematics in a galaxy, then of course one needs *observations of the stellar kinematics*. It is currently infeasible to measure stellar kinematics in galaxy halos outside the Local Group, so the only possibility is to use *planetary nebulae* (PNe) as stellar proxies—an approach which has taken off in recent years (e.g. [30–35]).

2.2 Observational Studies

To credibly measure rotation without assuming a preferred axis, or estimate mass without assuming some orbital anisotropy, one needs at least ~ 100 GC velocities. Many GC spectroscopic studies using 8-m-class telescopes have been mainly aimed at line-strengths, and so did not produce samples particularly suitable for dynamical studies (e.g. [36, 26, 37]). There are till now only ~ 10 galaxies with large GC velocity samples (see Fig. 1, right).

The first distant GCS studied in kinematical detail belongs to the Virgo Cluster central elliptical M87, with 288 velocities acquired out to galactocentric distances of 45 kpc (6 $R_{\rm eff}$)—mostly by Keck+LRIS and CFHT+MOS [7–11, 38, 39]. The *brightest* elliptical in Virgo, M49 (=NGC 4472) now surpasses M87 with 396 velocities to 90 kpc, mainly from CFHT+MOS, Keck+LRIS, and UT2+FLAMES [9, 40, 41, 12, 42].

The central Fornax Cluster elliptical, NGC 1399, has the largest data set of any galaxy, with $\sim$700 velocities to 90 kpc, mainly from UT2+FORS2 [13, 43]. NGC 1407, the central Eridanus A Group elliptical, has ~ 100 velocities from Keck+LRIS, UT2+FLAMES, and Clay+LDSS-3 [44].

These "high-end" ellipticals have been so extensively studied because of their very rich GCSs, but results for more ordinary ellipticals (with $\sim L^*$ luminosity) have been slow in coming. Even 8-m-class telescopes may have trouble getting large kinematic data sets in some of these galaxies because of their limited GC populations. An example is NGC 3379, which is a very interesting galaxy but unfortunately has a sparse GCS. Several observing campaigns have achieved only 49 GC velocities [26–28], although another recent UT2+FLAMES study should double or triple this number [42].

The first large GC kinematics study of an L^* elliptical involves NGC 4636, at the Virgo Cluster outskirts. Here UT2+FORS2 has been used to obtain 437 GC velocities out to 45 kpc [45, 43]. Further L^* elliptical studies are now underway with UT2+FORS2, Gemini+GMOS, and elsewhere.

Another extensively studied GCS is from the nearby (4 Mpc) disturbed early-type galaxy NGC 5128. Its ever-growing tally using 4-m telescopes is 343 GC velocities to 40 kpc [15, 46–48].

Spiral galaxies have received less attention than ellipticals. By far the largest study is of the M31 GCS, resulting in 321 velocities, largely from the MMT and the WHT+WYFFOS [49,50]. M33 has 143 velocities, mostly from WIYN+HYDRA [51,52].

3 GCS Kinematics in Spiral Galaxies

The available GCS kinematics data in spiral galaxies permit a brief summary, going from latest type to earliest. In M33, there is a rotating component of young GCs, and a weakly-rotating old GCS which may include halo and disk sub-components [52]. In the Milky Way, there is a rotating component of GCs associated with the stellar bar or bulge, and a hot, non-rotating halo GCS (see [53] for review). There are "old" and "young" halo GCs with differing kinematics [54,55], and the halo *stars* have a velocity dispersion that decreases with radius, similarly to the halo GCS [56]. In M31, there is rotation in both the metal-poor and metal-rich GCs [50], and a thin-disk GCS whose age is controversial [57,58]. There is again a metal-poor stellar halo with a decreasing dispersion profile [59]. In M81, there is a rotating metal-rich GC component and a non-rotating metal-poor component [60]. In M104, there appears to be little rotation regardless of metallicity, although the GC sample is still small [61,62]. Thus we see that a GCS kinematical structure like the Milky Way's may be common, but perhaps not universal, and also that there is more work to be done comparing the behavior of halo stars and GCs.

4 GCS Rotation in Elliptical Galaxies

While the luminous bodies of elliptical galaxies are observed to have strikingly lower angular momenta than spirals, the "missing" spin might reside in ellipticals' halos. This scenario is supported by simulations of elliptical formation via galaxy mergers, which predict high rotation ($v/\sigma \sim 1$) outside $\sim 2R_{\rm eff}$ [63–66]—with the caveat that non-gravitational baryonic effects are not yet well-studied (e.g. [67]). Although a galaxy's GCS would not directly contain much of its angular momentum, it could be used to infer the rotation of the halo stars and/or dark matter.

So far there is no obvious pattern to GCS rotational properties, except that dwarf ellipticals may be more rotationally dominated than giants [68]. There is no consistent trend of GC rotation with metallicity, nor a clear correlation between GC and stellar halo rotation fields (as inferred by PN kinematics). M87 and M49 may have large GCS kinematical twists, as evidence perhaps of GC accretion [39,12]. Further progress should come with more data, placed in the context of global dynamical models and more detailed theoretical predictions, with attention to the connections to galaxy sub-type and environment.

5 GCS Dynamical Modeling

The simplest ingredient for a GCS dynamical model is the projected velocity dispersion profile $\sigma_{\rm p}(R)$. Inferences from this are tricky because of the classic mass-anisotropy degeneracy, which can be lifted by attention to the *shape* of the line-of-sight velocity distribution (LOSVD). This requires ~ 1000 GC velocities even assuming spherical symmetry [69], but fortunately there are additional constraints that drastically reduce the uncertainties. The GCS surface density profile may be inferred from many more GCs than have measured velocities, and the central galaxy mass can be well-determined from stellar kinematics. One of the key potential advantages of GCS dynamics is less complicated geometry than stellar dynamics. In particular, the metal-poor GCs may lie in near-spherical distributions—although this supposition has had little empirical testing.

5.1 Elliptical Galaxy Halo Masses

The GCS $\sigma_{\rm p}(R)$ profiles of M87, M49, NGC 1399, and NGC 1407 are all constant or rising with the radius. This is clear evidence for massive dark matter halos, or else fairly pathological anisotropy would be required. For M87, "orbit modeling" has been carried out which includes non-parametric anisotropy profiles and LOSVD shape-fitting for the GCs and the stars [70]. A dark halo is found which appears to belong to the Virgo Cluster core, with a profile in encouragingly good agreement with independent X-ray-based mass results [71]. Another non-parametric study has found similar results [72]. In NGC 1399, a rising GCS $\sigma_{\rm p}(R)$ may also trace the Fornax Cluster core, or it may be a transient feature caused by galaxy interactions [73, 74].

NGC 4636 is a fairly typical L^* elliptical, except that its rich GC population and high X-ray luminosity suggest that it is at the center of a group halo. Its GCS $\sigma_{\rm p}(R)$ is roughly constant, and models do imply a dark matter halo of group scale—though not as massive as inferred from the (very disturbed) X-ray-emitting gas [45].

Studies of PN kinematics in ordinary ellipticals suggest that such galaxies do not have the centrally-concentrated massive dark matter halos expected from ΛCDM [75, 30, 31], which is supported by a meta-study of stellar, PN, and GC dynamics [76]. Other interpretations have been supplied [77, 78], and it would be invaluable to have independent GCS constraints on the same galaxies studied using PNe.

For the poster-child low-dark-matter galaxy, NGC 3379, there are not yet many GC velocities, but they do reach to much larger radii than the PNe. The GCS shows a much shallower decline in $\sigma_{\rm p}(R)$ than the PNe (see Fig. 2, left). This seems mostly due to very different spatial and anisotropy distributions—e.g., GCs and PNe at the *same projected radius* are sampling very different regions of the 3D potential. The implied mass profiles from stellar, PN, and GCS dynamics are consistent within the still-considerable

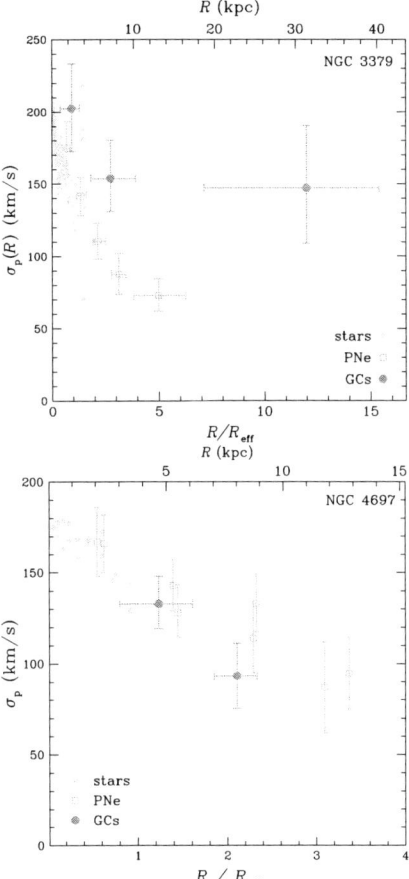

Fig. 2. Projected velocity dispersion profiles in nearby L^* ellipticals, for stars, planetary nebulae, and globular clusters. *Left:* Data from the round galaxy NGC 3379 [79, 31, 26–28, 80]. *Right:* Data from the flattened galaxy NGC 4697 [30, 20].

uncertainties. Intriguingly, this system also includes an HI gas ring [81] whose kinematics imply a low halo mass that is *not* consistent with the GCS results.

For another of the low-dark matter galaxies, NGC 4697, preliminary results from Gemini+GMOS indicate that the GCS $\sigma_p(R)$ does track the radial decline of the PNe (see Fig. 2, right). More data at larger radii are still needed from this ongoing program.

Further results from PN and GCS studies are awaited, but it is already clear that there is great diversity in the mass profiles of elliptical galaxies— which may be contrasted with the tight Tully-Fisher relations of spiral galaxies. Is this because the obvious dark matter halos in galaxies such as M49 are all examples of a group or cluster halo, while the individual halos of free-floating galaxies such as NGC 4697 are much scantier? Or is there a large amount of scatter in halo properties relating to the galaxies' collapse, merger, and interaction histories, coupled to normal scaling relations? Large, systematic dynamical surveys, as well as more progress with simulations, are needed to address such questions.

5.2 GCS Orbital Properties

The orbital properties of GCs can inform us about their formational histories, and their connections with their host galaxy's diffuse stellar populations. To determine these properties, it is very helpful to have independent constraints on galaxy mass profiles (e.g. from X-ray emission) so that all the GC kinematics information can be directed to determining the anisotropy.

Examples are M87 and M49, where the GCs turn out to have overall fairly isotropic orbit distributions [70,39,12,82]. In M87, the metal-poor GCs appear to have slightly tangential orbits inside $3R_{\rm eff}$, and slightly radial orbits in the outer parts, while the metal-rich GCs have slightly radial orbits everywhere. These orbital properties imply that dynamical processes could not have produced dramatic changes in the GC mass function [83].

More detailed analyses of large velocity data sets could cross-correlate GC and stellar kinematics, to test the notion that metal-rich GCs are coeval with their galaxy's old metal-rich field stars. One may also look for bimodality in dynamical phase-space that correlates with color, which could support the existence of metallicity bimodality. Further inferences about galaxy/GCS formation from GC orbital properties are in dire need of firmer theoretical predictions (see [63,66] for pioneering work). Some ΛCDM-based simulations do predict strong *radial anisotropy of the stars* in galaxy halos [84,85], which may not accord with the *observed isotropy of GCSs*. Orbit analyses can also find remnant substructures from accretion and interaction events, as may have already been identified in M31 and M49 (e.g. [86,12]).

5.3 Modified Newtonian Dynamics

An alternative to dark matter for explaining mass discrepancies is Modified Newtonian Dynamics (MOND), a theory with much success in modeling late-type galaxies. It has been little applied to elliptical galaxies, although it seems to be consistent with the PN data in ordinary ellipticals [87]. Perhaps the greatest observational challenge to MOND is the high apparent dark matter content in galaxy cluster cores. These cores are typically inhabited by bright ellipticals with rich GCSs whose dynamics can be used to test MOND—although the conclusions may not be as clear-cut as hoped [45,88].

6 Summary

GCS kinematics studies are now beginning to provide large data sets for a wide variety of galaxies. Such data are useful for identifying true GCS subcomponents. In elliptical galaxies, diverse GCS rotation profiles are seen, as well as typically constant dispersion profiles implying massive dark matter halos. However, these results come from massive group-central ellipticals, and new observations are addressing more ordinary ellipticals—in particular, to see how their mass distributions square with theoretical predictions.

References

1. R. Zinn: ApJ **293**, 424 (1985)
2. P. Côté: AJ **118**, 406 (1999)
3. B.J. Pritzl, K.A. Venn, M. Irwin: AJ **130**, 2140 (2005)
4. M.I. Wilkinson, N.W. Evans: MNRAS **310**, 645 (1999)
5. T. Sakamoto, M. Chiba, T.C. Beers: A&A **397**, 899 (2003)
6. W. Dehnen et al.: MNRAS **369**, 1688, (2006)
7. J.R. Mould, J.B. Oke, J.M. Nemec: AJ **92**, 53 (1987)
8. J. Huchra, J. Brodie: AJ **93**, 779 (1987)
9. J.R. Mould, J.B. Oke, P.T. de Zeeuw, J.M. Nemec: AJ **99**, 1823 (1990)
10. J.G. Cohen, A. Ryzhov: ApJ **486**, 230 (1997)
11. J.G. Cohen: AJ **119**, 162 (2000)
12. P. Côté, D.E. McLaughlin, J.G. Cohen, J.P. Blakeslee: ApJ **591**, 850 (2003)
13. B. Dirsch, T. Richtler, D. Geisler, K. Gebhardt et al.: AJ **127**, 2114 (2004)
14. B. Dirsch, T. Richtler, D. Geisler, J.C. Forte et al.: AJ **125**, 1908 (2003)
15. E.W. Peng, H.C. Ford, K.C. Freeman: ApJS **150**, 367 (2004)
16. K.L. Rhode, S.E. Zepf: AJ **127**, 302 (2004)
17. G.L.H. Harris, D. Geisler, W.E. Harris, B.P. Schmidt et al.: AJ **128**, 712
18. L.P. Bassino, T. Richtler, B. Dirsch: MNRAS **367**, 156 (2006)
19. K.L. Rhode: this vol.
20. G.R. Sivakoff, A. Jordán et al.: in prep; A.J. Romanowsky et al.: in prep
21. M. Rejkuba: A&A **369**, 812 (2001)
22. M. Gómez, D. Geisler, W.E. Harris, T. Richtler et al.: A&A **447**, 877 (2006)
23. D. Minniti, M. Kissler-Patig, P. Goudfrooij, G. Meylan: AJ **115**, 121 (1998)
24. P. Goudfrooij, J. Mack, M. Kissler-Patig et al.: MNRAS **322**, 643 (2001)
25. T. Richtler, B. Dirsch, K. Gebhardt, D. Geisler et al.: AJ **127**, 2094 (2004)
26. T.H. Puzia, M. Kissler-Patig, D. Thomas et al.: A&A **415**, 123 (2004)
27. M. Pierce, M.A. Beasley, D.A. Forbes et al.: MNRAS **366**, 1253 (2006)
28. G. Bergond, S.E. Zepf, A.J. Romanowsky et al.: A&A **448**, 155 (2006)
29. T. Richtler et al.: in prep
30. R.H. Méndez, A. Riffeser, R.-P. Kudritzki et al.: ApJ **563**, 135 (2001)
31. A.J. Romanowsky, N.G. Douglas, M. Arnaboldi et al.: Science **301**, 1696 (2003)
32. E.W. Peng, H.C. Ford, K.C. Freeman: ApJ **602**, 685 (2004)
33. A.M. Teodorescu, R.H. Méndez, R.P. Saglia et al.: ApJ **635**, 290 (2005)
34. C. Halliday, D. Carter, T.J. Bridges et al.: MNRAS **369**, 97 (2006)
35. H.R. Merrett, M.R. Merrifield, N.G. Douglas et al.: MNRAS **369**, 120 (2006)
36. M.A. Beasley, D.A. Forbes et al.: MNRAS **347**, 1150 (2004)
37. M. Pierce, T. Bridges, D.A. Forbes, R. Proctor et al.: MNRAS **368**, 325 (2006)
38. D.A. Hanes, P. Côté, T.J. Bridges, D.E. McLaughlin et al.: ApJ **559**, 812 (2001)

39. P. Côté, D.E. McLaughlin, D.A. Hanes, T.J. Bridges et al.: ApJ **559**, 828 (2001)
40. R.M. Sharples, S.E. Zepf, T.J. Bridges, D.A. Hanes et al.: AJ **115**, 2337 (1998)
41. S.E. Zepf, M.A. Beasley, T.J. Bridges, D.A. Hanes et al.: AJ **120**, 2928 (2000)
42. G. Bergond et al.: in prep
43. Y. Schuberth: *Globular Clusters-Guides to Galaxies* (Springer, Berlin 2009)
44. P. Sánchez-Blázquez et al.: in prep; A.J. Romanowsky et al.: in prep
45. Y. Schuberth, T. Richtler, B. Dirsch et al.: A&A **459**, 391 (2006)
46. E.W. Peng, H.C. Ford, K.C. Freeman: ApJ **602**, 705 (2004)
47. K.A. Woodley, W.E. Harris, G.L.H. Harris: AJ **129**, 2654 (2005)
48. K.A. Woodley: *Globular Clusters-Guides to Galaxies* (Springer, Berlin 2009)
49. J.P. Huchra, J.P. Brodie, S.M. Kent: ApJ **370**, 495 (1991)
50. K.M. Perrett, T.J. Bridges, D.A. Hanes, M.J. Irwin et al.: AJ **123**, 2490 (2002)
51. R.A. Schommer, C.A. Christian, N. Caldwell et al.: AJ **101**, 873 (1991)
52. R. Chandar, L. Bianchi, H.C. Ford, A. Sarajedini: ApJ **564**, 712 (2002)
53. W.E. Harris. In: *Star Clusters*, ed by L. Labhardt, B. Binggeli (Springer, Berlin 2001), p 223 (www.physics.mcmaster.ca/∼harris/Publications/saasfee.ps)
54. D.I. Dinescu, T.M. Girard, W.F. van Altena: AJ **117**, 1792 (1999)
55. A.D. Mackey, G.F. Gilmore: MNRAS **355**, 504 (2004)
56. G. Battaglia, A. Helmi, H. Morrison, P. Harding et al.: MNRAS **364**, 433 (2005)
57. H.L. Morrison, P. Harding, K. Perrett, D. Hurley-Keller: ApJ **603**, 87 (2004)
58. K.M. Perrett: *Globular Clusters-Guides to Galaxies* (Springer, Berlin 2009)
59. S.C. Chapman, R. Ibata, G.F. Lewis, et al.: ApJ **653**, 255, (2006)
60. L.L. Schroder, J.P. Brodie, M. Kissler-Patig et al.: AJ **123**, 2373 (2002)
61. T.J. Bridges, K.M. Ashman, S.E. Zepf, D. Carter et al.: MNRAS **284**, 376 (1997)
62. E.V. Held, A. Moretti et al. In: *Extragalactic Globular Cluster Systems*, ed by M. Kissler-Patig (Springer, Berlin 2003), pp 161–166 (astro-ph/0210687)
63. L. Hernquist, M. Bolte. In: *The Globular Cluster-Galaxy Connection*, ed by G.H. Smith, J.P. Brodie (ASP, San Francisco 1993), pp 788–799
64. M.L. Weil, L. Hernquist: ApJ **460**, 101 (1996)
65. G.J. Bendo, J.E. Barnes: MNRAS **316**, 315 (2000)
66. K. Bekki, M.A. Beasley, J.P. Brodie, D.A. Forbes: MNRAS **363**, 1211 (2005)
67. T. Naab, R. Jesseit, A. Burkert: MNRAS **372**, 839, (2006)
68. M.A. Beasley, J. Strader, J.P. Brodie et al.: AJ **131**, 814 (2006)
69. D. Merritt, B. Tremblay: AJ **106**, 2229 (1993)
70. A.J. Romanowsky, C.S. Kochanek: ApJ **553**, 722 (2001)
71. K. Matsushita, E. Belsole, A. Finoguenov, H. Böhringer: A&A **386**, 77 (2002)
72. X. Wu, S. Tremaine: ApJ **643**, 210 (2006)
73. N.R. Napolitano, M. Arnaboldi, M. Capaccioli: A&A **383**, 791 (2002)
74. K. Bekki, D.A. Forbes, M.A. Beasley, W.J. Couch: MNRAS **344**, 1334 (2003)
75. R. Ciardullo, G.H. Jacoby, H.B. Dejonghe: ApJ **414**, 454 (1993)
76. N.R. Napolitano, M. Capaccioli et al.: MNRAS **357**, 691 (2005)
77. G.A. Mamon, E.L. Łokas: MNRAS **363**, 705 (2005)
78. A. Dekel, F. Stoehr, G.A. Mamon, T.J. Cox et al.: Nature **437**, 707 (2005)
79. T.S. Statler, T. Smecker-Hane: AJ **117**, 839 (1999)
80. N.G. Douglas, N.R. Napolitano, A.J. Romanowsky: ApJ **664**, 257 (2007)
81. S.E. Schneider: ApJ **288**, L33 (1985)

82. D.E. McLaughlin: this vol.
83. E. Vesperini, S.E. Zepf, A. Kundu, K.M. Ashman: ApJ **593**, 760 (2003)
84. J. Diemand, P. Madau, B. Moore: MNRAS **364**, 367 (2005)
85. M.G. Abadi, J.F. Navarro, M. Steinmetz: MNRAS **365**, 747 (2006)
86. K.M. Perrett, D.A. Stiff, D.A. Hanes, T.J. Bridges: ApJ **589**, 790 (2003)
87. M. Milgrom, R.H. Sanders: ApJ **599**, L25 (2003)
88. T. Richtler et al.: *Globular Cluster-Guides to Galaxies*, (Springer, Berlin 2009)

Dark Matter in the Elliptical Galaxies NGC 1399 and NGC 4636

Ylva Schuberth[1,2], Tom Richtler[2], and Michael Hilker[1]

[1] Argelander–Institut für Astronomie, Universität Bonn, Germany
   ylva@astro.uni-bonn.de
[2] Departamento de Física, Universidad de Concepcíon, Chile

Abstract. We have used the multi–object spectrographs FORS2/MXU at the VLT and GMOS at Gemini South to obtain medium–resolution spectra of 625 and 437 globular clusters (GCs) around the giant elliptical galaxies NGC 1399 and NGC 4636, respectively. By measuring the radial velocities of a large sample of GCs, we determine the line–of–sight velocity dispersion as a function of projected galactocentric radius. These values are then compared to Jeans–models under different assumptions for the orbital anisotropy aswell.

1 NGC 4636 – In the Outskirts of Virgo

NGC 4636 is the southernmost bright elliptical in the Virgo cluster. Besides having a very populous globular cluster system (GCS) and an unusually high specific frequency [1], it is very luminous in the X–rays – which has earned this galaxy the reputation of being extremely dark matter dominated [2,3]. The central X–ray emission, however, arises from a region which is probably not in hydrostatic equilibrium [4]. Using GCs, we have an independent way to study the dark matter content of this galaxy, based on stellar dynamics.

In a recent study [5], based on 174 GC velocities, we showed that NGC 4636 does not seem to be extremely dark matter dominated. We have since obtained more data, and using the new sample of 437 GCs in total, we find that the velocity dispersion for both red (metal–rich) and blue (metal–poor) clusters decreases significantly for radii beyond 35 kpc, in accordance with the observed drop of the GC number density profiles in this region.

The left panel of Fig. 1 shows the line–of–sight velocity dispersion for the blue clusters (with velocity uncertainties below $50\,{\rm km\,s^{-1}}$) together with our Jeans–models (#1 – #4) from [5]. The corresponding circular velocity curves are shown in the right panel. The new data extend to larger radii, and are well described by our models calculated from the smaller dataset. The dark matter fraction at 30 kpc ($\simeq 4.5\,{\rm R_e}$) depends on the anisotropy parameter and lies between 37 and 52 percent, for $\beta = -2$ and $\beta = 0.5$, respectively.

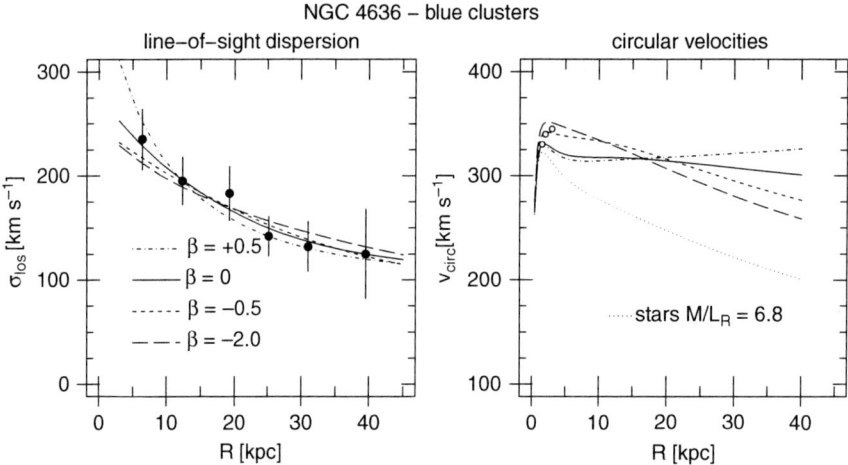

Fig. 1. NGC 4636. *Left panel*: line–of–sight dispersion of the blue GCs. The lines show models for different anisotropy parameters. *Right panel*: circular velocity curves for the sum of luminous and dark matter. The different line types correspond to those labeled in the left panel. The circular velocity of the stars alone is shown as dotted line. Circles indicate the values from the analysis by Kronawitter et al. [11].

2 NGC 1399 – At the Center of Fornax

NGC 1399, the central galaxy of the Fornax cluster, hosts a very rich and extended GC system (e.g. [6, 7]). The galaxy density in the central Fornax cluster is rather high, and the nearest (giant) neighbors of NGC 1399 are NGC 1404 and NGC 1387 (at 55 kpc and 105 kpc projected distance, respectively). This implies, that interactions between these galaxies are likely. Numerical simulations [8] show that GCs around early–type galaxies interacting with NGC 1399 are likely to be stripped. The blue GCs, which have shallower number density profiles, have an especially high chance of being affected by the interaction. In the simulated scenarios, these stripped GCs end up forming a population of intra–cluster GCs. It has indeed been shown that the GCS of NGC 1404 and NGC 1387, when compared to similar galaxies in Fornax, have low specific frequencies as well as unusually low fractions of blue GCs ([9] and references therein). Further, an excess of blue clusters between NGC 1387 and NGC 1399 is found – suggesting that this tail–like structure traces the interaction between the galaxies [7]. Given these findings, it is conceivable that, in our sample, the blue GCs are contaminated by these migratory objects. The red clusters, on the other hand, are more concentrated towards their host galaxies, and hence less prone to stripping. Moreover, the red population of NGC 1399 shows no sign of rotation [10], facilitating its use as dynamical tracers.

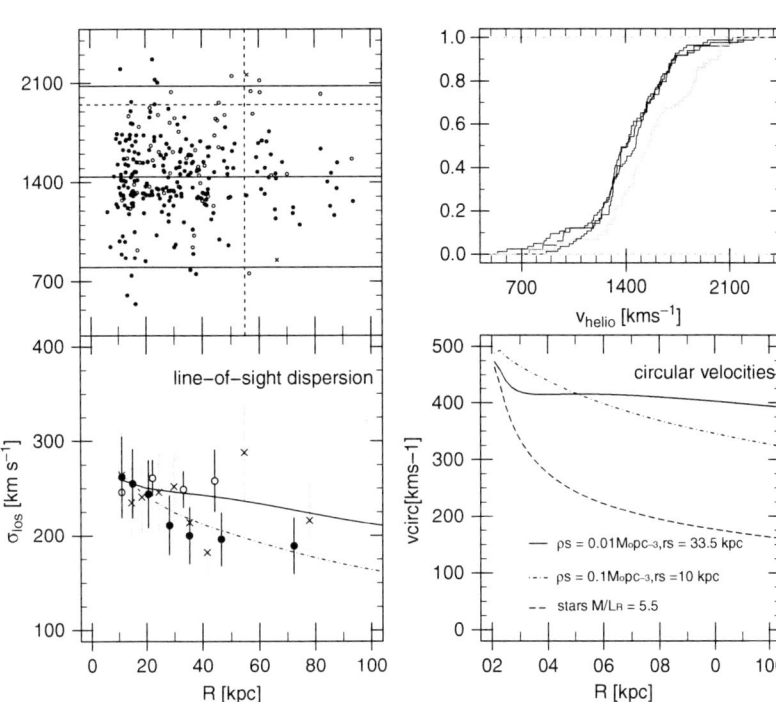

Fig. 2. NGC 1399. *Upper left panel*: Heliocentric velocity vs. galactocentric radius for the red GCs. The shaded region shows the velocity selection ($800 < v < 2080\,\mathrm{km\,s^{-1}}$) from [10]. The solid horizontal line at $1440\,\mathrm{km\,s^{-1}}$ indicates the systemic velocity of NGC 1399; the dashed line at $1950\,\mathrm{km\,s^{-1}}$ shows the velocity of NGC 1404. Open circles show GCs from the South–Eastern quadrant, filled circles show GCs from the other quadrants. Crosses indicate two extreme outliers which were excluded from the analysis. The line at 55 kpc shows the (projected) distance of NGC 1404. *Upper right panel*: Cumulative distribution of GC velocities. The thick grey line shows the red GCs from the SE quadrant. The red GCs in the other quadrants have significantly different distributions, and are shown with thin black lines. *Lower left panel*: Line–of–sight velocity dispersion of the red GCs. Open circles show the values from [10], crosses the dispersion obtained from the red sample (using the selection from [10], and 30 GCs per bin), and filled circles show the results when the SE quadrant is omitted. The solid line shows the model from [10]. The dash–dotted line indicates a model ($\beta = 0$) constructed to reproduce the declining dispersion. *Lower right panel*: Circular velocity curves. The solid line shows the circular velocity from [10]. The circular velocity corresponding to the declining velocity dispersion is shown as dot–dashed line. The circular velocity of the stars alone is shown as dashed line. The parameters for the NFW dark halos of the models are indicated in the plot.

Yet, when looking at local variations of the line–of–sight velocity dispersion of the red clusters, we found that the region in the vicinity of NGC 1404 is dynamically hotter than the surroundings. Whether this is an effect of the superposition of the two GCSs along the line–of–sight, or the result of an interaction is not clear. In any case, the velocity distribution of the South–Eastern quadrant, which contains NGC 1404, is significantly different from the other three quadrants. The upper right panel of Fig. 2 shows the cumulative distribution functions, and a two–sample KS–test yields a p–value of 0.02. Both mean velocity and standard deviation are higher towards NGC 1404. This prompted us to exclude the SE quadrant from our dynamical analysis. The upper left panel in Fig. 2 illustrates which GCs were omitted. We find a declining velocity dispersion for the red clusters. The values are lower than those found by [10], and significantly lower than the values found when using all four quadrants. The dispersion values are shown in the bottom left panel of Fig. 2, together with the model from [10], derived for a sample of GCs within 40 kpc of NGC 1399. A model with a declining dispersion is shown aswell. This model clearly needs to be refined, since it over–predicts the central circular velocity (see bottom right panel of Fig. 2), yet for larger radii, and assuming $\beta = 0$ (i.e. isotropy), the total enclosed mass is lower than what was found by [10]. To investigate whether the region near NGC 1404 is affected by dynamical heating or if the high dispersions found in this quadrant are merely due to projection effects, a more complete spatial coverage in this region is needed. Numerical simulations could help to determine to what degree the GCS of NGC 1399 is affected by the encounters with nearby massive Fornax members.

References

1. B. Dirsch, Y. Schuberth, T. Richtler: A&A **433**, 43 (2005)
2. K. Matsushita, K. Makishima, Y. Ikebe et al.: ApJ **499**, 13L (1998)
3. M. Loewenstein, F. Mushotzky: Nuclear Physics B Proc. Suppl **124**, 91 (2003)
4. C. Jones, W. Forman, A. Vikhlinin et al.: ApJ **567**, 115L (2002)
5. Y. Schuberth, T. Richtler, B. Dirsch et al.: A&A **459**, 391 (2006)
6. B. Dirsch, T. Richtler, D. Geisler et al.: AJ **125**, 1908 (2003)
7. L.P. Bassino, F.R. Faifer, J.C. Forte, B. Dirsch et al.: A&A **451**, 789 (2006)
8. K. Bekki, D.A. Forbes, M.A. Beasley, W.J. Couch: MNRAS **344**, 1334 (2003)
9. L.P. Bassino, T. Richtler, B. Dirsch: MNRAS **367**, 156 (2006)
10. T. Richtler, B. Dirsch, K. Gebhardt et al.: AJ **127**, 2094 (2004)
11. A. Kronawitter, R.P. Saglia, O. Gerhard, R. Bender: A&AS **144**, 53 (2000)

Ages, Abundances, and Kinematics of Globular Clusters in NGC 3379 and NGC 4649 with Gemini/GMOS

T. Bridges[1], M. Beasley[2], F. Faifer[3,4], D. Forbes[5], J. Forte[4], K. Gebhardt[6], D. Hanes[1], M. Norris[7], M. Pierce[5], R. Proctor[5], R. Sharples[7], and S. Zepf[8]

[1] Department of Physics, Queen's University, Kingston, ON, Canada
[2] Instituto de Astrofisica de Canarias, Spain
[3] Universidad Nacional de La Plata, La Plata, Argentina
[4] IALP - Conicet, Argentina
[5] Centre for Astrophysics & Supercomputing, Swinburne University, Australia
[6] Astronomy Department, University of Texas, Austin, Texas, USA
[7] Dept. of Physics, University of Durham, Durham, UK
[8] Dept. of Physics & Astronomy, Michigan State Univ., East Lansing, MI, USA

1 Introduction

Globular cluster (GC) spectroscopy is an excellent way to study GC kinematics, ages, and abundances, and galactic dark matter content [5, 2]. Our group is using the Gemini Multi-Object Spectrograph (GMOS) on the 8m Gemini telescopes to obtain photometry and spectroscopy of GCs in $\sim$ six early-type galaxies, covering a range of galaxy type, luminosity, and environment. Our photometric results for NGCs 3115, 3379, 3923, 4649, and 524 are presented in Faifer et al. 2006 (these proceedings). Here we discuss the ages, abundances, and kinematics of GCs in NGC 3379 and NGC 4649. NGC 3379 is an intermediate luminosity E in the Leo group at $\sim$ 11 Mpc, while NGC 4649 is a bright E in the Virgo cluster at $\sim$ 17 Mpc. (see [3]; and [4], [1] for further details about NGCs 3379 and 4649 respectively).

2 Observations and Data Analysis

Our data were obtained with GMOS on Gemini North. GMOS g', r', i' imaging was used to select targets by magnitude and colour in one 5.5 × 5.5 arcmin field per galaxy. In NGC 3379 the field was centered on the galaxy, while in NGC 4649 it was offset 1.9 arcmin NE. Spectra were obtained with the B600 grism, giving a resolution of 4–5.5 Å, and wavelength coverage from 3800–6600/3300–5900 Å (N3379/N4649). Exposure times were 10/8 hours for N3379/N4649, giving a S/N per Å of 20–60/5–20 (N3379/N4649). Data reduction was done using the Gemini/GMOS IRAF package, and velocities were determined via cross-correlation with Bruzual & Charlot (2003) model spectra. We have 22/38 confirmed GCs in N3379/N4649, with contamination rates < 10%.

3 Results

In NGC 3379, we find that all GCs are older than 10 Gyr and cover a range of metallicity. The [E/Fe] abundance ratio decreases with [Fe/H] (Fig. 1). The GC velocity dispersion is constant with radius, and isotropic, non-parametric models require that the M/L ratio increase with radius: in other words, a dark matter (DM) halo. This disagrees with Romanowsky et al. (2003), who found no evidence for a DM halo from planetary nebula velocities. No rotation is seen in either the NGC 3379 or NGC 4649 GC system.

In NGC 4649, we find that most GCs are old, but there is a group of 4 apparently young (2-3 Gyr), metal-rich GCs (Fig. 3). The origin of this young GC population is not clear, but they may have been accreted. As in NGC 3379, the α-element abundance ratio decreases with increasing metallicity. Figure 2 shows that the M/L ratio increases with radius (assuming isotropic orbits), and reaches $M/L_V = 20$ at 200 arcsec (16 kpc). There is excellent agreement between the M/L profiles for the GCs and the X-ray emission (Fig. 3), supporting the existence of a DM halo in this galaxy.

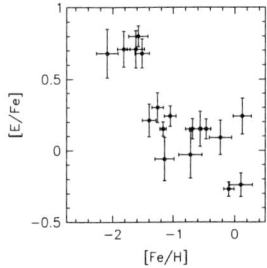

Fig. 1. [E/Fe] vs [Fe/H] for N3379 GCs.

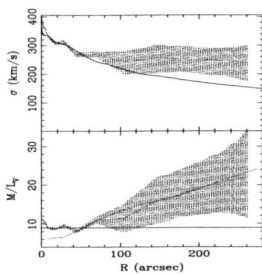

Fig. 2. *Top*: Velocity dispersion vs. radius for N4649 GCs. Upper *red line* and hatched region are the stellar and GC data with 1σ errors, and the lower *black line* is the dispersion profile for a constant M/L ratio model. *Bottom*: $(M/L)_V$ vs radius. *Red line* and hatched region as for top plot, while *blue line* is M/L profile from X-ray study by Humphrey et al. (2006).

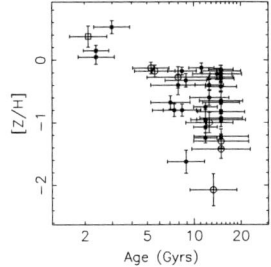

Fig. 3. (right): [Z/H] vs Age for N4649 GCs.

References

1. T. Bridges, K. Gebhardt, R. Sharples, F. Faifer, J. Forte, M. Beasley, S. Zepf, D. Forbes, D. Hanes, and M. Pierce, MNRAS, 373, 157 (2006)
2. J.P. Brodie & J. Strader, ARA&A, 44, 193 (2006)
3. M. Pierce, M. Beasley, D. Forbes, T. Bridges, K. Gebhardt, F. Faifer, J. Forte, S. Zepf, R. Sharples, D. Hanes, and R. Proctor, MNRAS, 366, 1253 (2006)
4. M. Pierce, T. Bridges, D. Forbes, R. Proctor, M. Beasley, K. Gebhardt, F. Faifer, J. Forte, S. Zepf, R. Sharples, and D. Hanes, MNRAS, 368, 325 (2006)
5. S. Zepf, "Extragalactic Globular Cluster Systems", ESO Workshop, p. 283 (2003)

The Dark Halo of NGC 1399 and MOND

Tom Richtler[1], Ylva Schuberth,[1,2], and A. Romanowsky[1,3]

[1] Depto. de Fisica, Universidad de Concepción, Chile
[2] Argelander Institut für Astronomie, Universität Bonn, Germany
[3] Lick Observatory, University of California, USA

The reigning CDM paradigm has been challenged for decades by alternative concepts without dark matter, most notably Modified Newtonian Dynamics (MOND) (Milgrom [2], Sanders and McGaugh [4]). While MOND is successfull in accounting for the kinematical behaviour of disk galaxies over a large range of masses, little is known about elliptical galaxies (see Romanowsky, this volume, for a review of early-type galaxy dynamics). NGC 4636, one of the few galaxies appearing in this context, seem to be consistent with MOND (Schuberth et al. [6]). However, MOND apparently does not explain the dynamics of galaxy clusters without invoking dark matter (e.g. Sanders [5]). It is therefore interesting to have a closer look at central cluster galaxies.

Schuberth et al., this volume, report on investigations of the dynamics of the globular cluster system of NGC 1399, the central galaxy of the Fornax cluster. See this contribution for information of the data set and the analysis. Here we focus on the MOND aspects.

Within the CDM framework, the dark halo of NGC 1399 can be best described by the NFW parameters $\rho_s = 0.0065 M_\odot/pc^3$ and $r_s = 50 pc$, which differs slightly from the values found by [3] inside 40 kpc (we assume isotropy). Figure 1 (left panel) shows the mass components in terms of circular velocities.

To evaluate the MOND circular velocity we apply the MOND recipe $g_N = \mu(g/a_0) \cdot g$, where g and g_N are there MOND acceleration and Newtonian acceleration, respectively. a_0 is a universal constant and $\mu(g/a_0)$ is a function interpolating between the Newtonian and the MONDian regime. The deep MOND regime is not yet reached at 80 kpc, so the exact form of the interpolation function is important. Following Famaey and Binney [1], we adopt $\mu = x/(1+x)$ with $x = g/a_0$. For a_0 we adopt the value 1.2×10^{-8} cm s^{-2}. The dashed line in Fig. 1 (left panel) is the circular velocity in MOND, resulting from the stellar mass alone (the gaseous mass is negligible within our radial range). In order to achieve consistency with the NFW halo, one has to add another (hypothetical) mass component. A possible circular velocity profile for this mass is shown in the right panel of Fig. 1. The total mass out to 80 kpc is about equal to the stellar mass, but has a shallower density profile.

The assumptions of isotropy and the interpolation function are caveats, but there are no indications of either a strong radial or tangential bias.

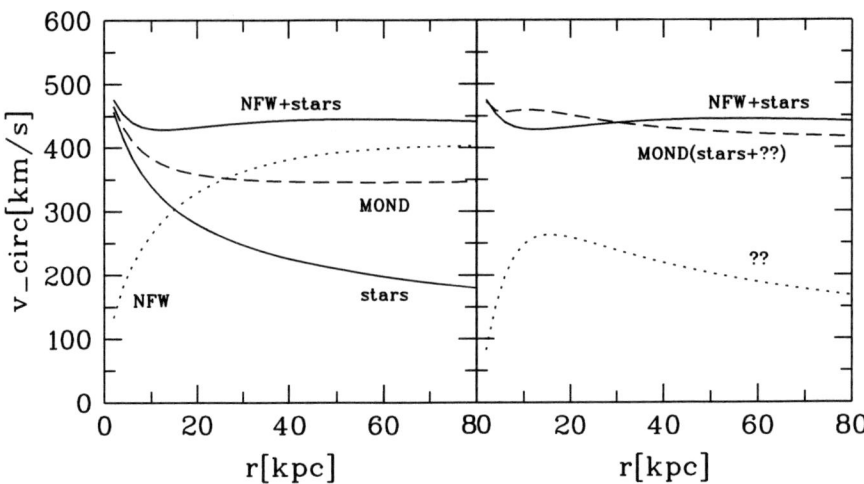

Fig. 1. *Left:* This graph shows the various circular velocities vs. galactocentric distance which represent the results of our analysis. The uppermost curve (*solid*), which is the sum of the stellar mass (*solid*) and a dark NFW halo (*dotted*) reproduces best the observed velocity dispersion. The MOND circular velocity (*dashed*) is too low. *Right:* The graph illustrates that MOND still needs an additional dark halo in order to be consistent with the observed velocity dispersion. The *solid line* is the same as in the *left panel*. The dashed line is the MOND circular velocity, now consistent with the *solid line*, if an hypothetical dark halo (*dotted line*) is assumed. This dark halo is of the order of the stellar mass.

Moreover, our adopted μ-function already gives a stronger MOND effect than the "classical" formula.

Our finding indicates that the central parts of a galaxy cluster still need dark matter in order for MOND to work. This mass has yet to be found.

References

1. B. Famaey, J. Binney: MNRAS **363**, 603 (2005)
2. M. Milgrom: Ap.J. **270**, 365 (1983)
3. T. Richtler, B. Dirsch, K. Gebhardt et al.: A.J. **127**, 2094 (2004)
4. R.H. Sanders, S.S. McGaugh: ARAA **40**, 263 (2002)
5. R.H. Sanders: MNRAS **342**, 901 (2003)
6. Y. Schuberth, T. Richtler, B. Dirsch, M. Hilker, et al. : A&A **459**, 391 (2006)

Dynamics of the Globular Cluster System of NGC 5128

Kristin A. Woodley

McMaster University, Department of Physics and Astronomy, Hamilton, ON, Canada L8S 4M1 woodleka@physics.mcmaster.ca

1 Introduction

NGC 5128 is a likely multiple-merger product, with currently 343 confirmed GCs. The kinematics of both the metal-rich and metal-poor globular clusters (GCs) within the galaxy can provide hints for signatures of mergers and/or accretion events producing the metallicity and kinematic structure.

2 Dynamics

2.1 Globular Cluster System of NGC 5128

The rotation amplitude and rotation axis (Fig. 1) of all 343 GCs, as well as the metal-poor, and metal-rich subpopulations were determined. The metal-poor population of GCs shows a change in rotation amplitude and axis with radius which could indicate a possible merger event disrupting the metal-poor population, while the metal-rich population shows radially consistent kinematics. The total mass of NGC 5128 (Fig. 1) was also determined as the rotationally supported mass determined by the Jeans equation plus the pressure supported mass using the Tracer Mass Equation [1]. The total mass of NGC 5128 is determined to be $1.3 \pm 0.5 \times 10^{12}$ $M_\odot$ out to a radius of 45 kpc.

2.2 Centaurus and M83 Groups

The kinematic analysis was extended to the nearest 42 galaxies in angular distance from NGC 5128 [2] located in the Centaurus and M83 groups (Fig. 2). The kinematics of the halo of NGC 5128 appears to extend to the satellite galaxies. The mean rotation amplitude and axis are 62 ± 5 km s^{-1} and 159 ± 4° E of N, respectively.

3 Conclusions

The kinematics of the outer halo of NGC 5128 matching the nearby satellites suggests the brightest group member and its satellite galaxies are part of the

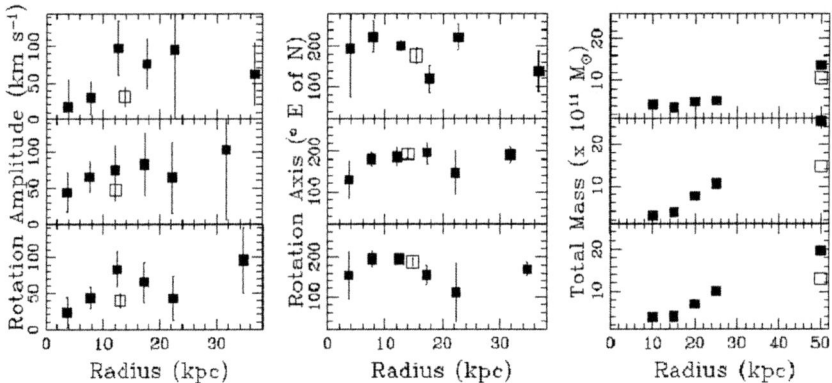

Fig. 1. Rotation amplitude (*left*) and rotation axis (*middle*) of the globular cluster system (GCS) and the total mass of NGC 5128 (*left*) as a function of radius were determined using the binned GCS (*solid squares*) and entire GCS (*open square*), for the metal-poor GCs (*top*), metal-rich GCs (*middle*), and all GCs (*bottom*).

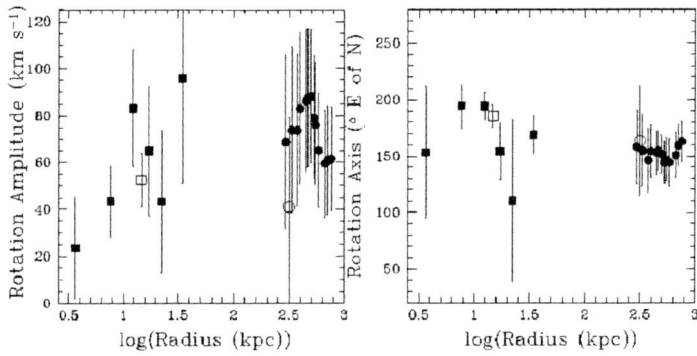

Fig. 2. Rotation amplitude (*right*) and rotation axis (*left*) as a function of radius from the center of NGC 5128 for the GCS (*squares*), the Centaurus group (*open circle*), and satellite galaxies (*closed circles*).

same large-scale structure, consistent with cold dark matter models. NGC 5128 could have, therefore, started as an initial "seed" galaxy and built up from the merger of many nearby small satellites. Full details on the dynamcis will be appear in Woodley et al. (2006) and Woodley (2006).

References

1. N.W. Evans, M.I. Wilkinson, K.M. Perrett, and T.J. Bridges: The Astrophysical Journal **583**, 752 (2003)
2. I.D. Karachentsev, M.E. Sharina, A.E. Dolphin, E.K. Grebel, D. Geisler, P. Guhathakurta, et al.: Astronomy & Astrophysics **385**, 21 (2002)